AF360713

Sustainable Supply Chains in Industrial Engineering and Management

Sustainable Supply Chains in Industrial Engineering and Management

Editors

Conghu Liu
Xiaoqian Song
Zhi Liu
Fangfang Wei

Basel • Beijing • Wuhan • Barcelona • Belgrade • Novi Sad • Cluj • Manchester

Editors

Conghu Liu
School of Mechanical and
Electronic Engineering
Suzhou University
Suzhou, China

Xiaoqian Song
China Institute of
Urban Governance
Shanghai Jiao Tong
University
Shanghai, China

Zhi Liu
College of Management
Engineering
Anhui Polytechnic University
Wuhu, China

Fangfang Wei
College of Economics
and Management
Shanghai Polytechnic
University
Shanghai, China

Editorial Office
MDPI
St. Alban-Anlage 66
4052 Basel, Switzerland

This is a reprint of articles from the Special Issue published online in the open access journal *Processes* (ISSN 2227-9717) (available at: https://www.mdpi.com/journal/processes/special_issues/A47ZFBWQ22).

For citation purposes, cite each article independently as indicated on the article page online and as indicated below:

Lastname, A.A.; Lastname, B.B. Article Title. *Journal Name* **Year**, *Volume Number*, Page Range.

ISBN 978-3-0365-9768-3 (Hbk)
ISBN 978-3-0365-9769-0 (PDF)
doi.org/10.3390/books978-3-0365-9769-0

Contents

About the Editors

Conghu Liu

Conghu Liu, Ph.D., Professor, Vice Dean of the School of Mechanical and Electronic Engineering at Suzhou University, Member of the Academic Committee of Suzhou University, Young Backbone Teacher of the Ministry of Education in Central and Western China, Visiting Scholar of Tsinghua University, Postdoctoral Fellow of Shanghai Jiao Tong University, Leader of Anhui Province's Mechanical Application Peak Discipline and Top Talents in Anhui Province's Universities, enjoys government special allowances. Dr. Liu has hosted multiple national/provincial level fund projects and published over 60 academic papers in domestic and foreign journals such as *IEEE Transactions on Industrial Informatics, Energy Conversion and Management, Energy and Computer Integrated Manufacturing Systems* (12 in Chinese Academy of Sciences journals, 5 highly cited ESI papers and 1 hot topic paper), as well as more than 50 papers indexed by SCI/EI. Dr. Liu has obtained three national invention patent authorizations and received three provincial and ministerial level awards.

Interests: sustainable manufacturing; industrial engineering.

Xiaoqian Song

Xiaoqian Song is an Associate Researcher in Resource and Environmental Management Studies at the School of International and Public Affairs (SIPA) of Shanghai Jiao Tong University. She finished her post-doctorate career in Shanghai Jiao Tong University and obtained a Ph.D. from the University of Science and Technology, Beijing. She held B.A. in Management from Xi'an University of Science and Technology, China.

Her current research interests include the low-carbon transition policy decision-making o resource-based cities in China, strategical resource management, urban governance, trash management, China's energy diplomacy, China's climate change policy, climate change governance, etc.

Dr. Song has been invited to deliver presentations at numerous conferences and workshops organized by government think tanks. She was invited as Visiting Scholar at the University of Dundee, UK, in 2011 and served as a visiting researcher at the National Institute for Environmental Studies (NIES) of Japan.

Interests: energy economics; energy policy; resource management; environmental management; urban governance; waste management; green transition; mineral resource material flow analysis.

Zhi Liu

Zhi Liu graduated from Hefei University of Technology with a Bachelor's and Master's degree in Industrial Engineering in 2009 and received a Ph.D. in Management Science and Engineering from Nanjing University of Aeronautics and Astronautics in 2020. Zhi Liu was a visiting scholar at Odette School of Business–University of Windsor, Canada, from June 2017 to June 2018. He is the Deputy Director of Industrial Engineering Department and the Director of Sustainable Operation Management Research Center at Anhui Polytechnic University. To date, Zhi Liu has published more than 40 academic papers in renowned journals, and 4 papers have been highly cited by ESI. He serves as a guest editor for the special issue of the SSCI journal *International Journal of Environmental Research and Public Health* and is an anonymous reviewer for over 30 journals, including *Transportation Research Part E: Logistics and Transportation Review* and the *International Journal of Production Research*.

Interests: sustainable operation management; closed-loop supply chain; corporate social responsibility.

Fangfang Wei

Fangfang Wei received her Ph.D from Yanshan University in China and served as a postdoctoral fellow at Shanghai Jiao Tong University. She is currently working at Shanghai Polytechnic University. Her research focuses on closed-loop supply chain issues, including recycling, remanufacturing and carbon emission reduction assessment. She has provided supervision for the projects of a vital soft science project under the Shanghai Science and Technology Innovation Action Plan 2021, with the title "Researching the recycling mechanism for incorporating informal collectors in the digital transformation context". In recent years, she has authored more than 20 papers in journals related to closed-loop supply chains.

Interests: sustainable supply chain management; remanufacturing; recycling; carbon emission; game theory.

Preface

Industrial engineering and management are of great importance in sustainable supply chains. The first reason is the optimization of resource consumption. Industrial engineering and management optimize production and logistics processes, allocate resources reasonably and reduce energy consumption and waste generation, thereby reducing negative impacts on the environment. Next is the improvement of efficiency and efficiency. By optimizing supply chain management and production processes, industrial engineering and management can improve production efficiency and efficiency, reduce costs and minimize negative impacts on the environment. Additionally, the application of industrial engineering and management in sustainable supply chains can promote environmental awareness among enterprises and consumers, create a good environmental atmosphere and promote sustainable development. It is important to ensure the stability of the supply chain. By allocating resources reasonably and optimizing supply chain management, industrial engineering and management can guarantee the stability of supply chains, ensuring the sustainable development and stable operation of various enterprises. The contributors to our Special Issue have conducted research on the above aspects, hoping to provide new ideas, theories, methods and technology for achieving the coordinated development of economic benefits, environmental protection and social benefits.

Conghu Liu, Xiaoqian Song, Zhi Liu, and Fangfang Wei
Editors

Editorial

Sustainable Supply Chains in Industrial Engineering and Management

Conghu Liu [1], Nan Wang [1,*], Xiaoqian Song [2], Zhi Liu [3] and Fangfang Wei [4]

[1] School of Mechanical and Electronic Engineering, Suzhou University, Suzhou 234000, China; lch339@126.com

[2] China Institute of Urban Governance, Shanghai Jiao Tong University, Shanghai 200030, China; songxq@sjtu.edu.cn

[3] College of Management Engineering, Anhui Polytechnic University, Wuhu 230009, China; liuzhi@nuaa.edu.cn

[4] College of Economics and Management, Shanghai Polytechnic University, Shanghai 201209, China; ffwei@sspu.edu.cn

* Correspondence: szxywn@126.com

Citation: Liu, C.; Wang, N.; Song, X.; Liu, Z.; Wei, F. Sustainable Supply Chains in Industrial Engineering and Management. *Processes* **2023**, *11*, 2280. https://doi.org/10.3390/pr11082280

Received: 25 June 2023
Revised: 15 July 2023
Accepted: 26 July 2023
Published: 28 July 2023

The integration of information technologies with the industry has marked the beginning of the Fourth Industrial Revolution and has promoted the development of industrial engineering. However, the depletion of resources and the creation of industrial waste caused by increasing industrial production pose a huge threats to nature. The application of sustainable supply chains in industrial engineering and management is one of the ways to balance the economy, society, and the environment. Therefore, it is a key concern for us to explore the construction of sustainable supply chains in industrial engineering and management. Moreover, to understand the impact of low-carbon, sustainable, and recycled supply chains on industrial engineering, we need more in-depth investigations. The Special Issue, entitled "Sustainable Supply Chains in Industrial Engineering and Management", has collected 18 recent works from relevant researchers, including research on sustainable supply chain technology in the fields of logistics, intelligent manufacturing (including remanufacturing), and management. The Special Issue is currently available online at: https://www.mdpi.com/journal/processes/special_issues/A47ZFBWQ22 (accessed on 15 July 2023).

1. Sustainable Supply Chain Design and Management

Gong et al. [1] established game models for a live-streaming supply chain and found optimal strategies for live-streaming members on streaming marketing modes, prices, and carbon emission reduction efforts.

Based on data-driven theory, Mu et al. [2] designed a sequence parameter index system for the logistics industry's ESE-B composite system, applied a Z-score to dimensionless data to standardize the original index data, and constructed a collaborative degree model to estimate the collaborative development between various subsystems of the logistics industry's ESE-B system.

Wang et al. [3] proposed a CNN-SA-NGU mixture model for forecasting silver closing prices, in which a CNN is used to extract the characteristics of the input data, SA is used to capture the correlation between different characteristic values, and the new NGU deep learning gating unit is used to predict the silver closing price. This system improves the ability to extract feature data and the non-linear fitting ability of the model.

Guo et al. [4] proposed a data-driven method that can be used to measure, evaluate, and identify the coupled and coordinated development (CCD) of the logistics industry (LI) and the digital economy (DE) to promote the integrated development of LI and DE.

Yang et al. [5] built a data-driven multimodel decision approach to calculate, assess, diagnose, and improve the regional innovation–economy–ecology (IEE) system. This method was used to test the coupling coordination degree of the Anhui IEE system and optimization measures were proposed.

2. Evaluation of Sustainable Supply Chains in Industrial Engineering

Huang et al. [6] reviewed the literature on the risk management of automotive supply chain interruption in recent years, listed the achievements in automotive supply chain interruption risk management, pointed out the problems in automotive supply chain management, and discussed the most concerned areas and development trends in automotive supply chain management.

Liu et al. [7] constructed an evaluation indicator system to measure organizational quality-specific immunity based on immune theory and introduced interval-valued hesitant fuzzy information and bidirectional projection technology to create bilateral matching evaluation and decision-making models in order to improve the bilateral matching decision making of manufacturing enterprises seeking partners in the manufacturing supply chain. The empirical analysis shows that this model can overcome the problem of information scarcity and help solve interval hesitation fuzzy decision-making problems.

Liu et al. [8] constructed a regression equation with employment in strategic emerging industries as the dependent variable, and the change direction, employment elasticity, and change speed of strategic emerging industries' structures as independent variables to measure the impact of structural changes in strategic emerging industries on employment in China. The research results indicate that the direction of changes in strategic emerging industries and employment elasticity is positively correlated with employment, while the change speed in strategic emerging industries is unstable and negatively correlated with employment.

Shi et al. [9] studied the role of carbon-abatement cost-sharing contracts in supply chains with a capital-constrained manufacturer. The results show that one-way cost-sharing contracts can improve manufacturers' carbon reduction levels, and two-way cost-sharing contracts have potential "economic-environmental" benefits.

Oršič et al. [10] used a new approach to achieve more sustainable deliveries through machine learning forecasts based on real-time data, different dynamic route planning algorithms, tracking logistics events, fleet capacities, and other relevant data. The developed model proposes to influence customers to choose a more sustainable delivery time window with important sustainability benefits using machine learning to predict accurate time windows with real-time data influence. This approach also leads to better vehicle utilization, less congestion, and fewer failures during home delivery.

Yang et al. [11] constructed a duopoly competition game model for channel competition between offline retailers and online retailers, studied the optimal pricing decisions in different scenarios, revealed the conditions for online retailers to provide return strategies and return insurance strategies, and provided management insights for online retailers.

3. Smart Manufacturing Process Monitoring and Control

Sun et al. [12] integrated blockchain and FL technologies in the intelligent manufacturing process, introducing the concept of Sustainable Production concerned with External Demands (SP-ED). The blockchain stores and manages detailed logs to identify defects, FL validates the sustainability and flaw detection for modifying the operations in consecutive operation cycles, which can improve sustainability by 11.48% and flaw detection by 14.65% and can reduce modifications by 11.11% and detection time by 10.46% for the varying energy supply-to-demand factor compared to DDSIM.

Chen et al. [13] proposed a control-centered data classification technology for the detection and analysis of emissions from industrial enterprises. Through intelligent hardware, the intensity, emission rate, and composition of emissions at different manufacturing intervals are obtained, and a repeated analysis is carried out via in-depth learning. Then, previous emission regulations and manufacturing guidelines can be improved to identify high emission intensities and dangerous components in gas emissions.

Bai et al. [14] developed the DECIA improvement model based on the internal quality control of water-saving products and external marketing policies, which promotes the upgrading of production processes or technologies to improve product quality and increase market penetration.

Taking into consideration the impact of lightweight quality on carbon emissions throughout the lifecycle of automobiles, Li et al. [15] proposed a more comprehensive lightweight design method. This method not only provides insight into the lightweight design of automobiles and other equipment against the background of low carbon but also offers a mean to calculate the carbon emission changes across the entire process after the implementation of the lightweight design.

4. Optimized Operation and Management of Remanufacturing Production System

Yang et al. [11] develop a game-theoretic model to examine the selection of different recycling strategies in the remanufacturing supply chain considering blockchain adoption and uncertain demand. Results show that the coefficient of collection investment costs determines the collection method and the incentive for collectors to participate in the blockchain.

Liu et al. [16] have established a remanufacturing supply chain recycling model based on the Bass innovation diffusion model. In this model, a single manufacturer takes the lead, while a single retailer follows. The retailer is responsible for the recycling aspect. The authors have determined the optimal wholesale price, retail price, and recovery effort path with the optimal control theory.

Chen et al. [17] studied the impact of subsidy policies on the donation strategy of the remanufacturing industry and found that the subsidy amount, first-mover advantage, and the form of the subsidy will affect the donation behavior of manufacturers and remanufacturers.

5. Conclusions

This Special Issue has published 18 papers on the sustainable development of supply chains in industrial engineering, including system modeling and simulation, supply chain management and evaluation, the combination of artificial intelligence and supply chains, and supply chain issues in remanufacturing. This Special Issue presents mathematical models for different industrial scenarios, improving supply chain management's efficiency in all aspects. Introducing artificial intelligence technology into industrial production promotes low-carbon and sustainable industrial development. We believe that integrating information technology and industry can promote sustainable supply chain development and accelerate the achievement of carbon neutrality goals in the manufacturing industry in the future [18,19].

We sincerely thank all the scientific contributors who submitted the papers in this special issue.

Author Contributions: Conceptualization, C.L. and N.W.; writing and editing—original draft preparation, X.S., Z.L. and F.W. All authors have read and agreed to the published version of the manuscript.

Funding: This research was funded by the Research Project on the Innovative Development of Social Sciences in Anhui Province (2021CX069, 2023CXZ018), the Anhui Provincial Natural Science Foundation General Project (2008085ME150) and the Academic Support Project for Top-notch Talents in Disciplines (Majors) in Colleges and Universities (gxbjZD2021083).

Conflicts of Interest: The authors declare no conflict of interest.

References

1. Gong, Y.; He, G. Research on Low-Carbon Strategies of Supply Chains, Considering Livestreaming Marketing Modes and Power Structures. *Processes* **2023**, *11*, 1505. [CrossRef]
2. Mu, W.; Xie, J.; Ding, H.; Gao, W. Data-Driven Evaluation of the Synergistic Development of Economic-Social-Environmental Benefits for the Logistics Industry. *Processes* **2023**, *11*, 913. [CrossRef]
3. Wang, H.; Dai, B.; Li, X.; Yu, N.; Wang, J. A Novel Hybrid Model of CNN-SA-NGU for Silver Closing Price Prediction. *Processes* **2023**, *11*, 862. [CrossRef]
4. Guo, Y.; Ding, H. Coupled and Coordinated Development of the Data-Driven Logistics Industry and Digital Economy: A Case Study of Anhui Province. *Processes* **2022**, *10*, 2036. [CrossRef]
5. Yang, Y.; Hu, F.; Ding, L.; Wu, X. Coupling Coordination Analysis of Regional IEE System: A Data-Driven Multimodel Decision Approach. *Processes* **2022**, *10*, 2268. [CrossRef]

6.   Huang, K.; Wang, J.; Zhang, J. Automotive Supply Chain Disruption Risk Management: A Visualization Analysis Based on Bibliometric. *Processes* **2023**, *11*, 710. [CrossRef]

7.   Liu, Q.; Sun, H.; He, Y. Bilateral Matching Decision Making of Partners of Manufacturing Enterprises Based on BMIHFIBPT Integration Methods: Evaluation Criteria of Organizational Quality-Specific Immunity. *Processes* **2023**, *11*, 709. [CrossRef]

8.   Liu, L.; Wu, C.; Zhu, Y. Employment Effect of Structural Changes in Strategic Emerging Industries. *Processes* **2023**, *11*, 599. [CrossRef]

9.   Shi, J.; Jiao, W.; Jing, K.; Yang, Q.; Lai, K. Joint Economic–Environmental Benefit Optimization by Carbon-Abatement Cost Sharing in a Capital-Constrained Green Supply Chain. *Processes* **2023**, *11*, 226. [CrossRef]

10.   Oršič, J.; Jereb, B.; Obrecht, M. Sustainable Operations of Last Mile Logistics Based on Machine Learning Processes. *Processes* **2022**, *10*, 2524. [CrossRef]

11.   Yang, T.; Li, C.; Bian, Z. Recycling Strategies in a Collector-Led Remanufacturing Supply Chain under Blockchain and Uncertain Demand. *Processes* **2023**, *11*, 1426. [CrossRef]

12.   Sun, F.; Diao, Z. Federated Learning and Blockchain-Enabled Intelligent Manufacturing for Sustainable Energy Production in Industry 4.0. *Processes* **2023**, *11*, 1482. [CrossRef]

13.   Chen, Z.; Chen, J. Control-Centric Data Classification Technique for Emission Control in Industrial Manufacturing. *Processes* **2023**, *11*, 615. [CrossRef]

14.   Bai, Y.; Liu, J.; Zhang, R.; Bai, X. Quality Control of Water-Efficient Products Based on DMAIC Improved Mode—A Case Study of Smart Water Closets. *Processes* **2023**, *11*, 131. [CrossRef]

15.   Li, Q.; Zhang, Y.; Zhang, C.; Wang, X.; Chen, J. Analysis Method and Case Study of the Lightweight Design of Automotive Parts and Its Influence on Carbon Emissions. *Processes* **2022**, *10*, 2560. [CrossRef]

16.   Liu, L.; Liu, Z.; Pu, Y.; Wang, N. Dynamic Optimal Decision Making of Innovative Products' Remanufacturing Supply Chain. *Processes* **2023**, *11*, 295. [CrossRef]

17.   Chen, X.; Li, Z.; Wang, J. Impact of Subsidy Policy on Remanufacturing Industry's Donation Strategy. *Processes* **2023**, *11*, 118. [CrossRef]

18.   Liu, C.; Chen, J.; Wang, W. Quantitative Evaluation Model of the Quality of Remanufactured Product. *IEEE Trans. Eng. Manag.* **2023**, 1–12. [CrossRef]

19.   Liu, C.; Chen, J.; Cai, W. Data-Driven Remanufacturability Evaluation Method of Waste Parts. *IEEE Trans. Ind. Inform.* **2021**, *18*, 4587–4595. [CrossRef]

Article

Research on Low-Carbon Strategies of Supply Chains, Considering Livestreaming Marketing Modes and Power Structures

Yonghua Gong [1,2,*] and Guangqiang He [1]

[1] School of Management, Nanjing University of Posts and Telecommunications, Nanjing 210003, China; 1021112620@njupt.edu.cn

[2] Information Industry Integration Innovation and Emergency Management Research Center, Nanjing University of Posts and Telecommunications, Nanjing 210003, China

* Correspondence: gongyh@njupt.edu.cn

Abstract: A livestreaming supply chain composed of a single manufacturer and a single streamer in the low-carbon market is examined. Motivated by the actual production and operation, both the manufacturer and the streamer have a chance to dominate the supply chain. Low-carbon strategies and livestreaming marketing modes of the supply chain are studied. The impacts of the consumer's price sensitivity coefficient, low-carbon preference, and streamer's promotion sensitivity coefficient on the equilibrium results are further studied. The results show that: the streamer achieves the optimal level of promotion effort in the resale mode under both power structures. The manufacturer achieves the optimal low-carbon level in the commission mode when the promotion sensitivity coefficient is smaller under both of two power structures. The streamer's profit is optimal in the resale mode, while the manufacturer's profit is optimal in the commission mode when under the streamer-led structure. Two parties' profits are optimal in the commission mode when the promotion sensitivity coefficient is smaller under the manufacturer-led structure. The low-carbon level, streamer promotion effort and selling price in two livestreaming marketing modes will increase when the streamer promotion sensitivity coefficient and consumer low-carbon preference increase and will decrease when consumer price sensitivity increases under two power structures. Lastly, the selling price in resale mode is always higher than that in commission mode under two power structures.

Keywords: low-carbon strategy; livestreaming marketing mode; consumer low-carbon preference; level of low-carbon promotion effort; power structures

Citation: Gong, Y.; He, G. Research on Low-Carbon Strategies of Supply Chains, Considering Livestreaming Marketing Modes and Power Structures. *Processes* **2023**, *11*, 1505. https://doi.org/10.3390/pr11051505

Academic Editors: Conghu Liu, Xiaoqian Song, Zhi Liu and Fangfang Wei

Received: 7 March 2023
Revised: 22 April 2023
Accepted: 6 May 2023
Published: 15 May 2023

1. Introduction

Global governments and environmental organizations have placed a high priority on the numerous climate and environmental issues caused by massive greenhouse gas emissions. In order to deal with this issue, the Chinese government has included carbon neutrality and peaking in its government work report for 2021, encouraging green and low-carbon development. Moreover, consumers' awareness of environmental protection has been continuously enhanced. Environmentally friendly, low-carbon products are more attractive to consumers [1–3]. With the development of online technology and the upgrading of shopping styles, consumers are increasingly turning to online shopping channels. According to Statista, global e-commerce sales will total about 5.2 billion U.S. dollars in 2021 and are expected to reach 8.1 billion U.S. dollars in 2026 [4]. To further cater to consumers, Facebook, TikTok, Jindong, Taobao, and others have started launching livestreaming marketing channels on their platforms. After the COVID-19 outbreak, livestreaming marketing has grown as a no-touch, interactive selling format. According to a report on livestreaming e-commerce in China, the number of livestreaming e-commerce users was 515 million as of December 2022 [5]. In that background, more manufacturers began to provide green products and adopt the popular livestreaming e-commerce channel to meet and promote consumers' low-carbon needs. The livestreaming e-commerce channel has been loved by a large number of consumers and manufacturers at present because it allows real-time interaction between consumers and steamers [6]. Then consumers can obtain professional explanations about the products or services, and manufacturers can obtain more sales. For the manufacturer, there are usually two main livestreaming marketing modes between manufacturers and streamers. One is the resale, and the other one is the commission. In the resale mode, the streamers have their own livestreaming room and operate the product sales channel; they purchase goods from manufacturers and resell them to consumers on the platform, such as Dongfangzhenxuan. Consumers buy the products in the livestreaming room, and the streamers gain sales revenue. In the commission mode, the manufacturers operate the product sales channel, and they hire the streamers to help them sell products in the streamers' live steaming room and pay a certain percentage of commission to the streamers, such as Li Jiaqi and Wei Ya. Consumers buy the products in manufacturers' online shops, and the manufacturers gain revenue. Streamers can stimulate consumers' willingness to buy low-carbon products.

However, the selection of the livestreaming marketing mode and low-carbon promotion effort is an important decision of the low-carbon livestreaming supply chain. Because although livestreaming marketing can attract more consumers to purchase low-carbon products through low-carbon promotion efforts made by streamers, many limitations remain. Under the commission mode, manufacturers are required to pay commissions to streamers, which typically account for 30–50% of sales [7], and the high commissions may squeeze manufacturers' profit margins. Under the resale mode, the manufacturer loses the power to price their products, and some streamers try to attract consumers by decreasing prices, which may create a psychology of fairness concern for consumers [8] and impact other sales channels of manufacturers. Moreover, different efforts will be made by the streamers in different livestreaming marketing modes. Then manufacturers need to choose the mode of cooperating with streamers, especially when the manufacturers or streamers play a leading role in the supply chain. What's more, manufacturers must also face the additional unit production cost of producing low-carbon goods. Obviously, high production costs often lead to higher product selling prices. The decision of low-carbon and selling price is also an important issue for manufacturers. Manufacturers' low-carbon strategy will also be affected by consumers' low-carbon preference and price sensitivity, streamer's promotion effort. Therefore, our study will also discuss how these factors affect the strategy of the livestreaming supply chain. We attempt to solve problems as follows:

(1) Optimal low-carbon strategies of the manufacturer and streamer under two livestreaming marketing modes when there are two different power structures, respectively, dominated by the manufacturer and the streamer.

(2) How do the marketing mode and power structure impact the optimal low-carbon strategies of the members? When to choose a "resale" or "commission" mode to obtain more profits?

(3) The impacts of the parameters, including the consumer price sensitivity coefficient, commission ratio, streamer's low-carbon promotion level coefficient and consumer's low-carbon preference on the optimal strategies of the manufacturer and streamer under the two livestreaming marketing modes.

To solve those issues, this paper establishes game models of a livestreaming supply chain and finds optimal strategies for livestreaming supply chain members on streaming marketing modes, prices and carbon emission reduction efforts. We discussed four supply chain structures: manufacturer-led in resale mode (model *wm*), streamer-led in resale mode (model *wl*), manufacturer-led in commission mode (model *am*), and streamer-led in commission mode (model *al*). Furthermore, the influences of model parameters on decision variables are analyzed. The results of our research will help the manufacturer and streamer manage and improve their strategies for livestreaming marketing mode selection and carbon emission reduction. The main contributions of this work include the following:

- This paper focuses on a livestreaming supply chain in which the manufacturer and the streamer need to decide both the low-carbon promotion level and optimal livestreaming marketing mode. Unlike previous research focused on the supply chain with manufacturers and e-commerce platforms [9,10], we considered the importance of streamers' low-carbon promotion which will influence consumers' willingness to purchase low-carbon products.
- We examine the influences of different power structures on manufacturers' and streamers' decisions to choose their livestreaming marketing mode. Studies on the choice of livestreaming marketing mode have only considered the case of power parity between manufacturers and streamer [6,7]. And interestingly, we found that the optimal selection of marketing mode of both the manufacturer and the streamer always depends on their position in the supply chain, and they will make the same selection in the manufacture-led structure and the opposite selection in the streamer-led structure.

This paper seeks the maximum low-carbon effort level. It is found that both the manufacturer's and streamer's optimal low-carbon effort levels do not depend on the power structure of the supply chain. The manufacturer's optimal low-carbon effort level only depends on the livestreaming marketing mode, while the streamer's low-carbon effort level depends on both the livestreaming marketing mode and the promotion sensitivity coefficient. The manufacturer always makes the maximal effort in the commission mode. The streamer makes the maximal promotion effort in the resale mode when the promotion sensitivity coefficient is smaller, and the streamer makes the maximal promotion effort in the commission mode when the promotion sensitivity coefficient is bigger. The rest part of the article is arranged as follows: In Section 2, related studies are offered. Section 3 is about model descriptions and assumptions. In Section 4, models of resale mode and commission mode are established, and the optimal decision is obtained. In Section 5, decision results are compared and analyzed. In Section 6, numerical experiments are conducted. In Section 7, a case is studied. Finally, the conclusions and limitations are in Section 8. All mathematical proofs are shown in Appendix A.

2. Related Works

2.1. Low-Carbon Strategies of Supply Chain

Consumer awareness of environmental protection is considered an important factor that can affect the emission reduction level of the supply chain. Zhang et al. [11] examined how consumers' low-carbon awareness and merchants' concerns about fairness in cost-sharing percentages and supply chain decisions. The research found that increasing consumers' environmental awareness and carbon emission reduction sharing ratio was beneficial to improving environmental quality. However, the more significant the retailers' fairness concern coefficient, the more unfavorable it was to environmental quality. Ji et al. [12] made a single-game model and a combined game model for reducing emissions. They found that the combined strategy for reducing emissions is good for both retailers and manufacturers. Wu et al. [13] created a differential game model with centralized versus decentralized decision-making. They discovered that consumer preferences for low-carbon products encouraged suppliers and manufacturers to reduce emission levels. The government's low-carbon policies also influence the emission reduction efforts of supply chain members [14,15]. Zhang et al. [16] studied the emission reduction of a dual-channel supply chain under the background of carbon quota. Chen et al. [17] compared manufacturers' channel choice and emission reduction under various government subsidies. Wang et al. [18] formulated a revenue-sharing contract with compensation. Wang et al. [19] considered retailers' altruistic preferences when making decisions and coordination. Gong et al. [20] studied the strategies of blockchain technology adoption and channel selection when consumers' willingness to pay for remanufactured products is low. Basiri et al. [21] proposed a supply chain coordination contract to improve product greenness while lowering retail prices. Zhang et al. [22] studied how an e-platform supply chain selects the optimal channel mode from platform mode and wholesale mode in the presence of a secondary marketplace. Dolai et al. [23] considered an EPQ model for screening imperfect products with GHG emission rates, and they concluded that higher screening rates lead to higher average profits when the screening rate is below a threshold, but higher screening rates may lead to lower profits when the screening rate is above a threshold. Manna et al. [24] proposes a two-plant production model that takes into account consumer demand. They believe that product warranty policies and rework policies are conducive to lowering greenhouse gas emissions and increasing average manufacturer profits and customer demand. Kumar et al. [25] developed a production inventory model with dynamic demand and found that reducing the cost of defective items facilitates manufacturers to increase their average profit. Manna et al. [26] studied the inventory problem of retailers with two warehouses based on the TDE algorithm.

Previous research on supply chain emission reduction mainly focused on manufacturers and retailers, but with the development of the platform economy, the mode of cooperation between manufacturers and streamers on the platform is becoming increasingly popular. Different from the previous research, we focus on the supply chain composed of manufacturers and streamers under the livestreaming marketing mode.

2.2. Livestreaming E-Commerce

There are few works of literature on livestreaming supply chains, and the existing research focuses primarily on manufacturers' livestreaming strategies and consumers' purchase behavior in live e-commerce. Zhang et al. [27] investigated the decision to implement live channels in multinational corporations while considering channel substitutability and tax differences. Gong et al. [28] discussed online retailers' live channel introduction strategy. Hao et al. [29] studied the effect of consumer returns and living sales on the selling mode chosen by e-commerce platforms and suppliers. Zhang et al. [30] found that live-streaming services always benefit e-commerce platforms, whereas manufacturers must consider production costs. Ma et al. [31] used an empirical approach to investigate the mechanism of streamer interaction's influence on consumer behavior and to analyze the factors influencing consumer purchase hesitation. Chen et al. [32] explored the mechanism of the role of weblebrities' traits and consumers' purchase intention and discovered that the more significant the weblebrities' traits, the stronger the consumers' purchase intention. Huang et al. [33] found that celebrity streamers selling hedonic goods and corporate streamers selling valuable goods are more likely to stimulate consumers' purchase desire. Xing et al. [34] examined the relationship between streamer commissions and streamer service quality efforts in both sign-up and non-sign-up scenarios. The research shows that the streamer's effort increases with the increase of streamer commission.

The preceding studies have empirically examined the impact of livestreaming marketing on consumers' purchase intention motivation and demonstrated the ability of streamers to stimulate consumers to purchase products. However, fewer studies have considered models to examine manufacturers' choices of livestreaming marketing mode and consider that manufacturers and streamers are in different power structures.

2.3. Power Structures of Supply Chain

Different power structures will lead to different decision results. Huang et al. [35] analyzed the supply chain members' optimal emission reduction decisions under different power structures, with manufacturers taking CSR and retailers taking social responsibility. It confirmed that the green degree and green publicity level of products are higher when the other party engages in CSR with both of the power structures. Li et al. [36] discovered that supply chain members always make more money when they have dominance. Lu et al. [37] studied the pricing problem with three structures: retailer-dominated, manufacturer-dominated, and power-parity situations. Chen et al. [38] explore the power structure's impact on the optimal O2O price and profit of retailers and suppliers. Generally, the manufacturer requires the dominant position between the streamer and the manufacturer, but not all streamers are subordinate, especially the celebrity streamers who can even take the initiative when cooperating with the manufacturer. Therefore, there are also different power structures between manufacturers and streamers under livestreaming marketing. Qiu et al. [39] considered a supply chain consisting of an online retail platform, a used sales platform, and a recycling platform and investigated the influence of different power structures on the integration strategies of used recycling platforms.

Most of the previous work considered the power imbalance between a manufacturer and an e-commerce platform and a manufacturer and a retailer in the supply chain. Our work differs from previous research in the following aspects. First, we consider the power imbalance between a manufacturer and a streamer in livestreaming marketing. Second, we study the impact of different power structures on the low-carbon strategies of the manufacturer and streamer.

To sum up, the existing literature has achieved specific results on low-carbon supply chain and livestreaming e-commerce, paying more attention to the low-carbon strategies between manufacturers and e-commerce platforms and the strategies about whether the manufacturer should choose a livestreaming channel, but the low-carbon problem of the livestreaming supply chain is rarely addressed. Not paying enough attention to the fact that the livestreaming platform is different from an e-commerce platform on the consumer's willingness to pay and the selection of the livestreaming marketing mode, especially when there are more than one livestreaming marketing mode and power structure. This paper will consider these facts. Our study can help manufacturers, and steamers determine the optimal livestreaming mode and emission reduction strategy.

3. Model Description and Assumption

3.1. Model Description

There is a manufacturer and a streamer. The decision objective of manufacturers and streamers is to maximize their own revenue. They have two livestreaming marketing modes to sell the products. One is resale mode, and the other one is commission mode. In the resale mode, the manufacturer sells the products to the streamer at a wholesale price, and the streamer reprices and sells them in the livestreaming room. Manufacturers gain income from wholesale products, and streamers gain income from product sales. In the commission mode, the streamer sells the products in the livestreaming room and draws a certain percentage of commission from each order, the selling price determined by the manufacturer. Manufacturers gain income from product sales.

There are two power structures between the manufacturer and the streamer: manufacturer-led and streamer-led. The research objective is to discuss the decision of low-carbon and livestreaming marketing mode of the manufacturer and streamer in four situations: manufacturer-led in resale mode (model wm), streamer-led in resale mode (model wl), manufacturer-led in commission mode (model am), and streamer-led in commission mode (model al). Then we build Stackelberg game models for four scenarios to analyze the optimal strategies of low-carbon efforts and livestreaming marketing modes.

3.2. Assumptions

Consumers are low-carbon conscious in the low-carbon supply chains; they are willing to pay higher prices for low-carbon products [40–43]. And consumers' purchasing needs in the livestreaming supply chain are also influenced by the promotion efforts of streamers [44]. Therefore, the consumer's demand in the low-carbon livestreaming supply chain is mainly influenced by the low-carbon level of the product e of manufacture, the low-carbon promotion effort level s of the streamer and the product price p. By increasing the low-carbon level of products, manufacturers attract low-carbon-preferring consumers to buy them [45], and streamers further strengthen consumers' low-carbon consciousness through product demonstration and narration, stimulating the growth of demand in the livestreaming room. According to [46–48], we have Assumption 1 about the demand function.

Assumption 1. *Consumers have a low-carbon preference, and more consumers are willing to pay higher prices for low-carbon products; manufacturers are willing to make low-carbon investments to meet consumers' low-carbon demand and gain more profits. The demand function of consumer's low-carbon products under the livestreaming marketing mode is*

$$d = a - \mu p + \eta e + \varphi s \tag{1}$$

where μ and η are the consumer's price sensitivity coefficient and low-carbon preference for the product, and the φ is the streamer low-carbon promotion sensitivity coefficient.

Assumption 2. *The manufacturer is responsible for improving the low-carbon level of the product, and the streamer is responsible for improving the low-carbon promotion effort level in the low-carbon supply chain. The manufacturer pays production and R&D costs for low-carbon products. The streamer pays promotion effort cost. Without losing generality, we assume the product's production cost is 0 without affecting the calculation result [21,49]. Refer to [50,51]. We assume the manufacturer's unit production cost is $c(e) = e^2/2$, and the streamer's unit sales cost is $c(s) = s^2/2$.*

Assumption 3. *Manufacturers pay the same percentage of commission to the same type of streamers in the industry in the commission mode. Because in actual operation, the manufacturer does not set a separate commission ratio for each streamer but often decides uniformly based on market conditions [52], and usually less than 0.5 [29], then we assume that the streamer commission ratio θ is an exogenous variable, $0 < \theta < 0.5$.*

The definitions of parameters in the paper are shown in Table 1.

Table 1. Parameters definition.

Symbol	Description
d	Demand function of consumers
a	Potential maximum demand
p	Product selling price
w	Wholesale price in resale mode
k	product unit profit in the resale mode
μ	Price sensitivity coefficient
e	Product low-carbon level
η	Consumer low-carbon preference
s	Streamer low-carbon promotion effort level
φ	Streamer low-carbon promotion sensitivity coefficient
θ	Commission ratio
π_m^i, π_m^j	Manufacturer's profit under two livestreaming marketing modes
π_l^i, π_l^j	Streamers' profit under two livestreaming marketing modes
i, j	$i = wm, wl$, wm: manufacturer-led in the resale mode, wl: streamer-led in the resale mode $j = am, al$, am: manufacturer-led in the commission mode, al: streamer-led in the commission mode

In the resale mode, the manufacturer produces at level e^i, and resells it to the streamer at price w^i. The streamer decides its promotion effort level s^i and selling price p^i, where $p^i = k^i + w^i$, k^i stands for the product's profit per unit; as a result, the streamer decides the profit per unit product k^i.

In the commission mode, the manufacturer produces at level e^j, and sells it to the consumer with the help of the streamer at price p^j. The streamer promotion effort is s^j. The structure of the models in resale and commission mode are shown in Figure 1.

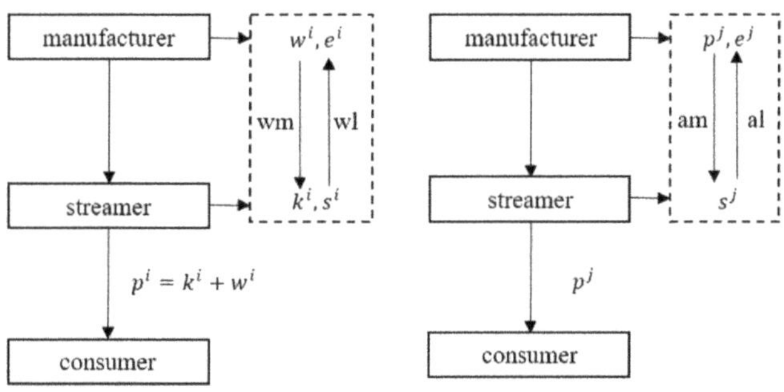

Figure 1. Model structure.

4. Modeling and Analysis

4.1. Resale Mode

The manufacturer's profit function in resale mode is:

$$\pi_m^i = w^i \left[a - \mu \left(k^i + w^i \right) + \eta e^i + \varphi s^i \right] - \frac{1}{2} (e^i)^2 \tag{2}$$

The streamer's profit function in resale mode is:

$$\pi_l^i = k^i \left[a - \mu \left(k^i + w^i \right) + \eta e^i + \varphi s^i \right] - \frac{1}{2} (s^i)^2 \tag{3}$$

4.1.1. Model *wm* (Manufacturer-Led)

In the model *wm*, the livestreaming marketing mode is the resale mode; the structure is manufacturer-led. The objective of the manufacturer and streamer is to maximize their own profit. The decision sequence is: (1) the manufacturer decides w and e, (2) the streamer decides k and s. Practice shows consumers' low-carbon preference and promotion sensitivity coefficient are not infinite, but there is a range. In order to make our study more realistic and have optimal solutions for functions (2) and (3), the following assumption about the model parameters are made regarding the studies of Wang et al. [19,48] scholars.

Assumption 4. *The parameters meet the following conditions*: the $\eta < \sqrt{-2\varphi^2 + 4\mu}$, $\varphi < \sqrt{-\frac{1}{2}\eta^2 + 2\mu}$ *in resale mode with a manufacturer-led structure.*

Lemma 1. *The optimal low-carbon strategies in resale mode with a manufacturer-led structure are as follows:*

$$e^{wm*} = -\frac{a\eta}{\eta^2 + 2\varphi^2 - 4\mu} \tag{4}$$

$$w^{wm*} = -\frac{a\left(-\varphi^2 + 2\mu\right)}{\mu(\eta^2 + 2\varphi^2 - 4\mu)} \tag{5}$$

$$k^{wm*} = \frac{a}{-\eta^2 - 2\varphi^2 + 4\mu} \tag{6}$$

$$p^{wm*} = \frac{a\left(-\varphi^2 + 3\mu\right)}{\mu(-\eta^2 - 2\varphi^2 + 4\mu)} \tag{7}$$

$$s^{wm*} = \frac{\varphi a}{-\eta^2 - 2\varphi^2 + 4\mu} \tag{8}$$

The following Propositions can be drawn from Lemma 1.

Proposition 1. $\frac{\partial e^{wm*}}{\partial \mu} < 0, \frac{\partial k^{wm*}}{\partial \mu} < 0, \frac{\partial s^{wm*}}{\partial \mu} < 0.$

From Proposition 1 we can find low-carbon level, unit profit and streamer's low-carbon promotion effort level are negatively related to the price sensitivity coefficient in resale mode with a manufacturer-led structure. That shows the higher the price sensitivity coefficient, consumers are easier influenced by the product's price, and the lower the selling price is. At the same time, the low-carbon product's selling price is always higher than the traditional product's price because of the higher R&D cost and streamer promotion cost. Consumers will be less interested in purchasing low-carbon products. The manufacturer will choose a lower low-carbon level; the streamer will choose a lower low-carbon promotion effort level.

Proposition 2. $\frac{\partial w^{wm*}}{\partial \eta} > 0, \frac{\partial e^{wm*}}{\partial \eta} > 0, \frac{\partial k^{wm*}}{\partial \eta} > 0, \frac{\partial p^{wm*}}{\partial \eta} > 0, \frac{\partial s^{wm*}}{\partial \eta} > 0.$

From Proposition 2, we can see wholesale price, low-carbon level, unit profit, selling price and streamer's low-carbon promotion effort are positively related to consumer's low-carbon preference in resale mode with manufacturer-led structure. The results are consistent with the reality. That shows both manufacturer and streamer will be motivated by consumers' low-carbon preference to pay more effort and costs for low-carbon promotion. The wholesale price, unit profit and selling price increase subsequently.

Proposition 3. $\frac{\partial w^{wm*}}{\partial \varphi} > 0, \frac{\partial e^{wm*}}{\partial \varphi} > 0, \frac{\partial k^{wm*}}{\partial \varphi} > 0, \frac{\partial p^{wm*}}{\partial \varphi} > 0, \frac{\partial s^{wm*}}{\partial \varphi} > 0.$

From Proposition 3 we can see wholesale price, low-carbon level, selling price, steamer's low-carbon promotion effort, and unit profit is positively related to the streamer's promotion sensitivity coefficient in resale mode with manufacturer-led structure. That means the streamer's low-carbon propaganda has positive feedback on the low-carbon level and product selling price, the product selling price has positive feedback on the streamer's low-carbon propaganda, then a feedback loop is formed.

4.1.2. Model wl (Streamer-Led)

In the model wl, the livestreaming marketing mode is the resale mode; the structure is streamer-led. The decision sequence is: (1) the streamer decides the value of k and s, and (2) the manufacturer decides the value of w and e.

Assumption 5. *The parameters meet the following conditions:* $\eta < \sqrt{\frac{-\varphi^2 + 4\mu}{2}}, \varphi < \sqrt{-2\eta^2 + 4\mu}$ *in resale mode with a streamer-led structure.*

The basis of Assumption 5 is similar to Assumption 4.

Lemma 2. *The optimal low-carbon strategies in resale mode with a streamer-led structure are as follows:*

$$e^{wl*} = \frac{a\eta}{-2\eta^2 - \varphi^2 + 4\mu} \tag{9}$$

$$w^{wl*} = \frac{a}{-2\eta^2 - \varphi^2 + 4\mu} \tag{10}$$

$$k^{wl*} = \frac{a(2\mu - \eta^2)}{(-2\eta^2 - \varphi^2 + 4\mu)\mu} \tag{11}$$

$$p^{wl*} = \frac{a(3\mu - \eta^2)}{(-2\eta^2 - \varphi^2 + 4\mu)\mu} \tag{12}$$

$$s^{wl*} = -\frac{a\varphi}{2\eta^2 + \varphi^2 - 4\mu} \tag{13}$$

The following Propositions can be drawn from Lemma 2.

Proposition 4. $\frac{\partial e^{wl*}}{\partial \mu} < 0, \frac{\partial w^{wl*}}{\partial \mu} < 0, \frac{\partial s^{wl*}}{\partial \mu} < 0.$

Proposition 4 shows low-carbon level, unit profit and streamer's low-carbon promotion effort level are negatively related to the price sensitivity coefficient in resale mode with a streamer-led structure. It is similar to Proposition 1.

Proposition 5. $\frac{\partial e^{wl*}}{\partial \eta} > 0, \frac{\partial w^{wl*}}{\partial \eta} > 0, \frac{\partial k^{wl*}}{\partial \eta} > 0, \frac{\partial p^{wl*}}{\partial \eta} > 0, \frac{\partial s^{wl*}}{\partial \eta} > 0.$

Proposition 5 means wholesale price, low-carbon level, unit profit, selling price and streamer's low-carbon promotion effort are positively related to consumer's low-carbon preference in resale mode with a streamer-led structure. It is similar to Proposition 2.

Proposition 6. $\frac{\partial e^{wl*}}{\partial \varphi} > 0, \frac{\partial w^{wl*}}{\partial \varphi} > 0, \frac{\partial k^{wl*}}{\partial \varphi} > 0, \frac{\partial p^{wl*}}{\partial \varphi} > 0, \frac{\partial s^{wl*}}{\partial \varphi} > 0.$

Proposition 6 shows that wholesale price, low-carbon level, selling price, steamer's low-carbon promotion effort, unit profit is positively related to a streamer's promotion sensitivity coefficient in resale mode to a streamer-led structure. Similar to Proposition 3.

4.2. Commission Mode

The manufacturer's profit function in commission mode is:

$$\pi_m^j = p^j (1 - \theta)\left(a - \mu p^j + \eta e^j + \varphi s^j\right) - \frac{1}{2}\left(e^j\right)^2 \tag{14}$$

The streamer's profit function in commission mode is:

$$\pi_l^j = p^j \theta \left(a - \mu p^j + \eta e^j + \varphi s^j\right) - \frac{1}{2}\left(s^j\right)^2 \tag{15}$$

4.2.1. Model *am* (Manufacturer-Led)

In the model *am*, the livestreaming marketing mode is commission mode, and the structure is manufacturer-led. The objective of the manufacturer and streamer is to maximize their own profit. The decision sequence is: (1) the manufacturer decides p and e, (2) the streamer decides s. In order to make our study more realistic and have optimal solutions for functions (14) and (15), we propose Assumption 6. The basis of Assumption 6 is similar to Assumption 4.

Assumption 6. *The parameters meet the following conditions:* $\eta < \sqrt{\frac{2(\mu - \theta \varphi^2)}{1 - \theta}}, \varphi < \sqrt{\frac{(\eta^2 \theta - \eta^2 + 2\mu)}{2\theta}}$ *in commission mode with manufacturer-led structure.*

Lemma 3. *The optimal low-carbon strategies in commission mode with a manufacturer-led structure are as follows:*

$$e^{am*} = \frac{\eta(1 - \theta)a}{\eta^2 \theta - 2\theta \varphi^2 - \eta^2 + 2\mu} \tag{16}$$

$$p^{am*} = \frac{a}{\eta^2 \theta - 2\theta \varphi^2 - \eta^2 + 2\mu} \tag{17}$$

$$s^{am*} = \frac{a\theta \varphi}{\eta^2 \theta - 2\theta \varphi^2 - \eta^2 + 2\mu} \tag{18}$$

The following Propositions can be drawn from Lemma 3.

Proposition 7. $\frac{\partial e^{am*}}{\partial \mu} < 0, \frac{\partial p^{am*}}{\partial \mu} < 0, \frac{\partial s^{am*}}{\partial \mu} < 0.$

From Proposition 7, it can be seen that the low-carbon level, selling price and streamer's low-carbon promotion effort level are negatively related to the price sensitivity coefficient in commission mode with manufacturer-led structure. It's similar to Propositions 1 and 4.

Proposition 8. $\frac{\partial e^{am*}}{\partial \eta} > 0$, $\frac{\partial p^{am*}}{\partial \eta} > 0$, $\frac{\partial s^{am*}}{\partial \eta} > 0$.

Proposition 8 shows that low-carbon level, selling price and streamer's low-carbon effort are positively related to consumer's low-carbon preference in commission mode with a manufacturer-led structure. It's similar to Propositions 2 and 5.

Comparing Proposition 8 with Proposition 2, we can find that the mode of cooperation between streamer and manufacturer does not change the impact of low-carbon preferences on the members' optimal strategies.

Proposition 9. $\frac{\partial e^{am*}}{\partial \varphi} > 0$, $\frac{\partial p^{am*}}{\partial \varphi} > 0$, $\frac{\partial s^{am*}}{\partial \varphi} > 0$.

Proposition 9 shows that the low-carbon level, selling price and streamer's low-carbon promotion effort are positively related to the steamer's promotion sensitivity coefficient in commission mode with a manufacturer-led structure. It's similar to Propositions 3 and 6. This indicates that if we want to improve the low-carbon level and streamer's promotion effort, we'd better improve the steamer's promotion sensitivity coefficient.

Proposition 10. $\frac{\partial e^{am*}}{\partial \theta} < 0$, $\frac{\partial s^{am*}}{\partial \theta} > 0$.

Proposition 10 reveals that the low-carbon level is negatively related to the commission ratio, streamer's low-carbon promotion effort level is positively related to the commission ratio in the commission mode with a manufacturer-led structure.

4.2.2. Model *al* (Streamer-Led)

In model *al*, the livestreaming marketing mode is commission mode; the structure is streamer-led. The decision sequence is: (1) the streamer decides s, (2) the manufacturer decides p and e.

Assumption 7. *The parameters meet the following conditions:* $\eta < \sqrt{\frac{-2\mu + \varphi\sqrt{2\mu\theta}}{1-\theta}}$, $\varphi < \frac{\sqrt{2}\left(\eta^2\theta - \eta^2 + 2\mu\right)}{2\sqrt{\mu\theta}}$ *in commission mode with a streamer-led structure.*

The basis of Assumption 7 is similar to Assumption 4.

Lemma 4. *The optimal low-carbon strategies in commission mode with a streamer-led structure are as follows:*

$$e^{al*} = \frac{(1-\theta)\eta a\left((\theta-1)\eta^2 + 2\mu\right)}{(-1+\theta)^2\eta^4 + 4\mu(-1+\theta)\eta^2 - 2\varphi^2\theta\mu + 4\mu^2} \tag{19}$$

$$p^{al*} = \frac{(-1+\theta)a\eta^2 + 2\mu a}{(-1+\theta)^2\eta^4 + 4\mu(-1+\theta)\eta^2 - 2\varphi^2\theta\mu + 4\mu^2} \tag{20}$$

$$s^{al*} = \frac{2a\theta\mu\varphi}{(-1+\theta)^2\eta^4 + 4\mu(-1+\theta)\eta^2 - 2\varphi^2\theta\mu + 4\mu^2} \tag{21}$$

The following Propositions can be drawn from Lemma 4. Due to the computational complexity, the effect of the consumers' low-carbon preference and commission ratio on optimal strategies will be discussed in the numerical study.

Proposition 11. $\frac{\partial e^{al*}}{\partial \mu} < 0$, $\frac{\partial p^{al*}}{\partial \mu} < 0$, $\frac{\partial s^{al*}}{\partial \mu} < 0$.

Proposition 11 shows that the low-carbon level, selling price and streamer's low-carbon promotion effort level is negatively related to the price sensitivity coefficient in commission mode with a streamer-led structure. It's similar to Propositions 1, 4 and 7. Comparing Propositions 1, 4, 7 and 11, it can be found that the impacts of the price sensitivity coefficient on the optimal strategies are not changed by both the power structure or livestreaming marketing mode.

Proposition 12. $\frac{\partial e^{al*}}{\partial \varphi} > 0$, $\frac{\partial p^{al*}}{\partial \varphi} > 0$, $\frac{\partial s^{al*}}{\partial \varphi} > 0$.

Proposition 12 shows that the low-carbon level, selling price and streamer's low-carbon promotion effort in commission mode with the streamer-led structure is positively related to the steamer's promotion sensitivity coefficient. It's similar to Propositions 3, 6 and 9. Comparing Propositions 3, 6, 9 and 12, it can be found that the impacts of the promotion sensitivity coefficient on the optimal strategies are not changed by both power structure or livestreaming marketing mode.

5. Results Comparison

Optimal results in resale mode and commission mode are compared on the basis of the above conclusions so that we can obtain the conclusions of the two parties' optimal choice about the livestreaming marketing mode under two different power structures.

Proposition 13. *When it's under the manufacturer-led structure: (1)* $e^{wm*} < e^{am*}$*, if* $0 < \varphi < \sqrt{\mu}$*;* $e^{wm*} > e^{am*}$*, if* $\varphi > \sqrt{\mu}$*. (2)* $s^{wm*} > s^{am*}$*.*

Proposition 13 suggests that the manufacturer's optimal low-carbon level in commission mode is higher than that in resale mode when the streamer's promotion sensitivity coefficient does not meet critical value. Otherwise, the low-carbon level in resale mode is higher. The streamer's optimal low-carbon promotion effort level in resale mode is always higher than that in commission mode. For the manufacturer, the commission mode is a better choice when the steamer's low-carbon promotion sensitivity coefficient is low. Otherwise, the resale mode is a better choice. The resale mode is always better for the steamer than the commission mode. Therefore, the choice of livestreaming marketing mode depends on the steamer's low-carbon promotion sensitivity coefficient under the manufacturer-led structure. Both parties will unanimously choose the resale mode when the steamer's low-carbon promotion sensitivity coefficient is big. When the steamer's low-carbon promotion sensitivity coefficient is small, the streamer can only choose the commission mode listening to the manufacturer in practice because the manufacturer has a stronger voice.

Proposition 14. *When it's under the streamer-led structure: (1)* $e^{wl*} > e^{al*}$*, if* $\theta > \theta'$*;* $e^{wl*} < e^{al*}$*, if* $\theta < \theta'$*, where* $\theta' = \frac{\sqrt{2\mu(2\mu-\varphi^2)(2\mu-\eta^2)(2\mu-\eta^2-\varphi^2)} + (2\mu-\eta^2)(2\mu-\eta^2-\varphi^2)}{\eta^2(4\mu-\eta^2-\varphi^2)}$*. (2)* $s^{wl*} > s^{al*}$*.*

Proposition 14 suggests that the manufacturer's optimal low-carbon level in commission mode is higher than that in resale mode when the streamer's commission ratio does not meet critical value. Otherwise, the low-carbon level in resale mode is higher. The streamer's optimal low-carbon promotion effort level in resale mode is always higher than that in commission mode. The resale mode is a better choice for the manufacturer than the resale mode when the commission ratio is low. Otherwise, the resale mode is the better choice. The resale mode is always better for the streamer than the commission mode. Therefore, the choice of livestreaming marketing mode depends on the commission ratio under the streamer-led structure. Both parties will unanimously choose the resale mode when the commission ratio is big. When the commission ratio is small, the manufacturer can only choose the resale mode listening to the streamer in practice because the streamer has a stronger voice.

Proposition 15. *(1) When it's under the manufacturer-led structure:* $p^{wm*} > p^{am*}$. *(2) When it's under the streamer-led structure:* $p^{wl*} > p^{al*}$.

Proposition 15 suggests the optimal product selling prices in resale mode are always higher than those in commission mode under both the manufacturer-led and streamer-led structures. Under the resale mode, the manufacturer and the streamer jointly determine the product's selling price, and both of them want to maximize their interests. Then the selling price is increased as much as possible.

Proposition 16. *(1) When it's under the manufacturer-led structure:* $\pi_m^{wm*} < \pi_m^{am*}$, *if* $0 < \varphi < \sqrt{\mu}$; $\pi_m^{wm*} > \pi_m^{am*}$, $\varphi > \sqrt{\mu}$. *(2) When it's under the streamer-led structure:* $\pi_l^{wl*} > \pi_l^{al*}$.

Proposition 16 suggests that the manufacturer can make more profit through the commission mode when the streamer promotion sensitivity factor is weak under the manufacturer-led structure. Otherwise, the manufacturer can make more profit through the resale mode. The streamer can always make more profit through the resale model. The comparison of the profits of the streamer under the manufacturer-led structure and the profits of the manufacturer under the streamer-led structure is too complex to calculate and is therefore analyzed in the numerical example.

6. Numerical Experiment Study

The effects of the parameters μ, η, φ and θ on the manufacturer's and streamer's optimal low-carbon strategy under two power structures have been simulated by numerical experiments. Referring to the numerical experiment studies of [12,19], it is assumed that $a = 40$, $\eta = 0.7$, $\mu = 0.5$, $\varphi = 0.4$, $\theta = 0.3$.

6.1. Impact of Coefficient μ

In this section, the values of parameter η, φ, θ are held constant, and the value of μ keeps increasing. Figure 2 shows that the low-carbon level, selling price, streamer's low-carbon promotion effort level, and profit of both parties all decrease with the increase of coefficient μ, which are consistent with Propositions 1, 4, 7, and 11. The selling price in resale mode is higher than that in commission mode, and Proposition 15 has the same conclusion. That shows when consumers' sensitivity coefficient to selling price increases, their willingness to buy decreases, then the selling price, low-carbon level, low-carbon promotion effort will be reduced. The profits of both parties will also decrease.

6.2. Impact of Preference η

In this section, the values of parameter μ, φ, θ are held constant, and the value of η keeps increasing. The results shown in Figure 3 are consistent with Propositions 2, 5, and 8. As can be seen in Figures 2b and 3a, in the resale mode, both the optimal low-carbon level of the product and the optimal low-carbon promotion effort level is higher when the streamer dominates the supply chain rather than under manufacturer domination. In the commission mode, even when streamers dominate the supply chain, the level of optimal low-carbon promotional effort is still higher than that under manufacturer domination. In addition, the more consumers that prefer low-carbon products in both the resale and commission models, the greater the influence of the power structure on the low-carbon strategies of manufacturers and streamers. It needs to be added that the power structure has less impact on the low-carbon level of the product in the commission model.

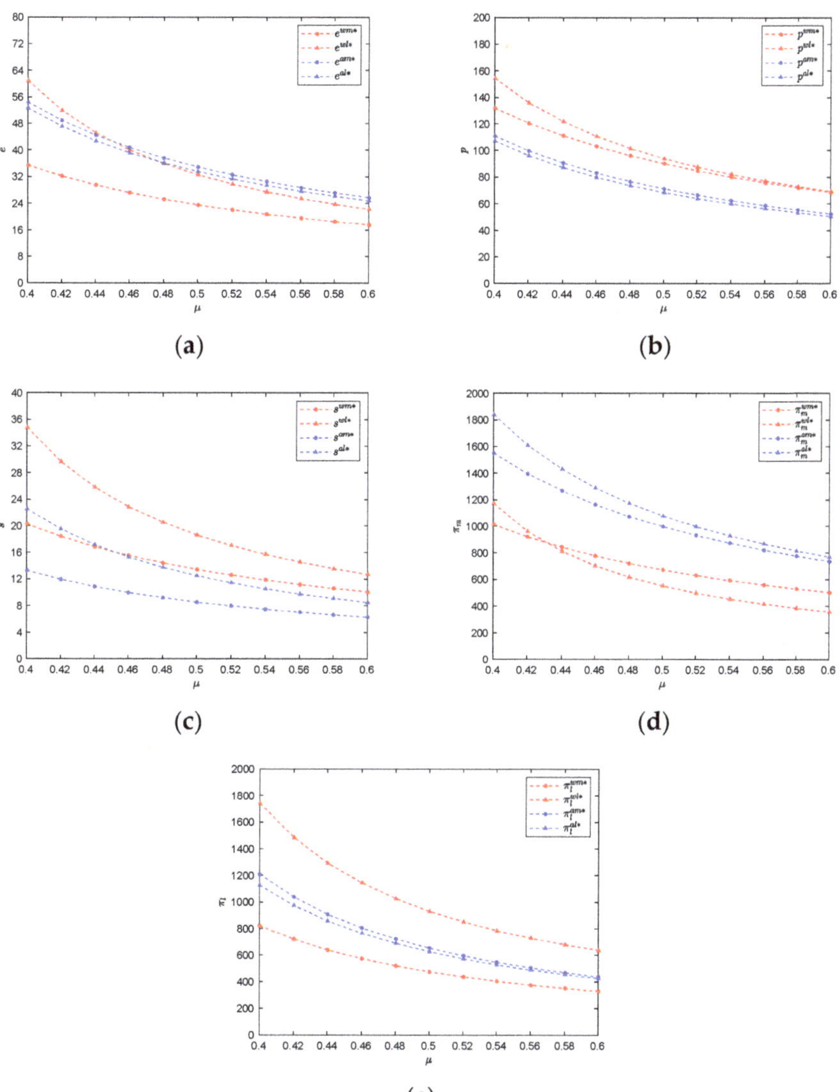

Figure 2. (**a**) changes of low-carbon level with μ; (**b**) changes of the product sale price with μ; (**c**) changes of low-carbon promotion efforts level with μ; (**d**) changes of manufacturer profits with μ; (**e**) changes of streamer profits with μ.

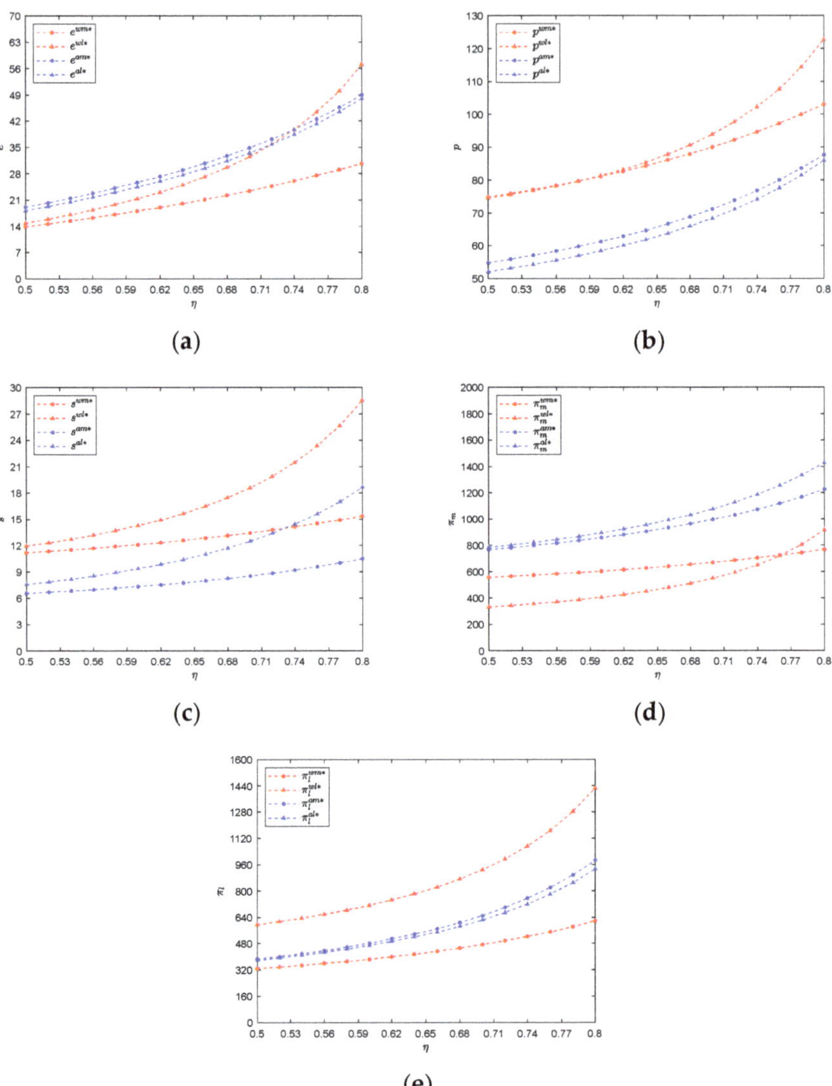

Figure 3. (**a**) changes of low-carbon level with η; (**b**) changes of the product sale price with η; (**c**) changes of low-carbon promotion efforts level with η; (**d**) changes of manufacturer profits with η; (**e**) changes of streamer profits with η.

6.3. Impact of Promotion Sensitivity Coefficient φ

In this section, the values of parameter η, μ, θ are held constant, and the value of φ keeps increasing. The results are shown in Figure 4a–c are consistent with Propositions 3, 6, 9, 12 and 15. (1) In the manufacturer-led structure. The manufacturer's optimal profit, shown in Figure 4d, and low-carbon level, shown in Figure 4a, in resale mode, are higher compared to the commission mode when φ is more than about 0.7. While it is better in commission mode when φ is less than about 0.7. The streamer's optimal profit in resale mode, shown in Figure 4e, is higher compared to the commission mode when φ is more than about 0.6. While it is higher in commission mode when φ is less than about 0.6. The streamer's optimal low-carbon promotion effort level in the resale mode shown in Figure 4c is always higher than that in commission mode regardless of the value of φ. (2) In the streamer-led structure. The manufacturer's optimal profit in commission mode, as shown in Figure 4d, is always higher than that in resale mode regardless of the value of φ. The low-carbon level shown in Figure 4a in resale mode is higher compared to the commission mode when φ is more than about 0.5. While it is better in commission mode when φ is less than about 0.5. The streamer's optimal profit and promotion effort level in the resale mode shown in Figure 4c,e is always higher compared to the commission mode, regardless of the value of φ. These results are consistent with Propositions 13 and 16.

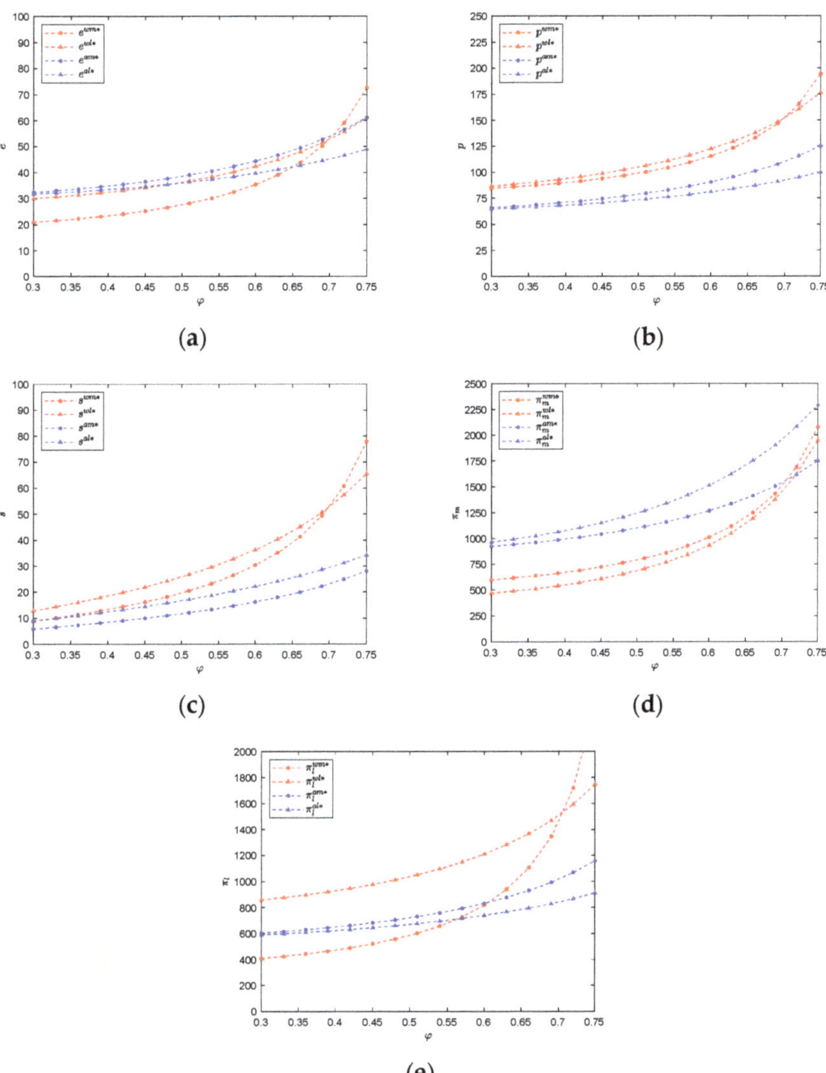

Figure 4. (**a**) changes of low-carbon level with φ; (**b**) changes of the product sale price with φ; (**c**) changes of low-carbon promotion efforts level with φ; (**d**) changes of manufacturer profits with φ; (**e**) changes of streamer profits with φ.

Thus, it can be observed that: (1) In the manufacturer-led structure. The manufacturer and streamer will choose the commission mode if the streamer's promotion sensitivity coefficient is smaller; otherwise, they will choose the resale mode if the streamer's promotion sensitivity coefficient is larger. In the above two cases, there is no disagreement between the manufacturer and streamer on the choice of livestreaming marketing mode. What's more, that choice will also make the manufacturer's low carbon level the highest, but the streamer's effort level is not optimal. (2) In the streamer-led structure. The streamer favors the resale mode, but the manufacturer prefers the commission mode, regardless of the value of the promotion sensitivity coefficient. This will need coordination between the manufacturer and the streamer. Generally, the weaker manufacturer will obey the streamer's decision to choose the resale mode in practice.

6.4. Impact of Commission Ratio θ

In this section, the values of parameters μ, φ, η are held constant, the value of θ keeps increasing. The results shown in Figure 5a,b are consistent with Propositions 10 and 14, respectively. We can find the manufacturer and streamer will make more efforts to low-carbon when they own the dominant position. The dominant player tends to be more active than that when they are a follower. It is also found from Figure 5a that when it is under the manufacturer-led structure, the optimal low-carbon level is always higher in commission mode, while the optimal promotion effort level is always higher in resale mode. From Figure 5b, when it is under the streamer-led structure, the optimal low-carbon level in commission mode is higher compared to the resale mode when θ is less than about 0.32. While it is higher in resale mode when θ is more than 0.32. The optimal promotion effort level is always higher in resale mode than that in the commission mode. It can be seen that regardless of the commission ratio and power structure, the resale mode has a higher incentive effect on the streamer's promotion effort.

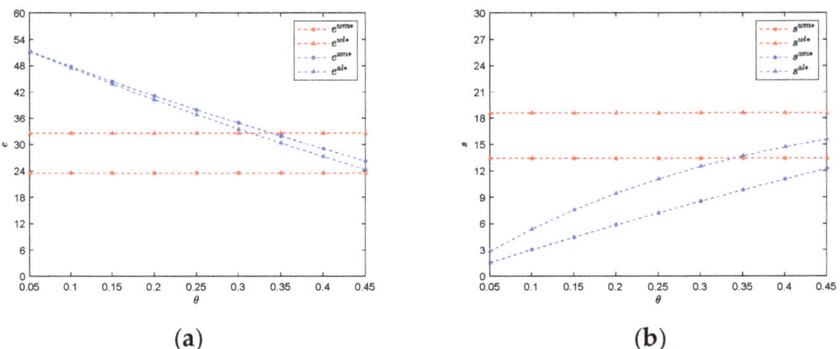

Figure 5. (**a**) changes of low-carbon level with θ; (**b**) changes of low-carbon promotion effort level with θ.

7. Case Study

In recent years, the Chinese government has vigorously implemented a "double carbon" policy to address climate change, which has been positively responded to by the Chinese home appliance brand, such as Midea and Gree, which have started to develop "low energy" products. At the beginning of 2020, the global COVID-19 pandemic greatly impacted the Chinese home appliance market, especially the large appliances mainly sold offline. According to the data from iimedia.cn, in the first quarter of 2020, the market share of Chinese home appliances was 117.2 billion CNY, down 36.1% year-on-year [53]. At the same time, the rapid development of "livestreaming + e-commerce", a new type of retail, has brought new development opportunities and sales chances for China's home appliance market. In order to adapt to market changes, some home appliance brands began to try livestreaming selling channels. For example, in May 2020, Dong Mingzhu of Gree Group livestreaming selling on the Kuaishou platform, and Midea Group combined with JD Five

Star Appliances to livestreaming selling; in June 2020, the AUX Group invited actor Lin Gengxin to livestream the selling of air conditioners. The common denominator in this series of livestreaming events is evident: livestreaming e-commerce has opened a new sales model for brands that mainly sell offline channels.

The most typical of these livestreaming events is the Gree Group, which had opened up new selling channels for Chinese manufacturing industries that stagnated during COVID-19. On 10 May 2020, Dong Mingzhu sold 44 million dollars in just three hours in a livestreaming event on the Kuaishou platform, attracting 16 million consumers to watch [54]. On 15 May, Dong conducted a livestreaming event on JD, generating 100 million dollars in sales in three hours, making it the highest sales in JD home appliance livestreaming marketing events [55]. The details of the two livestreaming events are shown in Table 2. In these two livestreaming events, the cooperation mode between Dong and Gree is different. In the Kuai livestreaming event, consumers buy products directly from Gree, so the cooperation mode can be regarded as commission mode, while in the JD livestreaming event, JD buys products from Gree and then resells them to consumers, so the cooperation mode can be regarded as resale mode. Comparing the two livestreaming events, we find that the sales in the JD are higher than those in the Kuaishou for the following reasons: First, in the JD livestreaming event, Dong introduced the technical features of the product from a professional perspective, with a higher level of low-carbon promotion effort, while in the Kuaishou livestreaming event, Dong was not yet skilled enough for livestreaming marketing, with a lower level of low-carbon promotion effort. Therefore, a higher level of low-carbon promotion can significantly increase consumer demand and improve manufacturer profits. Secondly, the Kuaishou livestreaming event mainly sells air purifiers, juicers and other small home appliances, and the low-carbon level of these products is relatively low; in JD livestreaming event, it mainly sells air conditioners and other large appliances, and the low-carbon level of these products is relatively high. The higher low-carbon level of products can increase the profits of manufacturers. Finally, we found that Dong, a socially well-known figure known as the "Queen of Home Appliances", dominated both the JD livestreaming event and Kuaishou livestreaming events. In addition, we know that the cooperation mode between Gree and Dong in Kuaishou livestreaming event is the commission model, and in JD livestreaming event is the resale model. It can be seen that Gree gained more profit in the resale mode when the streamer dominated. The above three points are reflected in our article, which fully illustrates the reasonableness and truthfulness of our proposed model.

Table 2. Comparison of Livestreaming Effect.

Gree Livestreaming	Commission Mode	Resale Mode
Time	11 May 2020	15 May 2020
Platform	Kuaishou	JD
Number of viewers	7.45 million	16 million
Power structure	streamer-led	streamer-led
Production low-carbon level	high	low
Low-carbon promotion effort level	high	low
Sales	$44 million	$100 million

Gree Group's success has caught the attention of other Chinese home appliance brand owners, driving livestreaming selling to become a virtual digital channel and development trend for Chinese home appliance sales. However, there are reasons behind Gree's success. There are two main reasons for Gree Group's success: First, as a company focused on home appliance manufacturing, Gree has been upholding the concept of low-carbon and environmental protection and constantly launching green and low-carbon home appliances. At the same time, Gree Group also actively promotes product innovation through the launch of the "photovoltaic air conditioning" project, successfully developed "zero carbon source" air conditioning technology; this advanced low-carbon technology has successfully attracted a large number of consumers. Secondly, Dong Mingzhu, the streamer of Gree, knows more about the products and has made more efforts to promote them, showing the features and advantages of Gree's low-carbon products to consumers and conveying the value of the products. At the same time, through the livestreaming Q&A, the recognition of Gree products was improved, further enhancing consumers' willingness to purchase low-carbon products and, thus, increasing product sales.

In general, with the awareness of green consumption has gradually gained popularity, and green consumption has become the mainstream trend. In order to adapt to the changing market trends, manufacturers must cater to consumer preferences and invest in the production of green, low-carbon products. In addition, consumers' consumption habits are also changing, and livestreaming e-commerce is highly sought after by consumers and has become one of the necessary channels for manufacturers to sell their products. By interacting with anchors in livestreaming, consumers will be more likely to recognize low-carbon products, leading to increased sales.

8. Conclusions

The optimal strategies of the low-carbon efforts and the livestreaming marketing mode of the manufacturer and streamer were studied. Two power structures are considered in the resale and commission cooperation modes. Members' profit functions are established under four scenarios: manufacturer-led in resale mode, streamer-led in resale mode, manufacturer-led in commission mode and streamer-led in commission mode. It also investigates the effects of supply chain power structure, and parameters concluding consumer's low-carbon preference and price sensitivity coefficient, streamer low-carbon promotion sensitivity coefficient, and commission ratio on the optimal product low-carbon level, low-carbon promotion effort level, profit, selling price and wholesale price. The significant findings can be summarized as follows.

(1) Consumers' price sensitivity coefficient is negatively related to the two parties' low-carbon efforts. Both the product's low-carbon level and promotion effort will decrease when the consumer's price sensitivity coefficient increases. That suggests that the more sensitive customers are to the price, the streamer and manufacturer are less motivated to promote low-carbon products. Therefore, it's necessary for the government to reduce taxes or provide financial subsidies to encourage the consumer and the manufacturer. For example, the Chinese government will subsidize consumers who buy new energy vehicles.

(2) Consumers' low-carbon preferences can deeply influence two parties' decisions. As long as it is improved, the low carbon efforts and profits of the two parties will increase, no matter in the resale mode or commission mode, or whether the manufacture-led structure or the streamer-led structure. So consumers' low-carbon preference should be improved through various ways, such as the manufacturer, the streamer and the government. For example, new energy vehicles can be exempted from restrictive measures, such as license plate auctions, lottery, and travel restrictions in China. This government policy has greatly improved the consumer's preference for new energy vehicles.

(3) The streamer's low-carbon effort coefficient is positively related to the manufacturer's profit and product low-carbon level, the streamer's promotion effort and profit, and the product sale price. This indicates that the more significant the impact of the streamer's low-carbon promotion on consumers, the more motivated the manufacturer and streamer are to make low-carbon efforts. So, the streamer can be encouraged to improve its influence of live broadcast skills through such means as cost-sharing with the manufacturer. So that the streamer's promotion effect can be improved, more consumers to buy. This is a win-win cooperation for the streamer and manufacturer.

(4) The commission ratio only affects the manufacturer's effort under the streamer-led structure. It is higher in commission mode when the commission ratio is smaller. The streamer's effort in resale mode is always higher than it is in the commission mode.

(5) The selection of the livestreaming marketing mode of the two parties is determined by the power structure and promotion sensitivity coefficient. When it is in the manufacturer-led structure, if the promotion sensitivity coefficient is smaller, the profits of two parties reach the maximization when choosing the commission mode, and the commission mode is the optimal choice for two parties; otherwise, the resale mode is the optimal choice. When in the streamer-led structure, the two parties' choices are opposite. The streamer prefers the resale mode, while the manufacturer prefers the commission mode.

(6) For the low-carbon effort level, the manufacturer makes the maximal effort in the commission mode, but the streamer makes the maximal effort in the resale mode. Accordingly, the streamer makes the maximal promotion effort in the resale mode, while the manufacturer makes the maximal effort in the commission mode if the promotion sensitivity coefficient is smaller.

Our study provides several managerial insights that may be useful to manufacturers, streamers, and governments.

Firstly, the manufacturer needs to continuously pay attention to consumers' low-carbon needs and market trends and adjust the low-carbon strategies in time. As for the government, it is necessary to take measures to make more consumers aware of low-carbon products, which also can motivate manufacturers to improve low-carbon technologies and achieve sustainable economic and social development. Secondly, manufacturers should focus on their position in the supply chain and the streamer's influence when choosing a cooperation mode. For example, when a manufacturer dominates the market, choosing the commission model is beneficial to increase profits if working with tail streamers; the resale model is the best choice if working with head streamers. Similarly, streamers should be concerned about their place in the supply chain. For example, the resale mode is best chosen when the streamer dominates the market.

Livestreaming marketing is now a new sales trend; our study provides theoretical support for strategies of low-carbon and the selection of livestreaming marketing modes. It is worth stating that there are some limitations to this study. Firstly, we assume that the commission ratio is given exogenously, but sometimes it is bargained by the two parties. Secondly, we find that in actual operation, besides livestreaming sales, manufacturers often have other sales channels, and the channels often have interactions with each other. Finally, corporate social responsibility(CSR) is also important for the low-carbon strategy. However, due to spatial constraints, we will further study the optimal low-carbon strategies of the manufacturer and the streamer when the manufacturer has two channels. And the cooperation mechanism between the manufacturer and the streamer when considering the CSR.

Author Contributions: Y.G.: conceptualization, revising and editing. G.H.: methodology, revising and editing. All authors have read and agreed to the published version of the manuscript.

Funding: This research was supported in part by the National Natural Science Foundation of China, No. 72202103.

Institutional Review Board Statement: Not applicable.

Informed Consent Statement: Not applicable.

Data Availability Statement: Not applicable.

Conflicts of Interest: The authors declare no conflict of interest.

Appendix A. Proofs

Proof of Lemma 1. With Equation (3), we can obtain $H_1 = \begin{bmatrix} -2\mu & \varphi \\ \varphi & -1 \end{bmatrix}$. From Assumption 4, H_1 is a negative definite matrix. Hence, π_l^{wm} is concave in k and s. Similarly, with

Equation (2), we can obtain $H_2 = \begin{bmatrix} -\dfrac{2\mu^2}{-\varphi^2+2\mu} & \dfrac{\mu\eta}{-\varphi^2+2\mu} \\ \dfrac{\mu\eta}{-\varphi^2+2\mu} & -1 \end{bmatrix}$. From Assumption 4, H_2 is also a negative definite matrix. Hence, π_m^{wm} is concave in w and e. Let $\frac{\partial \pi_l^{wm}}{\partial k} = 0$, $\frac{\partial \pi_l^{wm}}{\partial s} = 0$, then the best response of streamer is $k = \frac{\eta e - \mu w + \varphi s + a}{2\mu}$, $s = k\varphi$. Bring k and s to Equation (2), and let $\frac{\partial \pi_m^{wm}}{\partial w} = 0$, $\frac{\partial \pi_l^{wm}}{\partial e} = 0$, then we can obtain w^{wm*}, e^{wm*}, k^{wm*}, s^{wm*}, p^{wm*}. $\square$

Proof of Lemma 2, 3, 4. The process of proving Lemma 2, 3, and 4 is identical to that of Lemma 1.

Proof of Proposition 1. $\frac{\partial e^{wm*}}{\partial \mu} = -\frac{4a\eta}{\left(\eta^2+2\varphi^2-4\mu\right)^2} < 0$, $\frac{\partial k^{wm*}}{\partial \mu} = -\frac{4a}{\left(-\eta^2-2\varphi^2+4\mu\right)^2} < 0$, $\frac{\partial s^{wm*}}{\partial \mu} = -\frac{4\varphi a}{\left(-\eta^2-2\varphi^2+4\mu\right)^2} < 0$. $\square$

Proof of Proposition 2. $\frac{\partial e^{wm*}}{\partial \eta} = \frac{a\left(\eta^2-2\varphi^2+4\mu\right)}{\left(-\eta^2-2\varphi^2+4\mu\right)^2} > 0$, $\frac{\partial w^{wm*}}{\partial \eta} = \frac{2a\left(-\varphi^2+2\mu\right)\eta}{\mu\left(\eta^2+2\varphi^2-4\mu\right)^2} > 0$, $\frac{\partial k^{wm*}}{\partial \eta} = \frac{2a\eta}{\left(-\eta^2-2\varphi^2+4\mu\right)^2} > 0$, $\frac{\partial p^{wm*}}{\partial \eta} = \frac{2a\left(-\varphi^2+3\mu\right)\eta}{\mu\left(-\eta^2-2\varphi^2+4\mu\right)^2} > 0$, $\frac{\partial s^{wm*}}{\partial \eta} = \frac{2\varphi a\eta}{\left(-\eta^2-2\varphi^2+4\mu\right)^2} > 0$. $\square$

Proof of Proposition 3. $\frac{\partial e^{wm*}}{\partial \varphi} = \frac{4\varphi a\eta}{\left(\eta^2+2\varphi^2-4\mu\right)^2} > 0$, $\frac{\partial w^{wm*}}{\partial \varphi} = \frac{2a\varphi\eta^2}{\mu\left(\eta^2+2\varphi^2-4\mu\right)^2} > 0$, $\frac{\partial k^{wm*}}{\partial \varphi} = \frac{4\varphi a}{\left(-\eta^2-2\varphi^2+4\mu\right)^2} > 0$, $\frac{\partial p^{wm*}}{\partial \varphi} = \frac{2a\varphi\left(\eta^2+2\mu\right)}{\mu\left(\eta^2+2\varphi^2-4\mu\right)^2} > 0$, $\frac{\partial p^{wm*}}{\partial \varphi} = \frac{a\left(-\eta^2+2\varphi^2+4\mu\right)}{\left(-\eta^2-2\varphi^2+4\mu\right)^2} > 0$. $\square$

Proof of Proposition 4. $\frac{\partial e^{wl*}}{\partial \mu} = -\frac{4a\eta}{\left(-2\eta^2-\varphi^2+4\mu\right)^2} < 0$, $\frac{\partial w^{wl*}}{\partial \mu} = -\frac{4a}{\left(-2\eta^2-\varphi^2+4\mu\right)^2} < 0$, $\frac{\partial s^{wl*}}{\partial \mu} = -\frac{4a\varphi}{\left(2\eta^2+\varphi^2-4\mu\right)^2} < 0$. $\square$

Proof of Proposition 5. $\frac{\partial e^{wl*}}{\partial \eta} = \frac{a\left(2\eta^2-\varphi^2+4\mu\right)}{\left(2\eta^2+\varphi^2-4\mu\right)^2} > 0$, $\frac{\partial w^{wl*}}{\partial \eta} = \frac{4a\eta}{\left(-2\eta^2-\varphi^2+4\mu\right)^2} > 0$, $\frac{\partial k^{wl*}}{\partial \eta} = \frac{2a\eta\varphi^2}{\left(2\eta^2+\varphi^2-4\mu\right)^2\mu} > 0$, $\frac{\partial p^{wl*}}{\partial \eta} = \frac{2a\eta\left(\varphi^2+2\mu\right)}{\left(2\eta^2+\varphi^2-4\mu\right)^2\mu} > 0$, $\frac{\partial s^{wl*}}{\partial \eta} = \frac{4a\varphi\eta}{\left(2\eta^2+\varphi^2-4\mu\right)^2} > 0$. $\square$

Proof of Proposition 6. $\frac{\partial e^{wl*}}{\partial \varphi} = \frac{2a\eta\varphi}{\left(-2\eta^2-\varphi^2+4\mu\right)^2} > 0$, $\frac{\partial w^{wl*}}{\partial \varphi} = \frac{2a\varphi}{\left(-2\eta^2-\varphi^2+4\mu\right)^2} > 0$, $\frac{\partial k^{wl*}}{\partial \varphi} = -\frac{2a\left(\eta^2-2\mu\right)\varphi}{\left(2\eta^2+\varphi^2-4\mu\right)^2\mu} > 0$, $\frac{\partial s^{wl*}}{\partial \varphi} = \frac{a\left(-2\eta^2+\varphi^2+4\mu\right)}{\left(-2\eta^2-\varphi^2+4\mu\right)^2} > 0$, $\frac{\partial p^{wl*}}{\partial \varphi} = -\frac{2a\left(\eta^2-3\mu\right)\varphi}{\left(2\eta^2+\varphi^2-4\mu\right)^2\mu} > 0$. $\square$

Proof of Proposition 7. $\frac{\partial e^{am*}}{\partial \mu} = \frac{2\eta(\theta-1)a}{\left(\eta^2\theta-2\theta\varphi^2-\eta^2+2\mu\right)^2} < 0$, $\frac{\partial p^{am*}}{\partial \mu} = -\frac{2a}{\left(\eta^2\theta-2\theta\varphi^2-\eta^2+2\mu\right)^2} < 0$, $\frac{\partial s^{am*}}{\partial \mu} = -\frac{2a\theta\varphi}{\left(\eta^2\theta-2\theta\varphi^2-\eta^2+2\mu\right)^2} < 0$. $\square$

Proof of Proposition 8. $\frac{\partial e^{am*}}{\partial \eta} = -\frac{a(\theta-1)\left(-\eta^2\theta-2\theta\varphi^2+\eta^2+2\mu\right)}{\left(\eta^2\theta-2\theta\varphi^2-\eta^2+2\mu\right)^2} > 0$, $\frac{\partial p^{am*}}{\partial \eta} = \frac{2a\eta(1-\theta)}{\left(\eta^2\theta-2\theta\varphi^2-\eta^2+2\mu\right)^2} > 0$, $\frac{\partial s^{am*}}{\partial \eta} = \frac{2\eta a\theta\varphi(1-\theta)}{\left(\eta^2\theta-2\theta\varphi^2-\eta^2+2\mu\right)^2} > 0$. $\square$

Proof of Proposition 9. $\frac{\partial e^{am*}}{\partial \varphi} = \frac{4a\theta\varphi(1-\theta)\eta}{\left(\eta^2\theta-2\theta\varphi^2-\eta^2+2\mu\right)^2} > 0$, $\frac{\partial p^{am*}}{\partial \varphi} = \frac{4a\theta\varphi}{\left(\eta^2\theta-2\theta\varphi^2-\eta^2+2\mu\right)^2} > 0$, $\frac{\partial s^{am*}}{\partial \varphi} = \frac{a\theta\left(\eta^2\theta+2\theta\varphi^2-\eta^2+2\mu\right)}{\left(\eta^2\theta-2\theta\varphi^2-\eta^2+2\mu\right)^2} > 0$. $\square$

Proof of Proposition 10. According to Lemma 3, we can obtain $\eta^2\theta - \eta^2 - 2\theta\varphi^2 + 2\mu > 0$. Because $\eta^2\theta - \eta^2 < 0$, so $-\theta\varphi^2 + \mu > 0$. Hence, $\frac{\partial e^{am*}}{\partial \theta} = -\frac{2\eta a\left(-\varphi^2+\mu\right)}{\left(\eta^2\theta-2\theta\varphi^2-\eta^2+2\mu\right)^2} < 0$, $\frac{\partial s^{am*}}{\partial \theta} = \frac{a\varphi\left(-\eta^2+2\mu\right)}{\left(\eta^2\theta-2\theta\varphi^2-\eta^2+2\mu\right)^2} > 0$. $\square$

Proof of Proposition 11. $\frac{\partial e^{al*}}{\partial \mu} = \frac{\left[(-1+\theta)^2\eta^4+4\mu\eta^2(-1+\theta)+\varphi^2\theta\eta^2(1-\theta)+4\mu^2\right](-1+\theta)a\eta}{2\left[(-1+\theta)^2\eta^4+4\mu(-1+\theta)\eta^2+2\mu(-\varphi^2\theta+2\mu)\right]^2} < 0$, $\frac{\partial p^{al*}}{\partial \mu} = -\frac{a\left[(-1+\theta)^2\eta^4+4\mu\eta^2(-1+\theta)+\varphi^2\theta\eta^2(1-\theta)+4\mu^2\right]}{2\left[(-1+\theta)^2\eta^4+4\mu(-1+\theta)\eta^2+2\mu(-\varphi^2\theta+2\mu)\right]^2} < 0$, $\frac{\partial s^{al*}}{\partial \mu} = \frac{2a\theta\varphi\left[(-1+\theta)^2\eta^4-4\mu^2\right]}{\left[(-1+\theta)^2\eta^4+4\mu(-1+\theta)\eta^2+2\mu(-\varphi^2\theta+2\mu)\right]^2} < 0$. □

Proof of Proposition 12. $\frac{\partial e^{al*}}{\partial \varphi} = \frac{4\varphi\theta\mu\eta a(1-\theta)\left[(\theta-1)\eta^2+2\mu\right]}{\left[(-1+\theta)^2\eta^4+4\mu(-1+\theta)\eta^2-2\varphi^2\theta\mu+4\mu^2\right]^2} > 0$, $\frac{\partial p^{al*}}{\partial \varphi} = \frac{4a\varphi\theta\mu\left[(-1+\theta)\eta^2+2\mu)\right]}{\left[(-1+\theta)^2\eta^4+4\mu(-1+\theta)\eta^2-2\varphi^2\theta\mu+4\mu^2\right]^2} > 0$, $\frac{\partial s^{al*}}{\partial \varphi} = \frac{2a\theta\mu\left[(-1+\theta)^2\eta^4+4\mu(-1+\theta)\eta^2+2\varphi^2\theta\mu+4\mu^2\right]}{\left[(-1+\theta)^2\eta^4+4\mu(-1+\theta)\eta^2-2\varphi^2\theta\mu+4\mu^2\right]^2} > 0$. □

Proof of Proposition 13. $s^{wm*} - s^{am*} = \frac{\varphi a(1-2\theta)\left(-\eta^2+2\mu\right)}{(-\eta^2-2\varphi^2+4\mu)(\eta^2\theta-2\theta\varphi^2-\eta^2+2\mu)} > 0$, thus, $s^{wm*} > s^{am*}$.

$e^{wm*} - e^{am*} = \frac{2a\eta(2\theta-1)\left(-\varphi^2+\mu\right)}{(-\eta^2-2\varphi^2+4\mu)(\eta^2\theta-2\theta\varphi^2-\eta^2+2\mu)}$, since $0 < \theta < 0.5$, thus, when $\varphi < \sqrt{\mu}$, $e^{wm*} < e^{am*}$; otherwise, $e^{wm*} > e^{am*}$. □

Proof of Proposition 14. $e^{wl*} - e^{al*} = \frac{\eta a}{-2\eta^2-\varphi^2+4\mu} + \frac{a\eta(\theta-1)\left[(\theta-1)\eta^2+\mu\right]}{(\theta-1)^2\eta^4+4\mu(\theta-1)\eta^2-2\mu\theta\varphi^2+4\mu^2}$, let $e^{wl*} > e^{al*}$, then we can obtain $\frac{\eta a}{-2\eta^2-\varphi^2+4\mu} > -\frac{a\eta(\theta-1)\left[(\theta-1)\eta^2+\mu\right]}{(\theta-1)^2\eta^4+4\mu(\theta-1)\eta^2-2\mu\theta\varphi^2+4\mu^2}$. Thus, when $\theta > \theta'$, then $e^{wl*} > e^{al*}$, otherwise $e^{wl*} < e^{al*}$.

$s^{wl*} - s^{al*} = -\frac{a\varphi\left[(\theta-1)^2\eta^4+\mu(4\theta-1)\eta^2-\mu^2(4\theta-1)\right]}{2\left[(\theta-1)^2\eta^4+4\mu(\theta-1)\eta^2+\mu(-2\theta\varphi^2+4\mu)\right](-4\eta^2-2\varphi^2+8\mu)} > 0$, thus, $s^{wl*} > s^{al*}$. □

Proof of Proposition 15. $p^{wm*} - p^{am*} = \frac{a\left(-\varphi^2+3\mu\right)}{\mu(-\eta^2-2\varphi^2+4\mu)} - \frac{a}{\eta^2\theta-2\theta\varphi^2-\eta^2+2\mu} > 0$,

$p^{wl*} - p^{al*} = \frac{\left(\eta^2-3\mu\right)a}{(2\eta^2+\varphi^2-4\mu)\mu} - \frac{(\theta-1)a\eta^2+2a\mu}{(\theta-1)^2\eta^4+4\mu(\theta-1)\eta^2-2\mu\theta\varphi^2+4\mu^2} > 0$.

Thus, $p^{wm*} > p^{am*}$, $p^{wl*} > p^{al*}$. □

Proof of Proposition 16. $\pi_m^{wm*} - \pi_m^{am*} = \frac{a^2(2\theta-1)\left(-\varphi^2+\mu\right)}{(-\eta^2-2\varphi^2+4\mu)(\eta^2\theta-2\theta\varphi^2-\eta^2+2\mu)}$, thus, when $\varphi < \sqrt{\mu}$, then $\pi_m^{wm*} < \pi_m^{am*}$, otherwise, $\pi_m^{wm*} > \pi_m^{am*}$.

$\pi_l^{wl*} - \pi_l^{al*} = \frac{\left[(\theta-1)^2\eta^4+\mu(4\theta-1)\eta^2-\mu^2(4\theta-1)\right]a^2}{\left[(\theta-1)^2\eta^4+4\mu(\theta-1)\eta^2+\mu(-2\theta\varphi^2+4\mu)\right](-4\eta^2-2\varphi^2+8\mu)} > 0$, thus, $\pi_l^{wl*} > \pi_l^{al*}$. □

References

1. Moon, W.; Florkowski, W.J.; Brückner, B.; Schonhof, I. Willingness to Pay for Environmental Practices: Implications for Eco-Labeling. *Land Econ.* **2002**, *78*, 88–102. [CrossRef]
2. Adaman, F.; Karalı, N.; Kumbaroğlu, G.; Or, İ.; Özkaynak, B.; Zenginobuz, Ü. What determines urban households' willingness to pay for CO_2 emission reductions in Turkey: A contingent valuation survey. *Energy Policy* **2011**, *39*, 689–698. [CrossRef]
3. Zhao, R.; Yang, M.; Liu, J.; Yang, L.; Bao, Z.; Ren, X. University Students' Purchase Intention and Willingness to Pay for Carbon-Labeled Food Products: A Purchase Decision-Making Experiment. *Int. J. Environ. Res. Public Health* **2020**, *17*, 7026. [CrossRef]
4. Statista. Retail E-Commerce Sales Worldwide from 2014 to 2026. 2023. Available online: https://www.statista.com/statistics/379046/worldwide-retail-e-commerce-sales/ (accessed on 20 April 2023).
5. CNNIC. The 51st Statistical Report on Internet Development in China. 2023. Available online: https://www.cnnic.net.cn/NMediaFile/2023/0322/MAIN16794576367190GBA2HA1KQ.pdf (accessed on 20 April 2023).
6. Scheibe, K.; Fietkiewicz, K.J.; Stock, W.G. Information behavior on social live streaming services. *J. Inf. Sci. Theory Pract.* **2016**, *4*, 6–20. [CrossRef]
7. Ji, G.; Fu, T.; Li, S. Optimal selling format considering price discount strategy in live-streaming commerce. *Eur. J. Oper. Res.* **2023**, *309*, 529–544. [CrossRef]
8. Yi, Z.; Wang, Y.; Liu, Y.; Chen, Y.J. The impact of consumer fairness seeking on distribution channel selection: Direct selling vs. agent selling. *Prod. Oper. Manag.* **2018**, *27*, 1148–1167. [CrossRef]
9. Zhang, C.; Li, Y.; Ma, Y. Direct selling, agent selling, or dual-format selling: Electronic channel configuration considering channel competition and platform service. *Comput. Ind. Eng.* **2021**, *157*, 107368. [CrossRef]

10. Zhen, X.; Xu, S.; Li, Y.; Shi, D. When and how should a retailer use third-party platform channels? The Impact of spillover effects. *Eur. J. Oper. Res.* **2022**, *301*, 624–637. [CrossRef]

11. Zhang, L.; Liu, F.; Zhu, L.; Zhou, H. The Optimal Carbon Emission Reduction Strategy with Retailer's Fairness Concern and Advertising Effort Level. *Chin. J. Manag. Sci.* **2021**, *29*, 138–148. [CrossRef]

12. Ji, J.; Zhang, Z.; Yang, L. Carbon emission reduction decisions in the retail-/dual-channel supply chain with consumers' preference. *J. Clean. Prod.* **2017**, *141*, 852–867. [CrossRef]

13. Wu, D.; Wang, Y. Study on the differential game model for supply chain with consumers' Low carbon preference. *Chin. J. Manag. Sci.* **2021**, *29*, 126–137. [CrossRef]

14. Xia, L.; Bai, Y.; Qin, J.; Li, Y. Supply Chain's Emission Reduction and Lowcarbon Promotion Policies in Cap-and-trade System: The View of Information Asymmetry. *Oper. Res. Manag. Sci.* **2018**, *27*, 37–45.

15. Benjaafar, S.; Li, Y.; Daskin, M. Carbon footprint and the management of supply chains: Insights from simple models. *IEEE Trans. Autom. Sci. Eng.* **2012**, *10*, 99–116. [CrossRef]

16. Zhang, L.; Xu, H.; Li, Y. Research on emission reduction decision-making of dual-channel supply chain under the carbon cap-and-trade policy. *J. Ind. Eng. Eng. Manag.* **2022**, *37*, 1–9. [CrossRef]

17. Chen, K.; Wang, Y. Channel selection strategy of the green-product manufacturer under different forms of government subsidies. *Chin. J. Manag. Sci.* **2022**, *in press*. [CrossRef]

18. Wang, W.; Wang, F.; Zhang, S. Coordination Contract in a Dual Channel Supply Chain with Low Carbon Efforts. *Manag. Rev.* **2021**, *33*, 315–326. [CrossRef]

19. Wang, Y.; Yu, Z.; Jin, M.; Mao, J. Decisions and coordination of retailer-led low-carbon supply chain under altruistic preference. *Eur. J. Oper. Res.* **2021**, *293*, 910–925. [CrossRef]

20. Gong, B.; Zhang, H.; Gao, Y.; Liu, Z. Blockchain adoption and channel selection strategies in a competitive remanufacturing supply chain. *Comput. Ind. Eng.* **2023**, *175*, 108829. [CrossRef]

21. Basiri, Z.; Heydari, J. A mathematical model for green supply chain coordination with substitutable products. *J. Clean. Prod.* **2017**, *145*, 232–249. [CrossRef]

22. Zhang, Z.; Xu, H.; Chen, K.; Zhao, Y.; Liu, Z. Channel mode selection for an e-platform supply chain in the presence of a secondary marketplace. *Eur. J. Oper. Res.* **2023**, *305*, 1215–1235. [CrossRef]

23. Dolai, M.; Manna, A.K.; Mondal, S.K. Sustainable Manufacturing Model with Considering Greenhouse Gas Emission and Screening Process of Imperfect Items Under Stochastic Environment. *Int. J. Appl. Comput. Math.* **2022**, *8*, 93. [CrossRef]

24. Manna, A.K.; Benerjee, T.; Mondal, S.P.; Shaikh, A.A.; Bhunia, A.K. Two-plant production model with customers' demand dependent on warranty period of the product and carbon emission level of the manufacturer via different meta-heuristic algorithms. *Neural Comput. Appl.* **2021**, *33*, 14263–14281. [CrossRef]

25. Kumar, N.; Manna, A.K.; Shaikh, A.A.; Bhunia, A.K. Application of hybrid binary tournament-based quantum-behaved particle swarm optimization on an imperfect production inventory problem. *Soft Comput.* **2021**, *25*, 11245–11267. [CrossRef]

26. Manna, A.K.; Akhtar, M.; Shaikh, A.A.; Bhunia, A.K. Optimization of a deteriorated two-warehouse inventory problem with all-unit discount and shortages via tournament differential evolution. *Appl. Soft Comput.* **2021**, *107*, 107388. [CrossRef]

27. Zhang, T.; Tang, Z.; Han, Z. Optimal online channel structure for multinational firms considering live streaming shopping. *Electron. Commer. Res. Appl.* **2022**, *56*, 101198. [CrossRef]

28. Gong, H.; Zhao, M.; Ren, J.; Hao, Z. Live streaming strategy under multi-channel sales of the online retailer. *Electron. Commer. Res. Appl.* **2022**, *55*, 101184. [CrossRef]

29. Hao, C.; Yang, L. Resale or agency sale? Equilibrium analysis on the role of live streaming selling. *Eur. J. Oper. Res.* **2023**, *307*, 1117–1134. [CrossRef]

30. Zhang, X.; Chen, H.; Liu, Z. Operation strategy in an E-commerce platform supply chain: Whether and how to introduce live streaming services? *Int. Trans. Oper. Res.* **2023**, *in press*. [CrossRef]

31. Ma, X.; Zou, X.; Lv, J. Why do consumers hesitate to purchase in live streaming? A perspective of interaction between participants. *Electron. Commer. Res. Appl.* **2022**, *55*, 101193. [CrossRef]

32. Chen, H.; Zhang, Y.; Guo, W. A study on the impact of influencers on Fans' Purchase intention in live broadcasting platform. *China Bus. Mark.* **2020**, *34*, 28–37. [CrossRef]

33. Huang, M.; Ye, Y.; Wang, W. The interaction effect of broadcaster and product type on consumers' purchase intention and behaviors in livestreaming shopping. *Nankai Bus Rev.* **2022**, *in press*. Available online: https://kns.cnki.net/kcms/detail/12.1288.F.20210915.0954.002.html (accessed on 1 May 2023).

34. Xing, P.; You, H.; Fan, Y. Optimal quality effort strategy in service supply chain of live streaming e-commerce based on platform marketing effort. *Control Decis.* **2022**, *37*, 205–212. [CrossRef]

35. Huang, J.; Wang, X.; Luo, Y.; Yu, L.; Zhang, Z. Joint Green Marketing Decision-Making of Green Supply Chain Considering Power Structure and Corporate Social Responsibility. *Entropy* **2021**, *23*, 564. [CrossRef] [PubMed]

36. Li, Z.; Xu, Y.; Deng, F.; Liang, X. Impacts of Power Structure on Sustainable Supply Chain Management. *Sustainability* **2018**, *10*, 55. [CrossRef]

37. Lu, Q.; Liu, N. Pricing games of mixed conventional and e-commerce distribution channels. *Comput. Ind. Eng.* **2013**, *64*, 122–132. [CrossRef]

38. Chen, X.; Wang, X.; Jiang, X. The impact of power structure on the retail service supply chain with an O2O mixed channel. *J. Oper. Res. Soc.* **2016**, *67*, 294–301. [CrossRef]

39. Qiu, R.; Li, X.; Sun, M. Vertical integration of an online secondhand platform and a recycling platform under different power structures. *Eur. J. Oper. Res.* **2023**, *in press*. [CrossRef]

40. Tao, F.; Zhou, Y.; Bian, J.; Lai, K.K. Optimal channel structure for a green supply chain with consumer green-awareness demand. *Ann. Oper. Res.* **2023**, *324*, 601–628. [CrossRef]

41. Ghosh, D.; Shah, J.; Swami, S. Product greening and pricing strategies of firms under green sensitive consumer demand and environmental regulations. *Ann. Oper. Res.* **2020**, *290*, 491–520. [CrossRef]

42. Zhu, C.; Ma, J.; Li, J. Dynamic production and carbon emission reduction adjustment strategies of the brand-new and the remanufactured product under hybrid carbon regulations. *Comput. Ind. Eng.* **2022**, *172*, 108517. [CrossRef]

43. Sun, L.; Cao, X.; Alharthi, M.; Zhang, J.; Taghizadeh-Hesary, F.; Mohsin, M. Carbon emission transfer strategies in supply chain with lag time of emission reduction technologies and low-carbon preference of consumers. *J. Clean. Prod.* **2020**, *264*, 121664. [CrossRef]

44. Zhang, W.; Yu, L.; Wang, Z. Live-streaming selling modes on a retail platform. *Transp. Res. Part E Logist. Transp. Rev.* **2023**, *173*, 103096. [CrossRef]

45. Yi, Y.; Wang, Y.; Fu, C.; Li, Y. Taxes or subsidies to promote investment in green technologies for a supply chain considering consumer preferences for green products. *Comput. Ind. Eng.* **2022**, *171*, 108371. [CrossRef]

46. Zhang, L.; Wang, J.; You, J. Consumer environmental awareness and channel coordination with two substitutable products. *Eur. J. Oper. Res.* **2015**, *241*, 63–73. [CrossRef]

47. Li, Q.; Xiao, T.; Qiu, Y. Price and carbon emission reduction decisions and revenue-sharing contract considering fairness concerns. *J. Clean. Prod.* **2018**, *190*, 303–314. [CrossRef]

48. Wang, Y.; Fan, R.; Shen, L.; Jin, M. Decisions and coordination of green e-commerce supply chain considering green manufacturer's fairness concerns. *Int. J. Prod. Res.* **2020**, *58*, 7471–7489. [CrossRef]

49. Li, Q.; Chen, X.; Huang, Y. The stability and complexity analysis of a low-carbon supply chain considering fairness concern behavior and sales service. *Int. J. Environ. Res. Public Health* **2019**, *16*, 2711. [CrossRef]

50. Li, Z.; Pan, Y.; Yang, W.; Ma, J.; Zhou, M. Effects of government subsidies on green technology investment and green marketing coordination of supply chain under the cap-and-trade mechanism. *Energy Econ.* **2021**, *101*, 105426. [CrossRef]

51. Luo, R.; Zhou, L.; Song, Y.; Fan, T. Evaluating the impact of carbon tax policy on manufacturing and remanufacturing decisions in a closed-loop supply chain. *Int. J. Prod. Econ.* **2022**, *245*, 108408. [CrossRef]

52. Geng, X.; Tan, Y.; Wei, L. How Add-on Pricing Interacts with Distribution Contracts. *Prod. Oper. Manag.* **2018**, *27*, 605–623. [CrossRef]

53. iiMedia Research. Bimonthly Report on Market Operation and Analysis of Typical Enterprises in China's Home Appliance Industry, May–June 2020. 2020. Available online: https://www.iimedia.cn/c400/72452.html (accessed on 20 April 2023).

54. Xinhua. Chinese Air-Con Giant Reaps Big in Livestreaming Sales Attempt. 2020. Available online: http://www.xinhuanet.com/english/2020-05/11/c_139048374.htm (accessed on 20 April 2023).

55. JD Retail. Gree's Boss: Record Setter of Livestreaming on JD with Sales of $100 Million. 2020. Available online: https://jdcorporateblog.com/grees-boss-record-setter-of-livestreaming-on-jd-with-sales-of-100-million/ (accessed on 20 April 2023).

Article

Federated Learning and Blockchain-Enabled Intelligent Manufacturing for Sustainable Energy Production in Industry 4.0

Fanglei Sun [1],* and Zhifeng Diao [2],*

[1]   School of Creativity and Art, Shanghai Tech University, Shanghai 201210, China
[2]   College of Design and Innovation, Tongji University, Shanghai 200092, China
*   Correspondence: sun_fanglei@outlook.com (F.S.); diaozhifeng188@outlook.com (Z.D.)

Abstract: Intelligent manufacturing under Industry 4.0 assimilates sophisticated technologies and artificial intelligence for sustainable production and outcomes. Blockchain paradigms are coined with Industry 4.0 for concurrent and well-monitored flawless production. This article introduces Sustainable Production concerned with External Demands (SP-ED). This method is more specific about energy production and the distribution for flawless and outage-less supply. First, the energy demand is identified for internal and external users based on which sustainability is planned. Secondly, Ethereum blockchain monitoring for a similar production and demand satisfaction is coupled with the production system. From two perspectives, the monitoring and condition satisfaction processes are validated using federated learning (FL). The perspectives include demand distribution and production sustainability. In the demand distribution, the condition of meeting the actual requirement is validated. Contrarily, the flaws in internal and external supply due to production are identified in sustainability. The failing conditions in both perspectives are handled using blockchain records. The blockchain records reduce flaws in the new production by modifying the production plan according to the federated learning verifications. Therefore, the sustainability for internal and external demands is met through FL and blockchain integration.

Keywords: blockchain; federated learning; intelligent manufacturing; sustainable energy

Citation: Sun, F.; Diao, Z. Federated Learning and Blockchain-Enabled Intelligent Manufacturing for Sustainable Energy Production in Industry 4.0. *Processes* **2023**, *11*, 1482. https://doi.org/10.3390/pr11051482

Academic Editors: Conghu Liu, Xiaoqian Song, Zhi Liu and Fangfang Wei

Received: 10 February 2023
Revised: 13 April 2023
Accepted: 18 April 2023
Published: 12 May 2023

1. Introduction

Sustainability in various manufacturing aspects, such as energy, is achieved using intelligent processing in Industry 4.0. Sustainable energy comes from renewable sources such as wind power, water resources, and solar energy. Sustainable energy production is a crucial task to perform in industries [1]. Renewable energy applications and technologies are used in industries. Energy applications provide various services and policies to increase the sustainability range in energy production. An energy production scheme provides functions and services to perform specific tasks in industries [2]. Renewable energy production improves the energy-efficiency range of organizations. Sustainable energy production reduces the energy consumption ratio in Industry 4.0. Greenhouse gas (GHG) emission control is a complicated task to perform in industries. Greenhouse gas emission control uses various methods and techniques [3]. Artificial Intelligence (AI) technology is used in Industry 4.0—it increases the computational efficiency level of the systems. The AI method identifies the critical factors for the energy production process. The AI-based technique enhances industries' effectiveness and performance range [4]. Energy surplus or deficit may threaten the energy supply and demand security, leading to a demand–response issue in the industrial environment. It is becoming increasingly tricky to optimally schedule in a smart industry with varying energy consumption patterns and to engage in trustworthy energy trading due to potential privacy and security challenges in the distributed energy system.

Ethereum blockchain is the decentralized, open-source blockchain-based technology used for sustainable energy production in Industry 4.0. Blockchain-based techniques are mainly used in industries to detect problems and issues in the production process [5]. Blockchain techniques provide specific solutions to solve problems in production. For example, peer-to-peer transmission is carried out in manufacturing sectors, including effective assistance for sustainable energy production. Transmission and production contain various issues and threats that reduce the industry's production speed [6]. Essential qualities and attributes are identified in the database, yielding actionable data for multiple applications. The blockchain-based method identifies industries' potential benefits and features that provide necessary data for energy production [7]. A blockchain-based secure system is implemented in industries that increase the accuracy of sustainable energy production processes. The security system uses blockchain to initiate production based on certain conditions and functions. The Ethereum-blockchain-based data analysis method is also used in Industry 4.0, which analyzes the relevant datasets for the sustainable energy production process. The data analysis technique reduces the computation process' latency and energy consumption ratio [8,9]. Ethereum blockchain in industrial energy production allows for one to store the collected data (or proof of such data) to exchange them securely between entities that do not trust each other. Furthermore, blockchain technologies permit the creation of smart contracts, described as self-sufficient decentralized codes performed autonomously when certain conditions of an industry progression are met.

Machine learning (ML) models and techniques are widely used in various fields and applications. ML models are commonly used to improve production, computation efficiency, and feasibility range [10]. ML models are also used in Industry 4.0 for sustainable energy production. The convolutional neural network (CNN) algorithm is in the production model that performs specific industry tasks. CNN uses a feature extraction method that extracts the essential features and patterns from the database. CNN reduces the energy consumption range in computation, improving the efficiency ratio in the energy production process [11,12]. The support vector regression (SVR) technique is also used in sustainable production. SVR uses analysis that analyzes the data required for the production process. SVR increases the accuracy and performance range in Industry 4.0 [13,14]. The multiple linear regression (MLR) model is used for Industry 4.0, which implements a power forecasting system. The MLR model predicts the problems presented in the computation process, of which reduce the error range in sustainable energy production [15].

Sun et al. introduced a combined production scheduling model for sustainable manufacturing systems [16]. The primary goal of the presented model is to pinpoint the origin of the scheduling-related variation in resource use. The particle swarm optimization (PSO) algorithm is used here to analyze the data required for the scheduling process. PSO minimizes the overall time and energy consumption level in computation and scheduling processes. As a result, the proposed model enhances the performance and feasibility ratio of the manufacturing systems; however, the complex production efficiency modeling necessities need to be explored.

Li et al. introduced a digital twin-driven information mechanism for manufacturing systems [17]. A hierarchical analytic process analyzes the information relevant to scheduling and further processes. The digital twin mechanism uses evidence theory to build proper intelligent manufacturing techniques for the systems. The introduced method reduces the complexity and latency in the computation process. The presented strategy broadens the platforms' potential for efficiency and long-term sustainability. However, the incompleteness of primary data sources, the difficulty and uncertainty of actual indicators, and inaccuracy in human cognitive progression exist in the model procedure.

Majeed et al. developed an infrastructure for Sustainable and Smart Addiction Management (SSAM) using big data [18]. The proposed framework is mainly used for the decision-making process. Big data analytics identify the necessary data which are relevant for SSAM systems. The big data approach is mainly used for analyzing processes that reduce energy consumption in the identification process. The suggested architecture has been shown to improve the efficiency and functionality of SSAM systems in research conditions. However, due to the company's available capabilities and setup of IoT devices, the SSAM model can only be implemented in the first phases of a product's life cycle.

Psarommatis et al. presented a holistic approach to sustainable manufacturing systems [19]. Zero Defect Manufacturing (ZDM) is used here to improve the systems' efficiency and feasibility range. The suggested method's true motivation is to lessen the workload on the power grid during computing. ZDM is a required method since it supplies essential information and characteristics for production. In addition, the proposed approach increases the Quality of Service (QoS) in sustainable manufacturing systems. However, the suggested system data-driven model insufficiently attains the sustainable factor in the manufacturing process.

Ma et al. introduced a demand–response-based data-driven framework for a sustainable manufacturing system [20]. The goal of the proposed method is to manage multi-level requests which occur during the manufacturing process. The introduced framework reduces the computation cost and latency in manufacturing systems. The particle swarm optimization algorithm also manages the data required for various methods. The presented framework increases decision-making accuracy, enhancing the systems' performance. The suggested sustainable, innovative manufacturing model only considers the product lifecycle's manufacturing phase, disregarding other stages, such as operation, design, recycling, maintenance, and remanufacturing.

Tian et al. introduced dynamic evaluation based on correlation relationships for sustainable manufacturing in industrial cloud robots (ICR) [21]. Correlation relationships produce appropriate data which are related to assessments. The suggested technique may identify issues throughout the computation and provide a workable answer to fix them. The proposed approach has a lower energy usage ratio in the calculation compared to prior methods. The recommended strategy increases ICR's efficiency and dependability. Impacting sustainability objectives must be considered when developing a more comprehensive evaluation indicator.

Jasiulewicz-Kaczmarek et al. introduced a multiple-criteria approach for manufacturing systems [22]. The method offered is a continuous sustainability performance evaluation based on fuzzy set theory. The presented method uses a maintenance indicator that identifies the synthetic index and patterns necessary for the assessment process. The maintenance indicator reduces the time consumption ratio in both computation and identification processes. The introduced approach improves the manufacturing systems' overall sustainability and feasibility range. However, several aggregation functions have a limitation, primarily from their natural assumption that input criteria are independent.

Zimmermann et al. designed an action-oriented teaching approach for an intelligent precision manufacturing system [23]. The proposed method detects the exact demands and reasons for requests in manufacturing systems. Various machine tools are also used in a teaching approach that provides certain services in the decision-making process. An intelligent-based reduction strategy is used here that reduces the latency rate in assessment and scheduling processes. The systems' effectiveness and energy efficiency are improved by the proposed method. However, the limitation concerning the included thermal errors and the inadequate prediction accuracy makes an extensive industrial application unrealistic.

Wang et al. developed an energy consumption intelligent model for additive manufacturing (AM) systems [24]. A multisource fusion method is used in the proposed model that identifies the exact data from the database. The proposed model is mainly used for 3D printing (3DP), which detects the necessary pixels and features from the images. As the experiments show, the suggested model improves the AM devices' efficiency range. However, modeling energy consumption and forecasting with multiple source data are infrequent. The residues obscure how various sources can be leveraged to effectively learn a detailed depiction of the prediction task.

Favi et al. proposed an energy management framework for a sustainable life cycle in Industry 4.0 [25]. The proposed structural framework uses energy material flow analysis

(EMFA) for the data analysis. Key Performance Indicators (KPIs) are used in industries that identify the exact production performance ratio of the systems. KPI provides relevant data for decision-making and allocation processes. The proposed framework minimizes the material flow range in industries. Furthermore, the efficiency in supporting companies in the analysis, managing production plants' energy, and identifying criticalities and material flows have yet to be verified.

Pei et al. introduced an approximation algorithm for unrelated parallel machine scheduling in manufacturing systems [26]. The suggested method aims to reduce manufacturing systems' electric power consumption rate. The introduced algorithm identifies the regional problems that occurred during the manufacturing process. The proposed approach improves the precision of machine scheduling, which strengthens the systems overall. The new algorithm enhances both the speed and accuracy of existing systems. However, one commonly known disadvantage of the branch and bound technique is its time-consuming feature.

Cañas et al. designed a conceptual framework for Smart Production Planning and Control (SPPC) in Industry 4.0 [27]. Small- and medium-sized (SMEs) businesses are the primary users of the proposed framework. The proposed framework provides a systematic structure to analyze the relevant data for the SPPC process. As a result, the conceptual framework enhances the efficiency and accuracy of SPPC4.0. Furthermore, compared with other frameworks, the proposed framework achieves high performance in scheduling processes. However, it must be noted that multidisciplinary engineering is essential for establishing SPPC 4.0 models.

Friederich et al. introduced a standardized data-driven architecture for intelligent production systems [28]. The proposed framework aims to maintain the data in digital-twin-based systems. Machine learning (ML) and data mining techniques are used here to reduce the computation process' complexity. ML techniques are mainly used here for the validation and detection process. The introduced framework enhances the effectiveness and feasibility range of smart manufacturing systems. However, due to issues with devices, networks, etc., data may be incomplete.

Gu et al. developed a cyber–physical architecture for smart factories [29]. The architecture implements a deep reinforcement learning (DRL) algorithm that detects the relevant data for different processes. DRL selects the necessary information for decision-making, reducing the computation process' latency and energy consumption. The suggested strategy also improves the systems' effectiveness by boosting the precision of their decision-making and planning. However, a single scheduling rule cannot preserve high-quality scheduling efficiency in the face of orders of dissimilar sizes.

Liu et al. proposed product lifecycle management (PLM) infrastructure for Industry 4.0 based on blockchain technology [30]. The blockchain technique identifies the exact relationship among the nodes that provide optimal data for the scheduling process. Blockchain also detects the problems which occur during manufacturing. The conceptual approach reduces PLM systems' overall time-to-energy ratio. The proposed platform increases PLM's long-term viability and scope of industrial importance. This study still needs to implement the real-life case-study use fully; thus, existing outcomes supported the possibility of implementing this platform but cannot make a quantitative comparison with conventional PLM platforms.

Krithika L. B. [31] discussed the advancements in blockchain technology that have shown promising properties that might be useful in farming. There have been some helpful upheavals and progressive acceptance of blockchain in agribusiness owing to the development and rollout of blockchain, which has helped to modernize the sector. Decent quality development in farming has led to the use of blockchain technology at several stages of the process. This research comprehensively examines the existing research on the opportunities and threats posed by blockchain technology in the agricultural sector. Much of the study is in its infancy, the PoCs are based on outdated versions of blockchain, and the concept has undergone significant rehabilitation since its beginnings.

Guruprakash Jayabalasamy and Srinivas Koppu [32] suggested nonrepudiation in Internet of Things (IoT) apps developed on blockchain using High-Performance Edwards Curve Aggregate Signatures (HECAS). Compared to the standard digital signature model, the signing and verifying procedures created in the present study took 10% and 13% less time to process, respectively. Additionally, using HECAS in a blockchain context may reduce storage costs by 40%, improve transaction flows by 10%, and improve block validation by 10% compared to a system that does not use HECAS. Finally, the author tested their technology by simulating several blockchain-based Internet of Things systems. As a result of their efforts, blockchain-based technologies for the smart Internet of Things may produce effective, consistent results across various sensor types.

The leakage or collision of secret keys is possible if two identities are generated using the same randomized integer.

Zixiao Xu et al. [33] suggested a blockchain-based power trading and bidding mechanism for several microgrids. Consequently, to accomplish source–sale integration, a competitive spot market with scattered "multi-seller and multi-buyer" was constructed. Researchers in this study compared and contrasted traditional power trading with a blockchain-based alternative. In addition, an ant colony optimization technique was used for randomized bidding matching, and a blockchain-based multi-microgrid energy trading model was developed. This method integrates the transfer of energy, data, and money into a single procedure. Lastly, the efficient allocation of power resources was ensured by openness and transparency in power transactions. Nonetheless, business still needs to develop and improve the Energy Internet trading system.

Qu et al. [34] introduced federated learning and a blockchain-based distributed approach for Data-Driven Cognitive Computing (D2C). Federated learning's emphasis on privacy and efficiency makes it well-suited to address the "data island" issue. In contrast, blockchain's reward mechanism, completely decentralized nature, and resistance to poisoning assaults make it an attractive complement. Improvements in decision-making and data-driven intelligent manufacturing are already visible thanks to the development of different AI and machine learning technologies. Furthermore, rapid convergence may be achieved via sophisticated verifications and member choices made possible by blockchain-enabled federated learning. The results of a comprehensive review and assessment show that D2C is superior to the state-of-the-art in terms of efficiency.

Zhao et al. [35] proposed a federated learning (FL) system that uses a reputation mechanism to allow for home appliance makers to train a machine learning model using data from actual consumers to facilitate the development of an intelligent home system. The first step in the system's workflow involves users training the manufacturer-supplied baseline model on their mobile device and the mobile edge computing (MEC) server in addition to an incentive system to reward participants for enticing more consumers to participate in the crowd-sourced FL work.

Energy surplus or deficit may threaten the energy supply and demand security, leading to a demand–response issue in the industrial environment. It is becoming increasingly more work to optimally schedule in an intelligent industry with varying energy consumption patterns and to engage in trustworthy energy trading due to potential privacy and security challenges in the distributed energy system. Based on the survey, there are several challenges in existing methods in achieving high sustainability factors, attack detection time, modifications, and flaw detection for energy supply–demand. Hence, in this paper, Sustainable Production concerned with External Demands (SP-ED) has been proposed for practical energy production and distribution for flawless and outage-less supply.

2. Sustainable Production Concerned with External Demands

The features of Ethereum blockchain are used in Industry 4.0 for a synchronous and well-observed, flawless production. This method introduces Sustainable Production concerned with External Demands [SP-ED]. This method is used for energy production and the distribution of flawless outages. Industry 4.0 evolves many technologies, and blockchain is one of them. Blockchain enhances the Industry 4.0's security, privacy, and data transparency. This Industry 4.0 enables the manufacturers to achieve their goals in a more agile and quick way. Blockchain is used to attain more identification and improve the manufacturing environment. As blockchain is more straightforward and less intermediary, it is used to defend their inventions. Using this blockchain technology in Industry 4.0 enhances their competitiveness, which can access the world of copyrights. This unique technology eliminates transaction communication, effectively as a productive production flow. Decentralized energy trade and supply, the safe records of all industrial activities in energy generation, and the effective automated management of energy and storage flow via smart contracts are ways the Ethereum blockchain with FL might benefit the energy sector. This proposed SP-ED is portrayed in Figure 1.

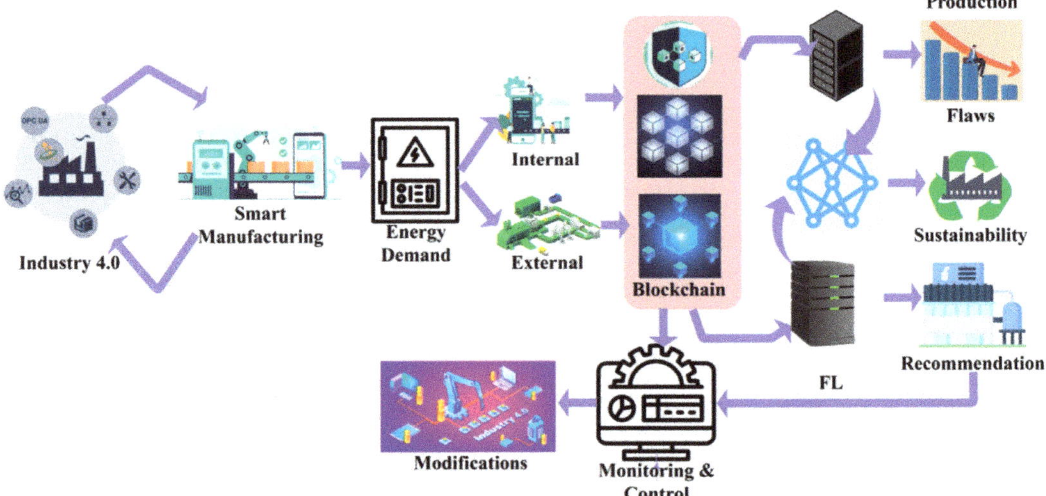

Figure 1. SP-ED illustration.

This method's manufacturing process is the modern version of automation. The manufacturing process depends on the energy demand to improve energy sustainability during the production of goods. The internal and external demands are identified from the energy demand. Internal demand is the one that has the energy used in the process of manufacturing. External demand is the one that contains the excess energy not used in the procedure of manufacturing. Blockchain technology observes these internal and external demands, which will be given as input to federated learning (FL). This FL checks the demands, distribution of energy, and energy consumption. From this FL, the production flaws, energy sustainability, and recommendations are obtained. These production flaws check the energy distribution and the internal usage of energy. If the energy is insufficient, it can be identified in this process. Energy sustainability checks how long the energy lasts to achieve the demand. A recommendation is used to recommend scheduling the energy and time depending on the output. From the blockchain, the monitor control takes place. Based on the input in the blockchain, the modification process is carried out to improve the manufacturing procedure. Ethereum blockchain technology can improve energy efficiency and give consumers more control over their utilities in the industrial environment.

Furthermore, the data on how much energy is used is updated securely and promptly due to an immutable ledger. Here, the intelligent manufacturing process is carried out in Industry 4.0 based on the energy demand. X_1, X_2 are the subset of variance to calculate the energy consumption. The process of enhancing the manufacturing procedure by the energy demand is explained by Equation (1), as given below:

$$a - b - c(X_1 + X_2), \; if \; 0 \leq X_1 + X_2 \leq b/c - 0, \tag{1}$$

where a is denoted as the measure of observation, b represents the data feature, c is the covariance of data, and X_1 and X_2 are characterized as the internal and external energy demand. Now, from the energy demand, the sustainability of the energy can be improved. Then, the internal demand and the external demand are identified. In this internal demand, the information on the energy used in the manufacturing process can be determined. This is used to obtain information on the energy that can enhance manufacturing in Industry 4.0. This also determines the energy utilized in the process and how much energy can be saved. Based on the demand, the energy can be accommodated for internal usage, and thus it will be helpful in flawless production. The internal demand accumulates the amount of energy that is needed in the manufacturing process. Additionally, it does not occupy unnecessary energy consumption, which will not be used during the process. This information can be observed by blockchain technology later, and it gives input for the upcoming process. Thus, the internal demand stores the needed energy during the manufacturing process. This eliminates the excess energy which is not required for the process. This procedure is used to enhance energy sustainability to produce flawless execution. The internal demand is extracted from the energy demand to improve manufacturing. This also helps identify the needed amount of energy for the process and helps make the energy last longer during flawless production. This process of internal demand usage is based on the sustainability plan. Equation (2) below explains the process of extracting the internal demand and its functions. d_1 and d_2 are the internal demand parameters, and the energy function is denoted as C.

$$\left. \begin{array}{l} \pi_1 - \pi_1(\gamma_1, \gamma_2) - (b - CX_1 - CX_2 - d_1)X_1 \\ \pi_2 - \pi_2(\gamma_1, \gamma_2) - (b - CX_1 - CX_2 - d_2)X_2 \end{array} \right\} \tag{2}$$

where (π_1, π_2) is denoted as the mean of all observations and (γ) is represented as the energy variance during the manufacturing process. Now, the external demand is extracted from the energy demand. In this external demand, the excess energy not used for the process is stored. Based on the energy demand, the amount of energy can be used for the process. Then, if an excessive amount of energy is occupied, it will be stored in the external demand. This extreme energy can be used for the upcoming process in production. This can be well monitored by blockchain technology and given as the input for federated learning. Based on the energy demand, the energy can be used for manufacturing. The energy utilized in the process is stored in the internal demand, and the energy not utilized will be stored in the external demand. Thus, the excessive amount of unused energy will be helpful in the other flawless production process. This will be helpful to the observation process, which is carried out by blockchain technology. Internal and external demands are based on the energy demand for the manufacturing process. The external demand is used to accumulate the excess energy which is not used in the manufacturing process. This excessive energy acquired during the process can be helpful in the upcoming flawless production process. Both internal and external energy is used to improve energy sustainability. This sustainable energy will last longer in the manufacturing process and production processes. Equation (3) below explains the process of obtaining the external demand from the energy demand for manufacturing.

$$\left. \begin{array}{l} L_1 - L_2(X_1, X_2) - \gamma_1\pi_1 + (1 - \gamma_1)Q_1 - (b - CX_1 - CX_2 - d_1)X_1, [\gamma_1 - 1] \\ M_1 - M_2(X_1, X_2) - \gamma_2\pi_2 + (1 - \gamma_2)Q_2 - (b - CX_1 - CX_2)X_2 - \gamma_2d_2X_2, [\{\gamma_2/\gamma_2 \leq 1\}] \end{array} \right\} \tag{3}$$

where (L_1, L_2) is denoted as the different energy demand levels and M_1 and M_2 are denoted as the excessive energy monitored in manufacturing and production levels. Now, the internal demand and the external demand are observed by blockchain technology. This observed information value is given as the input to the federated learning. Furthermore, this blockchain process can identify the energy used and available for the upcoming process. The blockchain implication for internal and external monitoring is illustrated in Figure 2.

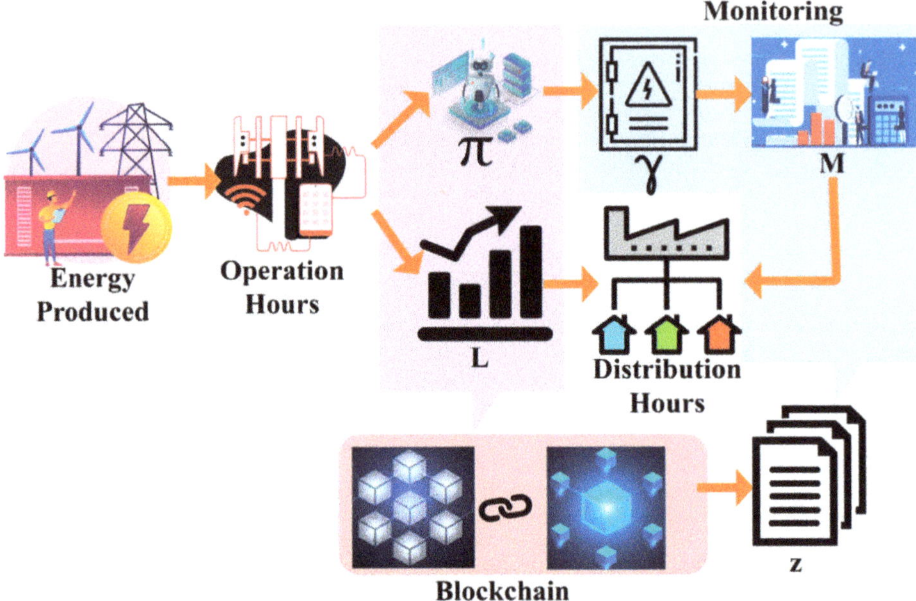

Figure 2. Blockchain implication.

Ethereum blockchain monitoring is used for similar products, and demand satisfaction is carried out in the production system. This process is used to check whether it satisfies the production system and whether it carries out the demanded process. This blockchain technology will obtain the energy used to fit the demand. It took care of both the demand and production satisfaction coextending. If there is any issue in the process, blockchain technology takes a further step to resolve the problems. It is used to check whether the demand satisfaction is met and the production flow. It also assumes energy and excessive energy consumption in the internal and external demand (Figure 2). Based on this input, federated learning is used for flawless production. In this technology, more identification is attained to produce the perfect deliverance without any flaws.

Ethereum blockchain technology is used to manage energy, which satisfies both the production and the demand. It is also used to observe the entire internal and external demand process extracted from the energy demand. It can be given as input to FL for production without flaws. It can also help monitor the control process. It has information about the used and unused energy for manufacturing and production. Therefore, it can be observed by satisfying both the presentation and demand during flawless execution. This process of observing blockchain technology's internal and external demand is explained by Equation (4) below:

$$\left. \begin{array}{l} \alpha(X_2) - X_1 b - \frac{d_1}{2a} - \frac{X_2}{2} \\ \beta(X_1, X_2) - X_2 - \frac{b - \gamma_2 d_2}{2a} - \frac{X_2}{2} \\ \alpha(\gamma_2) - \frac{b - 2d_1 + \gamma_2 d_2}{2b} \\ \beta(\gamma_2) - \frac{b - 2\gamma_2 d_2 + d_1}{2b} \end{array} \right\} \tag{4}$$

where α denotes the performance threshold of the blockchain technology, β is the energy evaluation function, γ is represented as the output of the internal and external demand from the energy demand, and d is the energy distribution. The observed information by the blockchain technology is sent as input to federated learning. The FL validates the monitoring and condition satisfaction processes from two perspectives. This perspective includes the distribution of the demand and the sustainability of the production. In the distribution of the demand perspective, the condition is validated to meet the actual requirement. FL is used for the intelligent manufacturing process by using sustainable energy. It uses the input given by the blockchain for flawless production. It is used to check the sustainability of production and demand distribution. It also verifies whether the energy can satisfy the demand and execute a perfect show. It also has information about the energy consumption and distribution rate used to fulfill the need. It is used to enhance energy sustainability during manufacturing and production processes. It is also used to validate whether the condition meets the actual requirement. The input given by the blockchain to FL is used in the monitoring control for further modification. Then, it is also used in FL for the perfect flawless production with good energy sustainability. The FL functions are explained in Figure 3.

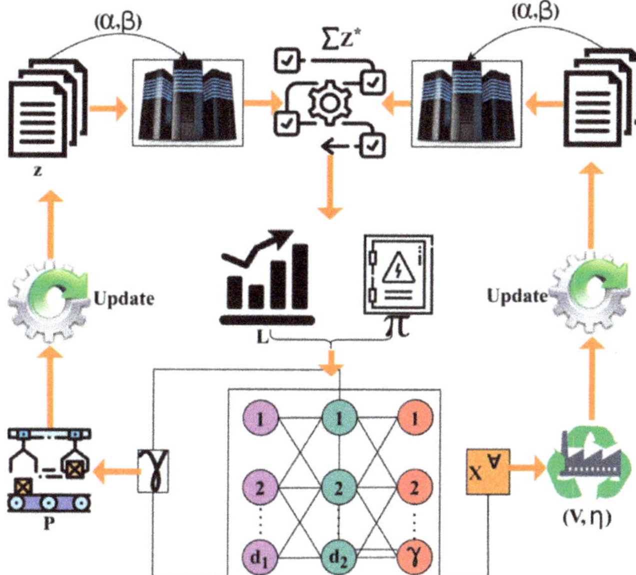

Figure 3. FL functions.

Federated learning is an ML method that trains an algorithm across servers holding local data samples and multiple decentralized edge devices. FL allows for ML to be used locally without transmitting data to a centralized server. The centralized storage permits the evaluation progression to work fully asynchronously. Since every FL training round creates a model for every user and a united model resulting from the merge procedure, the number of models grows fast. The blockchain and federated-learning-assisted solution deliver secure energy distribution between industrial applications. Federated learning is used to verify the demand distribution and the sustainability of the production with the two perspectives. The input given by the Ethereum blockchain to FL checks whether it satisfies the actual requirement and checks the availability and sustainability of the energy to meet the demand. The FL extracts the production flaws, energy sustainability, and recommendation. It can validate the energy state for the manufacturing process and flawless production (Figure 3). The distribution rate and energy consumption can be identified through FL. The method of FL using the input given by blockchain technology is explained by Equations (5) and (6), as given below:

$$\left. \begin{array}{l} \forall_1(\gamma_2) - \dfrac{b-(2d_1+\gamma_2 d_2)^2}{Z} \\[2mm] \forall_2(\gamma_2) - \dfrac{(b-d_2+d_1+\gamma_2 d_2)(b-\gamma_2 d_2+d_1)^2}{Z} \end{array} \right\} \tag{5}$$

$$\sum_z^* = \frac{d_2 - b_1 - a}{d_2} \tag{6}$$

where $(\forall)$ is denoted as the normalization vector of federated learning, $(\sum_z^*)$ is designated as the energy distribution rate, and Z represents the production flaws. Now, the production flaws, energy sustainability, and recommendation are extracted from FL. The production flaws verify the internal usage of energy. Additionally, they also check whether there needs to be more energy to lead the manufacturing process. Suppose there is an issue in the distribution process due to insufficient energy—in this case, the needed energy is given, and the redistribution process is carried out to the flawless production. It is used to check whether it satisfies the energy demand and sustainability. If there are any flaws during the show, further steps are taken to resolve the insufficient energy. After verifying the internal usage of energy, the needed amount is calculated for the other process. The redistribution process eliminates the flaws and makes the energy so long for the production process. The redistribution is made to improve the manufacturing process in the industry without any time delays and defects. It can also be used in the improvement in demand and product satisfaction. More energy can be identified, and further steps are taken to enhance the sustainability of energy. The production flaws were used to check whether the production rate met the demand satisfaction without flaws and insufficient energy. FL verifies production distribution and demand satisfaction, and the production flaws are obtained from that. If there is an inadequate amount of energy in the process, then the redistribution process is carried out with sufficient energy needed for the perfect manufacturing process. The method of production flaws obtained from FL is explained by Equation (7) below:

$$((\gamma_1^*, \gamma_2^*), (X_1(\beta_1, \beta_2), X_2(\gamma_1, \gamma_2)) - \left(\left(1, \frac{bd_2 - d_1 - b}{d_2} \right), \left(\frac{b - 2d_1 + \gamma_2 d_2}{b}, \frac{b - 2\gamma_2 c_2 + d_1}{b} \right) \right) \tag{7}$$

where (γ_1^*) is denoted as the rate of production flaws during manufacturing. Now, the energy sustainability is verified by FL. Here, it demonstrates how long the energy lasts to achieve the demands. Additionally, internal and external energy supply flaws during production are identified. They are used to verify the sustainability of the energy from the energy demand. The Ethereum blockchain makes the input for FL; thus, the flaws and production rate can be determined. These are the things that have information about the energy used in manufacturing. If there is excessive energy in the external demand, it will be used for the redistribution process when there is insufficient energy. The FL is used to enhance the sustainability of the energy in order to last a long time to meet the demand satisfaction. It also improves the production process without any flaws in it.

Energy sustainability checks whether energy can last far for the required demand and improves the production rate. The condition of meeting the requirement is also validated in this energy sustainability. From the input of the Ethereum blockchain, these features are extracted and used to improve the production rate and the manufacturing process in Industry 4.0. The sustainability of the energy helps in the elimination of flaws and increases the production rate. The energy sustainability process validates the appropriate value of energy consumption. It also makes the process effective and improves the satisfaction of the required demands during manufacturing. The method of energy sustainability verification from FL by the input of the blockchain is explained by Equation (8), as given below.

$$\left. \begin{array}{l} X_1^*(\gamma) = \frac{b - Vd_1 + d_2}{\eta} \\ X_2^*(P) = \frac{b - Vd_2 + d_1}{\eta} \end{array} \right\} \tag{8}$$

where $\left(X_1^*\right)$ is denoted as the sustainability of the energy, η is designated as the process carried out by the sustainable energy production, and V is the volume of the energy characterized as the output of FL. Now, the recommendation takes place from the FL processes. This recommendation gives information about sufficient and insufficient energy, scheduling the time and energy. By this, other methods can be carried out for the manufacturing process. It provides recommendations to alter the current approach to meet the demand after flawless production. This information makes changes for the successful production process without flaws and delays. The recommendation information is preferred for the following methods by changing accordingly with the perfect amount of energy and time. The recommendation flow based on decisions is presented in Figure 4.

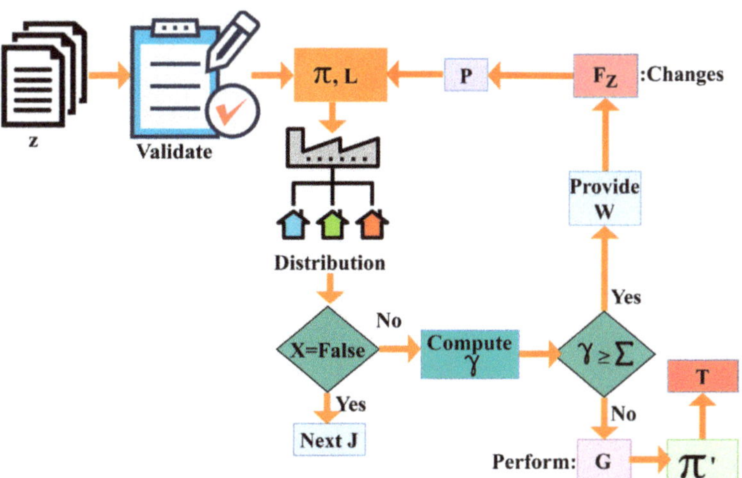

Figure 4. Recommendation flow.

The recommendation includes what to change and add for the upcoming processes. With this reference, the changes are made for the speedy manufacturing process in Industry 4.0. These changes can be stored in the blockchain records for further modification processes. In addition, it is given as input for the monitoring control process in the industry (refer to Figure 4). The recommendation process extracted from FL with the blockchain input is explained by Equation (9), as given below:

$$W(X_1, \gamma_2) - \frac{b - GX_1 - \gamma_2 d_2}{a} \tag{9}$$

where (W) is denoted as the weighted recommendation for the process and (G) is designated as the energy gain. Now, the monitoring control process takes place from the blockchain records. The observed information by the Ethereum blockchain modification is carried out accordingly. By monitoring the production plan based on the federated learning verification, the Ethereum blockchain records can be used to reduce the flaws in production. The monitoring control process is explained by Equations (10) and (11), as given below:

$$\pi'_1 - \pi'_1(X_1\gamma_2) - \left(\frac{b - GX_1 - \gamma_2 d_2 - 2J_1}{a}\right)X_1 \tag{10}$$

$$\pi'_2 - \pi^1_2(X_1\gamma_2) - \frac{(b - GX_1 + \gamma_2 d_2 - 2J_2)(b - CX_1 - \gamma_2 d_2)}{2a} \tag{11}$$

where $\left(\pi'_1\right)$ is denoted as the monitoring control process and (J) is represented as the flaw rate obtained during production. The modification is carried out for the new production process based on the monitoring control output. Improvement is made to improve the sustainability and production rate during manufacturing. The changes are made according to the records in the Ethereum blockchain and the validation of FL. These can help produce a productive manufacturing process in Industry 4.0. The method of the modification procedure carried out based on the Ethereum blockchain records and FL validation is explained by Equations (12) and (13), as given below:

$$\sum_{z}^{*} X_1(\gamma_2) - \frac{b + T_2 d_2 - 2d_1}{2a} \tag{12}$$

$$F_z(X_1) - 1 \tag{13}$$

where (F_z) is denoted as the functionality changes made to improve the manufacturing process and (T) is the required time in the manufacturing due to the Ethereum blockchain. Hence, this method uses Ethereum blockchain technology and federated learning to improve the production rate and manufacturing process. The sustainability for internal and external demand from the energy demand is met through FL and blockchain support integration. Here, the processing time is reduced and the flaw ratio is decreased. As a result, the sustainability of the energy is high during the manufacturing process. This method helps in the improvement in production and demand satisfaction. From the considered dataset, the monitoring control process is illustrated. The monitoring control process is shown in Figure 5.

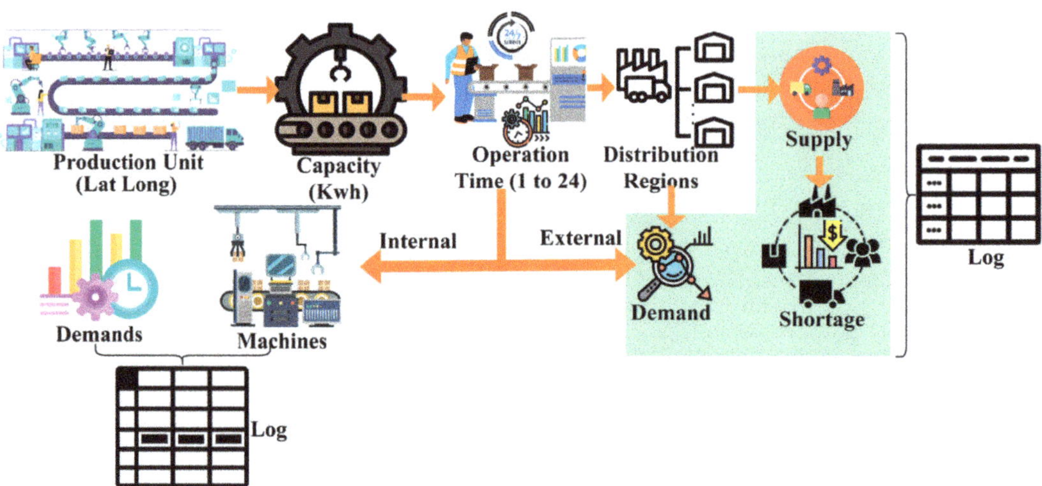

Figure 5. Monitoring control process using the considered data.

The production unit (industry) is located by its latitude (Lat) and longitude (long) markers. Based on the capacity, the distribution regions are organized. The generated energy is split into internal (machines) and external (public) distributions. The log post the single operation cycle provides the next cycle's shortage, consumption, and requirements. The T is predominantly performed for internal and external distributions (Figure 5).

3. Dataset Description

The dataset from [36] is used for validating the SP-ED and verifies sustainability. Data on greenhouse gas emissions supplied by significant emitters to the United States Environmental Protection Agency's (EPA) Greenhouse Gas Reporting Program (GHGRP) are used to calculate energy usage from commercial combustion at the plant level. Information on fuel usage is calculated using the EPA's standard pollutants factors. The values for the amount of energy required to burn fuel at a specific facility are calculated based on several factors, including sector (six-digit NAICS code), geographic coordinates (elevation, meridian, zip/postal code, county, and state), type of combustor, and name of the unit. The manufacturer's North American Industrial Classification System (NAICS) codes may further define combustion energy consumption by identifying energy end-use (e.g., conventional boiler use, co-generation/CHP utilization, process heating, and other facility support). The proportion of combustion fuel energy utilized for each end-use group in assembly plants may be calculated using data from the 2010 Manufacturing Energy Consumption Survey (MECS, produced by the Energy Information Administration), using the NAICS code and the stated fuel type. Based on industrial combustion energy, two observations are presented. The first observation is the attributes such as location, operation time, etc., as illustrated in Figure 5. The second observation is the utilization and sustainability of energy generation and distribution. In the first observation, a total of 20,118 (with average production) and 183 entries are jointly used for assessment in the second observation. The joint evaluation is performed according to 20 cycles for internal and external distribution. The number of nodes included in the eight industrial nodes and the amount of data available to each node in federated learning is two. The two blockchain networks were evaluated as potential components of the experimental setup's distributed ecosystem. Each benchmark included 1000 transactions sent at speeds ranging from 20 to 500 transactions per second to determine the maximum, average, and lowest transaction latency and throughput. Sustainability is accounted as the maximum possible distribution ratio that is consistently achieved in maximum operation cycles. Based on π and L, the actual representation in the dataset for sustainability is presented in Table 1.

Table 1. π & L representation from the data set.

Cycles	π (kWh)	Distribution (kWh)	M_1	L (kWh)	Distribution (kWh)	M_2
2	65.23	64.36	0.87	133.79	133.79	0
4	61.2	61.2	0	259.32	251.36	7.96
6	82.36	78.35	4.1	458.23	452.36	5.87
8	231.21	213.16	18.05	369.6	362.31	7.29
10	198.36	190.3	8.06	698.25	690.58	7.67
12	254.36	251.36	3.0	784.62	785.3	−0.68
14	274.63	269.54	5.90	1008.1	1008.1	0
16	289.46	285.34	4.12	985.36	878.36	107.0
18	303.05	301.21	1.84	874.23	870.69	3.54
20	298.25	298.25	0	963.21	962.21	1.0

The π and L different cycles are tabulated above, wherein π shows up more minor variations. Contrarily, L is different due to the distribution regions and unpredictable consumptions. In the π distribution, the fixed count of machines serves the purpose of planning the distribution beforehand. Therefore, the maximum utilization is improved, from which an excess is reported if the machinery is not functional. In this case, M_1 (extreme) is augmented for $L-$based distribution. However, M_2, a cycle modification, is comparatively high and therefore is required to prevent flaws. Pursued by the above represents the data recorded by the blockchain for identifying defects. For example, if the M_2 is vast even after M_1 scheduling, then the production needs to be modified. Therefore, the revisiting cycle across the different working machines is pursued under modifications. It is represented in Figure 6.

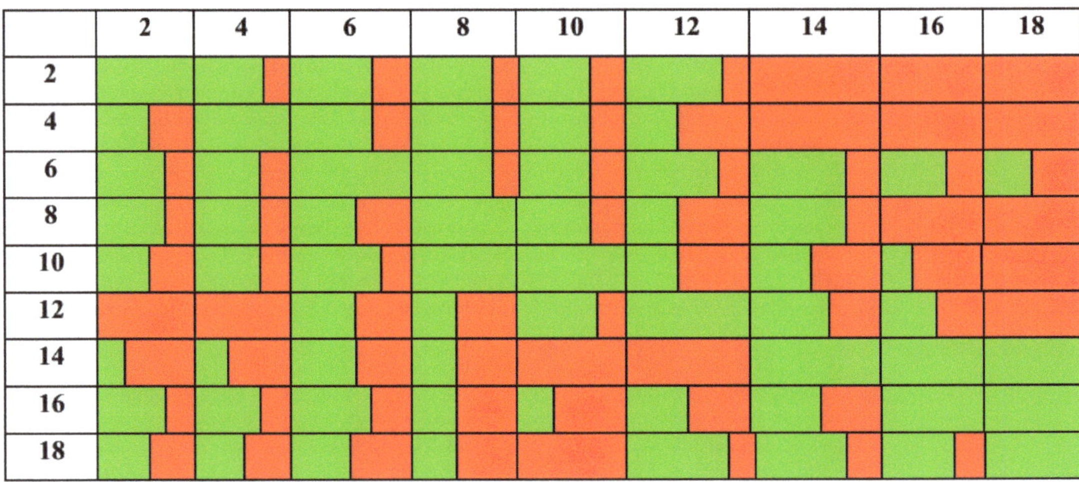

Figure 6. Revisited cycle for different hours.

The representation in Figure 6 presents four different combinations—namely, excessive (green), semi-demand (half red-half green), demand (red), and depleting (red concentration is high). Therefore, the depleting- and demand-based combinations are revisited to prevent distribution flaws. Based on (α, β) all depletion and demand, (α, β) is analyzed between successive cycles. In this process, γ^* due to a lack occurs, and hence, F_z is required. This represents the precise cycle for which modification is required. A total of 141 changes (internal 16, external 125) are observed in a given dataset. As the internal is significantly less, we discard it; the sustainability for 125 is analyzed in Table 2.

Table 2. (V, η) for 120 modifications.

Modifications	Flaw Detection (%)	X^*	Excessive Distribution (kWh)	(V, η)	Revisiting Required
20	13.06	0.365	98.56	0.658	1
40	21.36	0.263	137.604	0.462	3
60	25.69	0.458	121.36	0.541	2
80	29.69	0.547	69.25	0.745	2
100	32.45	0.619	12.39	0.883	0
120	36.46	0.587	58.25	0.851	1

The actual modifications are required to prevent frequent F_Z between the operating cycles. Therefore, sustainability is short-lived, and thus the new γ^* is identified. The $(M_1 + M_2) \, \forall \, (L)$ distribution is planned using G, such that π^* induces successful allocation. This is further studied using the blockchain output for the logs (α, β). In this process, W is the modification, and T for preventing sustainability falls (Table 2). Following this process, the flaw minimization duet to M and G is analyzed in Table 3 data.

Table 3. Flaws for varying $G = 1$ *to* 7.

G	W	(V,η)	Distribution	Flaws (/Cycle)
1	3	0.883	0.955	6
2	5	0.584	0.854	4
3	4	0.651	0.654	5
4	12	0.521	0.741	3
5	18	0.591	0.845	1
6	16	0.625	0.745	2
7	27	0.462	0.608	0

The flaw minimization is achieved by increasing the distribution and sustainability. W is segregated from multiple intervals (cycles) to improve the distribution. This is achieved by considering (α, β) various d_1, d_2 and γ of the FL process. Any flaws are tracked and addressed by providing precise G between T's, and hence, appropriate demands are satisfied. This generates no hassle in further distribution, thus preventing defects (Table 3).

4. Comparative Discussion

The comparative analysis uses the metrics sustainability factor, flaw detection, demand satisfaction, modifications, and detection time. The operating hours and supply-to-demand factors are modified accordingly. Alongside the proposed method, the existing MQIP-TOU [26], DDSIM [20], and GDDF [28] methods are considered. However, several current ways have limitations, such as complex production efficiency, a high time-consuming feature, and inadequate prediction accurateness. When compared to all of the existing methods, the proposed method has higher efficiency, which is discussed as follows:

4.1. Sustainability Factor

The efficacy of the sustainability factor is high in this method by using Ethereum blockchain technology and the federated learning technique. Based on the energy demand, the manufacturing process is carried out. From the energy demand, the internal and external demand is extracted. By using this demand, the use and the excessive amount of energy can be determined. Energy demand is used to enhance the sustainability of the energy during flawless production. Blockchain technology is used to observe the internal and external needs and provides input to FL. Monitoring control is also carried out based on blockchain records and FL validation. From this, modifications are made to improve the redistribution and satisfaction of production. This process helps to strengthen both demand and production satisfaction. Using Equation (5), the sustainability factor has been determined for the energy production system. By improving energy sustainability, the flaws can be reduced for the intelligent manufacturing process in Industry 4.0. With these methods, sustainability is high in this process (Figure 7).

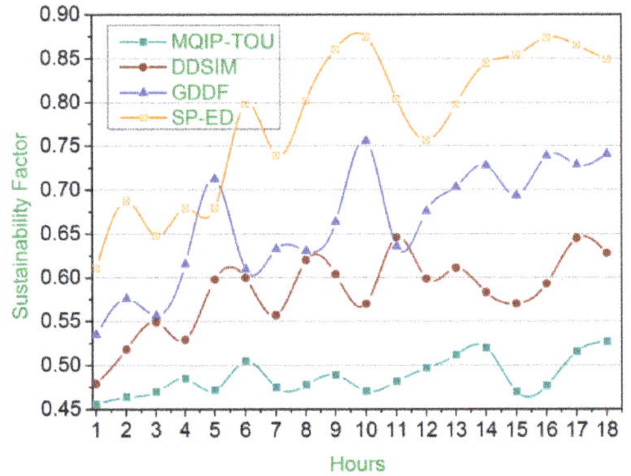
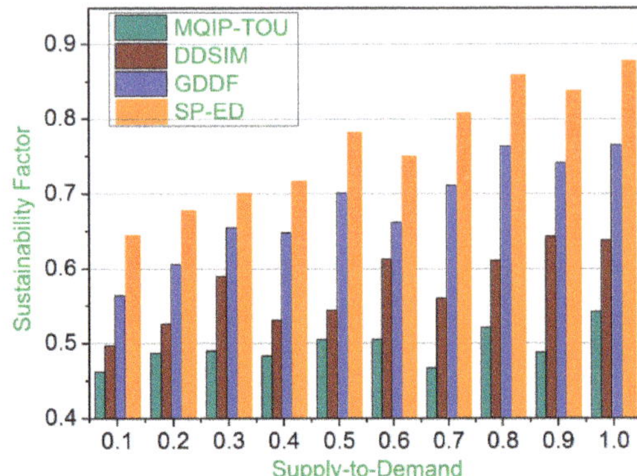

Figure 7. Sustainability factor.

4.2. Flaw Detection

Figure 8 depicts flaw detection during manufacturing and production in Industry 4.0 compared with the conventional method on which the proposed model has a high flaw detection rate. The flaw detection is high in this method using the FL technique, which uses the Ethereum blockchain records input. The production flaws are detected from FL, which verifies the consumption and distribution ratio during the manufacturing process. In this production flaw, if energy is inadequate, it can be verified, and further steps are taken to provide the needed energy. Thus, the redistribution process is carried out based on the blockchain records and the FL validations for the new manufacturing processes. The energy is redistributed based on its need, and a further process is carried out to modify the manufacturing procedure in the industry. The redistribution process is carried out to reduce the flaws and to make the energy last so long for the production process. Based on Equation (7), the production process flaws have been detected. This production flaw identifies the internal energy usage and the insufficiency of energy. It is used to check whether it satisfies the energy demand and sustainability. Through this FL validation process, the flaws in the production can be detected quickly, and further steps are taken to resolve them.

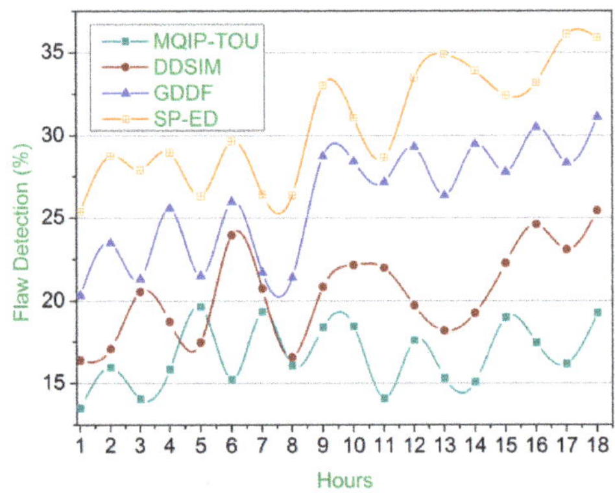
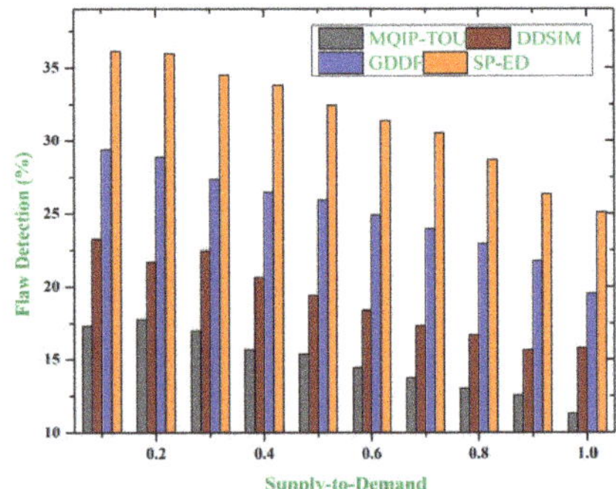

Figure 8. Flaw detection.

4.3. Demand Satisfaction

The demand satisfaction is high using Ethereum blockchain technology for manufacturing and flawless production. Ethereum blockchain is used to observe the internal and external demand obtained from the energy demand. It also checks whether the condition meets the required show and satisfies the needed directions. It is also used to enhance the energy's sustainability, which helps to satisfy the demand needed for flawless production. FL validates whether the sustainability of the energy lasts longer for the process based on the demand requirement. Both production distribution and demand satisfaction are identified by the FL technique, from which the flaws of the production are extracted during the manufacturing process in the industry. Ethereum blockchain monitoring is utilized for similar products, and demand satisfaction is completed in the production system. From Equation (2), demand satisfaction has been identified. The modification process is made from the blockchain records to improve the sustainability and production rate during manufacturing. The changes are made according to the blockchain records and FL validation (Figure 9).

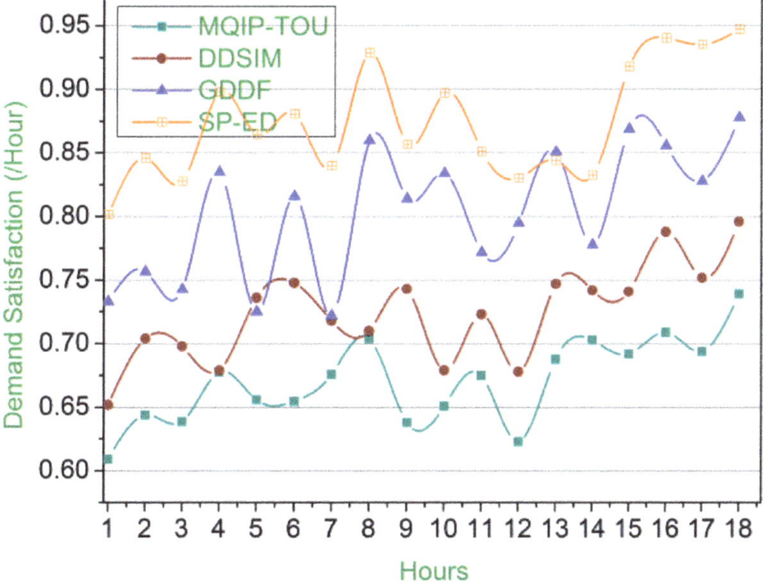

Figure 9. Demand satisfaction.

4.4. Modifications

The modifications are less in this method due to the usage of the required amount of energy for flawless production and the manufacturing process. Transformations are carried out to improve energy sustainability for the manufacturing process. Based on the blockchain records, the modification process is carried out. The modifications are carried out for the new production process based on the monitoring control output. It can help produce a productive manufacturing process in Industry 4.0. The improvements are made to the present method for the new manufacturing process, and the flaws are eliminated according to the demand requirement. This helps to enhance energy sustainability to meet production and demand satisfaction. The monitoring and condition satisfaction processes are identified using FL from two perspectives. The perspectives include demand allocation and production sustainability. From Equation (1), the modification process has been recognized. This process can be carried out to reduce the procedure of modification in the manufacturing procedure in Industry 4.0 (Figure 10).

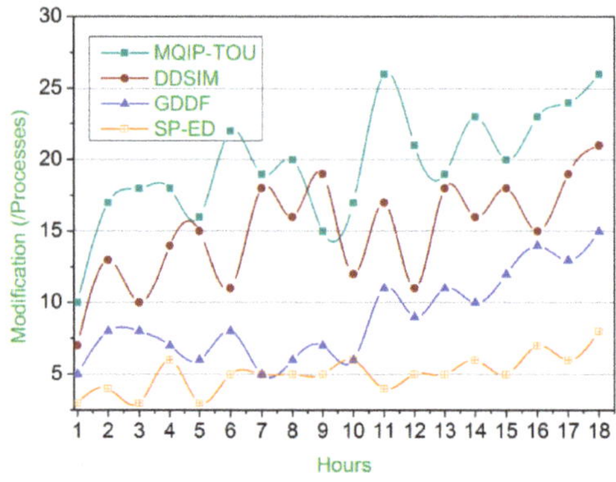
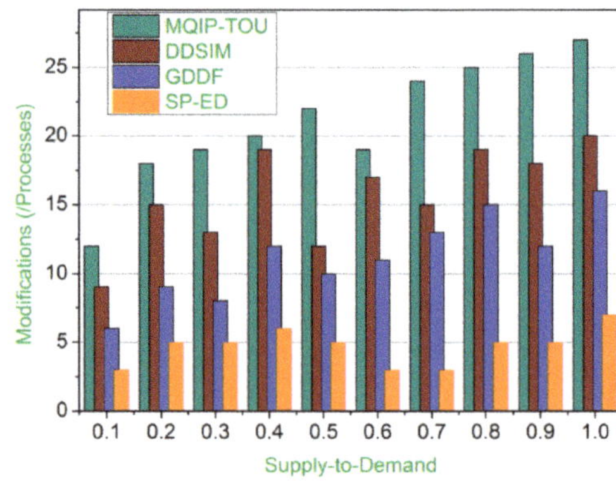

Figure 10. Modifications.

4.5. Detection Time

The time taken for detecting the flaws is recommended to be less in this method by using the Ethereum blockchain and FL techniques. By observing the internal and external demand, the blockchain technique uses this value as the input for FL. The production flaws are detected from FL; if there is insufficient energy, then it is said to be a flaw in the execution of the process. Energy sustainability checks how long the energy lasts to achieve the demand. The production system accompanies blockchain monitoring for lateral production and demand satisfaction. The observing and status gratification processes are approved using FL. The ineffectual conditions in prospects are manipulated using blockchain records. The Ethereum blockchain records reduce flaws in the new production by customizing the production plan according to the federated learning confirmations. Based on Equation (13), the detection time of marks has been identified. Using these processes, the time taken for the detection is less in this method for the flawless manufacturing procedures (Figure 11).

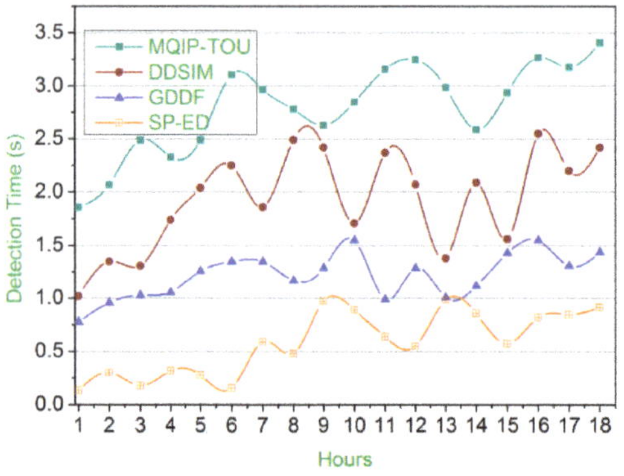
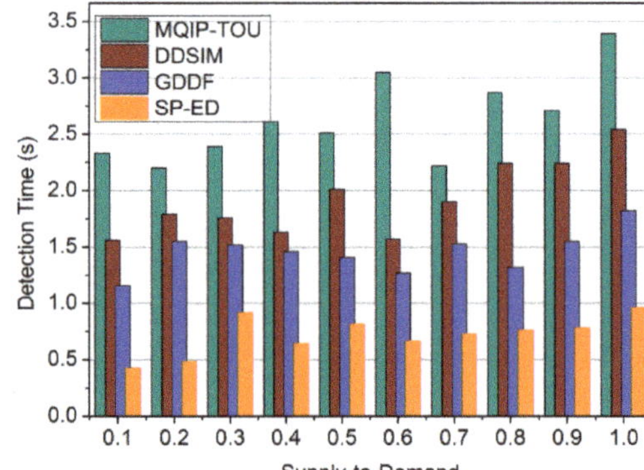

Figure 11. The detection time.

5. Conclusions

This article introduces and discusses the Sustainable Production concerned with External Demands method. This SP-ED method is designed to improve the efficacy of Industry 4.0 in energy production and distribution. The proposed method utilizes blockchain and federated learning concepts to improvise sustainability performances. The entire process is monitored, and the blockchain stores and processes detailed logs to identify production flaws. In this process, FL validates the sustainability and flaw detection for modifying the operations in consecutive operation cycles. The sustainability due to internal and external distribution demands is identified, and precise recommendations are provided. In the learning process, the maximum amount of achievable sustainability is predicted, and the performance is leveraged. The following learning process is instigated by considering the changes pursued in the production process using the recommendations. Therefore, the energy scheduling process is validated using joint blockchain and learning paradigms. Hence, sustainability is slowly leveraged across varying operation times and demands. The proposed federated learning and Ethereum blockchain model can achieve sustainability by 11.48%, flaw detection by 14.65%, and can reduce modifications by 11.11% and detection time by 10.46% for the varying energy supply-to-demand factor compared to DDSIM. The study's limitations are speed and scalability, a challenge identified for energy production and industrial applications. Future studies will examine the edge computing techniques for energy production for green innovation success in the industrial environment.

Author Contributions: F.S.—Data curation, writing—original draft preparation. Z.D.—Conceptualization, methodology, software, writing—reviewing and editing. All authors have read and agreed to the published version of the manuscript.

Funding: This research received no external funding.

Institutional Review Board Statement: Not available.

Informed Consent Statement: Not available.

Data Availability Statement: The data presented in this study are available on request from the corresponding author.

Conflicts of Interest: The authors declare no conflict of interest.

References

1. Rahmani, H.; Shetty, D.; Wagih, M.; Ghasempour, Y.; Palazzi, V.; Carvalho, N.B.; Correia, R.; Costanzo, A.; Vital, D.; Alimenti, F.; et al. Next-generation IoT devices: Sustainable eco-friendly manufacturing, energy harvesting, and wireless connectivity. *IEEE J. Microw.* **2023**, *3*, 237–255. [CrossRef]
2. Wang, Q.; Yang, H. Sensor-based recurrence analysis of energy efficiency in machining processes. *IEEE Access* **2020**, *8*, 18326–18336. [CrossRef]
3. Sarkar, B.; Mridha, B.; Pareek, S. A sustainable smart multi-type biofuel manufacturing with the optimum energy utilization under flexible production. *J. Clean. Prod.* **2022**, *332*, 129869. [CrossRef]
4. Chen, M.; Sinha, A.; Hu, K.; Shah, M.I. Impact of technological innovation on energy efficiency in industry 4.0 era: Moderation of shadow economy in sustainable development. *Technol. Forecast. Soc. Chang.* **2021**, *164*, 120521. [CrossRef]
5. Bauer, D.; Kaymakci, C.; Bauernhansl, T.; Sauer, A. Intelligent energy systems as enabler for increased resilience of manufacturing systems. *Procedia CIRP* **2021**, *104*, 217–222. [CrossRef]
6. Furstenau, L.B.; Sott, M.K.; Kipper, L.M.; Machado, E.L.; Lopez-Robles, J.R.; Dohan, M.S.; Cobo, M.J.; Zahid, A.; Abbasi, Q.H.; Imran, M.A. Link between sustainability and industry 4.0: Trends, challenges and new perspectives. *IEEE Access* **2020**, *8*, 140079–140096. [CrossRef]
7. Remme, A.M.R.; Stange, S.M.; Fagerstrøm, A.; Lasrado, L.A. Blockchain-enabled sustainability labeling in the fashion industry. *Procedia Comput. Sci.* **2022**, *196*, 280–287. [CrossRef]
8. Cui, X. Cyber-Physical System (CPS) architecture for real-time water sustainability management in manufacturing industry. *Procedia CIRP* **2021**, *99*, 543–548. [CrossRef]
9. Chen, J.; Zhang, W.; Wang, H. Intelligent bearing structure and temperature field analysis based on finite element simulation for sustainable and green manufacturing. *J. Intell. Manuf.* **2021**, *32*, 745–756. [CrossRef]
10. Chen, Y.; Lu, Y.; Bulysheva, L.; Kataev, M.Y. Applications of Blockchain in Industry 4.0: A Review. *Inf. Syst. Front.* **2022**, 1–15. [CrossRef]
11. Liu, C.; Luosang, R.; Yao, X.; Su, L. An integrated intelligent manufacturing model based on scheduling and reinforced learning algorithms. *Comput. Ind. Eng.* **2021**, *155*, 107193. [CrossRef]
12. Liu, Z.; Liu, Q.; Xu, W.; Wang, L.; Zhou, Z. Robot learning towards smart robotic manufacturing: A review. *Robot. Comput. Manuf.* **2022**, *77*, 102360. [CrossRef]
13. Aoun, A.; Ilinca, A.; Ghandour, M.; Ibrahim, H. A review of Industry 4.0 characteristics and challenges, with potential improvements using blockchain technology. *Comput. Ind. Eng.* **2021**, *162*, 107746. [CrossRef]
14. Mezair, T.; Djenouri, Y.; Belhadi, A.; Srivastava, G.; Lin, J.C.W. A sustainable deep learning framework for fault detection in 6G Industry 4.0 heterogeneous data environments. *Comput. Commun.* **2022**, *187*, 164–171. [CrossRef]

15. Ferreiro, S.F.S.; Larreina, J.; Tena, M.; Leunda, J.; Garmendia, I.; Arnaiz, A. Artificial intelligence methodology for smart and sustainable manufacturing industry. *IFAC-PapersOnLine* **2021**, *54*, 1041–1046. [CrossRef]
16. Sun, Z.; Dababneh, F.; Li, L. Joint energy, maintenance, and throughput modeling for sustainable manufacturing systems. *IEEE Trans. Syst. Man Cybern. Syst.* **2018**, *50*, 2101–2112. [CrossRef]
17. Li, L.; Qu, T.; Liu, Y.; Zhong, R.Y.; Xu, G.; Sun, H.; Gao, Y.; Lei, B.; Mao, C.; Pan, Y.; et al. Sustainability assessment of intelligent manufacturing supported by digital twin. *IEEE Access* **2020**, *8*, 174988–175008. [CrossRef]
18. Majeed, A.; Zhang, Y.; Ren, S.; Lv, J.; Peng, T.; Waqar, S.; Yin, E. A big data-driven framework for sustainable and smart additive manufacturing. *Robot. Comput. Manuf.* **2021**, *67*, 102026. [CrossRef]
19. Psarommatis, F.; Bravos, G. A holistic approach for achieving sustainable manufacturing using zero defect manufacturing: A conceptual framework. *Procedia CIRP* **2022**, *107*, 107–112. [CrossRef]
20. Ma, S.; Zhang, Y.; Liu, Y.; Yang, H.; Lv, J.; Ren, S. Data-driven sustainable intelligent manufacturing based on demand response for energy-intensive industries. *J. Clean. Prod.* **2020**, *274*, 123155. [CrossRef]
21. Tian, S.; Xie, X.; Xu, W.; Liu, J.; Zhang, X. Dynamic assessment of sustainable manufacturing capability based on correlation relationship for industrial cloud robotics. *Int. J. Adv. Manuf. Technol.* **2021**, *124*, 3113–3135. [CrossRef]
22. Jasiulewicz-Kaczmarek, M.; Żywica, P.; Gola, A. Fuzzy set theory driven maintenance sustainability performance assessment model: A multiple criteria approach. *J. Intell. Manuf.* **2021**, *32*, 1497–1515. [CrossRef]
23. Zimmermann, N.; Mayr, J.; Wegener, K. An action-oriented teaching approach for intelligent and energy efficient precision manufacturing. *Manuf. Lett.* **2022**, *33*, 961–969. [CrossRef]
24. Wang, K.; Yu, L.; Xu, J.; Zhang, S.; Qin, J. Energy consumption intelligent modeling and prediction for additive manufacturing via multisource fusion and inter-layer consistency. *Comput. Ind. Eng.* **2022**, *173*, 108720. [CrossRef]
25. Favi, C.; Marconi, M.; Mandolini, M.; Germani, M. Sustainable life cycle and energy management of discrete manufacturing plants in the industry 4.0 framework. *Appl. Energy* **2022**, *312*, 118671. [CrossRef]
26. Pei, Z.; Wan, M.; Jiang, Z.-Z.; Wang, Z.; Dai, X. An approximation algorithm for unrelated parallel machine scheduling under TOU electricity tariffs. *IEEE Trans. Autom. Sci. Eng.* **2020**, *18*, 743–756. [CrossRef]
27. Cañas, H.; Mula, J.; Campuzano-Bolarín, F.; Poler, R. A conceptual framework for smart production planning and control in Industry 4.0. *Comput. Ind. Eng.* **2022**, *173*, 108659. [CrossRef]
28. Friederich, J.; Francis, D.P.; Lazarova-Molnar, S.; Mohamed, N. A framework for data-driven digital twins for smart manufacturing. *Comput. Ind.* **2022**, *136*, 103586. [CrossRef]
29. Gu, W.; Li, Y.; Tang, D.; Wang, X.; Yuan, M. Using real-time manufacturing data to schedule a smart factory via reinforcement learning. *Comput. Ind. Eng.* **2022**, *171*, 108406. [CrossRef]
30. Liu, X.L.; Wang, W.M.; Guo, H.; Barenji, A.V.; Li, Z.; Huang, G.Q. Industrial blockchain based framework for product lifecycle management in industry 4.0. *Robot. Comput.-Integr. Manuf.* **2020**, *63*, 101897. [CrossRef]
31. Krithika, L.B. Survey on the Applications of Blockchain in Agriculture. *Agriculture* **2022**, *12*, 1333.
32. Jayabalasamy, G.; Koppu, S. High-performance Edwards curve aggregate signature (HECAS) for nonrepudiation in IoT-based applications built on the blockchain ecosystem. *J. King Saud Univ.-Comput. Inf. Sci.* **2022**, *34*, 9677–9687. [CrossRef]
33. Industrial Facility Combustion Energy Use. Available online: https://data.world/us-doe-gov/19c607fa-1687-4bc4-a6dd-74b8 8b849644 (accessed on 7 February 2022).
34. Xu, Z.; Wang, Y.; Dong, R.; Li, W. Research on multi-microgrid power transaction process based on blockchain Technology. *Electr. Power Syst. Res.* **2022**, *213*, 108649. [CrossRef]
35. Qu, Y.; Pokhrel, S.R.; Garg, S.; Gao, L.; Xiang, Y. A blockchained federated learning framework for cognitive computing in Industry 4.0 networks. *IEEE Trans. Ind. Inform.* **2020**, *17*, 2964–2973. [CrossRef]
36. Zhao, Y.; Zhao, J.; Jiang, L.; Tan, R.; Niyato, D.; Li, Z.; Lyu, L.; Liu, Y. Privacy-preserving blockchain-based federated learning for IoT devices. *IEEE Internet Things J.* **2020**, *8*, 1817–1829. [CrossRef]

Article

Recycling Strategies in a Collector-Led Remanufacturing Supply Chain under Blockchain and Uncertain Demand

Tianjian Yang [1], Chunmei Li [2,*] and Zijing Bian [3]

[1] School of Modern Post (School of Automation), Beijing University of Posts and Telecommunications, Beijing 100876, China
[2] School of Economics and Management, Beijing University of Posts and Telecommunications, Beijing 100876, China
[3] Department of Economic Theory Management, College of Social Sciences and Humanities, Moscow State Normal University, 119991 Moscow, Russia
* Correspondence: lcm@bupt.edu.cn

Abstract: Remanufacturing has been regarded as a key to the sustainable development of enterprises. However, collection strategies affect the remanufacturing and recycling of used products. Blockchain can ensure the authenticity of disclosed information and improve the consumer's trust in remanufactured products. Inspired by this, this paper develops a game-theoretic model to examine the selection of different recycling strategies in the remanufacturing supply chain considering blockchain adoption and uncertain demand. Incumbent collector 1 provides the manufacturer with used product 1 for remanufacturing product 1. For product 2, the manufacturer has two different collection strategies: in-house collection by the manufacturer or external collection by collector 2. The collectors act as the channel leader, and the manufacturer, who has private demand information, is the follower. Results show that collectors are incentivized to participate in the blockchain. If there is no blockchain, collector 1 prefers external collection. In the case of blockchain, the manufacturer prefers external collection when the demand variance is low. The manufacturer's decision on the in-house collection and external collection depends on the coefficient of collection investment costs.

Keywords: remanufacturing; blockchain; collection channel; recycling strategies; uncertain demand; game theory

Citation: Yang, T.; Li, C.; Bian, Z. Recycling Strategies in a Collector-Led Remanufacturing Supply Chain under Blockchain and Uncertain Demand. *Processes* **2023**, *11*, 1426. https://doi.org/10.3390/pr11051426

Academic Editors: Conghu Liu, Xiaoqian Song, Zhi Liu and Fangfang Wei

Received: 13 March 2023
Revised: 5 May 2023
Accepted: 7 May 2023
Published: 8 May 2023

1. Introduction

With the development of society, environmental deterioration and resource shortages are becoming more and more serious. As an effective way to protect the environment and save resources, remanufacturing has been recognized by enterprises [1,2]. Remanufacturing is a process in which used or underperforming products are collected through recycling channels and then remanufactured. Product recycling is an essential part of remanufacturing. Different recycling modes affect optimal decision-making and pricing in a remanufacturing supply chain. In practice, a manufacturer can collect used products in three main ways. One is that the manufacturer has in-house collection channels, such as Xerox and Fuji Films [3,4]. Another is that the manufacturer assigns product collection to its retailer, such as Kodak [4]. The third one is that the manufacturer outsources the collection activity to a dedicated collector. In the literature on collection mode selection, the manufacturer is usually the leader in the supply chain, while some powerful collectors have become upstream leaders of the manufacturers in recent years, for example, IBM's Global Asset Recovery Services, the world's largest mobile phone recycler ReCellular, and the world's largest metal electronics recycler SIMS Metal Management [5]. Therefore, we consider the situation that a third party performs the dedicated collection in which the collector is the leader and the manufacturer is the follower in this study.

In contrast to new products, however, consumers still have doubts about remanufactured products, such as uncertainty about the product's quality, low evaluations, or distrust [6–9], which decreases their willingness to purchase remanufactured products. In addition, the problem of uncertain demand in the remanufacturing supply chain has also attracted considerable theoretical and practical attention. In reality, upstream collectors are unable to obtain accurate market demand information because they are not familiar with the consumer market. Market information will have a certain impact on the operational efficiency of upstream companies [10]. Accurate market demand information can help upstream collectors adjust their inventories [11] and determine the transfer price of used products [5].

Blockchain adoption can help enable the information sharing of demand and increase consumer trust in remanufactured products. Blockchain technology ensures that the information recorded in a supply chain is transparent and unalterable for all stakeholders [12], which has been widely used in information traceability. In order to reduce consumers' concerns about quality issues, some second-hand trading platforms use blockchain technology to provide quality inspection information, such as Paipai, a second-hand trading platform of JD.com [1]. When blockchain exists, the downstream manufacturer records the demand information in the blockchain, and the upstream collector obtains full demand information. Moreover, the authenticity of disclosed information can improve consumer trust in remanufactured products. Hence, it naturally generates the following research problems:

(1) Does the manufacturer have the incentive to participate and record the demand information in blockchain?
(2) What are the effects of different blockchain scenarios on recycling decisions?
(3) How do demand variance and collection investment costs affect optimal decision-making in a collector-led remanufacturing supply chain?

To solve the above problems, we consider a collector-led remanufacturing supply chain comprised of a manufacturer and two collectors. The incumbent collector 1 recycles the used product 1 and sells them to the manufacturer. The manufacturer produces two new products with raw and used materials and sells them to the market. For product 2, the manufacturer has two different recycling strategies: in-house collection (Scenario A) or external collection (Scenario B). In-house collection denotes the manufacturer recycles the used product 2, and external collection denotes collector 2 recycles the used product 2. Because of familiarity with the market, the manufacturer has full knowledge of demand information, while collectors have to make predictions about market demand. In addition, the supply chain decides whether to adopt blockchain. In the case of blockchain, the manufacturer records the demand information in blockchain, and collectors can obtain full of market demand. As a result, four models are established depending on whether the supply chain adopts blockchain or not and whether product 2's collection is through in-house collection or through external collection (Model AN, Model BN, Model AB, and Model BB).

The main findings of this study are as follows. First, the unit transfer prices decrease with the increase in the collection rate in the four models. In Scenario B, the manufacturer's expected profits increase in the two collection rates when in-house collection exists. In other cases, the impact of collection rates on supply chain members' expected profits is related to the coefficient of collection investment costs. Second, the unit transfer prices and the selling quantities of product 1 in Scenario B are higher than that in Scenario A. Moreover, the unit transfer prices and selling quantities with blockchain are higher than without blockchain. However, the selling quantities of product 2 in Scenario B are lower than in Scenario A. Collectors prefer blockchain, but collector 1 is always inclined to the external collection under no blockchain. In the case of blockchain, collector 1's decision on the external collection depends on the demand variance and the coefficient of variance. Finally, the manufacturer's decision on the in-house collection and external collection depends on the coefficient of collection investment costs. The manufacturer may choose to implement the blockchain when the demand variance is less than a certain value.

The motivation for this research stems from the growing amount of empirical literature showing that consumers distrust the quality of remanufactured products [13,14]. Blockchain technology is an effective measure to increase consumer trust in the quality of remanufactured products. Moreover, recycling used products is a necessary link in the remanufacturing supply chain. Different recycling modes affect optimal decision-making and pricing in a remanufacturing supply chain. However, there is still a big research gap in the study of recycling models in collector-led remanufacturing supply chains, especially when considering the uncertain demand. Our findings provide new insights into blockchain applications and recycling strategies in the field of remanufacturing supply chains under uncertain demand.

The rest of this paper is organized as follows. In the next section, we briefly review the related literature. Problem formulation is listed in Section 3. Section 4 presents the collector-led supply chain models and equilibrium outcomes, respectively. In Section 5, we determine the supply chain members' preferences for collection scenarios and blockchain adoption by comparing four models. Section 6 summarizes and concludes this study.

2. Literature Review

This research is mainly related to studies about blockchain adoption in supply chains, demand uncertainty in supply chains, and remanufacturing collection modes.

2.1. Blockchain Adoption in Supply Chains

Blockchain technology has attracted considerable attention in supply chains. Considering the manufacturer's brand advantages and patent license fees, Yang et al. [1] studied the impact of blockchain on remanufacturing modes. Gong et al. [15] investigated the optimal strategies of the OEM regarding adopting blockchain technology and selecting distribution channels. Niu et al. [16] examined the supply chain members' preferences for blockchain adoption considering consumers' risk-aversion and quality distrust. Zhang et al. [17] analyzed the impact of three different blockchain adoption scenarios on the direct and retail channels of a dual-channel supply chain, where the three scenarios include both manufacturers and e-retailers adopting blockchain, manufacturers adopting blockchain, and e-retailers adopting blockchain. Cui et al. [18] used game theory to provide a theoretical investigation into the value and design of a traceability-driven blockchain under serial supply chains and parallel supply chains. Zheng et al. [19] studied the optimal blockchain-based traceability strategies in agricultural product supply chains under different strategic choices among multiple agents. Wang et al. [20] explored a three-echelon supply chain participants' motivation, condition, and roles by analyzing the game equilibrium of the no, upper-stream, lower-stream, and entire blockchain-driven accounts receivable chains. Zhang et al. [17] explored supply chain members' attitude towards three blockchain adoption scenarios (only manufacturer, only e-retailer, and both players) considering the direct sales channel and the retail channel. In contrast, our study focuses on the application of blockchain technology in collector-led remanufacturing supply chains. At the same time, we study the effect of blockchain adoption on the manufacturer's different collection scenarios considering uncertain demand.

2.2. Demand Uncertainty in Supply Chains

Most studies focus on the incentives for uncertain demand exchange among supply chain members. Uncertain demand can be categorized into two types: stochastic nature and fuzzy uncertainty. Currently, most studies with uncertain needs are stochastic nature. Cai et al. [5] examined how the manufacturer shares demand information and the effects of different demand-sharing strategies on collector-led CLSCs. Huang et al. [21] developed a win-win contract based on a revenue sharing and price markdown and studied how vendors and retailers share their risks and benefits under stochastic demand during the pandemic. Ji and Liu [22] studied how the two-part tariff and ZRS contract (zero wholesale price-revenue-sharing-plus-side-payment contract) affect risks and supply chain coordination when market demand and supplier yield are both uncertain. Zhang et al. examined partial demand information sharing from three sharing methods (neither, one, or both of the manufacturers) in a supply chain consisting of a single retailer and two competitive manufacturers. Garai and Paul [23] explored supply chain coordination in a closed-loop supply chain comprising one retailer, one main supplier whose demand is stochastic uncertain, and a backup supplier. Li et al. [24] built a two-stage stochastic program and investigated a comprehensive production planning problem considering uncertain demand and risk-averse. Some other literature has studied demand uncertainty in supply chains from the perspective of fuzzy uncertainty. For example, Pei et al. [25] investigated the pricing problem of dual-channel green supply chains based on fuzzy demand. Liu et al. [26] studied the closed-loop supply chain of second-hand products with ambiguous demand and different quality levels from the perspective of centralized and different authority structures. In this paper, we also set the demand as uncertain in stochastic nature. Differently, we examine asymmetric demand information in collector-led remanufacturing supply chains. Furthermore, this paper considers blockchain adoption and compares two collection modes that have been addressed in few previous studies.

2.3. Remanufacturing Collection Modes

The third related literature stream is about how manufacturers choose collection modes for remanufacturing. For example, Zheng et al. [27] investigated how the manufacturer and retailer choose the recycling cooperation modes between recycling alliance and cost-sharing and discovered that the optimal recycling cooperation option depends on the remanufacturing efficiency and the relative recycling cost efficiency. Considering the heterogeneity of willingness to pay, Long et al. [28] explored the optimal recycling and remanufacturing decisions by comparing four different remanufacturing modes. Yi et al. [29] examined the optimal decisions on a dual recycling channel in which the retailer and the third-party collector simultaneously collect the used products in the construction machinery industry. Huang et al. [30] further studied the optimal strategies for a triple recycling channel in a retailer-dominated closed-loop supply chain. Considering the retailer's bank loans or trade-credit financing, Zhang and Zhang [31] analyzed optimal equilibrium strategies of electric vehicle batteries in a closed-loop supply chain with a manufacturer or capital-constrained retailer recycling. He et al. [32] examined the competitive collection and channel convenience considering a manufacturer competing with a third-party collector. In the case of channel inconvenience, Guo et al. [33] investigated the optimal emission reduction strategy in three models with different recycling structures—manufacturer-led, retailer-led, and competitive under cap-and-trade regulation. Wan [34] investigated six game theory models which consist of different sales modes and recycling modes to explore the optimal pricing and recycling rate decisions under the discount coefficient of demand and the competing intensity of recycling. Some existing literature studies the collection strategies from different authority structures. For example, Cao and Ji [35] discussed the optimal recycling strategy by establishing three different Stackelberg leadership models in garment enterprises. Unlike previous literature, we investigate the optimal collection modes in a collector-led remanufacturing supply chain under demand uncertainty. Furthermore, this study also explores the impact of blockchain adoption on collection decisions.

3. Model

3.1. Problem Formulation

In our research, we consider a collector-led remanufacturing supply chain comprised of a manufacturer and two collectors. The manufacturer produces new products i ($i = 1, 2$) with raw and used materials and sells them to the market at a unit retail price p_i. Collector

i recycles the used product i and sells the used product i to the manufacturer at a transfer price b_i. For product 2, the manufacturer has two different recycling strategies: in-house collection through the manufacturer (Scenario A) or external collection through collector 2 (Scenario B). We use A and B to denote the in-house collection and external collection, respectively. Motivated by Cai et al. [5], we built the supply chain structure as illustrated in Figure 1. For simplicity, we assume the manufacturer (Scenario A) and collector 2 have the same collection rate. Collectors will invest in the collection channel at a cost of $k\lambda_i^2$, which quadratic form of the cost function is common in previous literature. $k > 0$ represents the coefficient of investment costs and $\lambda_i > 0$ denotes the collection rate of the collector i. The manufacturer has a less professional recycling network channel than collector 2, so the manufacturer's investment cost is $\phi k\lambda_2^2$ in Scenario A, where $\phi > 1$ represents the proportion of collector 2's investment costs. The unit production cost of producing a new product with raw/used materials are c_m, c_r, respectively, where $c_r < c_m$ represents the unit production cost with used materials is less than with raw materials. Denote $\Delta = c_m - c_r$, where $\Delta > b_i$ guarantees the manufacturer's positive profit from used products.

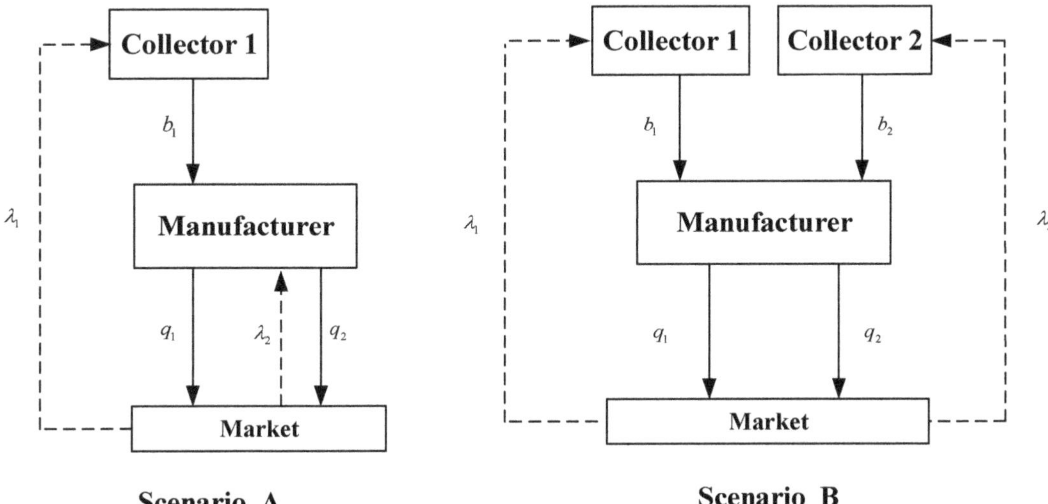

Figure 1. Supply chain structures.

Considering the uncertain market demand, the manufacturer has full knowledge of demand information since the manufacturer is closer to the market, while collectors have to predict the market demand. Consumers will have many uncertain concerns (including product function and product life) about remanufactured products when they know enterprises have collection channels. If blockchain is adopted, consumers can learn about the authenticity of remanufactured products and access the key information on remanufactured products from the blockchain database via cell phones. Moreover, blockchain technology can make sure the disclosed information is correct and improve the consumer's trust in remanufactured products, which will expand the consumer market. In addition, the manufacturer will log the sales information into the blockchain. Thus, the upstream collectors can obtain accurate demand information by accessing the blockchain platform. Note that blockchain adoption is a joint decision with the manufacturer and collectors, and this study does not consider the sunk costs associated with adopting blockchain technology [36,37]. Furthermore, we do not consider the unit collection cost and only consider the investment cost of collection channel [3,5,38].

Following Niu et al. [37] and Yang et al. [38], we introduce the inverse demand function given as:

$$p_i = a_j - q_i - \beta q_{3-i} + \varepsilon_i \tag{1}$$

where $j = \{B, N\}$ denotes the scenarios with/without blockchain technology and $i = \{1, 2\}$ denotes the product i. a_j stands for the deterministic market potential, where $a_B > a_N$ represents that the consumer market is expanded when blockchain exists. The random variable ε_i represents market demand uncertainty, which has a mean of zero and variance $Var[\varepsilon_1] = rV$, $Var[\varepsilon_2] = V$ where $0 < r < 1$ represents product 1 has a smaller demand variance than product 2. $\beta \in (0, 1)$ represents the competition coefficient between the two products. For simplification, we assume $\lambda_1 > \lambda_2 > 2\beta\lambda_1/(1 + \beta^2)$, which represents collector 2's collection rate is temperate. Four models are established depending on whether the supply chain adopts blockchain or not and whether product 2's collection is through in-house collection or external collection (Model AN, Model BN, Model AB, and Model BB). We characterize supply chain members' profits as π_h^l, where the subscript $h \in \{R1, R2, M\}$ stands for Collector 1, Collector 2, and the manufacturer, and the superscript $l \in \{AN, AB, BN, BB\}$ stands for the above four models. Table 1 shows the notations in this paper.

Table 1. Notations.

Notation	Definition
c_m / c_r	Unit production cost of producing a new product with raw/used materials
Δ	Unit saving cost of remanufacturing, $\Delta > b_i$
β	The competition coefficient between the two products
k	The coefficient of collection investment costs
ε_i	Random part of market potential for product i, $i = 1, 2$
rV / V	The variance of the random variable $\varepsilon_1 / \varepsilon_2$
a_N / a_B	The deterministic market potential without/with blockchain
λ_i	The collection rate of product i, $i = 1, 2$
p_i	Retail price of the product i, $i = 1, 2$
$\pi_M, \pi_{R1}, \pi_{R2}$	The profit functions of supply chain members
Decision Variables	
q_i	Selling quantity of the product i, $i = 1, 2$
b_i	Unit transfer price of used product i decided by collector i, $i = 1, 2$

The sequence of events is illustrated as follows. In stage 1, supply chain members decide whether to adopt blockchain technology. In stage 2, collectors decide the transfer price based on the demand information. In stage 3, the manufacturer determines the order quantities based on the transfer prices. Finally, the market demand will be realized.

3.2. Collector-Led Supply Chain Models

In this section, we investigate the four models (Model AN, Model BN, Model AB, and Model BB) and obtain the equilibrium outcomes through backward induction. To avoid trivial discussion and ensure that the equilibrium solutions are positive, we assume $(a_N - c_m)(1 - \beta) > \Delta\beta\lambda_2$. The equilibrium outcomes are summarized in Tables 2–5. The derivation and proof of this paper are in Appendix A. For ease of exhibition, we define some items in Table 6.

Table 2. Outcomes in Model AN.

Model AN ($i = 1, 2$)
$b_1^{AN} = ((1 - \beta)(a_N - c_m) + \Delta(\lambda_1 - \beta\lambda_2))/2\lambda_1$
$q_1^{AN} = ((1 - \beta)(a_N - c_m) + 2(\varepsilon_1 - \beta\varepsilon_2) + \Delta(\lambda_1 - \beta\lambda_2))/4(1 - \beta^2)$
$q_2^{AN} = ((1 - \beta)(\beta + 2)(a_N - c_m) + 2(\varepsilon_2 - \beta\varepsilon_1) - \Delta(\beta^2\lambda_2 + \beta\lambda_1 - 2\lambda_2))/4(1 - \beta^2)$
$E[\pi_{R1}^{AN}] = \left(((1 - \beta)(a_N - c_m) + \Delta(\lambda_1 - \beta\lambda_2))^2 - 8(1 - \beta^2)k\lambda_1^2\right)/8(1 - \beta^2)$
$E[\pi_M^{AN}] = ((1 - \beta)(a_N - c_m)((5 + 3\beta)(a_N - c_m) + 2\Delta(\lambda_1 + (4 + 3\beta)\lambda_2)) + 4(1 + r)V + F_1)/16(1 - \beta^2) - \phi k\lambda_2^2$
$F_1 = \Delta^2((\lambda_1 - 3\beta\lambda_2)(\lambda_1 + \beta\lambda_2) + 4\lambda_2^2)$

Table 3. Outcomes in Model BN.

Model BN ($i = 1, 2$)

$$b_1^{BN} = ((1-\beta)(2+\beta)(a_N - c_m) - \Delta(\beta^2\lambda_1 + \beta\lambda_2 - 2\lambda_1))/(4-\beta^2)\lambda_1$$

$$b_2^{BN} = ((1-\beta)(2+\beta)(a_N - c_m) - \Delta(\beta^2\lambda_2 + \beta\lambda_1 - 2\lambda_2))/(4-\beta^2)\lambda_2$$

$$q_i^{BN} = ((1-\beta)(2+\beta)(a_N - c_m) + (4-\beta^2)(\varepsilon_i - \beta\varepsilon_{3-i}) - \Delta(\beta^2\lambda_i + \beta\lambda_{3-i} - 2\lambda_i))/2(1-\beta^2)(4-\beta^2)$$

$$E[\pi_{Ri}^{BN}] = \left(((1-\beta)(2+\beta)(a_N - c_m) - \Delta(\beta^2\lambda_i + \beta\lambda_{3-i} - 2\lambda_i))^2 - 2(4-\beta^2)^2(1-\beta^2)k\lambda_i^2\right)/2(4-\beta^2)^2(1-\beta^2)$$

$$E[\pi_M^{BN}] = \left(2(1-\beta)(2+\beta)^2(a_N - c_m)(a_N - c_m + \Delta(\lambda_1 + \lambda_2)) + (4-\beta^2)^2(1+r)V + F_2\right)/4(4-\beta^2)^2(1-\beta^2)$$

$$F_2 = \Delta^2(\lambda_1^2 + \lambda_2^2)(4-3\beta^2) - 2\beta^3\Delta^2\lambda_1\lambda_2$$

Table 4. Outcomes in Model AB.

Model AB ($i = 1, 2$)

$$b_1^{AB} = ((1-\beta)(a_B - c_m) + \varepsilon_1 - \beta\varepsilon_2 + \Delta(\lambda_1 - \beta\lambda_2))/2\lambda_1$$

$$q_1^{AB} = ((1-\beta)(a_B - c_m) + \varepsilon_1 - \beta\varepsilon_2 + \Delta(\lambda_1 - \beta\lambda_2))/4(1-\beta^2)$$

$$q_2^{AB} = ((1-\beta)(\beta+2)(a_B - c_m) + (2-\beta^2)\varepsilon_2 - \beta\varepsilon_1 - \Delta(\beta^2\lambda_2 + \beta\lambda_1 - 2\lambda_2))/4(1-\beta^2)$$

$$E[\pi_{R1}^{AB}] = \left(((1-\beta)(a_B - c_m) + \Delta(\lambda_1 - \beta\lambda_2))^2 + (\beta^2 + r)V - 8(1-\beta^2)k\lambda_1^2\right)/8(1-\beta^2)$$

$$E[\pi_M^{AB}] = ((1-\beta)(a_B - c_m)((5+3\beta)(a_B - c_m) + 2\Delta(\lambda_1 + \lambda_2(4+3\beta))) + ((4-3\beta^2) + r)V + F_1)/16(1-\beta^2) - \phi k\lambda_2^2$$

Table 5. Outcomes in Model BB.

Model BB ($i = 1, 2$)

$$b_i^{BB} = ((1-\beta)(2+\beta)(a_B - c_m) + (2-\beta^2)\varepsilon_i - \beta\varepsilon_{3-i} - \Delta(\beta^2\lambda_i + \beta\lambda_{3-i} - 2\lambda_i))/(4-\beta^2)\lambda_i$$

$$q_i^{BB} = ((1-\beta)(2+\beta)(a_B - c_m) + (2-\beta^2)\varepsilon_i - \beta\varepsilon_{3-i} - \Delta(\beta^2\lambda_i + \beta\lambda_{3-i} - 2\lambda_i))/2(4-\beta^2)(1-\beta^2)$$

$$E[\pi_{R1}^{BB}] = \left(((1-\beta)(2+\beta)(a_B - c_m) - \Delta(\beta^2\lambda_1 + \beta\lambda_2 - 2\lambda_1))^2 + \left(\beta^2 + r(2+\beta)^2\right)V\right)/2(4-\beta^2)^2(1-\beta^2) - k\lambda_1^2$$

$$E[\pi_{R2}^{BB}] = \left(((1-\beta)(2+\beta)(a_B - c_m) - \Delta(\beta^2\lambda_2 + \beta\lambda_1 - 2\lambda_2))^2 + \left(r\beta^2 + (2+\beta)^2\right)V\right)/2(4-\beta^2)^2(1-\beta^2) - k\lambda_2^2$$

$$E[\pi_M^{BB}] = \left(2(1-\beta)(2+\beta)^2(a_B - c_m)((a_B - c_m) + \Delta(\lambda_1 + \lambda_2)) + (4-3\beta^2)(1+r)V + F_2\right)/4(4-\beta^2)^2(1-\beta^2)$$

Table 6. Definition.

k_i ($i=0,1,2,3,4,5,6,7$)

$$k_0 = \Delta((1-\beta)(a_N - c_m) + \Delta(\lambda_1 - \beta\lambda_2))/8\lambda_1(1-\beta^2)$$

$$k_1 = (2\Delta(1-\beta)(4+3\beta)(a_N - c_m) + \Delta^2((4-3\beta^2)\lambda_2 - \beta\lambda_1))/32\phi\lambda_2(1-\beta^2)$$

$$k_2 = \Delta(2-\beta^2)((1-\beta)(2+\beta)(a_N - c_m) + \Delta(2\lambda_1 - \beta^2\lambda_1 - \beta\lambda_2))/2\lambda_1(1-\beta^2)(4-\beta^2)^2$$

$$k_3 = \Delta(2-\beta^2)((1-\beta)(2+\beta)(a_N - c_m) - \Delta(\beta^2\lambda_2 + \beta\lambda_1 - 2\lambda_2))/2\lambda_2(1-\beta^2)(4-\beta^2)^2$$

$$k_4 = \Delta((1-\beta)(a_B - c_m) + \Delta(\lambda_1 - \beta\lambda_2))/8\lambda_1(1-\beta^2)$$

$$k_5 = (2\Delta(1-\beta)(4+3\beta)(a_B - c_m) + \Delta^2((4-3\beta^2)\lambda_2 - \beta\lambda_1))/32\phi\lambda_2(1-\beta^2)$$

$$k_6 = \Delta(2-\beta^2)((1-\beta)(2+\beta)(a_B - c_m) + \Delta(2\lambda_1 - \beta^2\lambda_1 - \beta\lambda_2))/2\lambda_1(1-\beta^2)(4-\beta^2)^2$$

$$k_7 = \Delta(2-\beta^2)((1-\beta)(2+\beta)(a_B - c_m) - \Delta(\beta^2\lambda_2 + \beta\lambda_1 - 2\lambda_2))/2\lambda_2(1-\beta^2)(4-\beta^2)^2$$

(1) In-house collection without blockchain (Model AN)

In this subsection, the manufacturer recycles the used product 2 through the in-house collection and there is no blockchain in the remanufacturing supply chain. In consequence, collector 1 only knows the expected value of the market demand. Thus, the supply chain members' expected profits are as follows:

$$\underset{q_1,\, q_2}{Max}\ E[\pi_M^{AN}] = (p_1 - c_m)q_1 + (\Delta - b_1)\lambda_1 q_1 + (p_2 - c_m)q_2 + \Delta\lambda_2 q_2 - \phi k\lambda_2^2 \tag{2}$$

$$\underset{b_1}{Max}\ E[\pi_{R1}^{AN}] = b_1\lambda_1\, q_1 - k\lambda_1^2 \tag{3}$$

where $p_1 = a_N - q_1 - \beta q_2 + \varepsilon_1$ and $p_2 = a_N - q_2 - \beta q_1 + \varepsilon_2$.

Thus, we obtain the optimal solutions from backward induction and the results are presented in Proposition 1.

Proposition 1. *In Model AN, the optimal transfer prices, selling quantities, and expected profits are summarized in Table 2.*

According to Table 2, Proposition 1 presents some important findings. (i) Unit transfer price is decreasing in λ_1 and λ_2. The higher the collection rate, the lower the unit transfer price. The selling quantity of product 1 is increasing in λ_1 and decreasing in λ_2, while the selling quantity of product 2 is decreasing in λ_1 and increasing in λ_2. The reason is that the higher collection rate of product i, the higher the selling quantity of product i. Conversely, it is unfavorable to the sales of the product if the collection rate of competing products is higher. (ii) Collector 1's expected profit is monotonically increasing if the collection investment costs coefficient k is lower than a threshold value (i.e., $k < k_0$) and decreasing in λ_2. In the market, product 1 and product 2 compete against each other. Thus, the collection rate of product 2 is not conducive to collector 1's profit. (iii) The manufacturer's expected profit is monotonically increasing in λ_2 if the coefficient of collection investment costs k is lower than a threshold value (i.e., $k < k_1$) and increasing in λ_1. For the manufacturer, the higher the collection rate λ_2, the higher the profit for the manufacturer when the collection investment costs are lower than the threshold value.

(2)   External collection without blockchain (Model BN)

In this subsection, the manufacturer agrees that collector 2 recycles the used product 2 through external collection. Two collectors also need to predict the uncertain demand since there is no blockchain in the remanufacturing supply chain. Therefore, the supply chain members' expected profits are as follows:

$$\underset{b_1}{Max}\ E[\pi_{R1}^{BN}] = b_1\lambda_1\,q_1 - k\lambda_1^2 \tag{4}$$

$$\underset{b_2}{Max}\ E[\pi_{R2}^{BN}] = b_2\lambda_2\,q_2 - k\lambda_2^2 \tag{5}$$

$$\underset{q_1,\,q_2}{Max}\ E[\pi_{M}^{BN}] = (p_1 - c_m + (\Delta - b_1)\lambda_1)q_1 + (p_2 - c_m + (\Delta - b_2)\lambda_2)q_2 \tag{6}$$

where $p_1 = a_N - q_1 - \beta q_2 + \varepsilon_1$ and $p_2 = a_N - q_2 - \beta q_1 + \varepsilon_2$.

Proposition 2. *In Model BN, the optimal policies can be formed in Table 3.*

Similar to Proposition 1, Proposition 2 also shows that the unit transfer prices decrease with the increase in the collection rate. An increased collection rate of a competitor's product is detrimental to the sales of its own product. Collector 1's expected profit is monotonically increasing in λ_1 if the coefficient of collection investment costs k is lower than a threshold value (i.e., $k < k_2$) and decreasing in λ_2. Collector 2's expected profit is monotonically decreasing in λ_1, and increasing in λ_2 if the coefficient of collection investment costs k is lower than a threshold value (i.e., $k < k_3$). As the collection rate of the two products increases, so does the manufacturer's expected profit.

(3)   In-house collection with blockchain (Model AB)

In this subsection, the remanufacturing supply chain consists only of the manufacturer and collector 1. The manufacturer recycles the used product 2 from the consumer market and records the demand information for the two products in the blockchain. Collector 1 can obtain accurate demand information. As a result, the supply chain members' expected profits are as follows:

$$\underset{q_1,\,q_2}{Max}\ E[\pi_{M}^{AB}] = (p_1 - c_m)q_1 + (\Delta - b_1)\lambda_1 q_1 + (p_2 - c_m)q_2 + \Delta\lambda_2 q_2 - \phi k\lambda_2^2 \tag{7}$$

$$\underset{b_1}{Max}\ E[\pi_{R1}^{AB}] = b_1\lambda_1\,q_1 - k\lambda_1^2 \tag{8}$$

where $p_1 = a_B - q_1 - \beta q_2 + \varepsilon_1$ and $p_2 = a_B - q_2 - \beta q_1 + \varepsilon_2$.

Proposition 3. *In Model AB, the optimal strategies can be given in Table 4.*

According to Table 4, in accordance with Proposition 1, we conclude that as collection rates increase, the unit transfer price decreases. The higher the collection rate, the greater the sales of the related product. When the collection rate of a competitor's product increases, the selling quantity of the own product also decreases. In addition, there is a monotonic increase in collector 1's expected profit in λ_1 if the coefficient of collection investment costs k is lower than a threshold value (i.e., $k < k_4$) and a monotonic decrease in λ_2. Similar to Model AN, high collection rates are not always advantageous for the manufacturer. The manufacturer's expected profit is monotonically increasing in λ_2 if the coefficient of collection investment costs k is lower than a threshold value (i.e., $k < k_5$) and increasing in λ_1.

(4) External collection with blockchain (Model BB)

In this subsection, collector 1 and collector 2 recycle the used products from the consumer market and sells them to the manufacturer. The manufacturer remanufactures the used products and records the demand information in the blockchain. Collectors can obtain accurate demand information. As a result, the supply chain members' expected profits are as follows:

$$\underset{b_1}{Max}\ E[\pi_{R1}^{BB}] = b_1\lambda_1\,q_1 - k\lambda_1^2 \tag{9}$$

$$\underset{b_2}{Max}\ E[\pi_{R2}^{BB}] = b_2\lambda_2\,q_2 - k\lambda_2^2 \tag{10}$$

$$\underset{q_1,\,q_2}{Max}\ E[\pi_{M}^{BB}] = (p_1 - c_m + (\Delta - b_1)\lambda_1)q_1 + (p_2 - c_m + (\Delta - b_2)\lambda_2)q_2 \tag{11}$$

where $p_1 = a_B - q_1 - \beta q_2 + \varepsilon_1$ and $p_2 = a_B - q_2 - \beta q_1 + \varepsilon_2$.

Proposition 4. *In Model BB, the optimal outcomes can be derived in Table 5.*

Based on Table 5, we can derive that the unit transfer prices are also decreasing in collection rates under the blockchain. The selling quantities are increasing in the corresponding collection rates while decreasing with the competitive product's collection rate. With an increase in λ_1, collector 1's expected profit increases if the coefficient of collection investment costs k is lower than a threshold value (i.e., $k < k_6$), and with an increase in λ_2, collector 1's expected profit decreases. Collector 2's expected profit increases as λ_2 increases if the coefficient of collection investment costs k is lower than a threshold value (i.e., $k < k_7$) and decreases with λ_1. The change in the manufacturer's expected profit with respect to the collection rate is similar to Model BN.

4. Analyses

Based on the aforementioned analyses, we further compare the equilibrium solutions in four models to gain the recycling strategies and blockchain preferences of the supply chain members.

4.1. Comparison of Different Recycling Strategies

In this subsection, we compare the equilibrium solutions of two recycling strategies with/without blockchain from the viewpoint of optimal recycling strategies. The results are summarized in Corollaries 1–3.

Corollary 1. *The optimal transfer prices and selling quantities satisfy the relations as follows:*
(i) For the transfer prices, we can get $b_1^{BN} > b_1^{AN}$ and $E[b_1^{BB}] > E[b_1^{AB}]$.

(ii) For the selling quantities of product 1, we can get $E[q_1^{BN}] > E[q_1^{AN}]$ and $E[q_1^{BB}] > E[q_1^{AB}]$. As for product 2, we have $E[q_2^{BN}] < E[q_2^{AN}]$ and $E[q_2^{BB}] < E[q_2^{AB}]$.

Corollary 1 clearly shows that collector 1's transfer prices and the selling quantities of product 1 are higher in Scenario B than in Scenario A. The reason is that when collector 2 enters the collection market, the manufacturer just needs to focus on the process of remanufacturing and selling, which leads to an increase in the selling quantities and improves the quality and recycling of remanufactured products. As a result, the selling quantities increase. Hence, in order to increase profitable profits, collector 1 has the incentive to improve the transfer prices as the selling quantities of product 1 increase in Scenario B. While the selling quantities of product 2 become lower in Scenario B than in Scenario A, the reason is that product 2's procurement cost increases in Scenario B. The manufacturer will need to pay an additional transfer price for product 2, which will reduce the incentive for the manufacturer to remanufacture product 2. Therefore, the order quantity of product 2 in the external collection mode is lower than in the in-house collection mode.

Corollary 2. *In the case of no blockchain, collector 1 is inclined to external collection (i.e., $E[\pi_{R1}^{BN}] - E[\pi_{R1}^{AN}] > 0$). When in the case of blockchain, collector 1 tends to the external collection if $r > r_0$ and $V > V_0$ (i.e., $E[\pi_{R1}^{BB}] - E[\pi_{R1}^{AB}] > 0$). Otherwise, collector 1 tends to the in-house collection. Here, $V_0 = \dfrac{\lambda_2(2-\beta^2)Eb_2^{BB}X_0}{r\beta(4-\beta)(2+\beta)^2 - \beta^2(6-\beta^2)(2-\beta^2)}$ and $r_0 = \dfrac{\beta(6-\beta^2)(2-\beta^2)}{(4-\beta)(2+\beta)^2}$, where $0 < r_0 < 1$. For ease of simplified calculation and exhibition, we define the items X_0 in Appendix A.*

According to Corollary 1, collector 1's transfer prices and selling quantities of product 1 are higher in Scenario B than in Scenario A. In the case of no blockchain, it is easy to conclude that collector 1 will be more profitable in the external collection scenario. Therefore, collector 1 is inclined to external collection under no blockchain. However, in the case of adopting blockchain, we cannot conclude from Corollary 1 that collector 1 is more profitable under the mode of external collection. This is because the transfer prices and order quantities contain random variables of demand in the blockchain scenario, and there are information values in the profit function of collector 1. Collector 1 can obtain accurate demand information through the blockchain platform. In addition, the selling quantities of product 2 decrease in Scenario B, thereby reducing market competition between the two products. When the variance of uncertain demand is higher (i.e., $V > V_0$) and the demand fluctuations of product 1 are higher (i.e., $r > r_0$), collector 1's profit is higher under external collection. At this point, the demand information value is larger, and the transfer prices and order quantities are higher. Thus, collector 1 prefers external collection. Otherwise, when the demand variance of product 1 is small, collector 1 is more focused on the market competition and does not want other collectors to enter the market. Therefore, collector 1 tends to in-house collection.

Corollary 3. *The manufacturer's attitude toward external collection or in-house collection depends on the following situations:*
(i) In the case of no blockchain, the manufacturer's expected profit in Model BN is higher than in Model AN if $k > k_8$ (i.e., $E[\pi_M^{BN}] - E[\pi_M^{AN}] > 0$). Otherwise, if $k < k_8$, we have $E[\pi_M^{BN}] - E[\pi_M^{AN}] < 0$.
(ii) In the case of adopting blockchain, the manufacturer's expected profit in Model BB is higher than in Model AB if $k > k_9$ (i.e., $E[\pi_M^{BB}] - E[\pi_M^{AB}] > 0$). Otherwise, if $k < k_9$, we have $E[\pi_M^{BB}] - E[\pi_M^{AB}] < 0$. Here, $k_8 = \dfrac{\lambda_2(4-\beta^2)b_2^{BN}X_1}{16\phi\lambda_2^2(1-\beta^2)(4-\beta^2)^2}$ and $k_9 = \dfrac{\lambda_2(2-\beta^2)Eb_2^{BB}X_2+X_3}{16\phi\lambda_2^2(1-\beta^2)(4-\beta^2)^2}$. For ease of simplified calculation and exhibition, we define the items X_1, X_2 and X_3 in Appendix A.

Corollary 3 demonstrates that the manufacturer will consider the coefficient of collection investment costs for external collection or in-house collection. In the case of no blockchain, the manufacturer can obtain a higher profit in Model BB if the coefficient of collection investment costs exceeds the threshold value (i.e., $k > k_0$). The reason is that the collection investment costs are larger than the purchase costs of product 2. At this time, the manufacturer is inclined to introduce collector 2 for product 2's collection. Conversely, when the coefficient of collection investment costs is smaller, the manufacturer prefers the in-house collection (see Figure 2 for illustration). In the case of adopting blockchain, similarly, external collection can provide the manufacturer with the most profit increase when the coefficient of collection investment costs is higher than the threshold value k_1 (see Figure 3 for illustration). From Figures 2 and 3, it can be seen that the collection rate affects the preference degree of the manufacturer's recycling decision. The higher the recycling rate, the more likely the manufacturer is to choose external collection.

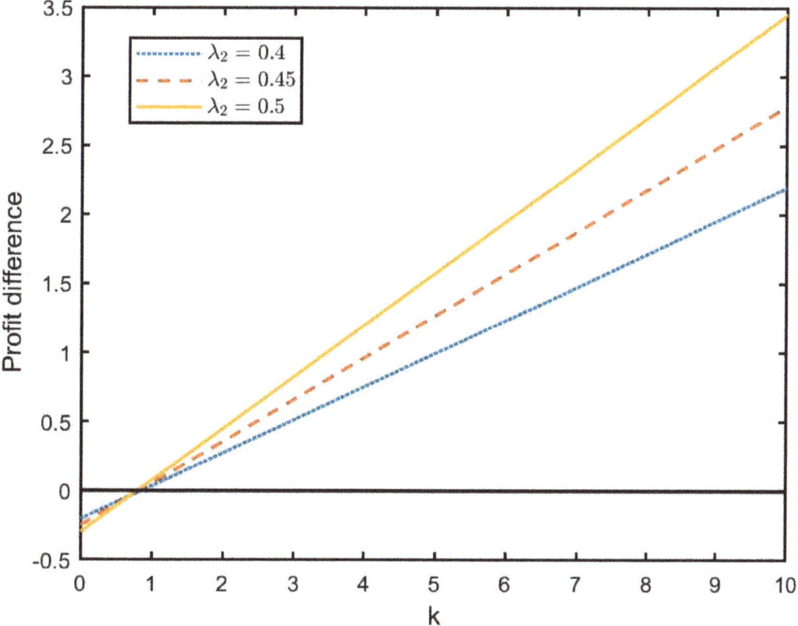

Figure 2. The impact of k on profit difference. ($a_N = 2$, $c_m = 1.5$, $\Delta = 0.5$, $r = 0.8$, $\beta = 0.5$, $\phi = 1.5$, $\lambda_1 = 0.6$).

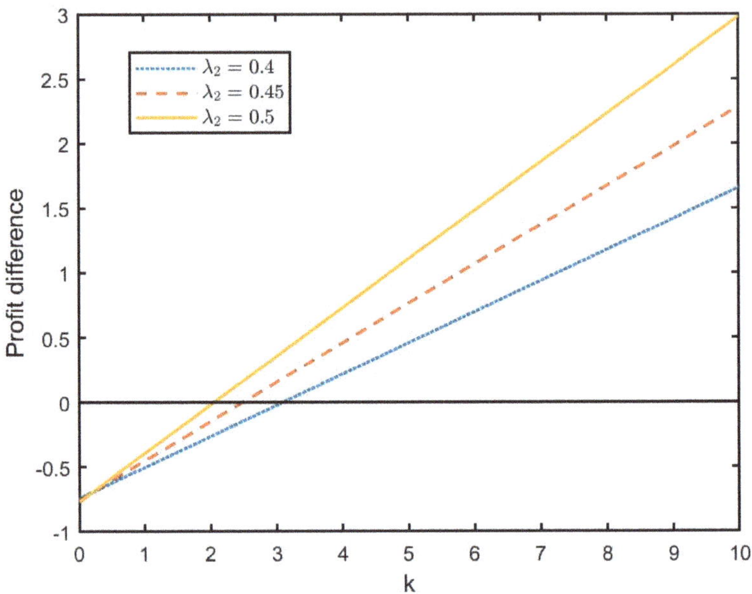

Figure 3. The impact of k on profit difference. ($a_B = 3$, $c_m = 1.5$, $\Delta = 0.5$, $r = 0.8$, $\beta = 0.5$, $\phi = 1.5$, $\lambda_1 = 0.6$).

4.2. Comparison of Different Blockchain Adoption

In this subsection, we compare the equilibrium outcomes of different blockchain scenarios under the same collection scenario. From the viewpoint of optimal blockchain adoption, we compare the equilibrium outcomes. The results are summarized in Corollaries 4–6.

Corollary 4. *The optimal transfer prices and selling quantities satisfy the relations as follows: (i) For the transfer price, we can get $E[b_1^{AB}] > b_1^{AN}$, $E[b_1^{BB}] > E[b_1^{BN}]$ and $E[b_2^{BB}] > E[b_2^{BN}]$. (ii) For the selling quantities, we can get $E[q_i^{AB}] > E[q_i^{AN}]$ and $E[q_i^{BB}] > E[q_i^{BN}]$ $(i = 1, 2)$.*

Corollary 4 articulates that the optimal transfer prices and selling quantities with blockchain are larger than that in the case of no blockchain. When the supply chain members introduce the blockchain, consumers can obtain the key information on remanufactured products, which improves the consumer's trust and expands the consumer market. Therefore, the selling quantities in the case of blockchain are larger than those without blockchain, as the increased selling quantities would stimulate collectors to increase the transfer prices. Hence, no matter the collection mode, the transfer prices and selling quantities with blockchain are higher than without blockchain. This means that the blockchain scenario benefits the improvement of both transfer prices and selling quantities.

Corollary 5. *By comparing the collectors' expected profits in the same collection scenario, we find that $E[\pi_{R1}^{AB}] > E[\pi_{R1}^{AN}]$, $E[\pi_{R1}^{BB}] > E[\pi_{R1}^{BN}]$ and $E[\pi_{R2}^{BB}] > E[\pi_{R2}^{BN}]$.*

Corollary 5 reveals that collectors can always benefit from blockchain technology. In the case of no blockchain, collectors do not have the full demand information and need to predict the uncertain demand based on the existing random information. When blockchain exists, collectors can obtain accurate demand from the blockchain. Thus, collectors have an extra demand information, which benefits the increase in collectors' profit. Furthermore, according to Corollary 4, the transfer prices and selling quantities with blockchain are larger than that without blockchain. It is easy to conclude that collector 1 and collector 2's profits in the case of blockchain are higher than that without blockchain.

Corollary 6. *By comparing the manufacturer's expected profits in four models, we find that the manufacturer's preference for blockchain depends on the variance of uncertain demand: (i) There exists a threshold V_1: if $V < V_1$, we have $E[\pi_M^{AB}] - E[\pi_M^{AN}] > 0$. Otherwise, if $V > V_1$, we have $E[\pi_M^{AB}] - E[\pi_M^{AN}] < 0$. (ii) There exists a threshold V_2: if $V < V_2$, we have $E[\pi_M^{BB}] - E[\pi_M^{BN}] > 0$. Otherwise, if $V > V_2$, we have $E[\pi_M^{BB}] - E[\pi_M^{BN}] < 0$.*

Corollary 6 illustrates that the variance in demand is one of the main factors that affect manufacturers in deciding whether to introduce blockchain technology. In Scenario A, the remanufacturing supply chain contains collector 1 and the manufacturer. When the demand variance is lower than a threshold value (i.e., $V < V_1$), the manufacturer is willing to introduce blockchain in order to expand the consumer market. When blockchain exists, the manufacturer records the demand information in the blockchain, which will provide collector 1 with accurate information about the market demand. Even though the manufacturer loses some demand information value, the increase in profits more than compensates for this loss. When the demand variance exceeds the threshold value (i.e., $V > V_1$), the demand information has a greater value than the profit increase. As a result, the manufacturer is unwilling to adopt blockchain (see Figure 4 for illustration). As shown in Figure 5, the manufacturer's blockchain decision is also influenced by a threshold value in Scenario B. When the demand variance exceeds the threshold value (i.e., $V > V_2$), the manufacturer has a higher demand information value than the improvement of profit and prefers no blockchain. When the demand variance is lower than a threshold value (i.e., $V < V_2$), blockchain adoption can make the manufacturer obtain more profit from market expansion. Based on Figures 4 and 5, it can be seen that there is a certain influence of the competition coefficient on the manufacturer's recycling decisions. The manufacturer is more likely to adopt blockchain technology as competition increases.

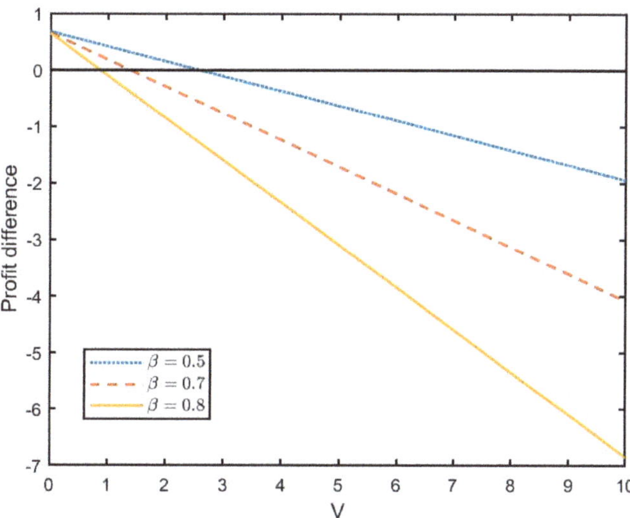

Figure 4. The impact of k on profit difference. ($a_B = 3$, $a_N = 2$, $c_m = 1.5$, $\Delta = 0.5$, $r = 0.8$, $\lambda_1 = 0.6$, $\lambda_2 = 0.5$).

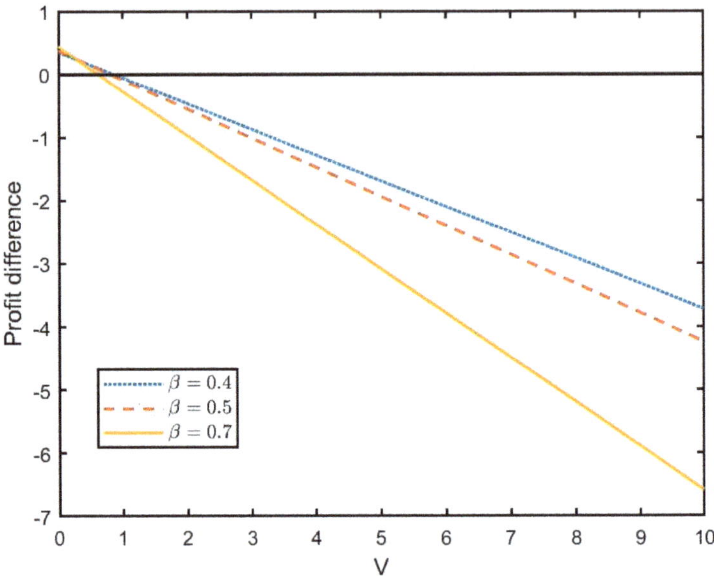

Figure 5. The impact of k on profit difference. ($a_B = 3$, $a_N = 2$, $c_m = 1.5$, $\Delta = 0.5$, $r = 0.8$, $\lambda_1 = 0.6$, $\lambda_2 = 0.5$).

5. Consumer Surplus

Consumer surplus is critically important for enterprises' sustainable development. In this section, this study analyzes consumer surplus in different recycling models without blockchain. Followed by Yang et al. [39] and Shen et al. [40], the consumer utility function and consumer surplus are as follows:

$$U(q_1^*, q_2^*) = \phi_1 q_1^* + \phi_2 q_2^* - \frac{\psi_1 q_1^{*2} + 2\gamma q_1^* q_2^* + \psi_2 q_2^{*2}}{2} \tag{12}$$

$$CS = U(q_1^*, q_2^*) - p_1^* q_1^* - p_2^* q_2^* \tag{13}$$

where the inverse demand functions are $p_1 = \phi_1 - \psi_1 q_1 - \gamma q_2$ and $p_2 = \phi_2 - \psi_2 q_2 - \gamma q_1$, respectively. Combining Equation (1) of the inverse demand function, we have $\phi_1 = \phi_2 = a_N, \psi_1 = \psi_2 = 1, \gamma = \beta$. Therefore, the consumer surplus equation of this chapter can be derived as follows:

$$CS = a_N(q_1^* + q_2^*) - \frac{q_1^{*2} + 2\beta q_1^* q_2^* + q_2^{*2}}{2} - p_1^* q_1^* - p_2^* q_2^* = \frac{1}{2}\left(q_1^{*2} + 2\beta q_1^* q_2^* + q_2^{*2} - 2q_1^* \varepsilon_1 - 2q_2^* \varepsilon_2\right) \tag{14}$$

Corollary 7. *For the consumer, the higher the recycling rate of the product, the more benefits the consumer can get, and the consumer surplus increases.*

Corollary 7 shows that the higher the collection rate, the more favorable the increase in consumer surplus. The optimal price of product 1 in Model AN is easily obtained $p_1^{AN} = \frac{(3-\beta)a_N + (1+\beta)c_m + 2\varepsilon_1 - \Delta(\lambda_1 + \beta\lambda_2)}{4}$ and $p_2^{AN} = \frac{1}{2}(a_N + c_m + \varepsilon_2 - \Delta\lambda_2)$. It is easy to get $\frac{\partial p_1^{AN}}{\partial \lambda_i} < 0$ ($i = 1, 2$) and $\frac{\partial p_2^{AN}}{\partial \lambda_2} < 0$. The first-order condition of the retail price in Model AN with respect to the collection rate shows that the retail price of the product decreases as the collection rate increases. The higher the product collection rate, the lower the retail price and the increase in consumer surplus. Therefore, for Model AN, the higher the collection rate, the higher the consumer surplus. In Model BN, the retail price of product 1 is $p_1^{BN} = \frac{(2+\beta)(3-2\beta)a_N+(2+\beta)c_m+\left(4-\beta^2\right)\varepsilon_1-\Delta(2\lambda_1+\beta\lambda_2)}{2(4-\beta^2)}$, the retail price of product 2 is $p_2^{BN} = \frac{(2+\beta)(3-2\beta)a_N+(2+\beta)c_m+\left(4-\beta^2\right)\varepsilon_2-\Delta(\beta\lambda_1+2\lambda_2)}{2(4-\beta^2)}$. The first order derivative of the collection rate for the optimal retail price are $\frac{\partial p_1^{BN}}{\partial \lambda_i} < 0$ and $\frac{\partial p_2^{BN}}{\partial \lambda_i} < 0$ $(i = 1, 2)$. It is easy to know that the higher the product collection rate, the lower the retail price of the product. For consumers, it is possible to purchase the remanufactured product at a lower retail price, which results in more benefits to consumers and an increase in consumer surplus. Therefore, for Model BN, the higher the product collection rate, the higher the consumer surplus.

6. Conclusions

In this paper, we examine the tradeoffs between different recycling strategies and blockchain adoption in a collector-led remanufacturing supply chain. The manufacturer has private demand information and has two different recycling strategies: in-house collection by the manufacturer or external collection by collector 2. There are four models for the recycling strategies and blockchain adoption: (1) Scenario A with no blockchain (Model AN); (2) Scenario B with no blockchain (Model BN); (3) Scenario A with blockchain (Model AB); (4) Scenario B with blockchain (Model BB). By comparing four models, we examine supply chain members' preferences for collection formats and blockchain adoption. The main findings of this paper are as follows.

First, our study investigates the effect of the collection rate on the equilibrium solutions in different models. The unit transfer prices decrease with the increase in the collection rate in the four models. The selling quantities of product 1 are monotonically increasing in the collection rate λ_1 and decreasing in the collection rate λ_2. Similarly, the selling quantities of product 2 increase with λ_2 and decrease with λ_1. Product 1 and product 2 compete with each other in the same market, and competitors' high collection rates are not conducive to their own selling quantities and profits. Collector 1's (Collector 2's) expected profits are increasing in λ_1 (λ_2) only when the coefficient of collection investment costs is lower than a threshold value. The manufacturer's expected profits are increasing in the two collection rates when external collection exists (Scenario B). However, In Scenario A, the manufacturer's expected profits are increasing in λ_2 only when the coefficient of collection investment costs is lower than a threshold value.

Based on the above findings, there are some discussions and implications for this study. Product 1 and Product 2 compete with each other in the market; collectors do not want competitors' collection rates to increase. However, it is not that the higher the collection rate of its own products, the higher its own profit value. Collectors need to make a cost investment in recycling channels. Only when the coefficient of collection investment cost is lower than a threshold value can a higher product collection rate increase the collectors' profits. Therefore, it is necessary for collectors to consider the improvement of product collection rate and control the cost of collection investment. For the manufacturer, the higher the product collection rate under external collection mode, the better the benefits. However, the manufacturer needs to control the cost of collection investment while considering increasing the collection rate of product 2 under in-house collection.

Second, we compare the differences in the equilibrium solutions of different models. The unit transfer prices and the selling quantities of product 1 in Scenario B are higher than in Scenario A. Moreover, the unit transfer prices and selling quantities of product 1 and product 2 with blockchain are higher than that without blockchain. However, the selling quantities of product 2 in Scenario B are lower than in Scenario A. Both collector 1 and collector 2 are willing to adopt blockchain in the remanufacturing supply chain. In the case of no blockchain, collector 1 is always inclined to the manufacturer to introduce external collection by collector 2. However, collector 1 prefers Scenario B with blockchain only when demand variance and the coefficient of variance are larger than a threshold value (i.e., $V > V_0$ and $r > r_0$). The manufacturer's decision on in-house collection and external collection depends on the coefficient of collection investment costs. Only when the coefficient of collection investment costs exceeds a certain value with/without blockchain will the manufacturer choose external collection; otherwise, the manufacturer prefers in-house collection. The manufacturer may choose to implement the blockchain when the demand variance is less than a certain value.

The findings of the second part have some inferences and implications. Collectors are looking to introduce blockchain technology to record information about remanufactured products. Not only can they gain accurate demand information value, but they can also obtain greater sales from blockchain technology that improves consumer trust in remanufactured products. However, the manufacturer is willing to introduce blockchain technology only when the variance of demand is smaller than a threshold value. Therefore, we recommend that the remanufacturing supply chain jointly introduce blockchain technology when the variance of demand is small so that a win-win situation can be achieved. In the case of blockchain, the manufacturer and collector 1 prefer external collection only if both the variance indicator of demand and the coefficient of collection investment costs meet a certain range.

Finally, our research expands the theoretical perspective and provides practical guidance. This study reveals the impact of demand randomization and blockchain technology on recycling strategies in a collector-led remanufacturing supply chain and enriches the theoretical research on the impact of uncertainties on the remanufacturing supply chain's operation decision. At the same time, it also provides a practical guide for the introduction of blockchain technology in the remanufacturing supply chain.

Furthermore, this paper has several limitations, which can be considered for future research. First, this study does not distinguish between new and remanufactured products. In some situations, remanufactured products do not meet the same quality standards as new products. Thus, it would be possible to study the scenario of remanufactured products that are differentiated in the future. Second, we assume that all participants are risk-neutral under uncertain demand and do not take into account their risk attitudes. The risk attitudes of heterogeneous players may vary in practice. Third, the product collection rate was not set as a decision variable in this study. In fact, product recovery rates are not fixed constants. Considering making the collection rate a decision variable is also a direction for future research. In the future, the above issues may prove to be interesting.

Author Contributions: Conceptualization, C.L. and T.Y.; methodology, C.L. and T.Y.; software, C.L. and Z.B.; formal analysis, C.L.; investigation, C.L., T.Y. and Z.B.; writing—original draft preparation, C.L.; writing—review and editing, C.L. and T.Y.; supervision, T.Y. All authors have read and agreed to the published version of the manuscript.

Funding: This research received no external funding.

Data Availability Statement: Not applicable.

Conflicts of Interest: The authors declare no conflict of interest.

Appendix A

Proof of Proposition 1. In Model AN, given the transfer price, taking the first-order condition of π_M^{AN} with respect to q_1 and q_2, respectively. Then, we can get the optimal response function: $q_1 = ((1-\beta)(a_N - c_m) + \varepsilon_1 - \beta\varepsilon_2 + (\Delta - b_1)\lambda_1 - \beta\Delta\lambda_2)/2(1-\beta^2)$. The collector 1 has no full of demand information, so taking $E[q_1]$ into the $E[\pi_{R1}^{AN}]$. Based on the first-order condition $\partial E[\pi_{R1}^{AN}]/\partial b_1 = 0$, we get the equilibrium outcomes b_1^{AN}, q_1^{AN}, and q_2^{AN}. Then we can get $E[\pi_{R1}^{AN}]$ and $E[\pi_M^{AN}]$. Taking the first-order condition of equilibrium outcomes with respect to λ_1 and λ_2, respectively, we have: $\frac{\partial b_1^{AN}}{\partial \lambda_2} = -\frac{\beta\Delta}{2\lambda_1} < 0$; $\frac{\partial E[q_1^{AN}]}{\partial \lambda_1} = \frac{\Delta}{4(1-\beta^2)} > 0$; $\frac{\partial E[q_1^{AN}]}{\partial \lambda_2} = -\frac{\beta\Delta}{4(1-\beta^2)} < 0$. Since $(a_N - c_m)(1-\beta) > \Delta\beta\lambda_2$, then $\frac{\partial b_1^{AN}}{\partial \lambda_1} = \frac{-((1-\beta)(a_N - c_m) - \beta\Delta\lambda_2)}{2\lambda_1^2} < 0$. And $\frac{\partial E[q_2^{AN}]}{\partial \lambda_1} = -\frac{\beta\Delta}{4(1-\beta^2)} < 0$: $\frac{\partial E[q_2^{AN}]}{\partial \lambda_2} = \frac{\Delta(2-\beta^2)}{4(1-\beta^2)} > 0$. We set $k_0 = \frac{\Delta(1-\beta)(a_N-c_m)+\Delta^2(\lambda_1-\beta\lambda_2)}{8\lambda_1(1-\beta^2)}$, then $\frac{\partial E[\pi_{R1}^{AN}]}{\partial\lambda_1} = 2\lambda_1(k_0-k)$. When $(i)k>k_0$, $\frac{\partial E[\pi_{R1}^{AN}]}{\partial\lambda_1}<0$; $(ii)k<k_0$, $\frac{\partial E[\pi_{R1}^{AN}]}{\partial\lambda_1}>0$. For the manufacturer, $\frac{\partial E[\pi_M^{AN}]}{\partial\lambda_1} = \frac{(1-\beta)\Delta(a_N-c_m)+\Delta^2(\lambda_1-\beta\lambda_2)}{8(1-\beta^2)}>0$ and we set $k_1 = \frac{2\Delta(1-\beta)(4+3\beta)(a_N-c_m)+\Delta^2\left((4-3\beta^2)\lambda_2-\beta\lambda_1\right)}{32\phi\lambda_2(1-\beta^2)}$, then $\frac{\partial E[\pi_M^{AN}]}{\partial\lambda_2} = 2\phi\lambda_2(k_1-k)$. When $(i)k>k_1$, $\frac{\partial E[\pi_M^{AN}]}{\partial\lambda_2}<0$; $(ii)k<k_1$, $\frac{\partial E[\pi_M^{AN}]}{\partial\lambda_2}>0$. $\square$

Proof of Proposition 2. Given the transfer price, taking the first-order condition of π_M^{BN} with respect to q_1 and q_2, respectively. Then, taking the optimal expected response functions of q_1 and q_2 into the collectors' profit functions, we get $E[\pi_{R1}^{BN}]$ and $E[\pi_{R2}^{BN}]$. Then, we have the first-order condition with respect to b_1 and b_2, and letting the derivative be zero, we have:

$$\begin{cases} \frac{\partial E[\pi_{R1}^{BN}]}{\partial b_1} = -\frac{b_1\lambda_1^2}{2(1-\beta^2)} + \frac{\lambda_1((1-\beta)(a_N-c_m)+(\Delta-b_1)\lambda_1-\beta(\Delta-b_2)\lambda_2)}{2(1-\beta^2)} = 0 \\ \frac{\partial E[\pi_{R2}^{BN}]}{\partial b_2} = -\frac{b_2\lambda_2^2}{2(1-\beta^2)} + \frac{\lambda_2((1-\beta)(a_N-c_m)-\beta(\Delta-b_1)\lambda_1+(\Delta-b_2)\lambda_2)}{2(1-\beta^2)} = 0 \end{cases}$$

Then we can get the optimal transfer prices b_1^{BN} and b_2^{BN}. Taking b_1^{BN} and b_2^{BN} into the optimal response functions of q_1 and q_2, we can get the optimal selling quantities q_1^{BN} and q_2^{BN}. Taking the optimal decision variables into the supply chain members' expected profits, we can get $E[\pi_{R1}^{BN}]$, $E[\pi_{R2}^{BN}]$ and $E[\pi_M^{BN}]$. Taking the first-order condition of equilibrium outcomes with respect to λ_1 and λ_2, respectively, we have: $\frac{\partial b_1^{BN}}{\partial\lambda_1} = \frac{-((1-\beta)(2+\beta)(a_N-c_m)-\beta\Delta\lambda_2)}{(4-\beta^2)\lambda_1^2}<0$; $\frac{\partial b_1^{BN}}{\partial\lambda_2} = \frac{-\beta\Delta}{(4-\beta^2)\lambda_1}<0$; $\frac{\partial b_2^{BN}}{\partial\lambda_1} = \frac{-\beta\Delta}{(4-\beta^2)\lambda_2}<0$; $\frac{\partial b_2^{BN}}{\partial\lambda_2} = \frac{-((1-\beta)(2+\beta)(a_N-c_m)-\beta\Delta\lambda_1)}{(4-\beta^2)\lambda_2^2}<0$ $\frac{\partial E[q_1^{BN}]}{\partial\lambda_1} = \frac{(2-\beta^2)\Delta}{2(1-\beta^2)(4-\beta^2)}>0$; $\frac{\partial E[q_1^{BN}]}{\partial\lambda_2} = \frac{-\beta\Delta}{2(1-\beta^2)(4-\beta^2)}<0$; $\frac{\partial E[q_2^{BN}]}{\partial\lambda_1} = \frac{-\beta\Delta}{2(1-\beta^2)(4-\beta^2)}<0$; $\frac{\partial E[q_2^{BN}]}{\partial\lambda_2} = \frac{(2-\beta^2)\Delta}{2(1-\beta^2)(4-\beta^2)}>0$. Define $k_2 = \frac{\Delta(2-\beta^2)\left((1-\beta)(2+\beta)(a_N-c_m)+\Delta(2\lambda_1-\beta^2\lambda_1-\beta\lambda_2)\right)}{2\lambda_1(1-\beta^2)(4-\beta^2)^2}$ and for collector 1's expected profit, the first-order condition is as follows: $\frac{\partial E[\pi_{R1}^{BN}]}{\partial\lambda_2} = -\frac{\beta\Delta\lambda_1 b_1^{BN}}{(1-\beta^2)(4-\beta^2)}<0$ and $\frac{\partial E[\pi_{R1}^{BN}]}{\partial\lambda_1} = 2\lambda_1(k_2-k)$. When $(i)k>k_2$, $\frac{\partial E[\pi_{R1}^{BN}]}{\partial\lambda_1}<0$ $(ii)k<k_2$, $\frac{\partial E[\pi_{R1}^{BN}]}{\partial\lambda_1}>0$.

Similarly, define $k_3 = \frac{\Delta(2-\beta^2)\lambda_2 b_2^{DN}}{2\lambda_2(1-\beta^2)(4-\beta^2)}$ and for collector 2's expected profit, the first-order condition is as follows: $\frac{\partial E[\pi_{R2}^{BN}]}{\partial\lambda_1} = \frac{-\beta\Delta\lambda_2 b_2^{BN}}{(1-\beta^2)(4-\beta^2)^2}<0$ and $\frac{\partial E[\pi_{R2}^{BN}]}{\partial\lambda_2} = 2\lambda_2(k_3-k)$. Thus, when $(i)k>k_3$, $\frac{\partial E[\pi_{R2}^{BN}]}{\partial\lambda_2}<0$; $(ii)k<k_3$, $\frac{\partial E[\pi_{R2}^{BN}]}{\partial\lambda_2}>0$. For the manufacturer, we have $\frac{\partial E[\pi_M^{BN}]}{\partial\lambda_1} = \frac{\Delta(1-\beta)(2+\beta)^2(a_N-c_m)+\Delta^2\left((4-3\beta^2)\lambda_1-\beta^3\lambda_2\right)}{2(1-\beta^2)(4-\beta^2)^2}$ $\frac{\partial E[\pi_M^{BN}]}{\partial\lambda_2} = \frac{\Delta(1-\beta)(2+\beta)^2(a_N-c_m)+\Delta^2\left((4-3\beta^2)\lambda_2-\beta^3\lambda_1\right)}{2(1-\beta^2)(4-\beta^2)^2}$. Because $\lambda_1 > \lambda_2 > \frac{2}{1+\beta^2}\beta\lambda_1 > \beta\lambda_1$ and $(4-3\beta^2)-\beta^3 = (1-\beta)(2+\beta)^2>0$, we have $\frac{\partial E[\pi_M^{BN}]}{\partial\lambda_1}>0$. $2(4-3\beta^2)\lambda_2 > (1+\beta^2)\lambda_2 > 2\beta\lambda_1$ and $(4-3\beta^2)\lambda_2-\beta^3\lambda_1>0$, so we get $\frac{\partial E[\pi_M^{BN}]}{\partial\lambda_2}>0$. $\square$

Proof of Proposition 3. Similar to Model AN, given the transfer price, taking the first-order condition of π_M^{AB} with respect to q_1 and q_2, respectively. Then, we can get the optimal response function: $q_1 = ((1-\beta)(a_B-c_m)+\varepsilon_1-\beta\varepsilon_2+(\Delta-b_1)\lambda_1-\beta\Delta\lambda_2)/2(1-\beta^2)$. Note that collector 1 has accurate demand information, so taking the optimal response function of q_1 into collector 1's profit function. Based on the first-order condition $\partial\pi_{R1}^{AB}/\partial b_1 = 0$, we get the equilibrium outcomes b_1^{AB}, q_1^{AB}, and q_2^{AB}. Then we can get $E[\pi_{R1}^{AB}]$ and $E[\pi_M^{AB}]$. Taking the first-order condition of equilibrium outcomes with respect to λ_1 and λ_2, respectively, we have $(a_B-c_m)(1-\beta) > (a_N-c_m)(1-\beta) > \Delta\beta\lambda_2$ and $\frac{\partial E[b_1^{AB}]}{\partial\lambda_1} = -\frac{(1-\beta)(a_B-c_m)-\beta\Delta\lambda_2}{2\lambda_1^2}<0$, $\frac{\partial E[b_1^{AB}]}{\partial\lambda_2} = -\frac{\beta\Delta}{2\lambda_1}<0$. The first-order condition of $E[q_1^{AB}]$ and $E[q_2^{AB}]$ in Model AB are the same as Model AN. $\frac{\partial E[\pi_{R1}^{AB}]}{\partial\lambda_2} = -\frac{\beta\Delta((1-\beta)(a_B-c_m)+\Delta(\lambda_1-\beta\lambda_2))}{4(1-\beta^2)}<0$ and $\frac{\partial E[\pi_M^{AB}]}{\partial\lambda_1} = \frac{\Delta(1-\beta)(a_B-c_m)+\Delta^2(\lambda_1-\beta\lambda_2)}{8(1-\beta^2)}>0$. Define $k_4 = \frac{\Delta((1-\beta)(a_B-c_m)+\Delta(\lambda_1-\beta\lambda_2))}{8\lambda_1(1-\beta^2)}$ and $k_5 = \frac{2\Delta(1-\beta)(4+3\beta)(a_B-c_m)+\Delta^2\left((4-3\beta^2)\lambda_2-\beta\lambda_1\right)}{32\phi\lambda_2(1-\beta^2)}$, we have $\frac{\partial E[\pi_{R1}^{AB}]}{\partial \lambda_1} = 2\lambda_1(k_4 - k)$ and $\frac{\partial E[\pi_M^{AB}]}{\partial \lambda_2} = 2\phi\lambda_2(k_5 - k)$, when $(i)k > k_4$, $\frac{\partial E[\pi_{R1}^{AB}]}{\partial \lambda_1} < 0$, $(ii)k < k_4$, $\frac{\partial E[\pi_{R1}^{AB}]}{\partial \lambda_1} > 0$; when $(i)\ k > k_5$, $\frac{\partial E[\pi_M^{AB}]}{\partial \lambda_2} < 0$, $(ii)k < k_5$, $\frac{\partial E[\pi_M^{AB}]}{\partial \lambda_2} > 0$. $\square$

Proof of Proposition 4. Similar to Model BN, given the transfer price, taking the first-order condition of π_M^{BB} with respect to q_1 and q_2, respectively. Then, taking the optimal response functions of q_1 and q_2 into the collectors' profit functions, we get the first-order condition with respect to b_1 and b_2, and letting the derivative be zero, we have:

$$\begin{cases} \frac{\partial \pi_{R1}^{BB}}{\partial b_1} = -\frac{b_1\lambda_1^2}{2(1-\beta^2)} + \frac{\lambda_1((1-\beta)(a_B-c_m)+\varepsilon_1-\beta\varepsilon_2+(\Delta-b_1)\lambda_1-\beta(\Delta-b_2)\lambda_2)}{2(1-\beta^2)} = 0 \\ \frac{\partial \pi_{R2}^{BB}}{\partial b_2} = -\frac{b_2\lambda_2^2}{2(1-\beta^2)} + \frac{\lambda_2((1-\beta)(a_B-c_m)+\varepsilon_2-\beta\varepsilon_1-\beta(\Delta-b_1)\lambda_1+(\Delta-b_2)\lambda_2)}{2(1-\beta^2)} = 0 \end{cases}$$

Then we can get the optimal transfer prices b_1^{BB} and b_2^{BB}. Taking b_1^{BB} and b_2^{BB} into the optimal response functions of q_1 and q_2, we can get the optimal selling quantities q_1^{BB} and q_2^{BB}. Taking the optimal decision variables into the supply chain members' expected profits, we can get $E[\pi_{R1}^{BB}]$, $E[\pi_{R2}^{BB}]$ and $E[\pi_M^{BB}]$. Taking the first-order condition of equilibrium outcomes with respect to λ_1 and λ_2, respectively, we have

$$\frac{\partial E[b_1^{BB}]}{\partial \lambda_1} = \frac{-((1-\beta)(2+\beta)(a_B-c_m)-\beta\Delta\lambda_2)}{(4-\beta^2)\lambda_1^2} < 0; \quad \frac{\partial E[b_1^{BB}]}{\partial \lambda_2} = -\frac{\beta\Delta}{(4-\beta^2)\lambda_1};$$

$$\frac{\partial E[b_2^{BB}]}{\partial \lambda_1} = -\frac{\beta\Delta}{(4-\beta^2)\lambda_2}; \quad \frac{\partial E[b_2^{BB}]}{\partial \lambda_2} = \frac{-(1-\beta)(2+\beta)(a_B-c_m)+\beta\Delta\lambda_1}{(4-\beta^2)\lambda_2^2} < 0.$$

The first-order condition of $E[q_1^{BB}]$ and $E[q_2^{BB}]$ in Model BB are the same as Model BN.

$$\frac{\partial E[\pi_{R1}^{BB}]}{\partial \lambda_2} = -\frac{\beta\Delta((1-\beta)(2+\beta)(a_B-c_m)+\Delta(2\lambda_1-\beta^2\lambda_1-\beta\lambda_2))}{(1-\beta^2)(4-\beta^2)^2} < 0 \quad \frac{\partial E[\pi_{R2}^{BB}]}{\partial \lambda_1} = -\frac{\beta\Delta((1-\beta)(2+\beta)(a_B-c_m)-\Delta(\beta^2\lambda_2+\beta\lambda_1-2\lambda_2))}{(4-\beta^2)^2(1-\beta^2)} < 0.$$

Define

$$k_6 = \frac{\Delta(2-\beta^2)((1-\beta)(2+\beta)(a_B-c_m)+\Delta(2\lambda_1-\beta^2\lambda_1-\beta\lambda_2))}{2\lambda_1(1-\beta^2)(4-\beta^2)^2} \text{ and } k_7 = \frac{\Delta(2-\beta^2)((1-\beta)(2+\beta)(a_B-c_m)-\Delta(\beta^2\lambda_2+\beta\lambda_1-2\lambda_2))}{2\lambda_2(1-\beta^2)(4-\beta^2)^2},$$

we have $\frac{\partial E[\pi_{R1}^{BB}]}{\partial \lambda_1} = 2\lambda_1(k_6 - k)$ and $\frac{\partial E[\pi_{R2}^{BB}]}{\partial \lambda_2} = 2\lambda_2(k_7 - k)$, when $(i)k > k_6$, $\frac{\partial E[\pi_{R1}^{BB}]}{\partial \lambda_1} < 0$, $(ii)k < k_6$, $\frac{\partial E[\pi_{R1}^{BB}]}{\partial \lambda_1} > 0$; when $(i)\ k > k_7$, $\frac{\partial E[\pi_{R2}^{BB}]}{\partial \lambda_2} < 0$, $(ii)k < k_7$, $\frac{\partial E[\pi_{R2}^{BB}]}{\partial \lambda_2} > 0$.

$$\frac{\partial E[\pi_M^{BB}]}{\partial \lambda_1} = \frac{\Delta(1-\beta)(2+\beta)^2(a_B-c_m)+\Delta^2((4-3\beta^2)\lambda_1-\beta^3\lambda_2)}{2(1-\beta^2)(4-\beta^2)^2},$$

$$\frac{\partial E[\pi_M^{BB}]}{\partial \lambda_2} = \frac{\Delta(1-\beta)(2+\beta)^2(a_B-c_m)+\Delta^2((4-3\beta^2)\lambda_2-\beta^3\lambda_1)}{2(1-\beta^2)(4-\beta^2)^2}.$$

According to the proof of Proposition 2, we have $(4-3\beta^2)\lambda_1 - \beta^3\lambda_2 > 0$ and $(4-3\beta^2)\lambda_2 - \beta^3\lambda_1 > 0$. Therefore, we have $\frac{\partial E[\pi_M^{BB}]}{\partial \lambda_1} > 0$ and $\frac{\partial E[\pi_M^{BB}]}{\partial \lambda_2} > 0$. $\square$

Proof of Corollary 1. Comparing Model BN and Model AN, Model BB and Model AB, the differences of the optimal transfer price and selling quantities are

$$b_1^{BN} - b_1^{AN} = \frac{2\beta(1-\beta^2)}{(4-\beta^2)\lambda_1}E[q_2^{AN}] > 0, \quad E[q_1^{BN}] - E[q_1^{AN}] = \frac{\beta}{2}E[q_2^{BN}] > 0,$$

$$E[q_2^{BN}] - E[q_2^{AN}] = \frac{\beta^2-2}{4-\beta^2}E[q_2^{AN}] < 0, \quad E[b_1^{BB}] - E[b_1^{AB}] = \frac{2\beta}{\lambda_1}E[q_2^{AB}] > 0,$$

$$E[q_1^{BB}] - E[q_1^{AB}] = \frac{\beta}{4-\beta^2}E[q_2^{AB}] > 0, \quad E[q_2^{BB}] - E[q_2^{AB}] = \frac{\beta^2-2}{4-\beta^2}E[q_2^{AB}] < 0. \quad \square$$

Proof of Corollary 2. Define $X_0 = (4-\beta)(1-\beta)(2+\beta)(a_B - c_m) + (8-3\beta^2)\Delta\lambda_1 - \Delta\lambda_2\beta(6-\beta^2)$. By comparing Model BN and Model AN, Model BB and Model AB, the differences of the collector 1's expected profit are

$$E[\pi_{R1}^{BN}] - E[\pi_{R1}^{AN}] = \frac{\beta\lambda_2(4-\beta^2)b_2^{BN}((4-\beta)(1-\beta)(2+\beta)(a_N-c_m)+(8-3\beta^2)\Delta\lambda_1-\beta\Delta\lambda_2(6-\beta^2))}{8(1-\beta^2)(4-\beta^2)^2}$$

$$E[\pi_{R1}^{BB}] - E[\pi_{R1}^{AB}] = \frac{(r\beta(4-\beta)(2+\beta)^2-\beta^2(6-\beta^2)(2-\beta^2))V-\lambda_2(2-\beta^2)E[b_2^{BB}]X_0}{8(4-\beta^2)^2(1-\beta^2)}.$$

Because $\lambda_1 > \lambda_2$ and $(8-3\beta^2) - \beta(6-\beta^2) > 0$, we have $(8-3\beta^2)\lambda_1 - \beta\lambda_2(6-\beta^2) > 0$ and $X_0 > 0$, thus we have $E[\pi_{R1}^{BN}] - E[\pi_{R1}^{AN}] > 0$. For Model BB and Model AB, we define $r_0 = \frac{\beta(6-\beta^2)(2-\beta^2)}{(4-\beta)(2+\beta)^2}$

and $V_0 = \frac{\lambda_2(2-\beta^2)E[b_2^{DB}]X_0}{r\beta(4-\beta)(2+\beta)^2-\beta^2(6-\beta^2)(2-\beta^2)}$, when $(1)r < r_0$, we have $E[\pi_{R1}^{BB}] - E[\pi_{R1}^{AB}] < 0$;

(2) $r > r_0$ and $V > V_0$, we have $E[\pi_{R1}^{BB}] - E[\pi_{R1}^{AB}] < 0$; $r > r_0$ and $V < V_0$, we have $E[\pi_{R1}^{BB}] - E[\pi_{R1}^{AB}] > 0$. $\square$

Proof of Corollary 3. Define

$$X_1 = (12 + 4\beta - 3\beta^2)(1 - \beta)(2 + \beta)(a_N - c_m) + (6 - \beta^2)(4 - 3\beta^2)\Delta\lambda_2 - (4 + \beta^2)\beta\Delta\lambda_1,$$

$$X_2 = (1 - \beta)(2 + \beta)(12 + 4\beta - 3\beta^2)(a_B - c_m) + (6 - \beta^2)(4 - 3\beta^2)\Delta\lambda_2 - (4 + \beta^2)\beta\Delta\lambda_1 \quad k_8 = \frac{\lambda_2(4 - \beta^2)b_2^{BN}X_1}{16\phi\lambda_2^2(1 - \beta^2)(4 - \beta^2)^2},$$

$$X_3 = ((6 - \beta^2)(2 - \beta^2)(4 - 3\beta^2) + \beta^2(4 + \beta^2)r)V, \quad \text{and} \quad k_9 = \frac{\lambda_2(2 - \beta^2)Eb_2^{BB}X_2 + X_3}{16\phi\lambda_2^2(1 - \beta^2)(4 - \beta^2)^2}.$$

By comparing Model BN and Model AN, Model BB and Model AB, the differences of the manufacturer's expected profit are $E[\pi_M^{BN}] - E[\pi_M^{AN}] = \phi\lambda_2^2(k - k_8)$ and $E[\pi_M^{BB}] - E[\pi_M^{AB}] = \phi\lambda_2^2(k - k_9)$.

Because

$$\lambda_1 > \lambda_2 > \tfrac{2}{1+\beta^2}\beta\lambda_1 > \beta\lambda_1,\ 12 + 4\beta - 3\beta^2 > 0 \text{ and } (6 - \beta^2)(4 - 3\beta^2) > (\beta^2 + 4),$$

we have $(6 - \beta^2)(4 - 3\beta^2)\lambda_2 - (\beta^2 + 4)\beta\lambda_1 > 0$. Thus, we have $X_1 > 0$ and $X_2 > 0$. When (i)$k > k_8$, $E[\pi_M^{BN}] - E[\pi_M^{AN}] > 0$; $k < k_8$, $E[\pi_M^{BN}] - E[\pi_M^{AN}] < 0$. (ii) $k > k_9$, we have $E[\pi_M^{BB}] - E[\pi_M^{AB}] > 0$; $k > k_9$, we have $E[\pi_M^{BB}] - E[\pi_M^{AB}] < 0$. $\square$

Proof of Corollary 4. Comparing Model AB and Model AN, Model BB and Model BN, the differences of the optimal transfer price and selling quantities are

$$E[b_1^{AB}] - b_1^{AN} = \frac{(a_B - a_N)(1 - \beta)}{2\lambda_1} > 0,\ E[b_1^{BB}] - b_1^{BN} = \frac{(a_B - a_N)(1 - \beta)}{(2 - \beta)\lambda_1} > 0,$$

$$E[b_2^{BB}] - b_2^{BN} = \frac{(a_B - a_N)(1 - \beta)}{(2 - \beta)\lambda_2} > 0,\ E[q_1^{AB}] - E[q_1^{AN}] = \frac{a_B - a_N}{4(1 + \beta)} > 0,$$

$$E[q_2^{AB}] - E[q_2^{AN}] = \frac{(a_B - a_N)(2 + \beta)}{4(1 + \beta)} > 0,\ E[q_1^{BB}] - E[q_1^{BN}] = \frac{a_B - a_N}{2(2 - \beta)(1 + \beta)} > 0,$$

$$E[q_2^{BB}] - E[q_2^{BN}] = \frac{a_B - a_N}{2(2 - \beta)(1 + \beta)} > 0. \quad \square$$

Proof of Corollary 5. Define

$$X_5 = (a_B - a_N)(1 - \beta)(2 + \beta)((1 - \beta)(2 + \beta)(a_N + a_B - 2c_m) - 2\Delta(\beta^2\lambda_1 + \beta\lambda_2 - 2\lambda_1)) \text{ and}$$

$$X_6 = (a_B - a_N)(1 - \beta)(2 + \beta)((1 - \beta)(2 + \beta)(a_B + a_N - 2c_m) - 2\Delta(\beta\lambda_1 - 2\lambda_2 + \beta^2\lambda_2)).$$

By comparing Model AB and Model AN, Model BB and Model BN, the differences of the collectors' expected profit are

$$E[\pi_{R1}^{AB}] - E[\pi_{R1}^{AN}] = \frac{(r + \beta^2)V + (a_B - a_N)(1 - \beta)((1 - \beta)(a_B + a_N - 2c_m) + 2\Delta(\lambda_1 - \beta\lambda_2))}{8(1 - \beta^2)} > 0 \quad E[\pi_{R1}^{BB}] - E[\pi_{R1}^{BN}] = \frac{(\beta^2 + r(2 + \beta)^2)V + X_5}{2(1 - \beta^2)(-4 + \beta^2)^2},$$

$$E[\pi_{R2}^{BB}] - E[\pi_{R2}^{BN}] = \frac{(r\beta^2 + (2 + \beta)^2)V + X_6}{2(1 - \beta^2)(-4 + \beta^2)^2}.$$

Since $a_B + a_N - 2c_m > 2(a_N - c_m) > 0$, we have $X_5 > 0$ and $X_6 > 0$. Thus, we have $E[\pi_{R1}^{BB}] - E[\pi_{R1}^{BN}] > 0$ and $E[\pi_{R2}^{BB}] - E[\pi_{R2}^{BN}] > 0$. $\square$

Proof of Corollary 6. Define

$$X_4 = (a_B - a_N)(1 - \beta)(5 + 3\beta)(a_B + a_N - 2c_m) + 2\Delta(a_B - a_N)(1 - \beta)(\lambda_1 + (4 + 3\beta)\lambda_2)$$

$$\text{and } X_7 = 2(a_B - a_N)(1 - \beta)(2 + \beta)^2(a_B + a_N - 2c_m + \Delta(\lambda_1 + \lambda_2)),$$

$$V_1 = \frac{X_4}{3(r + \beta^2)} \text{ and } V_2 = \frac{X_7}{(12 - 5\beta^2 + \beta^4)(1 + r)}.$$

By comparing Model AB and Model AN, Model BB and Model BN, the differences of the manufacturer's expected profit are $E[\pi_M^{BB}] - E[\pi_M^{BN}] = \frac{X_7 - (1 + r)(12 - 5\beta^2 + \beta^4)V}{4(4 - \beta^2)^2(1 - \beta^2)}$ and

$E[\pi_M^{IB}] - E[\pi_M^{IN}] = \frac{X_4 - 3(r + \beta^2)V}{16(1 - \beta^2)}$. When (i) $V < V_1$, $E[\pi_M^{AB}] - E[\pi_M^{AN}] > 0$; $V > V_1$, $E[\pi_M^{AB}] - E[\pi_M^{AN}] < 0$. (ii) $V < V_2$, $E[\pi_M^{BB}] - E[\pi_M^{BN}] > 0$; $V > V_2$, $E[\pi_M^{BB}] - E[\pi_M^{BN}] < 0$. $\square$

Proof of Corollary 7. Taking the first-order condition of the collection rate for the consumer surplus in Model AN and Model BN, respectively, are as follows: $\frac{\partial CS^{IN}}{\partial \lambda_1} = \frac{\Delta(1 - \beta)(a_N - c_m) + \Delta^2(\lambda_1 - \beta\lambda_2)}{16(1 - \beta^2)}$,

$$\frac{\partial CS^{IN}}{\partial \lambda_2} = \frac{2\left(1-\beta^2\right)(a_N - c_m + \Delta\lambda_2) + \lambda_2\left(4-\beta^2\right)b_2^{DN}}{16(1-\beta^2)}, \quad \frac{\partial CS^{DN}}{\partial \lambda_1} = \frac{(1-\beta)(2+\beta)^2\Delta(a_N - c_m) + \Delta^2\left(\left(4-3\beta^2\right)\lambda_1 - \beta^3\lambda_2\right)}{4(1-\beta^2)(4-\beta^2)^2},$$

$$\frac{\partial CS^{DN}}{\partial \lambda_2} = \frac{(1-\beta)(2+\beta)^2\Delta(a_N - c_m) + \Delta^2\left(\left(4-3\beta^2\right)\lambda_2 - \beta^3\lambda_1\right)}{4(1-\beta^2)(4-\beta^2)^2}.$$

Since $\lambda_1 > \lambda_2 > \beta\lambda_1 > \beta\lambda_2$ and $4 - 3\beta^2 > \beta^2$, we have $\left(4 - 3\beta^2\right)\lambda_1 > \beta^3\lambda_2$ and $\left(4 - 3\beta^2\right)\lambda_2 > \beta^3\lambda_1$. Therefore, we can obtain $\frac{\partial CS^{IN}}{\partial \lambda_1} > 0$, $\frac{\partial CS^{IN}}{\partial \lambda_2} > 0$, $\frac{\partial CS^{DN}}{\partial \lambda_1} > 0$, and $\frac{\partial CS^{DN}}{\partial \lambda_2} > 0$. Taking the first-order condition of retail prices with respect to λ_1 and λ_2, respectively, we have: $\frac{\partial p_1^{AN}}{\partial \lambda_1} = -\frac{\Delta}{4} < 0$, $\frac{\partial p_1^{AN}}{\partial \lambda_2} = -\frac{\beta\Delta}{4} < 0$, $\frac{\partial p_2^{AN}}{\partial \lambda_2} = -\frac{\Delta}{2} < 0$, $\frac{\partial p_1^{BN}}{\partial \lambda_1} = -\frac{\Delta}{4-\beta^2} < 0$, $\frac{\partial p_1^{BN}}{\partial \lambda_2} = -\frac{\beta\Delta}{2(4-\beta^2)} < 0$, $\frac{\partial p_2^{BN}}{\partial \lambda_1} = -\frac{\beta\Delta}{2(4-\beta^2)} < 0$, and $\frac{\partial p_2^{BN}}{\partial \lambda_2} = -\frac{\Delta}{4-\beta^2} < 0$. $\square$

References

1. Yang, L.; Gao, M.; Feng, L. Competition versus cooperation? Which is better in a remanufacturing supply chain considering blockchain. *Transp. Res. E-Logist. Transp. Rev.* **2022**, *165*, 102855. [CrossRef]
2. Han, X.; Yang, Q.; Shang, J.; Pu, X. Optimal strategies for trade-old-for-remanufactured programs: Receptivity, durability, and subsidy. *Int. J. Prod. Econ.* **2017**, *193*, 602–616. [CrossRef]
3. Wu, X.; Zhou, Y. The optimal reverse channel choice under supply chain competition. *Eur. J. Oper. Res.* **2017**, *259*, 63–66. [CrossRef]
4. Savaskan, R.C.; Van Wassenhove, L.N. Reverse Channel Design: The Case of Competing Retailers. *Manag. Sci.* **2006**, *52*, 1–14. [CrossRef]
5. Cai, K.; Zhang, Y.; Lou, Y.; He, S. Information sharing in a collectors-led closed-loop supply chain. *RAIRO-Oper. Res.* **2022**, *56*, 2329–2350. [CrossRef]
6. Yang, F.; Wang, M.; Ang, S. Optimal remanufacturing decisions in supply chains considering consumers' anticipated regret and power structures. *Transp. Res. E-Logist. Transp. Rev.* **2021**, *148*, 102267. [CrossRef]
7. Krikke, H.; Hofenk, D.; Wang, Y. Revealing an invisible giant: A comprehensive survey into return practices within original (closed-loop) supply chains. *Resour. Conserv. Recycl.* **2013**, *73*, 239–250. [CrossRef]
8. Ovchinnikov, A. Revenue and Cost Management for Remanufactured Products. *Prod. Oper. Manag.* **2011**, *20*, 824–840. [CrossRef]
9. Aydin, R.; Mansour, M. Investigating sustainable consumer preferences for remanufactured electronic products. *J. Eng. Res.* **2023**, *11*, 100008. [CrossRef]
10. Yue, X.; Liu, J. Demand forecast sharing in a dual-channel supply chain. *Eur. J. Oper. Res.* **2006**, *174*, 646–667. [CrossRef]
11. Huang, Y.; Wang, Z. Information sharing in a closed-loop supply chain with technology licensing. *Int. J. Prod. Econ.* **2017**, *191*, 113–127. [CrossRef]
12. Choi, T.M. Blockchain-technology-supported platforms for diamond authentication and certification in luxury supply chains. *Transp. Res. E-Logist. Transp. Rev.* **2019**, *128*, 17–29. [CrossRef]
13. Yanikoglu, I.; Denizel, M. The value of quality grading in remanufacturing under quality level uncertainty. *Int. J. Prod. Res.* **2021**, *59*, 839–859. [CrossRef]
14. Abbey, J.D.; Kleber, R.; Souza, G.C.; Voigt, G. Remanufacturing and consumers' risky choices: Behavioral modeling and the role of ambiguity aversion. *J. Oper. Manag.* **2019**, *65*, 4–21. [CrossRef]
15. Gong, B.; Zhang, H.; Gao, Y.; Liu, Z. Blockchain adoption and channel selection strategies in a competitive remanufacturing supply chain. *Comput. Ind. Eng.* **2023**, *175*, 108829. [CrossRef]
16. Niu, B.; Xu, H.; Chen, L. Creating all-win by blockchain in a remanufacturing supply chain with consumer risk-aversion and quality untrust. *Transp. Res. E-Logist. Transp. Rev.* **2022**, *163*, 102778. [CrossRef]
17. Zhang, T.Y.; Dong, P.W.; Chen, X.F.; Gong, Y. The impacts of blockchain adoption on a dual-channel supply chain with risk-averse members. *Omega-Int. J Manag. Sci.* **2023**, *114*, 102747. [CrossRef]
18. Cui, Y.; Hu, M.; Liu, J.C. Value and Design of Traceability-Driven Blockchains. *MSom-Manuf. Serv. Oper. Manag.* **2023**, 1–18, *Article in advance*. [CrossRef]
19. Zheng, Y.; Xu, Y.Q.; Qiu, Z.G. Blockchain Traceability Adoption in Agricultural Supply Chain Coordination: An Evolutionary Game Analysis. *Agriculture* **2023**, *13*, 184. [CrossRef]
20. Wang, C.F.; Chen, X.F.; Xu, X.; Jin, W. Financing and operating strategies for blockchain technology-driven accounts receivable chains. *Eur. J. Oper. Res.* **2023**, *304*, 1279–1295. [CrossRef]
21. Huang, Y.D.; Widyadana, G.A.; Wee, H.M.; Blos, M.F. Revenue and risk sharing in view of uncertain demand during the pandemics. *Rairo-Oper. Res.* **2022**, *56*, 1807–1821. [CrossRef]
22. Ji, C.Y.; Liu, X.X. Design of risk sharing and coordination mechanism in supply chain under demand and supply uncertainty. *Rairo-Oper. Res.* **2022**, *56*, 123–143. [CrossRef]
23. Garai, T.; Paul, A. The effect of supply disruption in a two-layer supply chain with one retailer and two suppliers with promotional effort under random demand. *J. Manag. Anal.* **2022**, *10*, 22–37. [CrossRef]

24. Li, Y.C.; Saldanha-da-Gama, F.; Liu, M.; Yang, Z.L. A risk-averse two-stage stochastic programming model for a joint multi-item capacitated line balancing and lot-sizing problem. *Eur. J. Oper. Res.* **2023**, *304*, 353–365. [CrossRef]
25. Pei, H.L.; Liu, Y.K.; Li, H.L. Robust Pricing for a Dual-Channel Green Supply Chain Under Fuzzy Demand Ambiguity. *IEEE Trans. Fuzzy Syst.* **2023**, *31*, 53–66. [CrossRef]
26. Liu, W.J.; Liu, W.; Shen, N.N.; Xu, Z.T.; Xie, N.M.; Chen, J.; Zhou, H.Y. Pricing and collection decisions of a closed-loop supply chain with fuzzy demand. *Int. J. Prod. Econ.* **2022**, *245*, 108409. [CrossRef]
27. Zheng, B.R.; Wen, K.; Jin, L.; Hong, X.P. Alliance or cost-sharing? Recycling cooperation mode selection in a closed-loop supply chain. *Sustain. Prod. Consum.* **2022**, *32*, 942–955. [CrossRef]
28. Long, X.F.; Ge, J.L.; Shu, T.; Liu, Y. Analysis for recycling and remanufacturing strategies in a supply chain considering consumers' heterogeneous WTP. *Resour. Conserv. Recycl.* **2019**, *148*, 80–90. [CrossRef]
29. Yi, P.X.; Huang, M.; Guo, L.J.; Shi, T.L. Dual recycling channel decision in retailer oriented closed-loop supply chain for construction machinery remanufacturing. *J. Clean. Prod.* **2016**, *137*, 1393–1405. [CrossRef]
30. Huang, M.; Yi, P.X.; Shi, T.L. Triple Recycling Channel Strategies for Remanufacturing of Construction Machinery in a Retailer-Dominated Closed-Loop Supply Chain. *Sustainability* **2017**, *9*, 2167. [CrossRef]
31. Zhang, W.S.; Zhang, T. Recycling channel selection and financing strategy for capital-constrained retailers in a two-period, closed-loop supply chain. *Front. Environ. Sci.* **2022**, *10*, 1–16. [CrossRef]
32. He, Q.D.; Wang, N.M.; Browning, T.R.; Jiang, B. Competitive collection with convenience-perceived customers. *Eur. J. Oper. Res.* **2022**, *303*, 239–254. [CrossRef]
33. Guo, Y.; Wang, M.M.; Yang, F. Joint emission reduction strategy considering channel inconvenience under different recycling structures. *Comput. Ind. Eng.* **2022**, *169*, 108159. [CrossRef]
34. Wan, N.A. Impacts of sales mode and recycling mode on a closed-loop supply chain. *Int. J. Syst. Sci.* **2023**, *54*, 1–22. [CrossRef]
35. Cao, H.Q.; Ji, X.F. Optimal Recycling Price Strategy of Clothing Enterprises based on Closed-loop Supply Chain. *J. Ind. Manag. Optim.* **2023**, *19*, 1350–1366. [CrossRef]
36. Yang, T.J.; Li, C.M.; Yue, X.P.; Zhang, B.B. Decisions for Blockchain Adoption and Information Sharing in a Low Carbon Supply Chain. *Mathematics* **2022**, *10*, 2233. [CrossRef]
37. Niu, B.Z.; Dong, J.; Liu, Y.Q. Incentive alignment for blockchain adoption in medicine supply chains. *Transp. Res. E-Logist. Transp. Rev.* **2021**, *152*, 32. [CrossRef]
38. Shi, C.-l.; Geng, W.; Sheu, J.-B. Integrating dual-channel closed-loop supply chains: Forward, reverse or neither? *J. Oper. Res. Soc.* **2020**, *72*, 1844–1862. [CrossRef]
39. Yang, X.D.; Cai, G.S.; Ingene, C.A.; Zhang, J.H. Manufacturer Strategy on Service Provision in Competitive Channels. *Prod. Oper. Manag.* **2020**, *29*, 72–89. [CrossRef]
40. Shen, B.; Zhu, C.; Li, Q.; Wang, X. Green technology adoption in textiles and apparel supply chains with environmental taxes. *Int. J. Prod. Res.* **2020**, *59*, 4157–4174. [CrossRef]

 *processes*

Article

Data-Driven Evaluation of the Synergistic Development of Economic-Social-Environmental Benefits for the Logistics Industry

Wei Mu [1], Jun Xie [1,*], Heping Ding [1,2,*] and Wen Gao [1]

[1] Business School, Suzhou University, Suzhou 234000, China
[2] Center for International Education, Philippine Christian University, Manila 1004, Philippines
* Correspondence: junxiesz@ahszu.edu.cn (J.X.); dingheping@ahszu.edu.cn (H.D.);
  Tel.: +86-18205575160 (H.D.)

Abstract: The receding globalization has reshaped the logistics industry, while the additional pressure of the COVID-19 pandemic has posed new difficulties and challenges as has the pressure towards sustainable development. Achieving the synergistic development of economic, social, and environmental benefits in the logistics industry is essential to achieving its high-quality development. Therefore, we propose a data-driven calculation, evaluation, and enhancement method for the synergistic development of the composite system of economic, social, and environmental benefits (ESE-B) of the logistics industry. Based on relevant data, the logistics industry ESE-B composite system sequential parametric index system is then constructed. The Z-score is applied to standardize the original index data without dimension, and a collaborative degree model of logistics industry ESE-B composite system is constructed to estimate the coordinated development among the subsystems of the logistics industry's ESE-B system. The method is then applied to the development of the logistics industry in Anhui Province, China from 2011 to 2020. The results provide policy recommendations for the coordinated development of the logistics industry. This study provides theoretical and methodological support for the sustainable development aspects of the logistics industry.

Keywords: data-driven; logistics industry; composite systems; synergy; sustainability

Citation: Mu, W.; Xie, J.; Ding, H.; Gao, W. Data-Driven Evaluation of the Synergistic Development of Economic-Social-Environmental Benefits for the Logistics Industry. *Processes* **2023**, *11*, 913. https://doi.org/10.3390/pr11030913

Academic Editors: Luis Puigjaner and Tsai-Chi Kuo

Received: 8 February 2023
Revised: 10 March 2023
Accepted: 15 March 2023
Published: 17 March 2023

1. Introduction

1.1. Background

Currently, the world is experiencing the largest changes in a century [1]. Further, the "second generation" open policy with regulatory convergence at its core has triggered a new round of global economic and trade rules games, and the global pattern is showing an evolutionary trend of accelerated multipolar development [2]. Given the reshaping of the global value chain, how can the logistics industry (LI) develop itself? In today's international multilateral environment, sustainable development should take precedence and the LI should achieve the synergistic development of economic, social, and environmental benefits (ESE-B) for sustainable development. As a link between production and consumption, the LI plays an important role in supporting and guaranteeing the regional economic development process and has become a new driving force for economic growth and social development [3]. For example, the LI provides a large number of employment opportunities and creates tax revenue but, at present, while the LI is playing an increasing role in promoting regional economic development, its negative impact on the ecological environment and society is also becoming increasingly obvious, and the rapidly developing LI will inevitably produce a series of environmental problems such as increased waste emissions, aggravated air pollution and increased energy consumption, as well as increased traffic accidents and noise pollution. For example, the average energy consumption of China's LI over 2009–2019 accounted for 8.44% of the total energy consumption [4].

Therefore, in the context of global sustainable development, it is the goal and motivation of this article to define the synergistic development of economic, social and environmental benefits of the logistics industry, evaluate the level of their synergistic development, realize the synergistic development of economic, social and environmental benefits of the logistics industry, propose suggestions for the government and practitioners to enhance the synergistic development of economic, social and environmental benefits of the logistics industry, and promote the sustainable development aspects of the logistics industry.

1.2. Research Overview

Achieving the synergistic development of economic, social, and environmental benefits of the logistics industry is essential to achieving its high-quality development [3,4]. Synergetics, founded by the German physicist Haken, is the study of the way in which subsystems within open systems work in concert and interact with each other to evolve the system from disorder to order [5,6]. The idea of sustainable development was first introduced in the World Conservation Strategy in 1980 and has since become a common worldwide effort [7]. Sustainable development chiefly alludes to ensuring that development of the population, economy, resources, society, and surrounding conditions, as well as the overall human development, does not deprive future generations of good living conditions [8–10]. Many studies on the synergistic development of the logistics industry are centered on the synergy with other industries; therefore, the literature mainly covers the following aspects [11–13].

(1)   Research on synergistic development of logistics industry

Scholars have mostly studied the coordinated development of the logistics industry with other industries. For instance, Gordon et al. considered that the productive service industries, including logistics, play an important role in the development of manufacturing industry [14]. Hu and Shu [15] studied the coordinated development of the LI and agricultural ecosystem, and the path selection for the coordinated development of agricultural logistics so that the ecosystem can be optimized. Liang and Gu [16] studied the level of synergistic development between agriculture and LI in China from 2010 to 2019 and its temporal trends, and measured the correlation between the two industries using a gray correlation model to explore the key factors affecting the synergistic development of the two industries. Scholars likewise focused on the coordinated development of the LI and the regional economy, with Klaus [17] arguing that the logistics expenditure of a country is proportional to its national economic wealth, finding that, with the increasing global integration, logistics has become the basis of wealth. Chen et al. [18] studied the coordinated development paths of logistics and the economy in metropolitan cities based on the coordinated development theory. The academic research on the relationship between the LI and economic system and other industries has achieved fruitful results, thus confirming that the LI and other industries, as well as the regional economy, can promote each other and develop together. However, there are fewer studies on the synergistic development of the subsystems within the logistics industry.

With increased financial globalization, deepening environmental degradation, and overexploitation of assets and energy, all modern logistics activities can have positive or bad impacts on the ecological surrounding conditions [19], and the relationship between LI and the environment is gradually becoming a research hotspot. Liu et al. [20] used data from 42 Asian countries between 2007 and 2016 to demonstrate that logistics performance is significantly related to environmental degradation. Zhou [21] took the regional "logistics-ecological environment" composite system as research object, and based on the synergy theory, studied the coordinated development of regional logistics and ecological environment from the perspective of low carbon. Evangelista [22] studied the environmental performance of third-party logistics service companies and proved that these companies assume a critical part in reducing carbon emissions and improving the ecological execution of logistics. Overall, the studies on the relationship between the LI and environment have concluded that the LI and environment can develop in a mutually supportive manner.

However, the studies treat the environment as an independent system, which cannot reflect the environmental impact of the logistics industry, and does not consider the social impact aspect of the logistics industry.

(2)   ESE-B synergy evaluation study in the logistics industry

When scholars study system coordination relationships, they generally construct evaluation index systems, collect data, and then use models to evaluate these systems. In constructing an evaluation index system, the LI is generally considered from the input-output perspective, the logistics economy from the growth level and foreign trade, and the logistics environment from pollution and carbon emissions [23–25]. These indicators do not fully reflect the benefits generated by the logistics industry, or are not closely related to the development of the logistics industry, and do not take into account the environmental factors such as resource utilization and energy consumption directly generated by logistics activities, not to mention the social factors. In terms of research methods, most studies used coupled coordination degree and composite system synergy models to study the coordinated development of two or more systems. Yu and Yin [26] studied the ESE-B of urban public transportation infrastructure using a coupled coordination degree model and investigated the influence of three benefits on its coordinated development level using a panel regression model. Li et al. [27] studied the growth and environmental impacts of green logistics performance in One Belt, One Road Initiative countries during 2007–2019 using least squares and generalized method of moments. There are also several other representative methods, such as data envelopment analysis to measure the low-carbon efficiency of LI [28,29] and the energy value approach to study the sustainability of logistics systems [30]. These methods have been proven to be better methods for studying synergistic relationships through theoretical and practical validation; however, development of more scientific measures for evaluating synergistic relationships is the direction of future research.

Based on these studies, scholars have each put forward policy recommendations to help increase the coordinated development of the LI and the economy or the environment, such as carrying out logistics infrastructure development, strengthening technological innovation, and creating a favorable environment [31,32].

1.3. Limitations of Prior Studies

In general, the extant research on coordinated development issues related to the LI has achieved clear results and has established a decent starting point for the study of the LI complex system, but due to the different perspectives and depths of scholars' understanding, the existing research needs further improvement as follows.

(1)   Most extant studies focus on the coordinated relationship between two systems: LI and ecological environment/regional economy/other industries, ignoring the social system. Further, most of them treated the ESE-B as independent systems instead of a single complex system; therefore, the research on LI's ESE-B composite system needs to be expanded.

(2)   In the construction of the ESE-B index system, the selected indicators are not comprehensive or not closely related to the development of the LI, and environmental factors such as resource utilization and energy consumption directly generated by logistics activities are not considered, so that the constructed index system cannot reflect well the essence and characteristics of the ESE-B composite system of the LI. As such, there is a need to accurately construct the LI ESE-B composite index system.

(3)   The CSSDM is generally utilized, which has laid a good foundation for this paper, but it is a challenge to measure and evaluate the synergy development level of LI's ESE-B composite system scientifically and accurately. To this end, we construct a data-driven LI's ESE-B composite system synergy degree model.

Therefore, our research objectives are: (1) to take the logistics industry as a total system and study the economic benefits, social benefits, environmental benefits and coordination of this total system; (2) to scientifically and objectively construct evaluation indicators for the ESE-B composite system of the logistics industry; (3) to scientifically and accurately measure and evaluate the level of synergistic development of the ESE-B composite system of the logistics industry using a data-driven approach, and on this basis, to propose targeted countermeasures and suggestions.

1.4. Manuscript Structure

This study adopts the composite system synergy model to study the ESE-B composite system of the LI and then considers Anhui Province as a case study. The evaluation index system of LI's ESE-B composite system is constructed, and the orderliness of ESE-B subsystems and the composite system synergy of LI in Anhui Province are evaluated and improved by collecting and calculating basic data on Anhui Province from 2011 to 2020. The remainder of this paper is organized as follows. The subsequent segment presents the way(s) of doing things, which introduces the data-driven methodological processes of data collection, data handling, data modeling and application. The third segment presents the case study, which estimates the level of coordinated development of the LI's ESE-B composite system in Anhui Province, China, and evaluates and seeks to enhance the coordinated development level. The final section offers the conclusions from the study.

2. Materials and Methods

To evaluate and improve the synergistic development level of LI's ESE-B composite system more objectively, comprehensively, and accurately, this section introduces the methods.

2.1. Method Flow

We promote the synergy of the ESE-B of LI, to achieve more efficient and sustainable development. Evaluating the relationship between these three systems from the systems theory viewpoint and formulating relevant policies from the worldwide optimization perspective are pressing requirements for SDLI research. However, the LI's ESE-B composite system includes multiple data types. As such, we need to determine how to apply effective models to quantify, assess, and recognize numerous pointers and whether the proposed strategy proposals can really uphold related government navigation. These conundrums present research challenges. To meet this difficulty and ameliorate the coordinated development of LI's ESE-B composite system, this paper proposes a data-driven method for measuring, evaluating, and enhancing coordinated development level of LI's ESE-B composite system. We collected economic, social, and environment-related data for LI, while data processing entailed turning the original data into measurement data in the evaluation index system [33]. The methodology is also depicted in Figure 1.

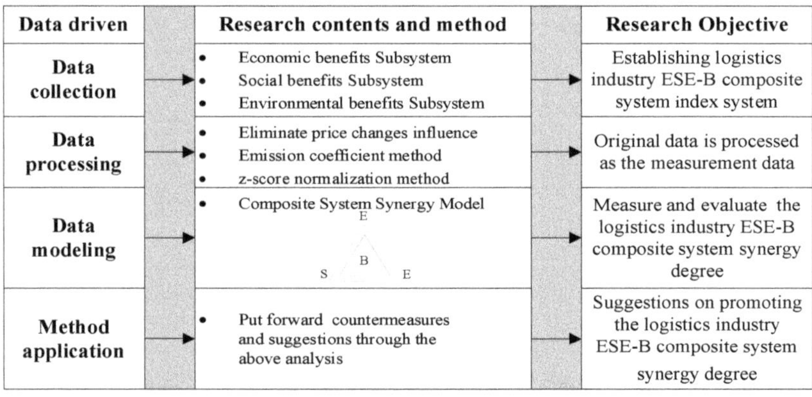

Data driven	Research contents and method	Research Objective
Data collection	• Economic benefits Subsystem • Social benefits Subsystem • Environmental benefits Subsystem	Establishing logistics industry ESE-B composite system index system
Data processing	• Eliminate price changes influence • Emission coefficient method • z-score normalization method	Original data is processed as the measurement data
Data modeling	• Composite System Synergy Model	Measure and evaluate the logistics industry ESE-B composite system synergy degree
Method application	• Put forward countermeasures and suggestions through the above analysis	Suggestions on promoting the logistics industry ESE-B composite system synergy degree

Figure 1. Research process.

The first step, data collection, is mainly to collect economic, social and environmental related data on the LI, including LI's cargo turnover, added value, energy consumption, carbon emissions, etc.

The second step, data processing, is to process the original data into the measurement data in the evaluation index system, including the elimination of price fluctuations, the conversion of standard coal, the use of z-score for standardization processing, etc.

The third step, data modeling, is to construct the composite system synergy model and correlation analysis model, including analyzing the sequential parameters, orderliness and synergy of the 3 subsystems of economic benefits, social benefits and environmental benefits of LI.

The fourth step, data application, is to put forward targeted policy recommendations to promote the coordinated development of an ESE-B composite system based on the research results.

2.2. Sequential Parametric Index System

In the related literature, the LI is characterized as three parts: transportation, warehousing, and postal industry. Since these three parts comprise over 85% of the gross value added of the LI [34], there is a certain reliability in adopting this definition. The review's motivation is to analyze the coordination relationship of the LI's ESE-B composite system according to the principle of covariance. The ordinal parameter is a parameter variable that influences the evolution direction of the system and characterizes the degree of ordering of the system; as long as the ordinal parameter is controlled, the development direction of the entire system can be grasped [35]. However, the selection of different indicators will produce different results of the ordinal parameter calculation. As such, the scientific and reasonable selection of indicators becomes key to the accuracy of the sequential parameter calculation results, which requires that the selected indicators should conform to the connotation and reflect the characteristics of sequential parameters. On the basis of the meaning of SDLI, referring to the existing research results on LI's coordinated development and guided by the sustainable development and green logistics theories, the typical order parameter index that can represent the ESE-B composite system of the LI is selected. We also construct a coordinated development index system of LI's ESE-B composite system consisting of a subsystem layer (ESE-B), element layer (input, output), and index layer. This evaluation index system considers the LI as a total system, and its system coordination degree and the orderliness of each subsystem reflect the SDLI level, being an accurate definition of the SDLI [13].

The assessment indexes of the LI's economic subsystem mainly contain the amount of investment in fixed assets and the added value of LI [36]. Compared with previous research indicators, this study adds "Internet broadband access port" [37] to reflect the information input of LI, because the input of information technology is necessary.

The evaluation indexes of LI's social subsystem mainly reflect the contribution and harm of the LI to the society, among which LI's added value and vehicle tax revenue are significant parts of the public economy, while t the freight turnover is LI's commitment to freight transportation. In this study, the indicator of "vehicle and vessel tax revenue" was added to reflect LI's contribution to the society, and the number of traffic fatalities and property damage were added to reflect the LI's harm to the society [38].

LI's environmental subsystem mainly measures the effect of the LI's development on the ecological environment from two aspects: pollution and governance. The unfriendly effect of LI through CO_2 emissions on the environment is climate warming, as LI's exhaust gas emissions will pollute the environment. The number of deaths and property loss in traffic accidents and road traffic noise reflect LI's adverse impact on society, which is also a new indicator added here based on previous studies [9,21]. The details are shown in Table 1.

Table 1. The index system of LI's ESE-B composite system.

Subsystem	Sequential Parametric Indicator Layer	Properties	References
LI Economic Benefits subsystem S_1	Investment in fixed assets (billion yuan) E11	Positive	[9]
	Number of cargo vehicles (million units) E12	Positive	[9]
	Internet broadband access ports (million) E13	Positive	[7]
	Cargo turnover (billion ton kilometers) E14	Positive	[21]
	Value added of logistics industry (billion yuan) E15	Positive	[21]
	Logistics industry contribution rate (%) E16	Positive	[21]
LI Social Benefits Subsystem S_2	Logistics network mileage (million km) E21	Positive	[13]
	Human resource input (10,000 people) E22	Positive	[13]
	Total wages of urban personnel (billion yuan) E23	Positive	[38]
	Vehicle tax revenue (billion yuan) E24	Positive	[38]
	Number of traffic fatalities (persons) E25	reverse	[38]
	Traffic accident property damage (million yuan) E26	reverse	[38]
LI Environmental Benefits Subsystem S_3	Energy consumption (million tons of standard coal) E31	reverse	[7]
	Electricity consumption (billion kWh) E32	reverse	[13]
	Greening coverage rate of built-up area (%) E33	Positive	[21]
	Carbon emissions (million tons) E34	reverse	[7]
	Exhaust gas emission (million tons) E35	reverse	[7]
	Road traffic noise (decibels) E36	reverse	[9]

2.3. Data Sources and Processing

(1) The data in Table 1 were predominantly acquired from the China Energy Statistical Yearbook, China Statistical Yearbook, and Anhui Statistical Yearbook for 2012–2021 [39,40]. To eliminate the price ups and downs effect, price-related factors, for example, property damage and vehicle charge were switched over completely to real values, with 2011 as the base period.

(2) LI's carbon emissions in Table 1 were calculated according to the carbon emission factors of 17 energy sources in the 2006 IPCC Guidelines for National Greenhouse Gas Inventories, and then calculated by the amount of energy consumed by the LI [41]. LI's exhaust gas emissions in Table 1 were determined by the emission factor method, drawing on EPA, AP-42, and Beijing emission factors [42] The emissions of NOX, PM10, PM2.5, SO_2 of the LI in Anhui Province from 2011 to 2020 were measured by the emission factor method in the light of the primary energy consumption of the LI in the energy balance sheet. Finally, the exhaust gas emissions were obtained by summing up and the missing energy data in 2020 was filled in by interpolation.

(3) Standardized data processing.

The standardization of raw index data is intended to transform raw index data into dimensionless index measurement values. Data standardization methods include, for example, min-max standardization or the z-score method. The z-score can reflect the original data values more accurately, so this paper uses it for standardization [43], as follows:

$$X_{ij}^* = \frac{X_{ij} - \overline{X}_j}{S_j} \tag{1}$$

$$\overline{X}_j = \frac{\sum_{i=1}^{n} X_{ij}}{n} \tag{2}$$

$$S_j = \sqrt{\frac{\sum_{i=1}^{n} \left(X_{ij} - \overline{X}_j \right)^2}{n - 1}} \tag{3}$$

where X_{ij}^* is the normalized data, X_{ij} the raw data, $\overline{X}_j$ the mean of X_{ij}, and S_j the standard deviation of X_{ij}.

2.4. Data Modeling

The complex system synergy model can scientifically calculate the level of complex system synergy and is now widely used to analyze the dynamic synergy evolution of a system [20,44]. We also use it to measure the dynamic synergy level of LI's ESE-B complex system as follows.

(1)   Orderliness of sequential parameters

LI's ESE-B composite system can be expressed as $S_j = \{S_1, S_2, S_3\}$, which represent the three subsystems of ESE-B, respectively. The order parameter of each subsystem is $e_j = (e_{j1}, e_{j2}, e_{j3} \ldots \ldots e_{jn})$, where $j = 1, 2, 3$, and $n \geq 1$ denotes the number of sequential parameters of each subsystem. $\alpha_{ji} \leq e_{ji} \leq \beta_{ji}, i = 1, 2, 3 \ldots \ldots n$. To ensure the stability of the system, α_{ji} and β_{ji} are the upper and lower limits of ordinal parametric components e_{ji} [41]. Each ordinal parametric component's orderliness is:

$$
\begin{aligned}
\mu_j(e_{ji}) &= \tfrac{e_{ji} - \alpha_{ji}}{\beta_{ji} - e_{ji}} \; e_{ji} \text{ is a positive indicator} \\
\mu_j(e_{ji}) &= \tfrac{\beta_{ji} - e_{ji}}{\beta_{ji} - e_{ji}} \; e_{ji} \text{ is a negative indicator}
\end{aligned}
\tag{4}
$$

From Equation (4), $\mu_j(e_{ji}) \in [0, 1]$; the larger the value is, the greater is the contribution of indicator e_{ji} to the ordering of the subsystem.

(2)   Subsystem orderliness

The degree of subsystem order reflects the sum of the contribution of all components of order parameter variables e_j to subsystem S_j. The orderliness of the environmental benefit subsystem S_3 reflects the contribution of ordinal variables E31, E32, $\cdots$,E36 to subsystem S_3. It can be obtained by integrating each $\mu_j(e_{ji})$. We use the more accurate geometric mean method to synthesize the orderliness of each subsystem, $\mu_j(e_j)$ [45,46]:

$$
u_j(e_j) = \sqrt[n]{\left| \prod_{i}^{n} u_j(e_{ji}) \right|}
\tag{5}
$$

where $\mu_j(e_j) \in [0, 1]$; the larger $\mu_j(e_j)$ is, the greater is the contribution that e_j makes to the orderliness of subsystem S_j and the higher is the degree of the subsystem orderliness, and vice versa.

(3)   Composite system synergy model

For a given initial moment t_0, let the degree of subsystem S_j order be $\mu_j^0(e_j), j = 1, 2, \ldots, k$ when the overall development of the composite system evolves to the moment t_1; at this point, the degree of subsystem S_j order is $\mu_j^1(e_j), j = 1, 2 \ldots, k$. Then, the coordination degree of composite system C(t) is [47]:

$$
C(t) = \theta \cdot \left[\left| \prod_{j}^{n} \left[u_j^t(e_j) - u_j^0(e_j) \right] \right| \right]^{1/k}
\tag{6}
$$

where $\theta = \dfrac{min\left[u_j^t(e_j) - u_j^0(e_j) \neq 0 \right]}{\left| min\left[u_j^t(e_j) - u_j^0(e_j) \neq 0 \right] \right|}, j = 1, 2, 3.$

In Equation (6), $u_j^t(e_j) - u_j^0(e_j)$ is the change in the magnitude of subsystem S_j from t_0 to t_1, and $u_j^t(e_j) - u_j^0(e_j) \in [-1, 1]$. As $C(t) \in [-1, 1]$, the larger the value is, the higher the degree of composite system coordinated development is, and vice versa. The significance of the introduction of parameter θ is that only condition $u_j^t(e_j) - u_j^0(e_j) > 0, \forall_j \in [1, k]$, and the synergy degree of the composite system is positive. If in the $[t_0, t_1]$ time period, the increase in the orderliness of one subsystem is larger and the increase in the orderliness of some other subsystems is smaller or even decreases, the coordinated development state of the entire complex system is bad or not coordinated at all, as shown by $C(t) \in [-1, 0]$. In this study, the LI ESE-B composite system has three subsystems. Thus, $k = 3$ and Equation (6) can be written as [48,49]:

$$C = \theta \cdot \sqrt[3]{\left[u_1^t(e_1) - u_1^0(e_1)\right] \cdot \left[u_2^t(e_2) - u_2^0(e_2)\right] \cdot \left[u_3^t(e_3) - u_3^0(e_3)\right]} \tag{7}$$

From the literature, there are two algorithms for Equation (6): one takes the same moment as the base period and the other takes the neighboring moments as the base period. The two algorithms provide useful ideas for the comprehensive analysis of the composite system synergy; specifically, the first one can well reflect the long-term evolution trend of the composite system and the second one can better reflect whether the composite system is in a stable state of synergistic evolution. As such, we use two methods to measure the LI's ESE-B composite system synergy. At present, most countries and international organizations generally use the same and adjacent base period synergy degree classification standard [50], as shown in Figure 2.

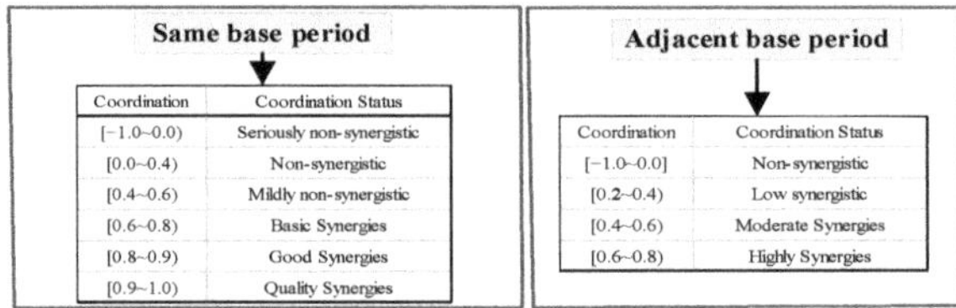

Coordination	Coordination Status
[−1.0~0.0)	Seriously non-synergistic
[0.0~0.4)	Non-synergistic
[0.4~0.6)	Mildly non-synergistic
[0.6~0.8)	Basic Synergies
[0.8~0.9)	Good Synergies
[0.9~1.0)	Quality Synergies

Coordination	Coordination Status
[−1.0~0.0]	Non-synergistic
[0.2~0.4)	Low synergistic
[0.4~0.6)	Moderate Synergies
[0.6~0.8)	Highly Synergies

Figure 2. Evaluation criteria for the synergy degree: same and adjacent base periods.

LI's ESE-B composite system synergy model can explore the synergistic development of LI's ESE-B and the classification of the coordination level can precisely define the level of coordination development, which is an important insight for the design of countermeasures to promote LI's synergistic development and indicates the direction of the SDLI.

2.5. Data Optimization for Decision Making Applications

In the context of the increasingly prominent contradiction between the economic development, resource shortage, and social demand of LI, promoting the synergistic development of ESE systems of LI is of great practical importance to enhance the SDLI. Since the ESE-B of LI are considered comprehensively, an index system containing these three subsystems is proposed here. Due to the multiple sources of data for the LI composite system, this paper puts into use a data-driven approach. The study thus aims to apply a data-driven method to precisely measure, evaluate, and identify the synergistic development level of the LI's ESE-B composite system, as well as to propose strategy proposals in view of the quantitative evaluation results and give a premise to decision making for LI specialists and chiefs. The specific applications are shown in Figure 3 below.

In the first step, the data related to the LI's ESE-B composite system are collected and the database established. As indicated by the standards of data accessibility and scientificity, the ESE-B composite system index system is constructed from three aspects, covering the ESE-B of LI. This step represents the extraction and organization of data.

In the second step, the raw data are standardized using the Z-score method, so that the raw data are all converted to dimensionless values; that is, the indicators are all at the same quantitative level and can be used for comprehensive measurement. This step represents the conversion of the data to make them comparable.

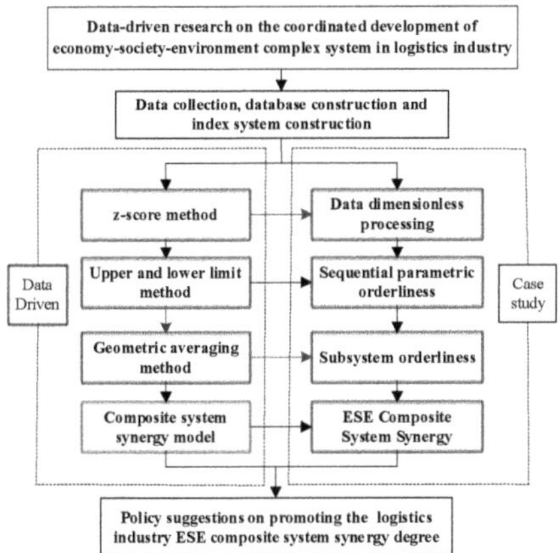

Figure 3. Data application diagram.

In the third step, the ordinal covariates orderliness is calculated using the upper and lower bound method, and the geometric mean method is applied to calculate LI's orderliness of ESE-B. This step represents the processing of standardized data to measure the contribution degree of ordinal covariates to the orderliness of subsystems.

In the fourth step, LI's ESE-B composite system synergy model is constructed, LI's ESE-B composite system synergy degree is calculated, and the coordination degree among LI's subsystems is evaluated. This represents is the modeling of the data, being used to solve the realistic problem of the synergy of LI's ESE-B.

In the fifth step, in view of the data-driven quantitative estimation and assessment results, the research conclusions are drawn and the shortcomings in LI's development identified; based on these, strategy proposals are proposed to promote the synergistic development of LI's ESE-B. This step represents the sublimation of the data and the presentation of the data modeling results to derive policy recommendations based on data.

Through the above data application, we can realize the measurement, evaluation. and optimization improvement of LI's ESE-B composite system synergy and provide theoretical and methodological support for the SDLI.

3. Case Study

Using the above research method, the study is carried out using the example of LI's ESE-B in Anhui Province, China. Section 3.1 presents the case background, Section 3.2 shows the data results calculated by using the above research method, and Section 3.3 provides suggestions to promote the synergy of LI's ESE-B complex system in Anhui Province based on the results.

3.1. Background

Anhui Province, as shown in Figure 4, is located in East China, connecting Jiangsu Province and Zhejiang Province to the east, Hubei Province and Henan Province to the west, Jiangxi Province to the south, and Shandong Province to the north. It is not only a national comprehensive transportation hub, but also a deep hinterland of the Yangtze River Delta region. The LI of Anhui Province has been developing well. In 2020, the total revenue of LI in Anhui Province was 495 billion yuan, increasing by 2% compared with the previous year. The ratio of total social logistics cost to GDP was 14.7% [13], being lower by 0.2 percentage points compared with the previous year, which was the same as the national level. In 2021, the total social logistics in Anhui Province maintained a steady growth, reaching 8.08 trillion yuan, with a year-on-year increase of 15.1% and being 5.9 percentage points higher than the national value. LI's added value was 98 billion yuan, with a year-on-year increase of 13.9%. The pace of transformation and upgrading is also accelerating. However, Anhui Province, as a latecomer in the Yangtze River Delta region, still requires improvements in terms of the ecological operation mode of LI and social impact. Therefore, through measuring, evaluating, and enhancing LI's ESE-B composite system in Anhui Province, this paper proposes policy recommendations to promote the coordinated development of LI's ESE-B composite system in this province in light of the appraisal results. This is of incredible importance to advance the SDLI of Anhui Province. However, this case study is restricted to Anhui Province only, and it can only be used as a reference by other provinces and cities. The condition underlying this case study is the uncoordinated development of ESE-B complex system of the logistics industry in Anhui Province.

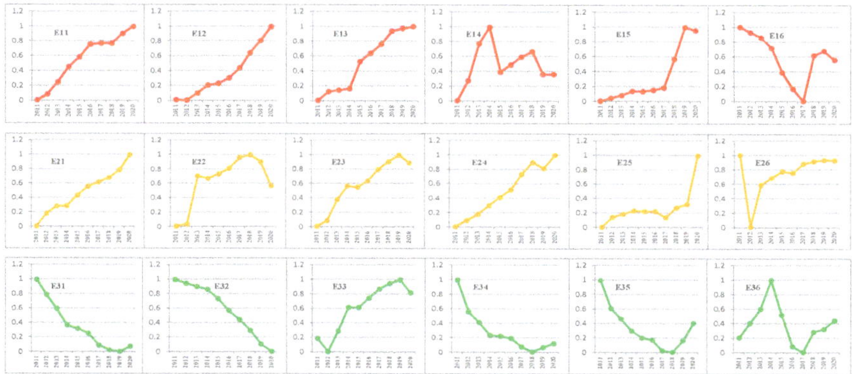

Figure 4. Orderliness of sequential parametric indicators of ESE-B composite system.

3.2. Data Analysis Results

(1)   Normalized values of sequential parameters

For the collected LI ESE-B composite system index data in Anhui Province, standardized values were calculated using Formulas (1)–(3) (i.e., the z-score method), as shown in Table A1. See Appendix A.

(2)   Orderliness of sequential parametric indicators

In this study, when calculating the orderliness of subsystem sequential parametric indexes, we use our own comparison method to set the lower limit of the index threshold as the minimum value and the upper limit as the maximum value of the same index. It is expressed as 1.01 times of the minimum and 1.01 times of the maximum values, respectively. According to Equation (4) and the data in Table A1, the orderliness of the sequential parameter indicators of the LI's ESE-B composite system in Anhui Province is calculated, as shown in Figure 4.

From Figure 5, the overall orderliness levels of the sequential covariates are low, but most of them show an upward trend from 2011 to 2020; the cargo turnover (E14), human resources input (E22), road traffic noise (E36) show an upward and then downward trend; the contribution rate of the LI (E16) and traffic accident property damage (E26) show a downward and then upward trend. In general, the degree of orderliness is improving. However, energy consumption (E31), electricity consumption (E32), carbon emissions (E34), and exhaust gas emissions (E35) show a decreasing trend, which means that these sequential parameters are areas that need to be improved (i.e., the orderliness of environmental benefits needs to be improved).

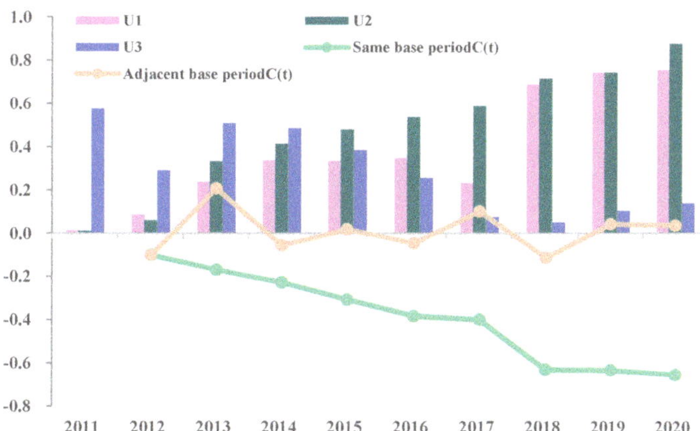

Figure 5. Orderliness of logistics industry ESE-B subsystems and synergy of composite systems.

(3)   Calculation results of the orderliness and synergy degree

According to Equation (5) and the data in Figure 5, the orderliness of the ESE-B subsystems of LI in Anhui Province during 2011–2020 are calculated, with 2011 as the base year. The coordination degree of the same and adjacent base periods of LI's ESE-B composite system in Anhui Province from 2012 to 2020 is calculated according to Equations (6) and (7), as shown in Figure 5.

Based on a single system, in terms of the economic efficiency of a subsystem, orderliness was only 0.012 in 2011 and reached 0.755 in 2020, with an average annual increase of 6.19%. Except for the decline in 2017 compared with 2016, the rest of the years showed an uptrend. The justification for this is that the orderliness of the LI's contribution (E16) decreased in 2017, because the country and provinces and cities began to promote the construction of ecological civilization and comprehensively deploy the green and low-carbon development of LI; as a result LI began to focus on environmental benefits, and relatively speaking, the economic development speed decreased. However, after two or three years of coordinated development, the economic benefits subsystem orderliness began to increase in 2018 [51]. The orderliness of social benefits was only 0.011 in 2011 and reached a peak of 0.877 in 2020, with an average annual growth of 8.66%, which shows that the LI of Anhui Province is relatively good in terms of social benefits. However, the orderliness of social benefits can be improved even further if human resource input is increased. The environmental benefit orderliness shows an opposite trend to those of economic and social benefits, being 0.575 in 2012, reaching the minimum of 0.051 in 2018, and increasing to 0.139 in 2020, which indicates that the LI in Anhui Province has begun to pay attention to environmental protection, actively adopt new technologies and equipment, change the previous rough development mode, and gradually move toward a new logistics mode of green and collaborative development.

Based on the overall trend, between 2012 and 2020, the synergy (using the same base period) of LI's ESE-B composite system in Anhui Province showed a decreasing trend year by year, reaching a minimum value of −0.655 in 2020, indicating a serious non-synergistic state. At the same time, the synergy degree of LI's ESE-B composite system in Anhui Province shows a jump forward with insignificant increase, except for 2013 when it reached a moderate synergistic evolution. The rest of the analyzed years are alternating between non-synergistic evolution and low synergistic evolution. This indicates that the synergy degree of LI's ESE-B composite system in Anhui Province needs to be improved, as does the stability degree. According to the order degree of each subsystem, the main reason for the serious non-synergy of LI's ESE-B composite system in Anhui Province is that the order degree of the environmental benefit subsystem has been decreasing in almost all years and the improvement of the environmental protection capability of the LI is the top priority in this province at present.

(4)   Calculation results of two-synergy degree of each subsystem

From Figure 6, the synergy of LI's ESE-B subsystems in Anhui is relatively stable in the same base period; basically, the synergy of two subsystems (economic and social benefits) increased year by year, but the synergy of two subsystems (environmental benefits and other subsystems) is not only negative, but also decreased year by year, which caused the synergy of the ESE-B composite system to decrease year by year. The synergy of LI's ESE-B subsystems in Anhui shows a repeated trend of "rise–fall" in the adjacent base period, which also causes the synergy of the ESE-B composite system to show almost the same characteristics. This indicates that the development of the environmental efficiency subsystem and other subsystems of the LI in Anhui Province is non-synergistic, which hinders the improvement of synergy, and indicates that environmental problems need to be solved urgently.

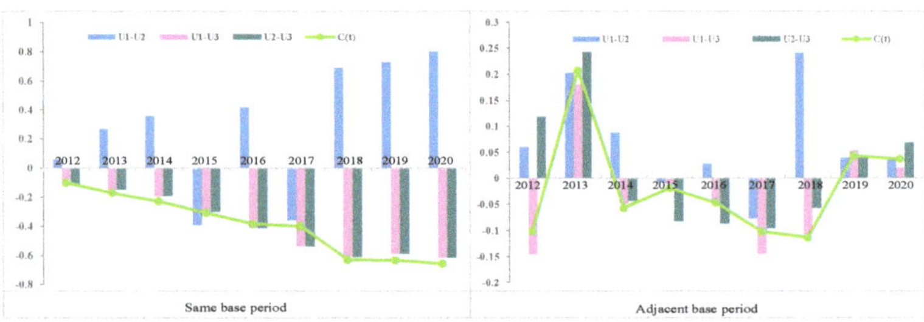

Figure 6. Synergy degree of two-two subsystem of logistics industry ESE-B.

3.3. Policy Recommendations

The synergistic development LI's ESE-B is of extraordinary and useful importance for sustainable development. The measurements and evaluation in Anhui Province from 2011 to 2020 continue to decline and the synergy degree within each subsystem varies and is unstable. The following suggestions are proposed to promote the coordinated development of LI's ESE-B complex system from three aspects: policy, intelligent systems, and energy, respectively. We thus propose the following countermeasures:

(1)   Formulate corresponding LI policies to promote the coordinated development of ESE-B

LI's ESE-B composite system relies upon the increase of each subsystem order degree and the order parameter. From the calculation results, problems arise due to the order degree decline of the environmental benefit subsystem. The synergy degree of the composite system is declining overall. Therefore, the government should give full play to its guiding role and formulate comprehensive policies for the development of the three aspects of the LI (ESE-B) to provide a solid policy platform for the synergistic development of LI's ESE-B complex system.

(2)   Promote the modernization of the LI and the development of intelligent logistics

The average orderliness of LI's social benefit subsystem in Anhui Province from 2011 to 2020 is 0.48, which is greater than those of the economic and environmental benefit subsystems, but the property loss from traffic accidents in the social benefit subsystem first decreased and then increased during 2011–2020, reaching the lowest value of 0.0739 in 2012. This result is mainly attributed to the lack of in-depth application of logistics information technology. As such, the construction of intelligent logistics is imperative. Anhui Province should establish a data processing center based on big data and cloud computing [51] to integrate the resources of the logistics operation process from the three aspects of supply, demand, and supervision to achieve the optimal fit of each subsystem in LI's ESE-B composite system.

(3)   Optimize the energy structure of the LI and enhance environmental orderliness

The orderliness of LI's environmental benefit subsystem in Anhui Province is much lower than that of the economic and social benefit subsystems, with an average value of only 0.29, and it plays a decisive role in the synergy with other subsystems and in the total system. Achieving carbon neutrality is a common development goal worldwide, meaning that environmental benefits are still something to which Anhui Province needs to continue to pay attention. The LI in Anhui Province should expand the application of new energy and clean energy such as electricity and natural gas; develop multimodal transportation and other ways to optimize transportation; support, cultivate, and introduce specialized and low-carbon logistics enterprises; promote the diversification and clean transformation of energy consumption in the LI; establish waste emission standards for the LI; force a low-carbon transformation; and promote the ecological development of the LI.

3.4. Discussion

Comparison with existing literature [21,22]: the present paper makes the following contributions. First, it constructs a more comprehensive evaluation index system of LI's ESE-B composite system from the viewpoint of synergy theory, systems theory, and input-output. Further, considering the three benefits of LI provides a more systematic and scientific approach. Second, the data-driven approach and the CSSMD are used to systematically and comprehensively consider the development of LI and more objectively quantify and evaluate the level of LI's ESE-B composite system synergy development. This approach provides a scientific decision-making framework for relevant subjects and practitioners. Finally, to achieve sustainable development, we propose targeted countermeasures and ideas to advance the synergistic development of LI's ESE-B composite system, which is of practical significance. However, the proposed method may not yield the obtained results in other provinces, cities or at other times, and the results require further temporal and spatial scaling.

Combining the findings, we can draw three management insights as follows:

First, to promote LI's ESE-B synergistic development according to the point of view of systems theory and overall optimization, we cannot depend on a single part of a larger system. Instead, it is necessary to bring into full play the back and forth support of the three parts of larger systems to effectively help increase the SDLI.

Second, with social, economic, and technological development, the foundation of a data-driven estimation and assessment method for the collaborative development of LI's ESE-B composite system can bring into full play of the advantages of each subsystem to achieve complementary strengths and weaknesses and collaborative development. As such, we can more effectively serve local government decision-making and local related enterprise management.

Finally, the synergistic development of LI's ESE-B composite system is crucial to promoting the sustainable development of the region and it is additionally important for the public authorities and significant divisions to mutually advance the joining and improvement of every subsystem and to understand the profoundly interacting and synergistic advancement of LI's ESE-B composite system.

4. Conclusions

Improving the coordinated development level of LI's complex system is important to ensure sustainable development and the SDLI is one of the important ways to protect the environment, save resources, and benefit society in today's world. To deal with the above research difficulties, this paper presents a data-driven research method and optimization countermeasure suggestions based on the data-driven coordination of LI's composite system, drawing on the exploratory thoughts of [52,53]. This approach has both theoretical and practical significance.

The theoretical significance of this study is as follows. (1) It enriches the connotation of the LI composite system index system and constructs a three dimensional index system of the ESE-B of the LI from the input-output perspective. When constructing the index system, social factors such as traffic property loss and road traffic noise and environmental factors such as carbon and waste emissions are considered. (2) Based on the data-driven theory, the composite system synergy model is used for analysis and the assessment results are accurate and objective, better reflecting the degree of interdependence and interaction among the ESE-B of the LI. (3) The internal mechanism of the economic, social, and environmental development of LI is explained from the synergy perspective, which improves the theoretical framework for the synergistic development of the LI's ESE-B. The practical significance of the results can be summarized as follows. (1) This study provides a quantitative basis for measuring, evaluating, and identifying the coordination level of the LI's ESE-B complex system. (2) It also provides reference for promoting the SDLI. (3) Finally, it offers new research ideas to researchers and policy makers in the LI field. This is also the innovation of this paper. The method may be used in the future to study the coordinated development of economic, environmental and social benefits of the LI and promote the future role of LI policies in supporting sustainable development.

The elements involved in LI's ESE-B composite system are complex, and the study of the synergistic development of each subsystem is complex. Hence, this study is not without limitations. (1) With the development of economy and society, the LI's composite system will include more elements. In future research, it is necessary to further make the index of the composite system more complete and accurate. (2) Further, the data capacity needs to be expanded to improve the representativeness of the sample by including more provinces and cities, as well as a longer sample period. (3) Finally, the mechanisms influencing the LI's complex system coordination need to be further explored. More effective countermeasures and suggestions need to be proposed to improve the coordination level, in support of regional sustainable development.

Author Contributions: Conceptualization, W.M.; methodology, W.M. and J.X.; investigation, W.G.; resources, W.G.; data curation, W.M.; writing—original draft preparation, W.M. and H.D.; writing—review and editing, J.X. and H.D. All authors have read and agreed to the published version of the manuscript.

Funding: This research is supported by key projects of Humanities and Social Sciences Research in Anhui Provincial Education Department (SK2016A1007 and SK2020A0539). Transportation rationalization consulting and transportation optimization services of Suzhou Chihan Transportation Co.(2022xhx150).

Data Availability Statement: Not applicable.

Conflicts of Interest: The authors declare no conflict of interest.

Appendix A

Table A1. Sequence parameters standardized values of ESE-B composite system.

Year	E11	E12	E13	E14	E15	E16	E21	E22	E23	E24	E25	E26	E31	E32	E33	E34	E35	E36
2011	−1.676	−1.127	−1.407	−1.827	−0.901	1.326	−1.664	−1.890	−1.781	−1.452	1.056	−0.913	−1.989	−1.207	−1.314	−2.486	−2.318	0.678
2012	−1.435	−1.142	−1.092	−0.811	−0.786	1.100	−1.051	−1.807	−1.530	−1.199	0.521	2.688	−1.347	−1.047	−1.864	−0.963	−0.977	−0.059
2013	−0.944	−0.842	−1.034	1.055	−0.686	0.871	−0.706	0.192	−0.639	−0.949	0.345	0.595	−0.765	−0.929	−0.999	−0.428	−0.458	−0.796
2014	−0.328	−0.517	−0.985	1.918	−0.530	0.407	−0.681	0.090	−0.052	−0.588	0.169	0.229	−0.039	−0.808	0.024	0.189	0.131	−2.270
2015	0.085	−0.443	0.001	−0.378	−0.537	−0.667	−0.170	0.283	−0.097	−0.252	0.201	−0.110	0.101	−0.432	0.024	0.240	0.463	−0.501
2016	0.601	−0.217	0.301	−0.012	−0.489	−1.381	0.247	0.519	0.177	0.058	0.201	−0.037	0.302	0.051	0.417	0.340	0.564	1.120
2017	0.641	0.198	0.630	0.384	−0.402	−1.912	0.469	0.968	0.671	0.717	0.529	−0.506	0.805	0.420	0.810	0.750	1.088	1.415
2018	0.651	0.832	1.108	0.661	0.681	0.085	0.686	1.076	1.001	1.212	0.002	−0.610	1.007	0.855	1.046	0.998	1.148	0.383
2019	1.057	1.337	1.210	−0.494	1.884	0.283	1.074	0.790	1.294	0.955	−0.182	−0.679	1.070	1.401	1.204	0.774	0.598	0.236
2020	1.347	1.920	1.269	−0.497	1.765	−0.112	1.797	−0.221	0.955	1.498	−2.841	−0.657	0.857	1.694	0.653	0.586	−0.240	−0.206

References

1. Chen, J.X.; Chen, J. From optimization to reinvention—High quality development of supply chain in the big change. *Syst. Eng. Theory Pract.* **2022**, *42*, 545–558. [CrossRef]
2. Peng, Z.W. Motivation and Countermeasures of Global Value Chain Adjustment, 20 January 2021. Available online: www.cssn.cn (accessed on 20 July 2022).
3. Li, M.J.; Wang, J. Spatial-temporal evolution and influencing factors of total factor productivity in China's logistics industry under low-carbon constraints. *Environ. Sci. Pollut. Res.* **2022**, *29*, 883–900. [CrossRef] [PubMed]

4. Dong, Q.L.; Bai, D.L.; Wang, D.F. Eco-Efficiency and pollution reducing potential of Logistics industry in the Yellow River Basin. *Ecol. Econ.* **2021**, *37*, 34–42.

5. Cooper, J. The paradox of logistics in Europe. *Int. J. Logist. Manag.* **1991**, *2*, 42–54. [CrossRef]

6. Quesada-Mateo, C.A.; Solís-Rivera, V. Costa Rica's national strategy for sustainable development: A summary. *Futures* **1990**, *22*, 396–416. [CrossRef]

7. Cao, C.L. Measuring Sustainable Development Efficiency of Urban Logistics Industry. *Math. Probl. Eng.* **2018**, *2018*, 9187541. [CrossRef]

8. Wu, H.J.; Dunn, S.C. Environmentally responsible logistics systems. *Int. J. Phys. Distrib. Logist. Manag.* **1995**, *25*, 20–38. [CrossRef]

9. Zaman, K.; Shamsuddin, S. Green logistics and national scale economic indicators: Evidence from a panel of selected European countries. *J. Clean. Prod.* **2017**, *143*, 51–63. [CrossRef]

10. Abareshi, A.; Molla, A. Greening logistics and its impact on environmental performance: An absorptive capacity perspective. *Int. J. Logist. Res. Appl.* **2013**, *16*, 209–226. [CrossRef]

11. Ramos, T.R.P.; Gomes, M.I.; Barbosa-Povoa, A.P. Planning a sustainable reverse logistics system: Balancing costs with environmental and social concerns. *Omega* **2014**, *48*, 60–74. [CrossRef]

12. Mota, B.; Gomes, M.I.; Carvalho, A.; Barbosa-Povoa, A.P. Towards supply chain sustainability: Economic, environmental and social design and planning. *J. Clean. Prod.* **2014**, *105*, 14–27. [CrossRef]

13. Ding, H.P.; Liu, Y.; Zhang, Y.; Wang, S.; Guo, Y.; Zhou, S.; Liu, C. Data-driven evaluation and optimization of the sustainable development of the logistics industry: Case study of the Yangtze River Delta in China. *Environ. Sci. Pollut. Res.* **2022**, *29*, 68815–68829. [CrossRef] [PubMed]

14. Gordon, I.R.; McCann, P. Industrial clusters: Complexes, agglomeration and/or social networks? *Urban Stud.* **2000**, *37*, 513–532. [CrossRef]

15. Hu, Y.; Shu, H. Synergetic mechanism of agricultural logistics ecosphere—The case study based on Jiangxi Taoxin. *Nankai Bus. Rev. Int.* **2022**; *ahead-of-print*. [CrossRef]

16. Liang, W.; Gu, D.C. Study on the synergistic development of agriculture and logistics industry and key influencing factors. *J. Yunnan Agric. Univ. Soc. Sci.* **2021**, *15*, 93–101. [CrossRef]

17. Klaus, P. Logistics research: A 50 years' march of ideas. *Logist. Res.* **2009**, *1*, 53–65. [CrossRef]

18. Chen, Y.; Lan, S.L.; Tseng, M.L. Coordinated development path of metropolitan logistics and economy in Belt and Road using DEMATEL-Bayesian analysis. *Int. J. Logist. Res. Appl.* **2018**, *22*, 1–24. [CrossRef]

19. Perez-Suarez, R.; Lopez-Menendez, A.J. Growing green? Forecasting CO_2 emissions with environmental Kuznets curves and logistic growth models. *Environ. Sci. Policy* **2015**, *54*, 428–437. [CrossRef]

20. Liu, J.; Yuan, C.; Hafeez, M.; Yuan, Q. The relationship between environment and logistics performance: Evidence from Asian countries. *J. Clean. Prod.* **2018**, *204*, 282–291. [CrossRef]

21. Zhou, T. Research on the coordinated development of regional logistics and ecological environment from the perspective of a low carbon. *Stat. Inf. Forum* **2021**, *36*, 62–72.

22. Evangelista, P. Environmental sustainability practices in the transport and logistics service industry: An exploratory case study investigation. *Res. Transp. Bus. Manag.* **2014**, *12*, 63–72. [CrossRef]

23. Lean, H.; Huang, W.; Hong, J.J. Logistics and economic development: Experience from China. *Transp. Policy* **2014**, *32*, 96–104. [CrossRef]

24. Chen, D.Q.; Ignatius, J.; Sun, D.Z.; Zhan, S.L.; Zhou, C.Y.; Marra, M.; Demirbag, M. Reverse logistics pricing strategy for a green supply chain: A view of customers' environmental awareness. *Int. J. Prod. Econ.* **2019**, *217*, 197–210. [CrossRef]

25. Shu, L.L.; Chen, Y.; Huang, G.Q. Data analysis for metropolitan economic and logistics development. *Adv. Eng. Inform.* **2017**, *32*, 66–76. [CrossRef]

26. Yu, S.; Yin, C. Evaluating the coordinated development of economic, social and environmental benefits of urban public transportation infrastructure: Case study of four Chinese autonomous municipalities. *Transp. Policy* **2018**, *66*, 116–126. [CrossRef]

27. Li, X.L.; Sohail, S.; Majeed, M.T.; Ahmad, W. Green logistics, economic growth, and environmental quality: Evidence from one belt and road initiative economies. *Environ. Sci. Pollut. Res.* **2021**, *28*, 30664–30674. [CrossRef] [PubMed]

28. Long, R.Y.; Ouyang, H.Z.; Guo, H.Y. Super-slack-based measuring data envelopment analysis on the spatial-temporal patterns of logistics ecological efficiency using global malmquist index model. *Environ. Technol. Innov.* **2020**, *18*, 100770–100784. [CrossRef]

29. Deng, F.M.; Xu, L.; Fang, Y.; Gong, Q.; Li, Z. PCA-DEA-Tobit Regression Assessment with Carbon Emission Constraints of China's Logistics Industry. *J. Clean. Prod.* **2020**, *271*, 12548. [CrossRef]

30. Müller, E.; Stock, T.; Schillig, R. A method to generate energy value-streams in production and logistics in respect of time- and energy-consumption. *Prod. Eng. Res. Devel.* **2014**, *8*, 243–251. [CrossRef]

31. Sun, Q. Empirical research on coordination evaluation and sustainable development mechanism of regional logistics and new-type urbanization: A panel data analysis from 2000 to 2015 for Liaoning Province in China. *Environ. Sci. Pollut. Res.* **2017**, *24*, 14163–14175. [CrossRef]

32. Yang, J.N.; Tang, L.; Mi, Z.F.; Liu, S.; Li, L.; Zheng, J.L. Carbon emissions performance in logistics at the city level. *J. Clean. Prod.* **2019**, *231*, 1258–1266. [CrossRef]

33. Alarcón, F.; Cortés-Pellicer, P.; Pérez-Perales, D.; MengualRecuerda, A. A Reference Model of Reverse Logistics Process for Improving Sustainability in the Supply Chain. *Sustainability* **2021**, *13*, 10383. [CrossRef]

34. Jiang, X.H.; Ma, J.X.; Zhu, H.Z.; Guo, X.C.; Huang, Z.G. Evaluating the Carbon Emissions Efficiency of the Logistics Industry Based on a Super-SBM Model and the Malmquist Index from a Strong Transportation Strategy Perspective in China. *Int. J. Environ. Res. Public Health* **2020**, *17*, 8459. [CrossRef] [PubMed]
35. Wu, C.X. Synergistic effects of low-carbon economic development in China. *Manag. World* **2021**, *8*, 105–116.
36. Feng, J.H.; Zhang, J.L.; Tang, L. Study on the coupling and coordinated development of agricultural economy ecology society composite system: Take Shaanxi Province as an example. *J. Syst. Sci.* **2021**, *29*, 92–96.
37. Cao, B.R.; Kong, Z.Y.; Deng, L.J. A study on provincial logistics efficiency and spatial and temporal evolution in the Yangtze River Economic Belt. *Geoscience* **2019**, *39*, 1841–1848.
38. Wang, Y.F. Research on the Evaluation of High-Quality Development of the Logistics Industry. Master's Dissertation, Henan University of Technology, Zhengzhou, China, 2020. [CrossRef]
39. Energy Statistics Division of the National Bureau of Statistics. *China Energy Statistical Yearbook*; China Statistics Press: Beijing, China, 2012–2021.
40. Anhui Provincial Bureau of Statistics. *Anhui Provincial Statistical Yearbook*; Anhui Statistics Publishing House: Anhui, China, 2012–2021.
41. Sun, H.; Hu, X.Y.; Nie, F.F. Spatio-temporal Evolution and Socio-economic Drivers of Primary Air Pollutants from Energy Consumption in the Yangtze River Delta. *China Environ. Manag.* **2019**, *11*, 71–78. [CrossRef]
42. IPCC. 2006 IPCC Guidelines for National Greenhouse Gas Inventories. Available online: http://www.ipcc-nggip.iges.or.jp/public/2006gl/index.html (accessed on 1 August 2022).
43. Ramanathan, R. The moderating roles of risk and efficiency on the relationship between logistics performance and customer loyalty in e-commerce. *Transp. Res. Part E* **2010**, *46*, 950–962. [CrossRef]
44. Zhao, J.Q.; Xiao, Y.; Sun, S.Q.; Sang, W.G.; Axmacher, J.C. Does China's increasing coupling of "urban population" and "urban area" growth indicators reflect a growing social and economic sustainability? *J. Environ. Manag.* **2022**, *301*, 113932. [CrossRef]
45. Liu, X.L.; Guo, P.B.; Yue, X.H.; Zhong, S.C.; Cao, X.Y. Urban transition in China: Examining the coordination between urbanization and the eco-environment using a multi-model evaluation method. *Ecol. Indic.* **2021**, *130*, 108056. [CrossRef]
46. Cheng, X.; Long, R.Y.; Chen, H.; Li, Q.W. Coupling coordination degree and spatial dynamic evolution of a regional green competitiveness system–A case study from China. *Ecol. Indic.* **2019**, *104*, 489–500. [CrossRef]
47. Fang, X.; Shi, X.Y.; Phillips, T.K.; Du, P.; Gao, W.J. The coupling coordinated development of urban environment towards sustainable urbanization: An empirical study of Shandong Peninsula, China. *Ecol. Indic.* **2021**, *129*, 107864. [CrossRef]
48. Wu, X.; Wenqi, C.; Xinyu, D. Coupling and coordination of coal mining intensity and social-ecological resilience in China. *Ecol. Indic.* **2021**, *131*, 108167. [CrossRef]
49. Ariken, M.; Zhang, F.; Liu, K.; Fang, C.L.; Kung, H.T. Coupling coordination analysis of urbanization and eco-environment in Yanqi Basin based on multi-source remote sensing data. *Ecol. Indic.* **2020**, *114*, 106331. [CrossRef]
50. Li, J. Reconstruction of intelligent logistics model based on big data and cloud computing. *China Circ. Econ.* **2019**, *33*, 20–29. [CrossRef]
51. Zheng, H.; Khan, Y.A.; Abbas, S.Z. Exploration on the coordinated development of urbanization and the eco-environmental system in central China. *Environ. Res.* **2021**, *204*, 112097. [CrossRef] [PubMed]
52. Liu, C.; Gao, M.; Zhu, G.; Zhang, C.; Cai, W. Data driven eco-efficiency evaluation and optimization in industrial production. *Energy* **2021**, *224*, 120170. [CrossRef]
53. Wang, C.; Zhang, C.; Hu, F.; Wang, Y.; Yu, L.; Liu, C. Emergy-based ecological efficiency evaluation and optimization method for logistics park. *Environ. Sci. Pollut. Res.* **2021**, *28*, 58342–58354. [CrossRef]

Article

A Novel Hybrid Model of CNN-SA-NGU for Silver Closing Price Prediction

Haiyao Wang [1], Bolin Dai [2], Xiaolei Li [2], Naiwen Yu [3,*] and Jingyang Wang [2,4]

1. School of Ocean Mechatronics, Xiamen Ocean Vocational College, Xiamen 361100, China
2. School of Information Science and Engineering, Hebei University of Science and Technology, Shijiazhuang 050018, China
3. FedUni Information Engineering Institute, Hebei University of Science and Technology, Shijiazhuang 050018, China
4. Hebei Intelligent Internet of Things Technology Innovation Center, Shijiazhuang 050018, China
* Correspondence: naiwenyu_hebust@163.com

Abstract: Silver is an important industrial raw material, and the price of silver has always been a concern of the financial industry. Silver price data belong to time series data and have high volatility, irregularity, nonlinearity, and long-term correlation. Predicting the silver price for economic development is of great practical significance. However, the traditional time series prediction models have shortcomings, such as poor nonlinear fitting ability and low prediction accuracy. Therefore, this paper presents a novel hybrid model of CNN-SA-NGU for silver closing price prediction, which includes conventional neural networks (CNNs), the self-attention mechanism (SA), and the new gated unit (NGU). A CNN extracts the feature of input data. The SA mechanism captures the correlation between different eigenvalues, thus forming new eigenvectors to make weight distribution more reasonable. The NGU is a new deep-learning gated unit proposed in this paper, which is formed by a forgetting gate and an input gate. The NGU's input data include the cell state of the previous time, the hidden state of the previous time, and the input data of the current time. The NGU learns the previous time's experience to process the current time's input data and adds a *Tri* module behind the input gate to alleviate the gradient disappearance and gradient explosion problems. The NGU optimizes the structure of traditional gates and reduces the computation. To prove the prediction accuracy of the CNN-SA-NGU, this model is compared with the thirteen other time series forecasting models for silver price prediction. Through comparative experiments, the mean absolute error (MAE) value of the CNN-SA-NGU model is 87.898771, the explained variance score (EVS) value is 0.970745, the r-squared (R^2) value is 0.970169, and the training time is 332.777 s. The performance of CNN-SA-NGU is better than other models.

Keywords: NGU; self-attention; CNN; silver prediction; deep learning

Citation: Wang, H.; Dai, B.; Li, X.; Yu, N.; Wang, J. A Novel Hybrid Model of CNN-SA-NGU for Silver Closing Price Prediction. *Processes* **2023**, *11*, 862. https://doi.org/10.3390/pr11030862

Academic Editors: Conghu Liu, Xiaoqian Song, Zhi Liu, Fangfang Wei and Tsai-Chi Kuo

Received: 7 February 2023
Revised: 7 March 2023
Accepted: 11 March 2023
Published: 14 March 2023

1. Introduction

In recent years, investors began to notice silver, and silver investment has become a means of financial management. Silver remains an essential part of financial markets, often playing dual roles as an investment product and an industrial metal. However, the epidemic has recently affected the economy, resulting in volatile silver prices. Therefore, accurate prediction of silver prices is significant to economic development.

Silver price prediction is a time series problem [1], which predicts the possible future price of silver according to the actual price data of the silver market. The change in silver price is relevant to the formulation of laws, the development of the world economy, political events, investors' psychology, etc. These factors lead to the fluctuation of the price of silver, which makes it difficult to accurately predict silver price [2,3]. Traditional machine learning methods, such as decision trees [4], support vector machines (SVM) [5], and genetic algorithms [6], are applied to time series data prediction. However, there are problems with all of these approaches, such as poor processing of special values in time series data and the poor nonlinear fitting ability of data [7]. As technology develops, more and more deep learning methods are applied to time series prediction. Deep learning algorithms can better fit the changes of nonlinear time series data [8].

The importance of different feature data is different in the actual training process. Because different eigenvalues have different influences on the prediction results, some important eigenvalues should be given greater training weight in the training process [9,10]. The SA mechanism is added to the silver price forecasting model, which can select a small number of important eigenvalues from feature data. The process of selecting is reflected in the calculation of the weight coefficient. The greater the influence of characteristic data on prediction results, the bigger the weight coefficient. The weight coefficient represents the importance of feature data. After introducing the SA mechanism, it is easier to capture the interdependent features among different characteristic data, thus improving the sensitivity of the model to important eigenvalues [11].

The NGU includes the forgetting gate and the input gate. The forgetting gate determines how much cell state from the previous time is retained in the current cell state. The input gate determines how much input data at the current time can be saved to the current cell state. The input data of each gate include the hidden state of the previous time, the cell state of the previous time, and the input data of the current time. The forgetting gate and the input gate learn the experience of the previous time to process the input data of the present time, which improves the prediction accuracy of the model. The *Tri* conversion module processes the data of the input gate, which significantly changes the output data value and alleviates the problems of gradient disappearance and gradient explosion.

Therefore, this paper presents a new neural network model to predict the closing price of silver. The CNN-SA-NGU time series prediction model is constructed by a CNN, SA mechanism, and NGU. The CNN processes the input data and extracts the features of the data. The SA mechanism is used to compute the importance of different feature data. Additionally, the important features are assigned larger weight coefficients so that the weight distribution is more reasonable. The NGU is used to forecast the silver closing price. To certify the validity of CNN-SA-NGU, this model is compared with the prediction results of Prophet, support vector regression (SVR), multi-layer perceptron (MLP), autoregressive integrated moving average mode (ARIMA), long short-term memory (LSTM), bi-directional long short-term memory (Bi-LSTM), gated recurrent unit (GRU), NGU, CNN-LSTM, CNN-GRU, CNN-NGU, CNN-NGU, CNN-SA-LSTM, and CNN-SA-GRU. The innovations and main contributions of this paper are as follows:

(1) This paper presents a new neural network NGU, which includes a forgetting gate and an input gate. The input data of each gate includes the hidden state of the previous time, the cell state of the previous time, and the input data of the current time. The NGU learns the previous moment's experience to process the current moment's input data, which improves the prediction accuracy of the model. The *Tri* data conversion module in the NGU alleviates the problems of gradient disappearance and gradient explosion. The NGU has a simple structure and few parameters to be calculated, so the training time is short. The NGU is mainly used to predict time series.

(2) In the silver price prediction experiment, the SA mechanism is applied to the model, which can improve the unreasonable distribution of weights and facilitate the gate unit to learn the law of silver price data.

(3) This paper presents a new silver price forecasting model: CNN-SA-NGU. Under the same experimental conditions and data, the silver price forecasting results of CNN-SA-NGU are better than other models.

2. Related Work

Yuan et al. [12] predicted the gold future price using least square support vector regression improved by the genetic algorithm. The SVR is unsuitable for large data sets. Additionally, when time series data sets have noise, the problem of overlapping target classes will occur. Aksehir et al. [13] put forward a prediction model of the Dow Jones index stock trend based on a CNN, which achieved good results in predicting stock trends. The study showed that the CNN algorithm performed well in extracting data features. However, the performance of the CNN is poor on small data sets [14].

Chen et al. [15] put forward a new model combining SVM and LSTM. This model used entropy space theory and price factors that may affect the gold price to predict the gold price. The experimental results show the price prediction of gold is good. There are too many parameters in the LSTM, leading to much calculation [16]. Combined LSTM and CNN can enhance the prediction of gold volatility [17]. By inputting time series data into the convolution layer, the features of data can be extracted better.

E et al. [18] presented a combination technique based on independent component analysis (ICA) and gate recurrent unit neural network (GRUNN), called ICA-GRUNN, to forecast the gold price. ICA is a multi-channel mixed signal analysis technology. The original time series data are decomposed into virtual multi-channel mixed signals by variational mode decomposition (VMD) technology. Comparative experiments show that ICA-GRUNN has higher prediction accuracy.

The attention mechanism was applied to image classification for the first time and achieved good results [19]. In 2017, the Google machine translation team abandoned recurrent neural networks (RNNs) and CNNs. The team implemented the translation task only using the attention mechanism, achieving an excellent translation effect. The attention mechanism can effectively capture the semantic relevance between all the words in context. To pursue better performance, Liu et al. [20] proposed a model based on a weighted pure attention mechanism. The authors introduced weight parameters into the artificially generated attention weight and transferred attention from other elements to key elements according to the setting of weight parameters. If the attention mechanism is not applied, long-distance information is weakened. The attention mechanism can give a higher weight to the feature data, which significantly influences the prediction results.

SA is also called intra-attention. Kim et al. [21] proposed a SAM-LSTM prediction model based on SA, which is composed of multiple LSTM modules and an attention mechanism. The SA mechanism gives different weight information to different parts of the input data. The change point detection technique is used to achieve the stability of prediction in the invisible price range. Finally, the model's effectiveness in cryptocurrency price prediction is impressive. To solve the problem that a fully connected neural network cannot establish a correlation for multiple related inputs, SA is used to make the machine notice the correlation between different parts of the data. After introducing SA, it is easy to capture the long-distance interdependent features in sentences. Wang et al. [22] presented a sentence-to-sentence attention network (S2SAN) using multi-threaded SA and carried out several emotion analysis experiments in specific fields, cross-fields, and multi-fields. Experimental results show that S2SAN is superior to other advanced models. Li et al. [23] improved the existing SA with the hard attention mechanism. The addition of the SA mechanism improves the autonomous learning ability of the model. The improved SA fully extracts the text's positive and negative information for emotion analysis. The improved SA can enhance the extraction of positive information and make up for the problem that the value in the traditional attention matrix cannot be negative. An RNN or LSTM needs to be calculated in sequence. For long-distance interdependent features, many calculations are needed to connect them. The farther the distance between features, the less likely it is to capture effectively [24]. SA connects any two words in a sentence directly through one calculation. Therefore, the distance between long-distance dependent features is shortened [25].

3. Models

3.1. SA

The SA mechanism determines the weight coefficients of different eigenvalues by calculating the relationship between different eigenvalues of a piece of data. Additionally, the SA mechanism obtains new eigenvectors by recalculating. The new eigenvector takes more information into account and assigns higher weight coefficients to the eigenvalues that significantly influence the prediction results. The SA mechanism is beneficial to the NGU's prediction of the silver closing price. The principle of the SA mechanism is shown in Figure 1.

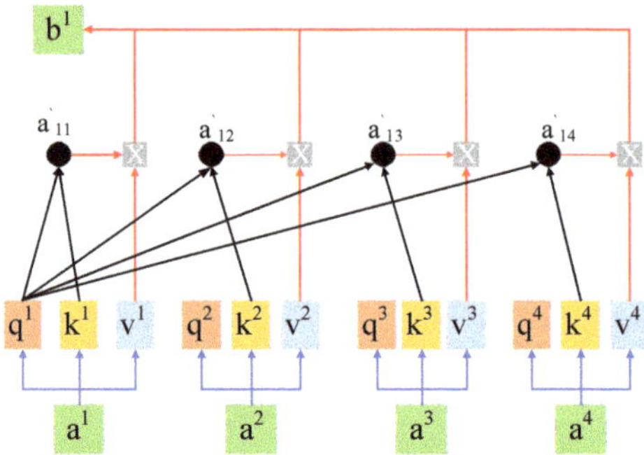

Figure 1. The principle of the SA mechanism.

An encoder encodes the feature data and the eigenvector a^i of the eigenvalue is obtained by nonlinear operation. The eigenvector is multiplied by the weight matrices of w^q, w^k, and w^v obtained by training to obtain query vector, key vector, and value vector, respectively. The calculation formulas are shown in (1), (2), and (3), respectively.

$$q^i = w^q \cdot a^i \tag{1}$$

$$k^i = w^k \cdot a^i \tag{2}$$

$$v^i = w^v \cdot a^i \tag{3}$$

where q^i is a query vector with the i-th eigenvalue; k^i is the key vector of the i-th eigenvalue; v^i is the value vector of the i-th eigenvalue. w^q, w^k, and w^v are the parameters obtained by model training. a_i is the eigenvector obtained by the encoder operation of the i-th eigenvalue.

a_{ij} is the similarity between the i-th and the j-th eigenvalues. The query vector of the i-th eigenvalue is multiplied by the key vector of the j-th eigenvalue, and the inner product of the two vectors is obtained. d is the dimension of the i-th eigenvalue key vector. After each element of the vector a^i divides by $\sqrt{d}$, the variance distribution becomes 1. Therefore, the gradient value in the training process remains stable. The formula for calculating a_{ij} is shown in (4).

$$a_{ij} = \frac{q^i \cdot k^j}{\sqrt{d}} \tag{4}$$

where k^j is the key vector of the j-th eigenvalue.

a'_{ij} is the weight coefficient between the i-th and the j-th eigenvalues. The weights between the i-th and other different eigenvalues need to be normalized to obtain their similarity. After the weight coefficients are normalized, the sum of the weight coefficients is 1. The calculation formula for calculating a'_{ij} is shown in (5).

$$a'_{ij} = \frac{\exp(a_{ij})}{\sum_{j=0}^{n} \exp(a_{ij})} \tag{5}$$

where $\exp(a_{ij})$ represents the exponential operation of e for a_{ij}. $\sum_{j=0}^{n} \exp(a_{ij})$ is the sum of the exponential power of e of all a_{ij} to obtain the sum of the weight coefficients of different eigenvalues. The weight coefficient vector of the i-th eigenvalue is obtained by division operation.

b^i is the output of the SA layer. The weight coefficient vector a'_{ij} of the i-th eigenvalue is multiplied by the v^i vector of the i-th eigenvalue to obtain the eigenvector. As the input of the NGU, b^i improves the model's sensitivity to important eigenvalues, thus improving the accuracy of forecasting the closing price of silver. The formula for calculating b^i is shown in (6).

$$b^i = \sum_{i=0}^{n} a'_{ij} \cdot v^i \tag{6}$$

3.2. NGU

Based on the in-depth study of the principle and structure of LSTM [26,27] and GRU [28,29], a new gated unit (NGU) is proposed in this paper. The NGU has a simple structure, including a forgetting gate and an input gate, and adds a *Tri* module. The structure diagram of the NGU is shown in Figure 2.

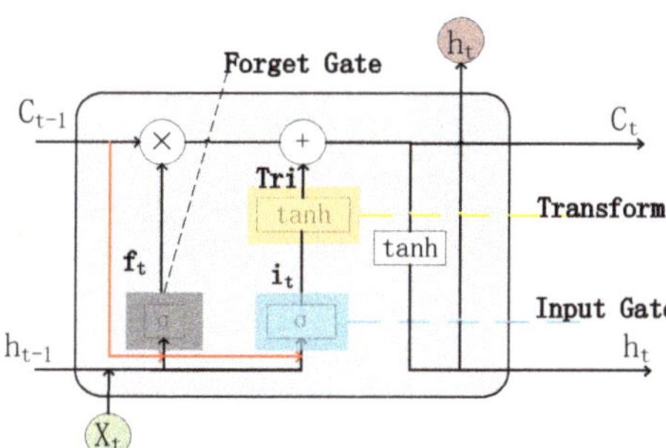

Figure 2. NGU structure diagram.

The function of the forgetting gate in the NGU determines how much cell state information can be kept from the previous time to the current time. The input data of the forgetting gate include the cell state of the previous time, the hidden state of the previous time, and the input data of the current time. The forgetting gate processes the input data through the sigmoid function, thus outputting the operation value. The sigmoid function's output value determines how much cell state information is retained from the previous time to the current time. The output value of the sigmoid function is 0 ~ 1, 0 means completely discarding the cell state at the previous time, and 1 means completely retaining the cell state from the previous time to the current time. The calculation formula of the forgetting gate is shown in Formula (7).

$$f_t = \sigma\left(w_{fh} \cdot h_{t-1} + w_{fx} \cdot x_t + w_{fc} \cdot c_{t-1} + b_f\right) \tag{7}$$

where σ represents the sigmoid activation function, h_{t-1} represents the hidden state at the previous time, x_t represents the input data at the current time, c_{t-1} represents the cell state at the previous time, and b_f is the bias vector. w_{fh}, w_{fx}, and w_{fc} correspond to the weight vectors obtained by training h_{t-1}, x_t, and c_{t-1}, respectively. The purpose of network training many times is to continuously adjust the values of these parameter vectors.

The function of the input gate in the NGU determines how much input data x_t can be saved to the cell state at the current time. The input data of the input gate include the cell state of the previous time, the hidden state of the previous time, and the input data of the current time. The input gate processes the input data through the sigmoid function, thus outputting the operation value. The sigmoid function's output value determines how much input data x_t is retained in the cell state at the current time. The calculation formula of the input gate is shown in Formula (8).

$$i_t = \sigma(w_{ih} \cdot h_{t-1} + w_{ix} \cdot x_t + w_{ic} c_{t-1} + b_i) \tag{8}$$

where b_i is the bias vector; w_{ih}, w_{ix} and w_{ic} correspond to the weight vectors obtained by training h_{t-1}, x_t, and c_{t-1}, respectively.

The input gate sigmoid function outputs data after the *Tri* conversion module operation as the output data to conduct output. After the input data are operated by the sigmoid function, the output result is between 0 and 1. When the input data of the sigmoid function is $(-\infty, 5)$ or $(5, \infty)$, the small variation of function value easily causes the problem of disappearing gradient, which is not conducive to the feedback transmission of deep neural networks. After the output data of the sigmoid function are processed by the tanh function, the output value will change significantly so as to improve the sensitivity of the model and alleviate the problem of gradient disappearance. The *Tri* calculation formula of the conversion module is shown in Formula (9).

$$Tri = \tan h(i_t) \tag{9}$$

The cell state c_t at the current time is the product of the output value of the forgetting gate and the cell state at the previous time plus the output value of the *Tri* module. The calculation of the cell state at the current time includes the cell state at the previous time. By learning the cell state at the previous time, the input data at the current time is processed by using the experience of historical data processing. The learning ability and nonlinear fitting ability of the NGU are improved. The formula for calculating c_t is shown in Formula (10).

$$c_t = f_t \cdot c_{t-1} + Tri \tag{10}$$

The calculation formula of the hidden state h_t at the current time is shown in Equation (11).

$$h_t = \tan h(c_t) \tag{11}$$

where h_t is also the current output of the NGU.

3.3. CNN-SA-NGU

The integral structure of the CNN-SA-NGU model for silver closing price prediction is shown in Figure 3.

Data preprocessing layer: Delete the data not needed for training (including trade_date, duplicate data, invalid data, and so on) in the original data set. Standardize the data in the data set, and convert the data of different specifications to the same value interval so as to reduce the influence of distribution difference on model training.

CNN layer: By convolution operation on the input data, the data's characteristics are extracted. The output of the CNN layer is passed to the SA layer as new input data.

SA layer: By calculating the feature data transmitted from the CNN layer, the weight coefficients are allocated, and new feature vectors are obtained.

NGU layer: This layer learns the law of silver price change and predicts silver's closing price.

Output layer: Through the inverse normalization operation of the data output from the NGU layer, the silver price prediction results of this model are output.

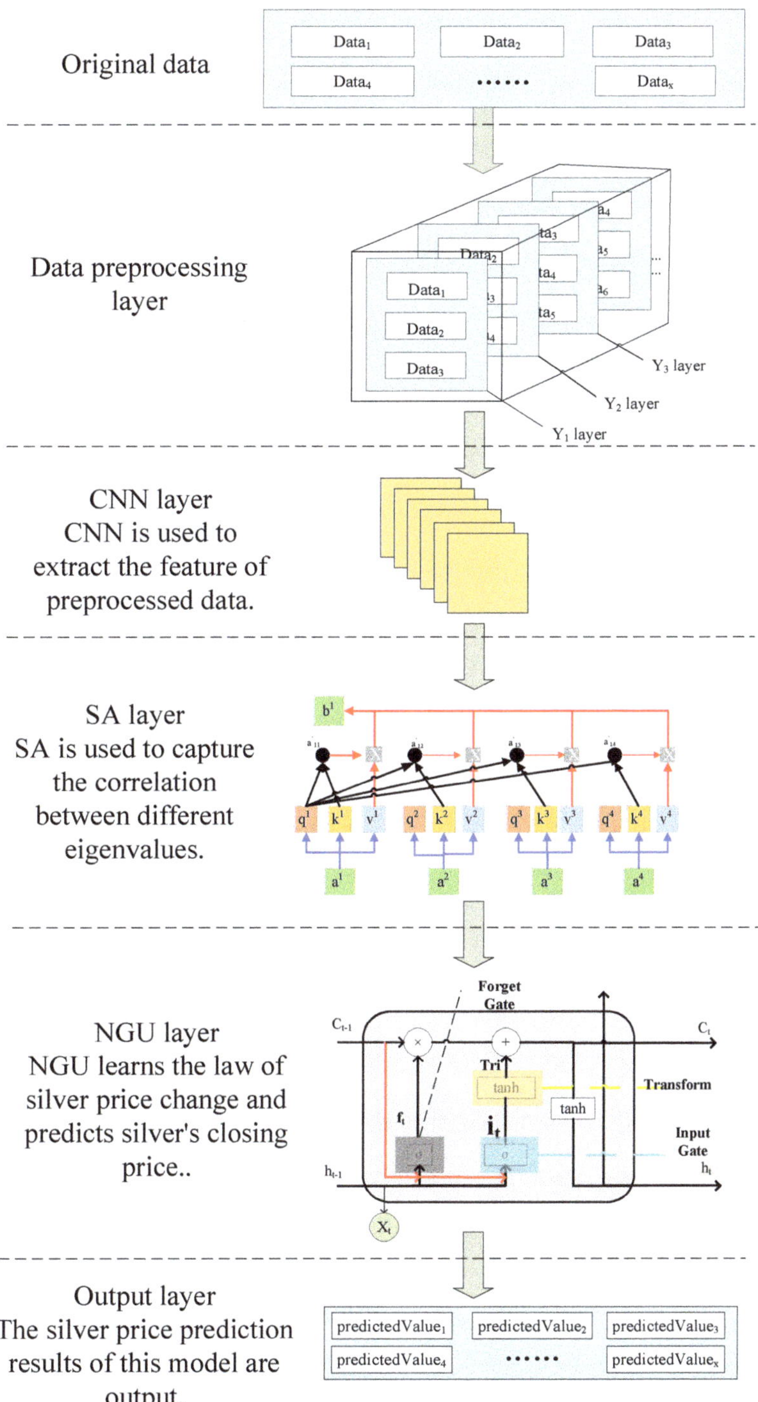

Figure 3. CNN-SA-NGU structure diagram.

4. Experiment

4.1. Experimental Environment

The hardware environment and software environment of this experiment are shown in Table 1.

Table 1. Experimental environment.

Environment Type	Project Name	Value
Hardware environment	Operating system	Windows 11
	CPU	Intel i7-12700H 2.30 GHz
	Memory	16GB
	Graphics card	RTX 3070Ti
Software environment	Development tools	PyCharm 2020 1.3
	Programming language	Python3.7.0
	Basic platform	Anaconda4.5.11
	Learning framework	keras2.1.0 and TensorFlow 1.14.0

4.2. Data Acquisition

In this experiment, the silver futures trading data of the Shanghai futures exchange from 5 January 2015 to 30 November 2022 are selected as experimental data. A total of 1925 pieces of data were collected. All the data collected in this experiment are obtained from the third-party data interface of the Tushare website, which is a data service platform. Silver futures price data are shown in Table 2.

Table 2. Silver futures price data items.

Trade_Date	Open	High	Low	Close	Change	Settle	Vol	Oi
5 January 2015	3498	3516	3478	3507	−17	3500	51379800	41042000
6 January 2015	3490	3566	3462	3554	54	3517	219997200	45015800
7 January 2015	3544	3596	3530	3554	37	3556	186587600	40197400
8 January 2015	3540	3578	3537	3548	−8	3558	143412200	41246400
9 January 2015	3568	3586	3544	3555	−3	3562	141589000	40017000

The trade_date in the table indicates the opening time; the open represents the silver opening price; the high represents the highest silver price; the low represents the lowest silver price; the close represents the silver closing price; the change represents a rise or fall in value; the settle represents the settlement price; the vol represents volume; the oi represents operating income.

We select the S&P 500 index (SPX), the Dow Jones industrial average (US30), the Nasdaq 100 index (NAS100), the U.S. dollar index (USDI), the gold futures (AU), Shanghai stock index (SSI) as factors affecting silver price. The original data of silver price impact factors are shown in Table 3.

Table 3. Original data of silver price impact factors.

Trade_Date	SPX	US30	NAS100	USDI	AU	SSI
5 January 2015	2000.63	17362	4102.8999	11648	242.15	3350.519
6 January 2015	2026.38	17590	4155.8999	11655	244.45	3351.446
7 January 2015	2060.1299	17881	4236.8999	11684	245.25	3373.9541
8 January 2015	2041.88	17720	4207.6001	11690	244.5	3293.4561
9 January 2015	2041.88	17720	4207.6001	11633	245.15	3285.4121

4.3. Data Preprocessing

The silver price data selected in this paper come from the trading data of Shanghai futures trading. The Shanghai futures exchange suspends trading on Saturdays and Sundays and on corresponding Chinese legal holidays. Therefore, there is no silver trading data on the corresponding date. SPX, US30, NAS100, USDI, and AU, which affect silver prices, are international market trading data. The legal working days of international exchanges are different from those in China. Therefore, there will be silver trading data on a certain day, but there are no corresponding impact factor data. For the missing impact factor data of a certain day, take the average value of the data of the previous day and the previous two days to fill in the missing data value. If the impact factor's trading data of a particular day exist, but the silver trading data do not, the impact factor's trading data will be deleted. The first duplicate data will be deleted when two experimental data are duplicated. If invalid trading data exist, they will be deleted.

The trade_date has no training significance in the original data, so the column data are deleted. In the original data of silver, the sample values of different characteristics are quite different. When the features have different value ranges, it will take a long time to reach the optimal local value or the optimal global value when the model is updated by the gradient. Data standardization refers to scaling the original data to eliminate the dimensional difference of the original data. That is, each index value is in the same quantity level to reduce the impact of excessive differences of orders of magnitude on model training. Z-score normalization is used for preprocessing the original data. After the standardization of the data, all the feature data sizes are in the same specific interval. Therefore, it is convenient to compare and weigh the characteristic data of different units or orders of magnitude and accelerate the convergence of the training model.

In this experiment, the first 1155 data are selected as training data, 385 data from 1155 to 1540 are used as verification data, and the remaining 385 data are used as test data.

4.4. Model Parameters

In this paper, the parameters of different models are determined by using the grid search method. By comparing the performance results obtained from different parameters, the optimal parameter combination is finally determined. In this experiment, fourteen models are compared. The important parameters of the fourteen models are shown in Table 4.

Table 4. Model parameters.

Model	Layer	Parameters
Prophet	Prophet	interval_width = 0.8
SVR	SVR	kernel = 'linear', epsilon = 0.07, C = 4
MLP	MLP	activation = "tanh"
ARIMA	ARIMA	dynamic = false
LSTM	LSTM	activation = 'tanh', units = 128
Bi-LSTM	Bi-LSTM	activation = 'tanh', units = 128
GRU	GRU	activation = 'tanh', units = 128
NGU	NGU	activation = 'tanh', units = 128
CNN-LSTM	Conv1D LSTM	filters = 16, kernel_size = 3, activation = 'tanh', units = 128
CNN-GRU	Conv1D GRU	filters = 16, kernel_size = 3, activation = 'tanh', units = 128
CNN-NGU	Conv1D NGU	filters = 16, kernel_size = 3, activation = 'tanh', units = 128

Table 4. *Cont.*

Model	Layer	Parameters
CNN-SA-LSTM	Conv1D SA LSTM	filters = 16, kernel_size = 3, initializer = 'uniform', activation = 'tanh', units = 128
CNN-SA-GRU	Conv1D SA GRU	filters = 16, kernel_size = 3, initializer = 'uniform', activation = 'tanh', units = 128
CNN-SA-NGU	Conv1D SA NGU	filters = 16, kernel_size = 3, initializer = 'uniform', activation = 'tanh', units = 128

4.5. Model Comparison

To certify the validity of the CNN-SA-NGU, the prediction results of this model are compared with those of other models. The evaluation indexes of the experiment are MAE, EVS, R^2, and training time. The results show that the CNN-SA-NGU is better than other models. The experimental results are shown in Table 5.

Table 5. Experimental results.

Model	MAE	EVS	R^2	Training Time (t/s)
Prophet	176.829765	0.899582	0.864999	73.432
SVR	182.038698	0.928241	0.903835	50.824
MLP	190.168172	0.848885	0.837680	5.598
ARIMA	168.655063	0.907159	0.907148	24.946
LSTM	116.539392	0.940564	0.940126	450.684
Bi-LSTM	119.670333	0.941758	0.941239	1306.247
GRU	118.748377	0.939636	0.936895	334.112
NGU	103.960158	0.955276	0.953869	278.847
CNN-LSTM	113.772953	0.956882	0.944692	398.622
CNN-GRU	108.031883	0.947018	0.944206	272.832
CNN-NGU	97.277688	0.965663	0.963685	253.501
CNN-SA-LSTM	102.664546	0.954118	0.952839	428.042
CNN-SA-GRU	97.424566	0.960515	0.957547	328.642
CNN-SA-NGU	87.898771	0.970745	0.970169	332.777

(1) Comparison of Prophet, SVR, ARIMA, MLP, LSTM, Bi-LSTM, GRU, and NGU

The fitting degrees of traditional machine learning algorithms SVR, ARIMA, and MLP in silver price prediction are only 0.903835, 0.907148, and 0.837680, respectively, which are poorer compared with other deep learning models. Traditional machine learning methods have poor nonlinear fitting ability. The processing of special values in data sets is not good enough, which leads to poor prediction results of the silver closing price. LSTM, Bi-LSTM, and GRU are variants of the RNN. The structure of the NGU is simple, and the training parameters are few, so the training time is greatly shortened. NGU learns the experience of the previous time to process the input data of the current time, which improves the prediction accuracy of the model. The *Tri* conversion module behind the NGU input gate changes the output value to alleviate the gradient disappearance and gradient explosion problems. The fitting degree of the NGU is 0.013743 higher than LSTM and 0.016974 higher than GRU. In terms of training time, NGU is 171.837 s faster than LSTM. NGU is 55.265 s faster than GRU. The comparison between the true values and the predicted results of Prophet, SVR, MLP, ARIMA, LSTM, Bi-LSTM, GRU, and NGU is shown in Figure 4.

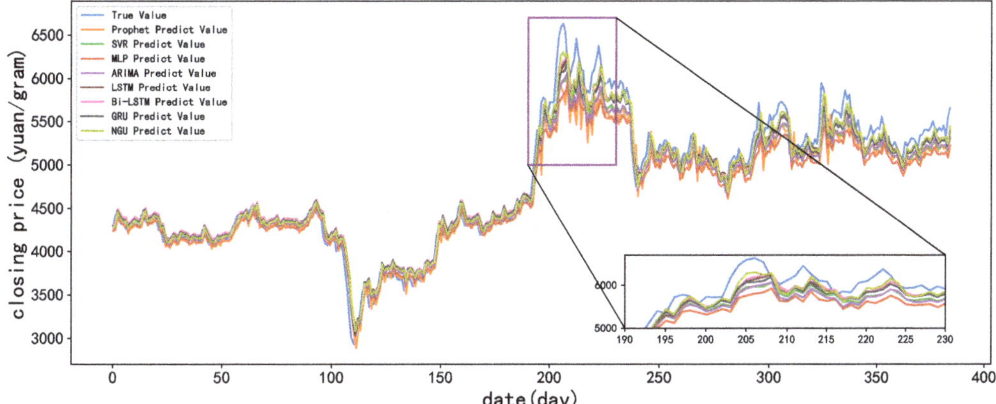

Figure 4. Comparison of true values with Prophet, SVR, ARIMA, MLP, LSTM, Bi-LSTM, GRU, and NGU prediction results.

The CNN extracts the features of silver data and outputs the convolution results to the NGU for learning. Through the convolution of the CNN layer, we can better extract the features from the original data. It is beneficial to the learning of the NGU and improves the model's prediction accuracy. After the convolution operation, the NGU directly learns the feature data without learning the rules from the original data. It shortens the training time to a certain extent. The CNN is combined with LSTM, GRU, and NGU to form a new silver forecasting hybrid model. The prediction results of CNN-LSTM, CNN-GRU, and CNN-NGU are much better than those without the CNN. The fitting degree of CNN-NGU is 0.009816 higher than the NGU. The comparison between the true values and the predicted results of LSTM, GRU, NGU, CNN-LSTM, CNN-GRU, and CNN-NGU is shown in Figure 5.

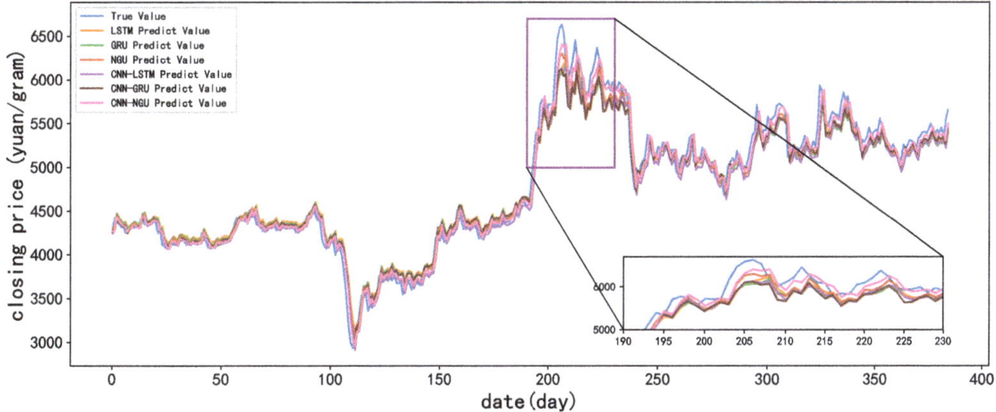

Figure 5. Comparison between true values and predicted results of LSTM, GRU, NGU, CNN-LSTM, CNN-GRU, and CNN-NGU.

(2) Comparison of CNN-LSTM, CNN-GRU, and CNN-NGU

The prediction fitting degree of CNN-NGU is 0.018993 higher than CNN-LSTM and 0.019479 higher than CNN-GRU. CNN-NGU's training time is 145.121 s faster than CNN-LSTM. The comparison of prediction results of CNN-LSTM, CNN-GRU, and CNN-NGU models is shown in Figure 6.

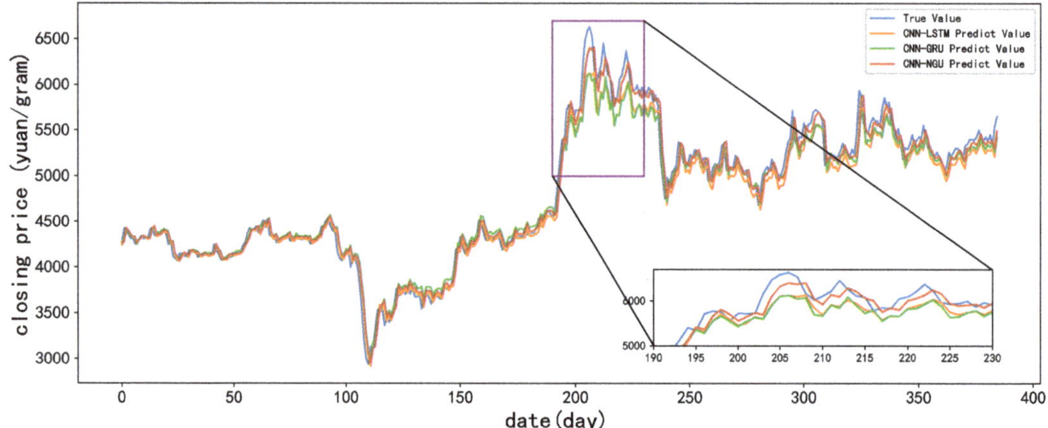

Figure 6. Comparison of true values with CNN-LSTM, CNN-GRU, and CNN-NGU prediction results.

The SA mechanism processes the feature data after the convolution of the CNN convolution layer. The SA layer determines the importance of different feature data by calculation. The characteristic data that have a great influence on the prediction results are given a larger weight factor. Feature data with less influence on prediction results are given smaller weight factors. Through the treatment of SA, different eigenvalues are given different weight factors. By reassigning different weight coefficients to different data, the subsequent gated unit can learn which data have a greater impact on the prediction result. It is beneficial to NGU to learn so as to better predict the closing price of silver. The fitting degree of CNN-SA-LSTM is 0.008147 higher than CNN-LSTM, and the MAE value is 11.108407 lower. The fitting degree of CNN-SA-GRU is 0.013341 higher than CNN-GRU, and the MAE value is 10.607317 lower. The fitting degree of CNN-SA-NGU is 0.006484 higher than CNN-NGU, and the MAE value is 9.378917 lower. The comparison between the true values and the predicted results of CNN-LSTM, CNN-GRU, CNN-NGU, CNN-SA-LSTM, CNN-SA-GRU, and CNN-SA-NGU is shown in Figure 7.

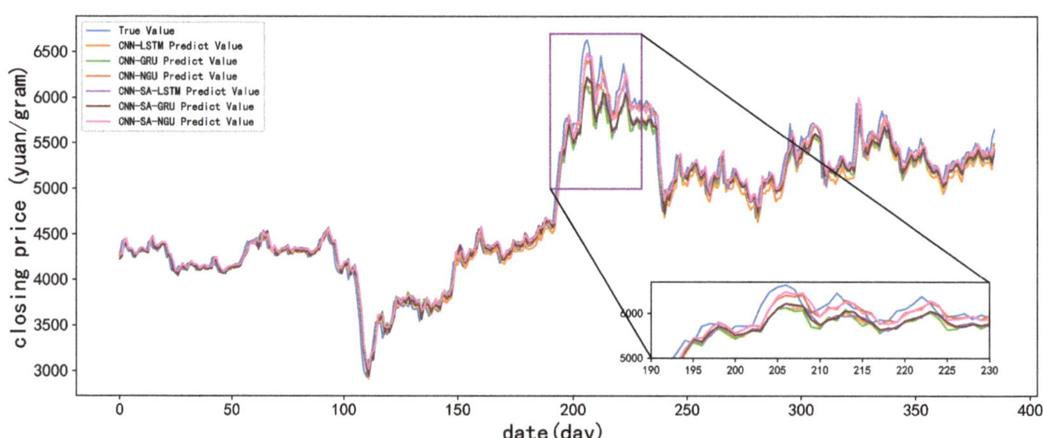

Figure 7. Comparison of true values with CNN-LSTM, CNN-GRU, CNN-NGU, CNN-SA-LSTM, CNN-SA-GRU, and CNN-SA-NGU predictions.

(3)   Comparison of CNN-SA-LSTM, CNN-SA-GRU, and CNN-SA-NGU

Among the three models of CNN-SA-LSTM, CNN-SA-GRU, and CNN-SA-NGU, the performance of CNN-SA-NGU is the best. The fitting degree of CNN-SA-NGU is 0.01733 higher than CNN-SA-LSTM and 0.012622 higher than CNN-SA-GRU. The training time of CNN-SA-NGU is 95.265 s shorter than CNN-SA-LSTM. The true values are compared with the predicted results of CNN-SA-LSTM, CNN-SA-GRU, and CNN-SA-NGU models, as shown in Figure 8.

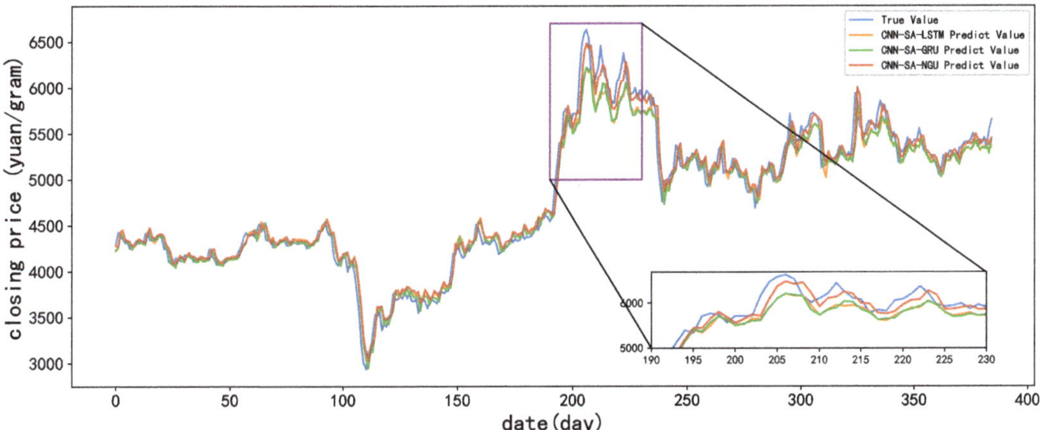

Figure 8. Comparison of true values with CNN-SA-LSTM, CNN-SA-GRU, and CNN-SA-NGU predictions.

4.6. Generalization Ability of Model

CNN-SA-NGU model has good generalization ability. It shows good performance in silver price prediction and is suitable for forecasting other time series data such as ETFs, gold futures, and stocks. The following experiments are carried out with gold futures and Shanghai stock composite index data. The experimental results of forecasting gold futures prices are shown in Table 6. The experimental results of forecasting the Shanghai stock composite index are shown in Table 7.

Through the above two tables, we can see that the CNN-SA-NGU model has good generalization ability.

Table 6. The experimental results of the forecasting table of gold futures prices.

Model	MAE	EVS	R^2	Training Time (t/s)
Prophet	7.328386	0.889623	0.849623	61.752
SVR	5.767165	0.935615	0.915764	45.185
MLP	7.012249	0.901912	0.861166	9.969
ARIMA	6.757669	0.941629	0.898242	28.905
LSTM	4.871939	0.942918	0.939975	479.996
Bi-LSTM	4.855796	0.944959	0.943941	1098.907
GRU	4.736281	0.946534	0.944854	323.123
NGU	4.814574	0.972503	0.955799	279.747
CNN-LSTM	4.625511	0.962245	0.951108	465.302
CNN-GRU	4.528336	0.959778	0.953008	306.796
CNN-NGU	4.032819	0.971674	0.966912	257.907
CNN-SA-LSTM	4.264018	0.960852	0.956380	483.592
CNN-SA-GRU	4.185553	0.959046	0.959038	374.097
CNN-SA-NGU	3.628549	0.972574	0.971670	367.560

Table 7. The experimental results of forecasting the Shanghai stock composite index.

Model	MAE	EVS	R^2	Training Time (t/s)
Prophet	47.545572	0.902315	0.901356	63.558
SVR	31.234645	0.959948	0.958887	36.740
MLP	40.541882	0.942551	0.933907	8.011
ARIMA	39.251600	0.955644	0.955076	25.791
LSTM	28.944838	0.968379	0.967678	466.249
Bi-LSTM	28.409177	0.969849	0.968916	1327.013
GRU	27.071643	0.971395	0.970907	304.327
NGU	26.573712	0.979191	0.975161	290.925
CNN-LSTM	28.279052	0.977564	0.972431	489.508
CNN-GRU	25.767393	0.978223	0.975720	273.203
CNN-NGU	23.452398	0.979790	0.979602	256.740
CNN-SA-LSTM	27.767957	0.979228	0.978946	495.647
CNN-SA-GRU	25.957919	0.983123	0.980307	386.600
CNN-SA-NGU	22.639894	0.984826	0.984815	377.040

5. Discussion

Compared with ten other silver price prediction models, the performance of the CNN-SA-NGU is the best. Compared with SVR, MLP, LSTM, and GRU, the NGU presented in this paper has a better performance in MAE, EVS, R^2, and training time. Adding a CNN to the model improves the ability to extract feature data. The SA layer is added to the model to redistribute the weights of different feature data. It is beneficial for NGU learning. The NGU learns from the previous training experience to deal with the input data at the current time, which improves the nonlinear fitting ability of the model. The CNN-SA-NGU model can achieve higher prediction accuracy for the following reasons:

(1) The NGU uses the original learning experience fully to enhance the processing ability of the input data at the current time, thus improving the nonlinear fitting ability of the model. The *Tri* conversion module changes the range of output value by processing the output data of the input gate, thus alleviating the problems of gradient disappearance and gradient explosion.

(2) With the addition of the SA mechanism, the feature data that significantly influence the prediction results can be well identified. The SA mechanism reallocates the weights of different feature data through calculation. Additionally, a higher weight factor is assigned to the feature data, which benefits the NGU's learning.

(3) By adding the CNN convolution layer, the model's feature extraction ability is improved. The hidden features between data can be mined by the CNN.

6. Conclusions

This paper presents a novel hybrid model of CNN-SA-NGU for silver closing price prediction. The CNN convolution layer solves the problem of incomplete feature data extraction in traditional models to some extent. After introducing the SA mechanism, the relationship between different feature data can be learned, thus increasing the sensitivity of the model to feature data. The structure of the NGU is simple, and the training parameters are few, greatly reducing the training time. The *Tri* conversion module of the NGU deals with the output data of the input gate, which ameliorates the problems of gradient disappearance and gradient explosion. NGU fully learns the experience of the previous time and deals with the input data at the current time, which improves the model's nonlinear fitting ability and improves its prediction accuracy. The comparative experiments show that the performance of CNN-SA-NGU is better than other models, but the model has the shortcoming of not fitting some extreme values in the data set well.

Our future research directions are as follows:

(1) Currently, the model only takes scalar data such as SPX, US30, NAS100, USDI, AU, and SSI as the influencing factors of the silver price. However, some factors still affect the silver price, such as investors' psychology, the formulation of laws, and political events. In future research, we should use natural language processing technology to quantify political events such as policy changes and wars as influencing factors and input them into the prediction model to improve prediction accuracy.

(2) We will further attempt to improve the SA model to make the weight coefficient allocation of the importance of feature data more reasonable.

Author Contributions: Conceptualization, H.W.; methodology, H.W. and J.W.; software, B.D. and X.L.; validation, N.Y.; investigation, N.Y. and B.D.; writing—original draft preparation, B.D. and N.Y. writing—review and editing, J.W. and H.W.; visualization, B.D. and X.L.; supervision, H.W. and J.W.; project administration, B.D. and N.Y.; funding acquisition, H.W. and J.W. All authors have read and agreed to the published version of the manuscript.

Funding: This research was funded by the Scientific Research Project Foundation for High-level Talents of the Xiamen Ocean Vocational College under Grant KYG202102, and Innovation Foundation of Hebei Intelligent Internet of Things Technology Innovation Center under Grant AIOT2203.

Institutional Review Board Statement: Not applicable.

Informed Consent Statement: Not applicable.

Data Availability Statement: Data are available on request due to restrictions privacy. The data presented in this study are available on request from the corresponding author.

Conflicts of Interest: The authors declare no conflict of interest.

Abbreviations

NO.	Abbreviation	Full Name
1	ARIMA	autoregressive integrated moving average
2	AU	gold futures
3	Bi-LSTM	bi-directional long short-term memory
4	CNN	conventional neural network
5	EVS	explained variance score
6	GRU	gated recurrent unit
7	GRUNN	gate recurrent unit neural network
8	ICA	independent component analysis
9	LSTM	long short-term memory
10	MAE	mean absolute error
11	MLP	multi-layer perceptron
12	NAS100	Nasdaq 100 index
13	NGU	new gated unit
14	R2	r squared
15	RNN	recurrent neural network
16	S2SAN	sentence-to-sentence attention network
17	SA	self-attention
18	SPX	S& P 500 index
19	SSI	Shanghai stock index
20	SVM	support vector machine
21	SVR	support vector regression
22	US30	Dow Jones industrial average
23	USDI	U.S. dollar index
24	VMD	variational mode decomposition

References

1. Kim, T.; Kim, H.Y. Forecasting stock prices with a feature fusion LSTM-CNN model using different representations of the same data. *PLoS ONE* **2019**, *14*, e0212320. [CrossRef] [PubMed]
2. Cunado, J.; Gil-Alana, L.A.; Gupta, R. Persistence in trends and cycles of gold and silver prices: Evidence from historical data. *Phys. A Stat. Mech. Its Appl.* **2019**, *514*, 345–354. [CrossRef]

3. Apergis, I.; Apergis, N. Silver prices and solar energy production. *Environ. Sci. Pollut. Res.* **2019**, *26*, 8525–8532. [CrossRef]
4. Kotsiantis, S.B. Decision trees: A recent overview. *J. Artif. Intell. Rev.* **2013**, *39*, 261–283. [CrossRef]
5. Heo, J.; Jin, Y.Y. SVM based Stock Price Forecasting Using Financial Statements. *J. Korea Ind. Inf. Syst. Soc.* **2015**, *21*, 167–172. [CrossRef]
6. Liu, R.; Liu, L. Predicting housing price in China based on long short-term memory incorporating modified genetic algorithm. *Soft Comput.* **2019**, *23*, 11829–11838. [CrossRef]
7. Yu, Y.; Si, X.S.; Hu, C.H.; Zhang, J.X. A Review of Recurrent Neural Networks: LSTM Cells and Network Architectures. *Neural Comput.* **2019**, *31*, 1235–1270. [CrossRef] [PubMed]
8. Li, G.N.; Zhao, X.W.; Fan, C.; Fang, X.; Li, F.; Wu, Y.B. Assessment of long short-term memory and its modifications for enhanced short-term building energy predictions. *J. Build. Eng.* **2021**, *43*, 103182. [CrossRef]
9. Liu, Y.; Zhang, X.M.; Zhang, Q.Y.; Li, C.Z.; Huang, F.R.; Tang, X.H.; Li, X.J. Dual Self-Attention with Co-Attention Networks for Visual Question Answering. *Pattern Recognit.* **2021**, *117*, 107956. [CrossRef]
10. Humphreys, G.W.; Sui, J. Attentional control and the self: The Self-Attention Network (SAN). *Cogn. Neurosci.* **2015**, *7*, 5–17. [CrossRef]
11. Lin, Z.H.; Li, M.M.; Zheng, Z.B.; Cheng, Y.Y.; Yuan, C. Self-Attention ConvLSTM for Spatiotemporal Prediction. In Proceedings of the AAAI Conference on Artificial Intelligence, New York, NY, USA, 7–12 February 2020; pp. 11531–11538. [CrossRef]
12. Yuan, F.C.; Lee, C.H.; Chiu, C.C. Using Market Sentiment Analysis and Genetic Algorithm-Based Least Squares Support Vector Regression to Predict Gold Prices. *Int. J. Comput. Intell. Syst.* **2020**, *13*, 234–246. [CrossRef]
13. Aksehir, Z.D.; Kilic, E. How to handle data imbalance and feature selection problems in CNN-based stock price forecasting. *IEEE Access* **2020**, *10*, 31297–31305. [CrossRef]
14. He, C.M.; Kang, H.Y.; Yao, T.; Li, X.R. An effective classifier based on convolutional neural network and regularized extreme learning machine. *Math. Biosci. Eng.* **2019**, *16*, 8309–8321. [CrossRef] [PubMed]
15. Chen, W.B.; Lu, Y.; Ma, H.; Chen, Q.L.; Wu, X.B.; Wu, P.L. Self-attention mechanism in person re-identification models. *Multimed. Tools Appl.* **2022**, *81*, 4649–4667. [CrossRef]
16. Yan, R.; Liao, J.Q.; Yang, J.; Sun, W.; Nong, M.Y.; Li, F.P. Multi-hour and multi-site air quality index forecasting in Beijing using CNN, LSTM, CNN-LSTM, and spatiotemporal clustering. *Expert Syst. Appl.* **2021**, *169*, 114513. [CrossRef]
17. Vidal, A.; Kristjanpoller, W. Gold volatility prediction using a CNN-LSTM approach. *Expert Syst. Appl.* **2020**, *157*, 113481. [CrossRef]
18. Jianwei, E.; Ye, J.M.; Jin, H.H. A novel hybrid model on the prediction of time series and its application for the gold price analysis and forecasting. *Phys. A Stat. Mech. Its Appl.* **2019**, *527*, 121454. [CrossRef]
19. Liu, Y.; Zhang, Z.L.; Liu, X.; Wang, L.; Xia, X.H. Deep Learning Based Mineral Image Classification Combined with Visual Attention Mechanism. *IEEE Access* **2021**, *9*, 98091–98109. [CrossRef]
20. Liu, J.J.; Yang, J.K.; Liu, K.X.; Xu, L.Y. Ocean Current Prediction Using the Weighted Pure Attention Mechanism. *J. Mar. Sci. Eng.* **2022**, *10*, 592. [CrossRef]
21. Kim, G.; Shin, D.H.; Choi, J.G.; Lim, S. A Deep Learning-Based Cryptocurrency Price Prediction Model That Uses On-Chain Data. *IEEE Access* **2022**, *10*, 56232–56248. [CrossRef]
22. Wang, P.; Li, J.N.; Hou, J.R. S2SAN: A sentence-to-sentence attention network for sentiment analysis of online reviews. *Decis. Support Syst.* **2021**, *149*, 113603. [CrossRef]
23. Li, Q.B.; Yao, N.M.; Zhao, J.; Zhang, Y.A. Self attention mechanism of bidirectional information enhancement. *Appl. Intell.* **2022**, *52*, 2530–2538. [CrossRef]
24. Liang, Y.H.; Lin, Y.; Lu, Q. Forecasting gold price using a novel hybrid model with ICEEMDAN and LSTM-CNN-CBAM. *Expert Syst. Appl.* **2022**, *206*, 117847. [CrossRef]
25. Chen, W.J. Estimation of International Gold Price by Fusing Deep/Shallow Machine Learning. *J. Adv. Transp.* **2022**, *2022*, 6211861. [CrossRef]
26. Chen, S.; Ge, L. Exploring the attention mechanism in LSTM-based Hong Kong stock price movement prediction. *Quant. Financ.* **2019**, *19*, 1507–1515. [CrossRef]
27. Zeng, C.; Ma, C.X.; Wang, K.; Cui, Z.H. Parking Occupancy Prediction Method Based on Multi Factors and Stacked GRU-LSTM. *IEEE Access* **2022**, *10*, 47361–47370. [CrossRef]
28. Deng, L.J.; Ge, Q.X.; Zhang, J.X.; Li, Z.H.; Yu, Z.Q.; Yin, T.T.; Zhu, H.X. News Text Classification Method Based on the GRU_CNN Model. *Int. Trans. Electr. Energy Syst.* **2022**, *2022*, 1197534. [CrossRef]
29. Sun, W.W.; Guan, S.P. A GRU-based traffic situation prediction method in multi-domain software defined network. *PeerJ Comput. Sci.* **2022**, *8*, e1011. [CrossRef]

Article

Automotive Supply Chain Disruption Risk Management: A Visualization Analysis Based on Bibliometric

Kai Huang [1,2], Jian Wang [1,2] and Jinxin Zhang [3,*]

[1] School of Economics and Management, Fuzhou University, Fuzhou 350001, China
[2] Logistics Research Center, Fuzhou University, Fuzhou 350001, China
[3] Business School, Hubei University, Wuhan 430062, China
* Correspondence: zhangjinxin999@foxmail.com

Abstract: The automobile industry is the pillar industry of the national economy. The good operation of the automobile supply chain is conducive to the sustainable development of the economy and social economy. In recent years, the popular research of automotive supply chain disruption risk management has been widely of concern by both business and academic practitioners. It is observed that most of the literature has focused only on a particular journal or field; there is a distinct lack of comprehensive bibliometric review of two decades, of research on automotive supply chain disruption risk management. This paper delivers a comprehensive bibliometric analysis that provides a better understanding not previously fully evaluated by earlier studies in the field of automotive supply chain disruption risk management. We used the 866 journal article during the period between 2000 and 2022 from the WOS database as sample data. Highlights research topics and trends, key features, developments, and potential research areas for future research. The research problems we solved are as follows: (1) Over time, how does the research in the field of automotive supply chain disruption risk management progress? (2) Which research areas and trends are getting the most attention in the field of automotive supply chain disruption risk management? (i) to recognize the scholarly production; (ii) the most productive authors; (iii) the most productive organization; (iv) the most cited articles; and (v) the most productive countries. (3) What is the research direction of automotive supply chain disruption risk management in the future? Also discusses the shortcomings of literature and bibliometric analysis. These findings provide a potential road map for researchers who intend to engage in research in this field.

Keywords: automotive supply chain disruption; supply chain resilience; disruption risk; bibliometric analysis; co-citation analysis

Citation: Huang, K.; Wang, J.; Zhang, J. Automotive Supply Chain Disruption Risk Management: A Visualization Analysis Based on Bibliometric. *Processes* **2023**, *11*, 710. https://doi.org/10.3390/pr11030710

Academic Editors: Conghu Liu, Xiaoqian Song, Zhi Liu and Fangfang Wei

Received: 6 January 2023
Revised: 21 February 2023
Accepted: 21 February 2023
Published: 27 February 2023

1. Introduction

As an industry indispensable to the national economy, the automobile industry has developed into a strategic leading industry with high interconnectivity and a long industrial chain. It is easy to evaluate both the software and hardware strengths of a country from the automobile industry. From an economic perspective, automobile supply chain risk management can benefit all supply chain participants. The supply chain of automobile enterprises includes production links such as raw material supply, parts manufacturing, vehicle production, and assembly, and jointly completes various tasks in the whole life cycle of automobile products such as procurement, production, sales, and service. With the increasing complexity of the technical and production links of the automobile commodity chain, it stands to reason that the automotive supply chain should be more elastic. A series of "supply interruption" events of auto parts manufacturers show that as supply chain risks emerge, automobile supply chain management has become increasingly difficult. Without effective risk management, it is hardly possible to see the automobile supply chain operate normally. Therefore, we need to have an in-depth discussion on the risks brought by supply disruption to the automotive supply chain.

Affected by COVID-19 and other emergencies, the development of the supply chain of the automobile industry slowed down, stirring a wide concern over the stability and continuity of China's manufacturing supply chain. According to a McKinsey & Company consulting report, the Chinese manufacturing sector contributes more than 35% to the world, and the contribution rate of the automobile industry is as high as 75%. According to customs data, in 2019, China's top five exporters of automobiles and their key parts and components are: the United States, Japan, Mexico, Germany, and Russia. Auto manufacturing in four of the five exporting countries is mired in production due to COVID-19. Chinese parts manufacturers "supply disruption" directly led to many foreign car production enterprises being in a state of suspension. Due to its globalized industrial characteristics and supply chain system, the automobile industry suffered the most serious impact caused by COVID-19 among the secondary industries. The COVID-19 pandemic is comparable to a magnifying glass, helping us to clearly realize that in the automotive industry, there are shortcomings during the current supply chain system construction, i.e., less toughness with more flexibility.

For the automobile industry, the development of its supply chain is characterized by regionalization, localization, and antiglobalization. In the global supply chain, participating countries gradually quit the supply chain; as a result, the upstream and downstream countries become less cooperative in production, causing the breakup of the supply chain. The resultant supply shock will lead to the reduction of the production capacity of each country. Anti-globalization can resist the risk of a global supply chain "a ripple effect on the whole body".

To help activate the new energy vehicle industry, in recent years, many countries have emphasized energy conservation and emission reduction, and have issued a number of support policies and subsidy policies to help produce and market more new energy vehicles. In addition, as unceasing progress has been made on clean energy technology and commercialization has bn a mainstream, hydrogen energy and fuel cell industries are also setting off a commercial upsurge around the world. At the same time, many local governments have clearly issued corresponding hydrogen energy subsidy policies. With the new energy vehicle expanding and the rise of the hydrogen energy vehicle industry, there is a big difference between the production of new energy vehicles and traditional vehicles, which is bound to tremendously affect the parts supply chain of the traditional automobile industry.

Taking the vehicle parts supply chain as an example, from the perspective of the vehicle industrial chain, these impacts will directly affect the fuel vehicle powertrain and vehicle manufacturing and maintenance industries. Compared with traditional fuel vehicles, electric vehicles are much simpler in respect of structure. Electric vehicles have no engines, and many original auto parts disappear. Consequently, the related industrial chain suffered a serious impact, directly changing the pattern of the traditional automobile industry.

Automobile commodity is one of the typical producer-driven commodity value chains, each car has about more than 30,000 parts. In the automotive industry, parts manufacturing and supply in the supply chain are different. For complex components, there are complex supply relationships between different parts enterprises and different vehicle enterprises. For auto parts products, its supply chain is operated under a strict grade system, and its suppliers are divided into one, two, and three, forming a pyramid structure of the multi-level division of labor. The OEMs are at the top of the pyramid, and the first-tier suppliers directly supply products to the OEMs, forming a direct cooperative relationship between the two sides, and the second-tier suppliers provide products to the first-tier suppliers. First-tier suppliers can control the manufacturing right of the core system, first-tier suppliers have strong competitiveness in a specific field, and first-tier suppliers lack core competitiveness due to their large number. In addition, the current expansion of automobile parts production enterprises is leading to changes in the structural relationship between "vehicle manufacturers" and "first-tier suppliers" in the pyramid structure, mainly reflected in the thinning automobile parts suppliers. With the deepening of the group of auto parts enterprises, the auto industry will form an hourglass structure, that is, a few enterprises monopolize the production of a certain part, and provide it to many enterprises. Suppliers of different levels have different positions in the supply chain of automobile enterprises, and they have different risk resistance abilities. These risk factors are bound to affect the resilience characteristics and resilience ability of automobile supply.

At present, for the automobile supply chain interruption risk identification, in a broad perspective, supply chain risk is defined as the occurrence of unexpected events (or conditions) at the macro or micro level, adversely affecting any part of the supply chain, leading to failures or violations at the operational, tactical or strategic levels [1]. Based on this perspective, the risks of the supply chain can be divided into two types: macro risk and micro risk. Macro risks are adverse and relatively rare external events or phenomena that may have a strong negative impact on an automobile company or its supply chain (e.g., unexpected shocks such as COVID-19); Micro risk refers to relatively periodic events, directly from the internal activities of the enterprise or the relationship between partners in the supply chain network (e.g., supply shortage, supply quality, supply delay, etc.).

From a narrow perspective, through combing the existing literature, supply chain risk can be summarized as external risk, time risk, information risk, financial risk, supply risk, operation risk, and demand risk.

External risks. External risk refers to the risk from outside the supply chain, mainly caused by economic, social political, or geographical reasons. These include fire accidents, natural disasters, economic downturns, external legal problems, corruption, and cultural differences [2–5]. These risks are more threatening and may directly or indirectly cause disruptions to the supply chain. For example, Hansen et al. found that recessions changed market demand and financial policy [6], and even broke the relationship between supply and demand [7].

Time risk. Time risk refers to delays in supply chain activities [8]. Failure to meet target time requirements can lead to risks in information, operations, demand, and supply chain performance. Manavazi et al. demonstrated that delays in the procurement of materials and equipment are often a major cause of cost overruns in construction projects [9]. Angulo et al. demonstrate that time-related risks can cause dissatisfaction among all partners in the supply chain. For example, information delays may disrupt communication among supply chain members [10]. Delays in delivering products to customers can cause the partner to go bankrupt [11], Late payment, however, leads to more disputes in all parts of the supply chain [12].

Information risk. Information risk refers to the risk caused by poor communication, the complexity of information infrastructure, information distortion, and information leakage in each link of the supply chain [13–15]. For example, excessive inventory investment, poor customer service, large revenue loss, wrong capacity planning, ineffective shipping, etc. [16].

Financial risk. Financial risks include inflation, interest rates, currency fluctuations, and the requirements of stakeholders [17–19]. This type of risk creates price volatility in supply activities, operational planning, labor disputes, demand changes, and supply chain disruptions. For example, inflation leads to rising prices, which irritate consumers, who then blame producers, which is one reason for changing demand [20]. In terms of interest rates, Zhi believed that with the increase of interest rates, the increase in fees charged by banks for corporate loans will reduce the ability of customers to buy products and services and increase demand risk, a phenomenon that will lead to price fluctuations in supply activities [21]. In addition, currency fluctuations have an impact on output and prices [22]. Yeo and Tiong's requirement for stakeholders is also a financial risk. Stakeholders participate in the daily operations of the enterprise, influence key decisions of operational planning activities, and have the right to supervise the selection of suppliers [23].

Supply risk. Supply risk is related to upstream activities in the supply chain [24]. Examples include: bankruptcy of suppliers, price fluctuations, unstable quality and quantity of inputs, etc. [8,25]. These risks can lead to the failure of the delivery of goods or services to the purchasing company, and the subsequent risk of goods propagation affects the downstream of the supply chain [26].

Operational risk. Operational risk refers to supply chain disruptions caused by problems within the enterprise organization, such as design and technological changes, accidents, and labor disputes [2,4,27]. The study of Williams et al. shows that design change will increase the production cost of enterprises, and technological change will destroy business activities, resulting in a decline in the expected return on investment. However, this study did not clarify the mechanism of device influence [28]. David pointed out that employees taking more than three days off may affect employees' ability to perform their daily duties, affecting supply chain stability [29].

Bibliometric method combines mathematical and statistical methods to summarize the research status of related topics and a specific research field, and mainly analyzes the research status, research hotspots, and research trends. Zhou et al. used VOS viewer, Sci2 and Cite Space to explore the publication trend, author and journal distribution, research topics, and hot spots of the research topic of the green supply chain, and predict the research frontier. However, it can be seen from its research that the factors considered have certain limitations [30]. Fan and Stevenson conducted a systematic review of the existing literature on supply chain risk management and proposed a new and comprehensive conceptual framework for supply chain risk management [31]. Moosavi et al. found through bibliometric analysis that blockchain technology can not only improve the transparency and traceability of supply chain management, but also improve the efficiency and information security of supply chain management. The application of blockchain technology in supply chain management is relatively lacking, and there are great research opportunities in future research [32]. Iftikhar et al. used bibliometric analysis and systematic review methods to review the relevant literature on technology innovation, data innovation, and supply chain resilience, in order to timely review the key research clusters, the evolution of research over time, the knowledge trajectory and the research methodology development in this field. The bibliometric method can effectively analyze the current research situation in the research object field, but in this study, the research team's application of the bibliometric method is still not deep enough [33]. Roblek et al. applied bibliometric methods to explore the overall development trend and importance of agile related to management and organization [34]. In order to track the research frontiers and hot spots of closed-loop supply chains, Guan et al. used visualization software VOS viewer and Cite Space for analysis. A descriptive analysis was first conducted to identify trends in the number of publications, major journals, top authors, and regions. A thematic cluster analysis was then performed to identify the research areas. Subsequently, hot issues and research trends are summarized based on co-keywords, clustering, and co-citation analysis. "Game theory" and "remanufacturing" are new trends in closed-loop supply chain research. "Dual channel", "quality" and "circular economy" have become hot topics. Although this research has achieved the summary of the current situation of supply chain tracking research, it has not made effective prediction and put forward effective suggestions for its development trend [35].

In summary, many researchers have studied automotive supply chain disruption risk management from multiple perspectives, but it is not clear how the research on automotive supply chain disruption risk management develops. We do not know the evolution characteristics of the research topics in this research area, and these questions deserve our further exploration. These subjects are worth exploring in depth because they provide clues to how this field of research will develop in the future. In view of the lack of in-depth research on the risk management of automobile supply chain disruption, we designed the present study to identify the research hotspots and future development trends in this field. To do so, we conducted a bibliometric analysis of automotive supply chain disruption risk management and performed a qualitative analysis of the results.

This paper mainly discusses these research questions as follows: (1) What are the most representative articles, the number of articles published in top journals, the co-cited articles, and the authors in the field of automotive supply chain disruption risk research?

(2) Which organizations, institutions, and countries contribute to the research of automotive supply chain disruption risk? (3) What are the main research clustering topics? (4) What are the future research trends and research hotspots in the field of automotive supply chain sustainability?

The rest of this paper is organized as follows: Section 2 presents the data source and search methodology. Section 3 presents the results of the bibliometric analysis and science mapping. The analysis includes the most representative scientific production, journals, authors, countries, institutions, and organizations. The science mapping shows the citation analysis, by articles, authors, and terms. Section 4 includes a discussion of the research, findings, analysis, and implications of the scientific research on the relevant aspects of automotive supply chain disruption risk management. Section 5 provides the conclusions, limitations, and future research concludes the paper.

There are two innovative points in the study. First, the study learned about the interruption management of the automobile supply chain in recent years through literature measurement, and then analyzed the interruption risk of the automobile supply chain through the results of literature measurement, without providing corresponding suggestions for the development and management of the automobile supply chain.

2. Data Source and Methodology

For the collection of article information, the source of data is very important. We used SCI, SSCI WOS and EI databases as sources of scientific literature. Web of Science (WOS) is one of the important database platforms for obtaining global academic information, including more than 20,000 international authoritative and influential academic journals, covering social sciences, humanities and arts, natural sciences, engineering and technology, and other disciplines. The database is considered the most trustworthy and thorough source of data [36,37] and is frequently used in bibliometric research on the progress and evaluation of various scientific fields [38].

SCI database is currently the most authoritative large-scale multidisciplinary comprehensive retrieval tool in the world, so it is regarded as one of the main source databases of foreign literature. In the SCI database, the search term is limited to "auto supply chain" and "supply chain interruption risk", the language is "English", the publication year is limited to "after 2000", and the document type is "Article". Under this condition, a total of 1024 documents are screened.

The SSCI WOS database mainly includes SSCI journal papers. The research takes WOS as the search source, takes "auto supply chain", "supply chain interruption risk" and "risk management" as the search keywords, limits the search language to "English", and sets the document type to "article" and "review". In order to obtain more effective supply chain information, the research selects the time node when the auto industry starts to develop as the initial search time, so set the time span = '2000–2022', Databases = 'WOS Core Collection'. A total of 1003 articles were retrieved.

The search time for this study began on 7 January 2023. After searching in the WOS database, a total of 1084 literatures were obtained. All the screening was conducted in one day to prevent data bias resulting from daily database updates. This is just a preliminary simple screening. Considering the defects in the retrieval technology, we will filter and clean the original literature data by means of manual screening to remove irrelevant and duplicate literature data. Finally, a total of 866 articles are selected as sample data for further detailed analysis.

The dataset we obtained includes important information such as author, title, year, journal source, affiliation, abstract, etc. The dataset was managed in Excel, and we used version 5.7R2 of the Cite Space software for bibliometric visualization analysis [39].

3. Current Status of the Field of Automotive Supply Chain Disruption Risk Management

This section may be divided into subheadings. It should provide a concise and precise description of the experimental results, their interpretation, as well as the experimental conclusions that can be drawn.

Next, this part focuses on the research status of the field of automotive supply chain disruption risk management, including the trend of the number of published articles, publication types, top journals, etc.

Figure 1 shows the number of publications and citations from 2000 to 2022 in the field of automotive supply chain disruption risk management. The overall trend of the publications follows an exponential growth ($y = 0.2409x^2 - 1.4415x + 8.9311$; $R^2 = 0.9$) after mathematical exponential adjustment. The result accords with a price curve and it demonstrates that the field of automotive supply chain disruption risk management has received increasing attention. Similar verifications are also presented in other literature [40,41]. In addition, from 2000 to 2011, the growth rate of the number of published articles was relatively flat. Since 2012, the number of articles published has increased significantly. Especially after 2015, the number of articles published showed a sharp upward trend. The reason may be that the Tohoku earthquake in March 2011 temporarily shut down Japan's auto industry, forcing European and North American manufacturers to halt production because their inventories in Japan had been depleted. Goldman Sachs estimates that Japanese carmakers are losing $200 millionm a day. A one-third drop in global daily car production has led to a total global loss of five million vehicles, against a planned loss of 72 million in 2011 (a loss of about 7%), and months of stagnation in the global car industry's supply chain. As well, floods in Thailand disrupted supply chains for electronics and auto parts. In 2012, an explosion at a major Ford supplier, Evonik, disrupted production. In order to improve the competitive advantage of the automotive supply chain, the research topic of automotive supply chain management has been widely discussed by many researchers. Therefore, more and more scholars focus on how to reduce the disruption risk of the automotive supply chain after these emergencies. In terms of the number of citations of articles, the number of citations has continued to rise. Since the data statistics of this study are up to 26 August 2022, the number of citations in Figure 1 shows a downward trend in 2022. It can be seen that more and more researchers have been involved in the study of automotive supply chain disruption risk in recent years.

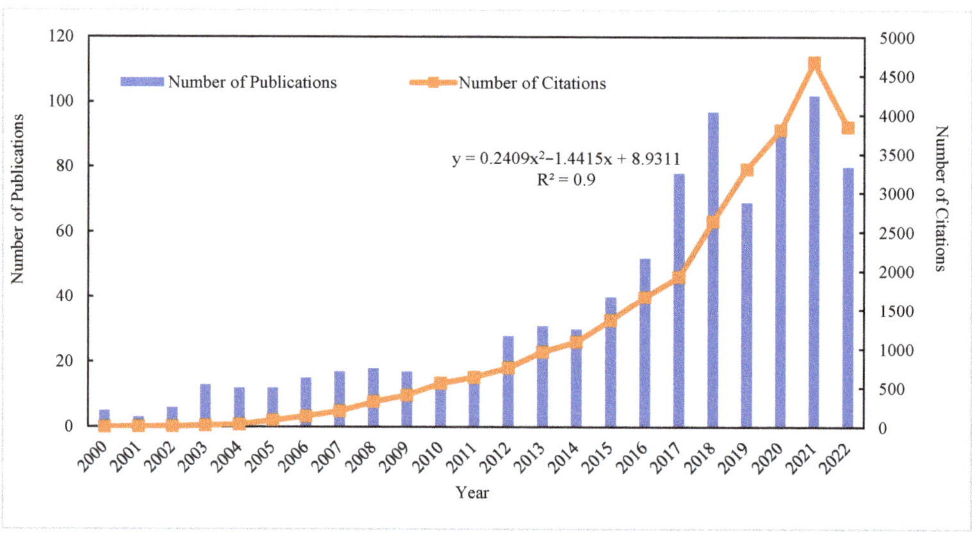

Figure 1. Number of publications and citations from 2000 to 2022.

According to the results reported in Table 1, the top 20 journals published about 40% of the total number of articles in the field of automotive supply chain disruption risk management. The Journal of Cleaner Production ranked first, with 56 papers published. This journal plays an important role in the display of research results in the field of Automotive Supply Chain Disruption risk management. Then, the top five remaining journals are International Journal of Production Economics, International Journal of Production Research, Supply Chain Management-An International Journal, Production Planning & Control. The top five journals account for about 20% of the total publications.

Table 1. Top 20 journals in the field of Automotive Supply Chain Disruption risk management.

Rank	Journal	Total Number	Percentage (%)	Cumulative Percentage (%)
1	Journal of Cleaner Production	56	4.77%	4.77%
2	International Journal of Production Economics	54	4.60%	9.37%
3	International Journal of Production Research	53	4.52%	13.89%
4	Supply Chain Management-An International Journal	37	3.15%	17.04%
5	Production Planning & Control	35	2.98%	20.02%
6	Sustainability	29	2.47%	22.49%
7	International Journal of Operations & Production Management	26	2.22%	24.71%
8	Computers & Industrial Engineering	25	2.13%	26.84%
9	Benchmarking-An International Journal	21	1.79%	28.63%
10	Industrial Management & Data Systems	17	1.45%	30.08%
11	Journal of Manufacturing Technology Management	16	1.36%	31.44%
12	Journal of Operations Management	15	1.28%	32.72%
13	International Journal of Logistics Management	14	1.19%	33.91%
14	European Journal of Operational Research	13	1.11%	35.02%
15	IFAC Papers OnLine	13	1.11%	36.13%
16	Business Strategy and the Environment	12	1.02%	37.15%
17	IFIP Advances in Information and Communication Technology	12	1.02%	38.17%
18	IEEE Transactions on Engineering Management	10	0.85%	39.02%
19	International Journal of Productivity and Performance Management	10	0.85%	39.87%
20	Procedia CIRP	10	0.85%	40.72%

The number of citations of a paper can indicate the research contribution and importance of the paper. According to the results reported in Table 2, Vickery et al. published the title of the effects of an integrative supply chain strategy on customer service in the Journal of Operations Management in 2003 and financial performance: an analysis of direct versus indirect relationships, which has been cited 691 times [42]. Zhu et al. published in the Journal of Cleaner Production in 2007 titled Green supply chain management: pressures, practices, and performance within the Chinese automobile industry, has been cited 637 times [43]. In 2011, published in the Journal of Operations Management, the contingency effects of environmental uncertainty on the relationship between supply chain integration and operational performance [44], Structural investigation of supply networks: A social network analysis approach published in the Journal of Operations Management [45], It can be seen that in the beginning, scholars mainly focused on the theoretical analysis of automotive supply chain operation. With the continuous changes and development of the automotive industry, scholars have also changed their research perspectives to supplier selection, green supply chain, supply chain optimization, and other aspects.

Table 2. Top 20 most cited articles in the field of Automotive Supply Chain Disruption risk management.

Rank	Article Titles	Year	Source Title	Cited by
1	The effects of an integrative supply chain strategy on customer service and financial performance: an analysis of direct versus indirect relationships	2003	Journal of Operations Management	691
2	Green supply chain management: pressures, practices, and performance within the Chinese automobile industry	2007	Journal of Cleaner Production	637
3	Gaining from vertical partnerships: Knowledge transfer, relationship duration, and supplier performance improvement in the US and Japanese automotive industries	2003	Strategic Management Journal	561
4	The contingency effects of environmental uncertainty on the relationship between supply chain integration and operational performance	2011	Journal of Operations Management	556
5	A comparison between Fuzzy AHP and Fuzzy TOPSIS methods to supplier selection	2014	Applied Soft Computing	458
6	Structural investigation of supply networks: A social network analysis approach	2011	Journal of Operations Management	365
7	Supplier evaluations: communication strategies to improve supplier performance	2004	Journal of Operations Management	365
8	Sourcing by design: Product complexity and the supply chain	2001	management science	357
9	An empirical analysis of supply chain risk management in the German automotive industry	2011	International Journal of Production Economics	323
10	An integrated green supplier selection approach with analytic network process and improved Grey relational analysis	2015	International Journal of Production Economics	319
11	The influence of green practices on supply chain performance: A case study approach	2011	Transportation Research Part E: Logistics and Transportation Review	284
12	Supply chain flexibility and firm performance—A conceptual model and empirical study in the automotive industry	2005	International Journal of Operations & Production Management	281
13	Incorporating sustainability into supply management in the automotive industry—the case of the Volkswagen AG	2007	Journal of Cleaner Production	268
14	Intuitionistic fuzzy-based DEMATEL method for developing green practices and performances in a green supply chain	2015	Expert Systems with Applications	257
15	Defining the concept of supply chain quality management and its relevance to academic and industrial practice	2005	International Journal of Production Economics	251
16	Greening the automotive supply chain: a relationship perspective	2007	Journal of Cleaner Production	238
17	A system dynamics model based on evolutionary game theory for green supply chain management diffusion among Chinese manufacturers	2014	Journal of Cleaner Production	226
18	Designing an integrated AHP-based decision support system for supplier selection in the automotive industry	2016	Expert Systems with Applications	194
19	Green innovation adoption in automotive supply chain: the Malaysian case	2015	Journal of Cleaner Production	185
20	Environmental management system certification and its influence on corporate practices Evidence from the automotive industry	2008	International Journal of Operations & Production Management	183

4. Bibliometric Analysis

4.1. Influence of Authors

Figure 2 presents the analysis of co-authorship in the field of Automotive Supply Chain Disruption risk management by Cite Space software. It shows the mutual cooperation and correlation between different authors. Among them, the authors with the highest correlation are mainly the following groups: The first group is MB Frazer, MS Kerr, WP Neumanm, RW Norman, RP Wells; The second group is Lin Zhou/Fulin Zhou/Yandong He/Yun Lin/Xu Wang; The third group is LC Brown/Dm Gute/Hj Cohen/Amc Desmars; The fourth group is Fabio Sgarbossa/Alena Otto/Alessandro Persona/Martina Calzavara; The fifth group is Hamzeh Soltanali/Abbas Rohani/Jose Torres Farinha; The sixth group is Shukriah Abdullah/Nor Kamaliana Khamis/Jaharah Ghani. From the above results, it can be seen that a large number of scholars currently use space software to analyze the risk of disruption in the automobile supply chain and propose management strategies accordingly. In addition, according to the correlation evaluation in Figure 2, there is a significant correlation between a large number of authors, which indicates that most of the authors have a consistent main direction in the study of automobile supply chain disruption risk.

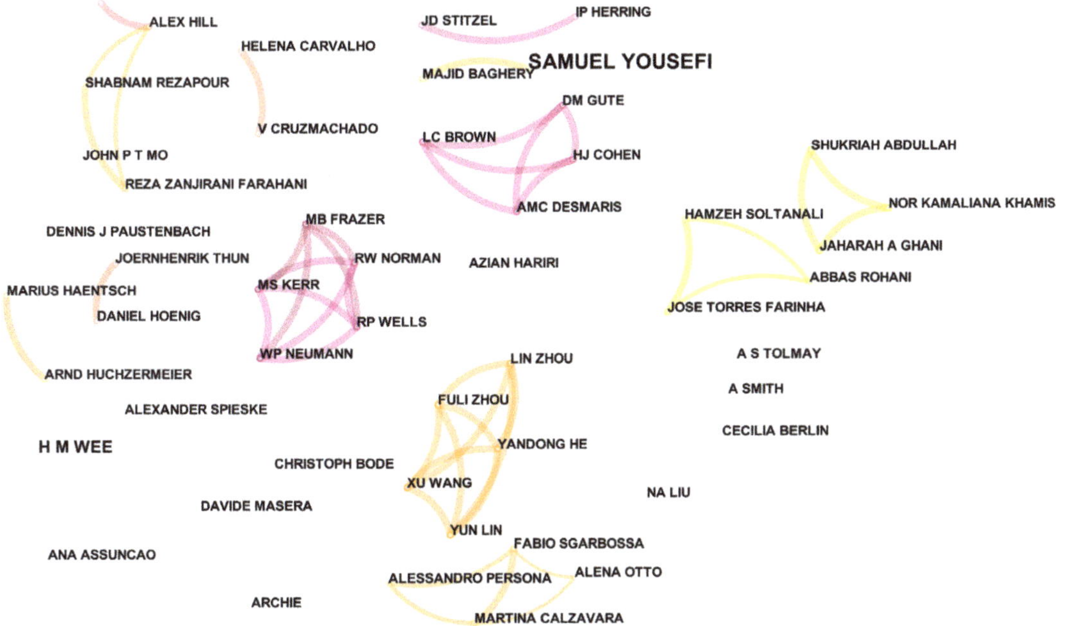

Figure 2. Author cooperation network in the field of automotive supply chain disruption risk management.

4.2. Affiliation Statistics and Analysis Countries, Institutions, Organizations

According to the national distribution of the number of papers published, the country with the largest number of papers is the United States (Figure 3). So far, a total of 162 papers have been published on the subject of automotive supply chain disruption risk management. It accounted for 18.7% of the total number of published papers. The second is Germany, which has published a total of 77 articles on the subject of automotive supply chain disruption risk management, accounting for 8.9%. Next is China. So far, a total of 62 articles have been published on the subject of automotive supply chain disruption risk management, accounting for 7.2%. The top ten countries accounted for 67.2% of the total number of published papers. In addition, from the distribution of papers published in Figure 3, it can be seen that the research related to the risk of automobile supply chain disruption is mainly concentrated in some countries. As shown in Figure 3, the number of papers published in the United States, Germany, China, India, Canada, and other countries accounts for a large proportion, which further indicates that the automobile industry in the above countries has developed more rapidly, increasing the number of scholars' research samples. Therefore, in this study, we will focus on the contents of papers from the United States, Germany, China, India, Canada and other countries, and use the method of literature measurement to summarize the risk of automobile supply chain interruption, so as to put forward universal risk management opinions.

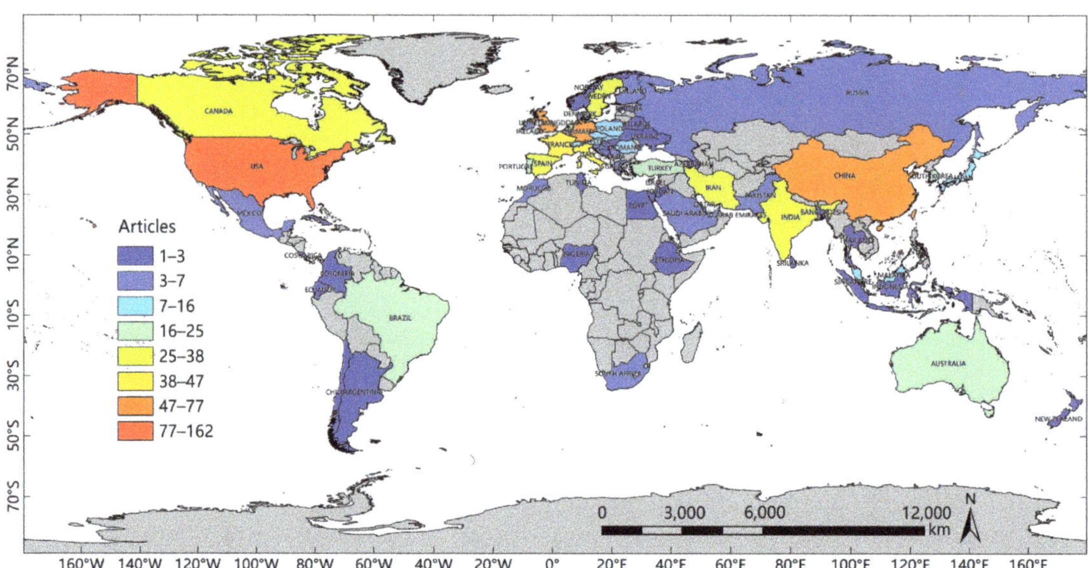

Figure 3. Number of countries' publications in the field of Automotive Supply Chain Disruption risk management.

According to the statistics of the number of papers published by each country, it can be seen from the results reported in Figure 4 that the United States ranks first with 162 papers published and shows the strongest centrality, with a centrality index of 0.38. In second place is Germany with a centrality index of 0.22, and in third place is China with a centrality index of 0.18. In addition, the number of published papers in the United Kingdom, France, Italy, Canada, Iran and other countries is above 35, and their centrality is above 0.15. The cooperative research among these countries is relatively sufficient, and they can use each other's resources to conduct cooperative research and jointly contribute to the research on automotive supply chain disruption risk. At the same time, according to the published statistical results of the papers in Figure 4, the development status of each country can be further analyzed in depth, and the risk factors of automobile supply chain interruption can be proposed from the national level. However, from the differences in the central evaluation results of various countries, it can be seen that most of the current research is still based on the United States. Therefore, if it is necessary to put forward the risk management strategy for domestic automobile supply chain disruption, the follow-up research still needs to focus on the domestic development status.

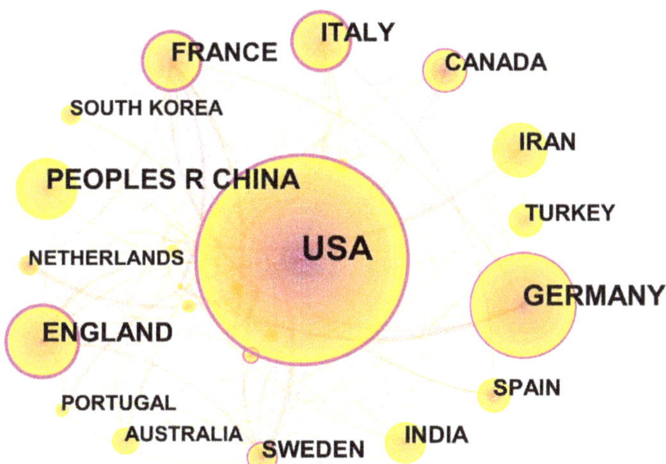

Figure 4. Countries working together in the field of automotive supply chain disruption risk management.

The United States, Germany, the United Kingdom, and China represent the study groups with the highest correlations. The institutions or universities with the most published articles are all located in the United States. At the top of the top 10 is Univ Michigan, which has published 11 papers. Univ Michigan's research on automotive supply chain risk management is relatively mature and widely recognized. Tied for second place are Islamic Azad Univ, Tech Univ Munich, and Chinese Acad Sci, all of which have published seven papers. Among the remaining top 10 institutions, Univ Waterloo, Politecn Torino, and Urmia Univ Technol each published six papers. Chongqing University, Charles Univ Prague, and Tech Univ Kosice each published five papers. The number of articles published by the top 10 universities and institutions accounts for 10% of the total number of articles published by all institutions worldwide. The remaining 90% are publications with between one and four articles per institution.

Among the many research institutions, Univ Waterloo, Univ Toronto, McMaster Univ, Univ Western Ontario, Inst Work and Health and NIOSH have the strongest cooperative research ability and the strongest correlation. By sharing teaching resources and scientific research results, they promote research collaboration between institutions and achieve significant research results in the field of automotive supply chain risk management. This was followed by Univ Tehran Med Sci, Islamic Azad Univ and Cranfield Univ, Also available are Urmia Univ Technol, Allameh Tabatabai Univ, Univ New Haven, Birla Inst Technol & Sci, Michigan State Univ. Further analysis from Figure 5 shows that the cooperation network of research institutions is relatively simple, only some universities and research structures are designed, and there are few studies in which enterprises participate. At present, there are more theoretical research works than practical ones in the research of automotive supply chain risk management. There is not a stable core research group in the study of automotive supply chain disruption risk management, which needs more researchers to participate in and carry out more in-depth research in this field. According to the research, the reason why theoretical research is more than practical research in the current research is that the communication between various research institutions is mostly online academic discussion, so the importance of offline practice is ignored. Therefore, under this background, the research will be based on the current theoretical research situation of the risk management of automobile supply chain interruption, and will be carried out with practice as the core of the follow-up research.

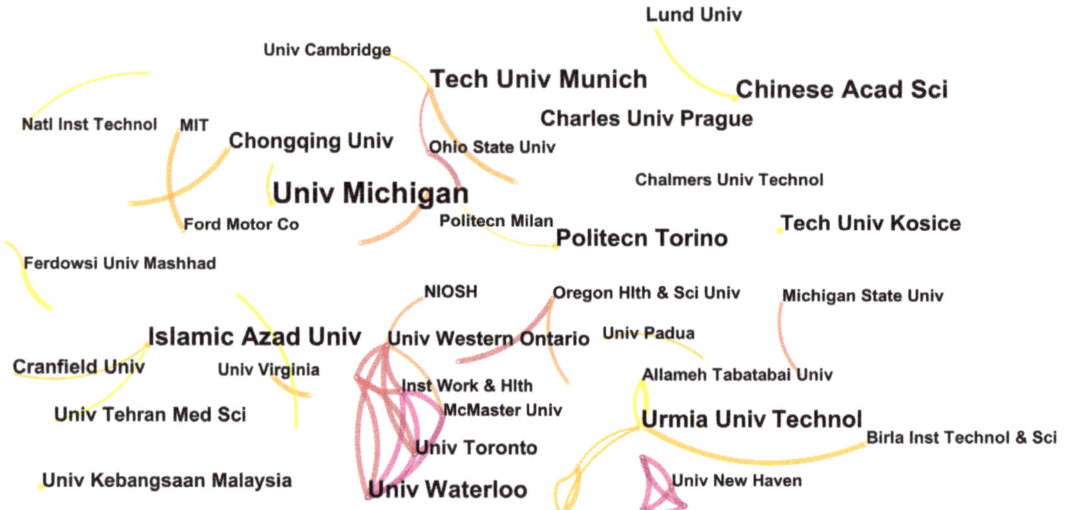

Figure 5. Institutions working together in the field of automotive supply chain disruption risk management.

Based on the sample data collected in this paper, Cite Space software was used to detect the strongest citation bursts on research institutions to identify active institutions in automotive supply chain disruption risk management research in a specific period. According to Figure 6, it can be seen that the research field of automotive supply chain disruption risk management has been the focus of the following top 15 institutions since 2000, among which the first institutional mutation in 2001 was related to Inst Work & Hlth, Univ Waterloo and Univ Toronto. In 2002, scholars at Univ New Haven paid great attention to the field of automotive supply chain disruption risk research. In 2003, Michigan State University began to focus on the research perspective on the disruption risk of the automotive supply chain. Among the top 15 institutions are research institutes in the United States and China. The reason may be that the development level of the automobile manufacturing industry in China and the United States is at the forefront of the world, and it is easier for scholars to start with the industry with certain development advantages in their own countries. By collecting research data and materials and using research methods, the development of the automobile industry is becoming better and better.

4.3. Keyword Analysis

To fully understand the history and development of automotive supply chain disruption risk management research, we used Cite Space software to visually analyze the keywords of the papers in the sample data and identify the research hotspots in this field. Figure 7 shows the network visualization of the keyword co-occurrence map from 2000 to 2022. With the continuous progress and development of the automotive industry, automotive supply chain risk management has gradually become an important strategic means to promote the core competitiveness of automotive enterprises. The node size in Figure 7 is proportional to the frequency of the keyword.

Top 15 Institutions with the Strongest Citation Bursts

Institutions	Year	Strength	Begin	End	2000–2022
Inst Work & Hlth	2000	1.52	**2001**	2022	
Univ Waterloo	2000	1.66	**2001**	2022	
Univ Toronto	2000	1.57	**2001**	2022	
Univ New Haven	2000	1.66	**2002**	2022	
Michigan State Univ	2000	1.44	**2003**	2022	
Ohio State Univ	2000	1.8	**2004**	2022	
Univ Western Ontario	2000	1.6	**2005**	2022	
Lund Univ	2000	1.62	**2009**	2022	
MIT	2000	1.56	**2012**	2022	
Islamic Azad Univ	2000	1.92	**2015**	2022	
Birla Inst Technol & Sci	2000	1.53	**2016**	2022	
Chongqing Univ	2000	1.52	**2016**	2022	
Urmia Univ Technol	2000	2.08	**2017**	2022	
Tech Univ Munich	2000	1.89	**2017**	2022	
Univ Michigan	2000	1.7	**2018**	2022	

Figure 6. Top 15 institutions with the strongest citation bursts.

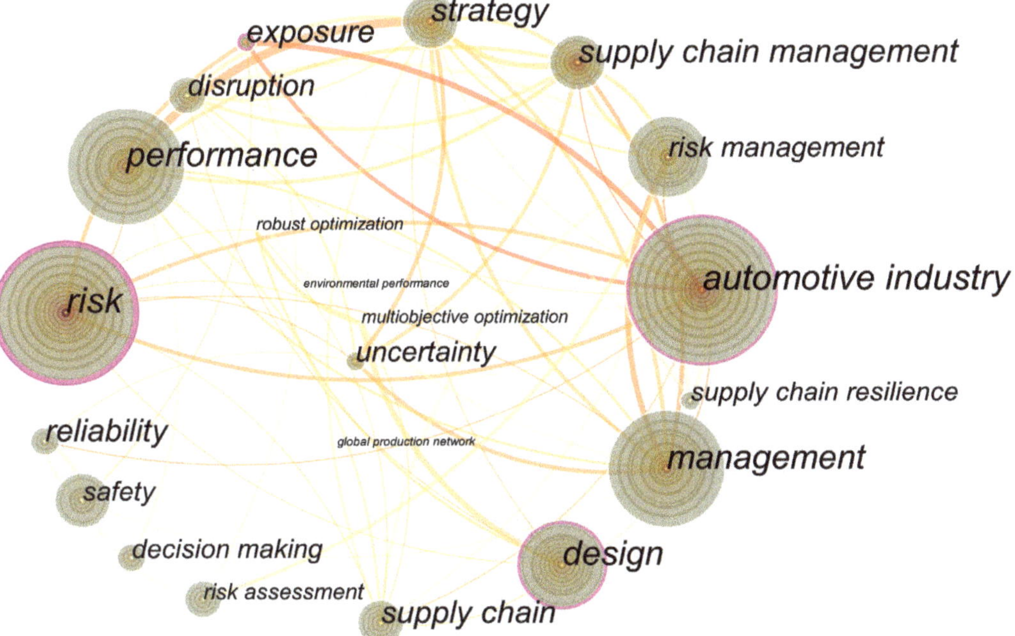

Figure 7. Network visualization of keyword co-occurrence analysis.

In the research of automotive supply chain risk management, the most important keyword is automotive industry (102), risk (81), performance (73), model (64), design (50), strategy (47), safety (39), reliability (39), simulation (38) and optimization (35). In general, the research in this field mainly involves supply chain performance improvement, supply chain operation, supply chain optimization and simulation, and supply chain sustainable development, indicating that with the increasing market competition between enterprises, researchers are trying to improve the level of automotive supply chain disruption risk management through the above methods. It makes the automobile enterprises maximize their profit while maximizing the cooperation and synergy effect among enterprises, realizes the optimal development path of the automobile supply chain system, and helps and promotes the sustainable development of the automobile enterprises in a benign competitive environment.

5. Citation Analysis and Reference Co-Citation Analysis

Ivanov et al. have made the most prominent contribution in the field of supply chain disruption risk research, He and his research team applied mathematical programming, simulation, control, and fuzzy theory methods to study the dynamics of complex networks in production, logistics, and supply chains, exploring supply chain structure dynamics and control, with an emphasis on global supply chain design and disruption considerations, distribution planning and dynamic rescheduling, among others [46–48]. Kamalahmadi et al. were cited seven times [49]. Zimmer et al. were cited five times. Zimmer et al. developed a model combined with the fuzzy AHP method to effectively evaluate the supplier selection of the German automobile industry in the global supply chain [50]. Munir et al. were cited five times.

Munir et al. explored the relationship between supply chain integration and supply chain risk management to improve supply chain operation performance, and studied the mediating role of supply chain risk management between supply chain integration and enterprise operation performance. The study shows that SCM plays a partial mediating role between internal integration and operational performance, and a fully mediating role between supplier and customer integration and operational performance [51]. Govindan et al. were cited five times [52]. Tomlin was cited four times. Tomlin believed that the ratio of supplier uptime to disruption frequency is the key factor of the optimal strategy. By comparing different types of suppliers, the author proposes a hybrid procurement strategy to alleviate the supply disruption risk of automobile enterprises [53]. Fan and Stevenson were cited four times. Fan and Stevenson conducted a systematic literature review on supply chain risk management, mainly including risk identification, risk assessment, and risk detection [31]. The network of the co-cited reference in the field of automotive supply chain disruption risk management is shown in Figure 8. It can be seen from the reference network diagram in Figure 8 that in the current research, the discussion is mainly based on the previous research, that is, the discussion on the risk of interruption of the automobile supply chain is still more traditional. For this reason, the research believes that the main research directions of the risk of automobile supply chain interruption can be screened out by analyzing the number of references cited in the past, providing theoretical support for the follow-up research.

Analyzing the keywords in the title and abstract and performing cluster analysis can help to determine the research trend of this topic. In order to identify the hot topics in the research field of automotive supply chain disruption risk, Cite Space was used to visually analyze the dynamic evolution path of research topics in this research field. The horizontal line in Figure 9 represents the change in research hotspots over time. The number one theme was supply chain resilience (#0) which exploded around 2001. This shows that there may have been many disruption risk events in the automobile industry around 2001, and the research on the improvement of supply chain resilience in this field began to attract the attention of scholars. Then, the clustering themes are respectively: network design (#1), game theory (#2), stochastic demand (#3), supply chain coordination (#4), green supply chain (#5), supply chain performance (#6), disruption management (#7), mixed integer programming (#8), risk management (#9). Green supply chain (#5), supply chain performance (#6), and disruption management (#7) may be the research direction that scholars focus on recently.

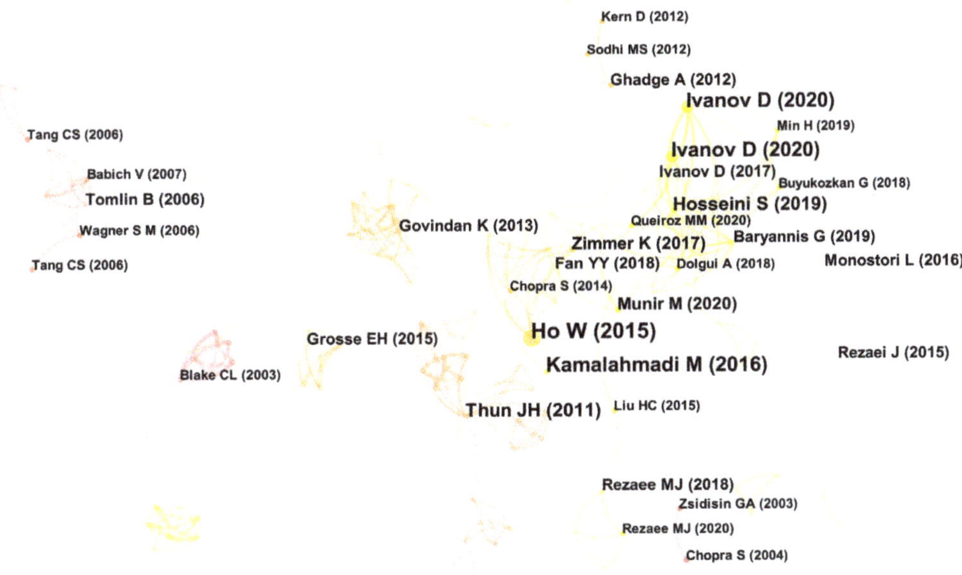

Figure 8. Network of the co-cited reference in the field of automotive supply chain disruption risk management.

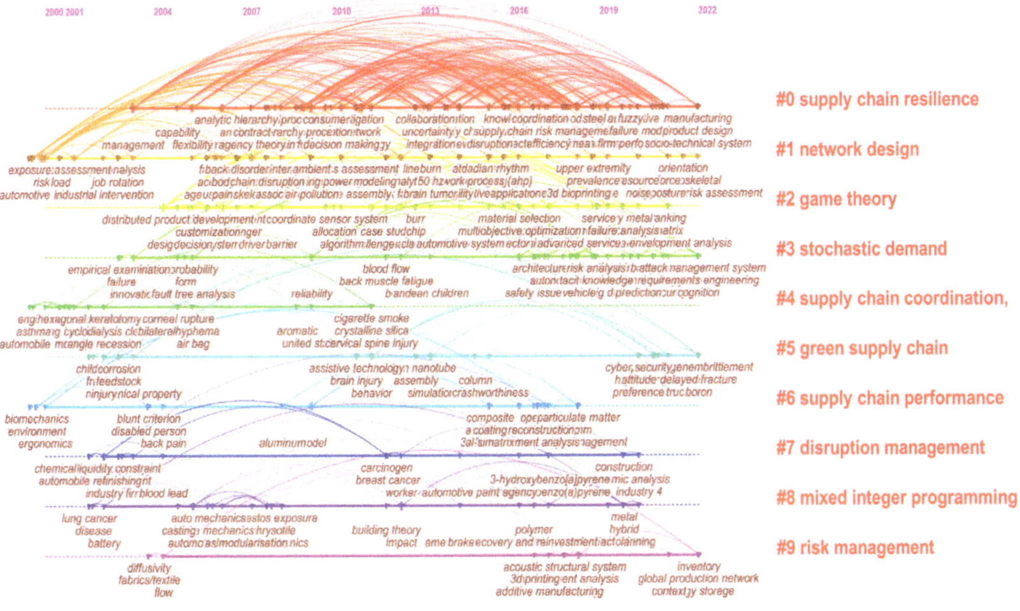

Figure 9. Timeline view of reference clusters in the field of automotive supply chain disruption risk management.

Supply chain resilience (#0): The word Resilience first comes from the concept of resilience in engineering. Its meaning is "returning to the original state", which refers to the ability of an object to recover to its original state after deformation under the action of external forces, namely "engineering resilience". In 1973, Holling, a Canadian ecologist, applied the idea of resilience to ecology for the first time and put forward the concept of "ecological resilience", which considered the ecosystem as a dynamic system with multiple stable states, overturning the traditional ecology view that the ecosystem has a single equilibrium state [54]. Since then, scholars' research on resilience has gradually spread from natural ecology to ecology, engineering, economy, and society. Alberti believed that resilience is the ability of a system to absorb and resolve changes before restructuring after a series of structural and process changes [55]. Rose believed that resilience is the inherent adaptive ability of individuals or enterprises to avoid potential losses in the face of disasters. It is usually expressed in two types, one emphasizing its inherent characteristics, and the other emphasizing its adaptive characteristics [56]. With the continuous infiltration of the concept of resilience, resilience has been given connotation interpretation in different fields and dimensions [57]. Quang and Hara proposed that supply chain resilience can be measured by five key dimensions, including: supplier performance, internal business, innovation and learning, customer service, and finance [58].

Although the focus of research in different fields is different, the relevant research on the connotation of comprehensive resilience generally believes that the most basic meaning of resilience is the ability of the system to resolve external shocks and maintain its main functions when crises occur.

Network design (#1): In view of the discussion of supply chain structure in the existing literature, Cox et al. studied the concept of supply chain management by focusing on the linear relationship between buyers and suppliers. While this approach is useful for understanding the relationship between buyers and suppliers, it does not capture the complexity of a firm's strategy, which relies on a larger network of firms that are embedded together to form a complex network [59]. Croxton [60] and Carter [61] have shown that the supply chain is a complex network that exhibits a similar organizational pattern to other network types. The company's "supply network" includes the connection with direct suppliers and customers and the connection between subsequent suppliers and customers, which is essentially a recursive relationship.

In addition, the researchers also explored the connections between the layers of the supply chain network. Kito et al. used the data in the Mark Line database to construct the supply chain network of Toyota Motor, and the study determined that the layered structure of Toyota was "barrel shaped", which was contrary to the pyramid structure assumed before. However, the authors also did not find the topology to be scale-free [62]. Brintrup et al. studied the automotive industry to explore robustness properties. While these studies discussed part of the topological structure features, the study did not evaluate the dynamic nature of these relationships [63].

Game theory (#2): Existing research mainly uses game theory to solve the following two problems:

1. Delay in quick response. Xiao et al. by delaying the final assembly of finished products until the actual demand is observed, can effectively avoid losses and risks caused by overproduction under the inventory-based production strategy. However, in the future with limited capacity, the producer must assemble part of the finished product in advance to maximize its own profit [64]. Liu et al. considered the impact of a supplier's delivery delay on the demand of end customers, established a supply chain ordering model considering delivery delay based on repurchase contract, and analyzed the sharing of delivery delay risk and demand risk [65].

2. Mechanism and compensation contract mechanism. In the supply chain environment with uncertain demand, distributors, and retailers have convenient conditions to get close to the market and customers, and can truly understand the whole market demand. If the manufacturers in the upstream of the supply chain can take incen-

tives and compensation measures to stimulate the sales of distributors and retailers and eliminate their worries, not only can distributors and retailers obtain more profits [66], but manufacturers' profits will also increase; on the other hand, incentives and compensation measures taken by manufacturers to achieve benefit sharing and risk sharing can maintain a good cooperative relationship among members of the supply chain, increase the elasticity of the supply chain, and reduce the occurrence of disruption risk. The commonly used incentive and compensation strategies are: wholesale price discounts, revenue-sharing strategies, buyback strategies, quantity flexibility contracts, reward and penalty strategies, lead time adjustments, order options, etc. [67].

Stochastic demand (#3): In the supply chain system, uncertainty is common due to the complexity of the supply chain network structure and the interaction between enterprises in the supply chain. These uncertainties come not only from the uncoordinated operation between internal suppliers and manufacturers, but also from the external environment such as natural disasters, policies, and regulations. Uncertainty will be transmitted step by step with information flow, logistics, and capital flow, and ultimately affect the overall operation and profitability of the supply chain.

There are many and mixed factors affecting demand, and the reasons for random demand are as follows: the operation mode and organizational structure of each member enterprise in the supply chain are different, and the way to deal with problems in the cooperation process is different. If not handled properly, the interests of some member enterprises can not only be guaranteed, but also cause vicious competition and other bad situations, and the needs of customers can not be met at all [68]; The opacity of market information. There are many factors that affect the quantity demanded, which also makes it difficult to predict the quantity demanded, thus leading to the deviation of the demand [69]; Supply chain members amplify demand information step by step based on maximizing their own interests [70]; The occurrence of an emergency. Emergencies are the product of the complexity and uncertainty of the supply chain system. The particularity of the automobile supply chain structure determines that it is more fragile and has a weak ability to resist the risk of emergencies [71].

Supply chain coordination (#4): Adhikari et al. considered that different enterprises in the supply chain have different risk resistance capabilities, and set up different risk-sharing mechanisms to improve the profitability of enterprises. At the same time, they designed joint pricing and order quantity decisions, and finally designed a two-way repurchase sales rebate punishment contract to coordinate the supply chain [72].

Some research works put forward that members of the supply chain share risks to promote good cooperation among the member enterprises in the supply chain. Obviously, insufficient supply can not fully reflect the characteristics of the automobile supply chain network only from the perspective of supply chain equilibrium. Therefore, it is particularly critical to improve the elasticity and stability of the automobile supply chain, reduce the damage of risks to the automobile supply chain network, and effectively prevent the occurrence of risks [73]. The Ivanov team presented the results of a simulation study that revealed some new research on the impact of COVID-19 on global supply chain risk. The specific features that frame epidemic outbreaks as a distinct type of supply disruption risk are first clarified. Secondly, the coronavirus and supply chain optimization software are used as examples to demonstrate how simulation-based methods can be used to examine and predict the impact of the outbreak on supply chain performance. The main observation of the simulation experiment is that important factors for risk propagation include lead time, the speed of epidemic spread, and the duration of upstream and downstream disruptions in the supply chain [74–76].

Green supply chain (#5): For the research on the green supply chain in the field of automotive supply chain research, the design of a reverse recycling network is the most discussed research issue. Many countries have introduced policies to minimize waste after vehicle scrap, which has led to a special focus on reverse supply chain research in the automotive industry in product design, electrical and electronic equipment, and battery industries. Mansour and Zarei proposed a network design and mathematical formula for recycling ELVs from the perspective of manufacturers to obtain maximum economic benefits and fulfill relevant legal provisions. The model focuses on establishing optimal locations for collection centers and demolition personnel, as well as material flows between facilities [77].

Żach proposed the EV recycling network design and its mathematical model [78]. Minimization of the total cost of the system including the costs of setting up the network and transportation was aimed in the model. Jonrinaldi and Zhang proposed an integrated production and inventory control model for supply chains containing multiple products. They considered 3PL providers collecting used products from end customers and dismantling used products [79].

Supply chain performance (#6): The occurrence of supply chain risk will have a significant impact on supply chain performance [80,81]. Different risks have different attributes and different impacts on supply chain performance. For example, Rice and Caniato proposed that core risks related to product processes may directly affect the day-to-day operations of the supply chain, and that these risks have a high probability of occurring but a lower impact on supply chain performance than external supply chain risks [82]. External supply chain risks, on the other hand, are rare but can disrupt supply chain activities directly or indirectly (e.g., the shutdown of Hyundai Motor in South Korea due to COVID-19). The interruption of information, finance, time, and other factors may have a serious impact on the supply chain process (especially the supply, manufacturing, and downstream links) [1]. Quang and Hara constructed a hypothetical model of the relationship between risk and supply chain performance. The core of the model is three core risks related to product process: supply risk, operation risk, and demand risk, which have a direct impact on supply chain performance [58]. Thun and Hoenig pointed out that the impact of internal and external supply chain risks on supply chain performance is significantly different [83]. Altay and Ramirez studied the impact of natural disasters on enterprises, analyzing the impact of more than 3500 natural disaster events (including fast-occurring natural disasters such as typhoons, floods, and earthquakes) on the performance of more than 100,000 enterprises, including financial leverage, total asset turnover, and operating cash flow [84]. Bergmann et al. believed that the excessive consumption of these natural resources leads to the escalation of climate change, and this exogenous risk can further significantly affect the operation and financial performance of organizations [85].

For the research on the relationship between operational risk and supply chain performance, most of the existing studies believe that the successful implementation of the organizational strategy has a positive role in promoting supply chain performance [86], however, the research on interruption risk management has not received enough attention. Cuthbertson identifies the broader consequences of failing to manage risk effectively. These include not only financial losses, but also reduced product quality, damage to property and equipment, loss of reputation in the eyes of customers, suppliers, and the wider public, and delivery delays [87]. Based on the background of promoting the safe and efficient flow of goods, Bueno et al. proposed a system dynamics model to analyze the impact of the spread of disruption risk caused by terrorist acts on the performance of global supply chains [88].

Disruption management (#7): Wagner et al. studied the data of several automobile enterprises, discussed the differences between just-in-time sequential delivery and other delivery concepts of automobile manufacturing enterprises, and then proposed a method to improve supply chain resilience based on JIS [89].

The problem of supply chain disruption management is studied. The supply chain dynamic simulation model is considered as a suitable method to observe and predict the changing behavior of the supply chain over time. On the basis of the original model, Torabi and Sadghiani et al. added additional dynamic characteristics to simulate the impact of risk on supply chain performance [90,91]. Most of the existing studies use discrete event simulation methods [92,93], some researchers choose to use heuristics to study supply chain optimization problems [94], and several studies also use methods based on multi-agent simulation [95,96]. Few studies have incorporated supply chain simulation studies into simulations and transportation disruptions during epidemic crises [97]. In the study of the impact of disruptive events on supply chain performance, Klibi and Ivanov et al. believe that adding some parameter conditions that change according to time dependence in the model will be more conducive to simulation model calculation [98,99]. In addition, Li and Pavlov et al. analyze detailed control policies based on various financial, customer, and operational performance indicators [100,101]. The simulation model considers logical and stochastic constraints such as disruptions, stochastics of inventory, production, procurement, and transportation control policies, and stepwise capacity degradation and restoration [102–105]. Simulation studies play an important role in model simulation studies because they involve time-varying parameters, duration of recovery measures, and capacity degradation and recovery. The advantage of simulation is that it allows for further optimization studies by extending complex problems through time-varying situational behavior in the system.

Mixed integer programming (#8): Spengler et al. focused on the MILP model of product recycling in the steel industry [106]. Munoz et al. argue that in the present work, LCA was used to evaluate an existing automotive component, a plastic facade panel, and compare it with a recyclable prototype panel based on a compatible polyolefin design. From an environmental point of view, LCA has proven to be a very useful tool for validating redesigned automotive components; Moreover, it allows one to identify not only the benefits of increased recyclability, but also the improvement of other stages of the life cycle. From this case study, some recommendations are made for companies in order to design eco-friendly components for automotive interiors [107]. Zarei et al. proposed a mathematical model in which new vehicle distributors are responsible for ELV collection, and they use joint potential facilities for distribution and collection. The effectiveness of the GA-based model and solution method is verified by generating several test problem instances [108]. Ozceylan and Paksoy proposed an integrated, multi-echelon, multi-period, and multi-part MILP model to optimize production and distribution plans for CLSC networks. They presented computational results for many scenarios [109]. Farel et al. proposed a mathematical model for an ELV glazing recycling network including car manufacturers, dismantlers, shredders, collection and transportation facilities, and glass treatment facilities [110].

Risk management (#9): Through the literature review, the existing research mainly uses the fault tree analysis method for supply chain risk assessment research [111], risk matrix method [112], fuzzy comprehensive evaluation method [113], and other methods. Kolotzek et al. used the analytic hierarchy process to rank 11 indicators from the perspectives of political risk, demand increase, and supply decrease to evaluate the supply risk of manufacturing enterprises [114]. Abdel et al. constructed an evaluation index system considering demand, supply, environment, and business risks, and introduced the ideal solution similarity order preference technology TOPSIS method to evaluate supply chain risks [115]. In addition, some studies use probability theory and survey methods to evaluate supply risk. For example, Vilko and Hallikas used probabilities and losses to characterize supply risk and assessed the risk of multi-modal supply chains [116]. Defining demand as a probabilistic parameter, Nooraie and Mellat Parast constructed a multi-objective approach to assess supply risk [117].

In addition to the overall assessment of risk factors, existing studies also discuss the causal relationship between risk factors in the supply chain. Shin used quantitative methods to map the causal relationship between supply chain security risks [118]. Studies have found that quantitative methods have limited interdependence ability to compare and distinguish risks, such as the petri net model [119], and coupled with a lack of quantitative data or inappropriate model parameters, quantitative methods do not always work [120]. Therefore, in the existing literature, such as explanatory structural modeling [121] and Bayesian networks are used to evaluate causality in supply chain risk. Among them, when using the Bayesian network to analyze the risk of a supply chain system, it can identify the causal relationship between the key factors and the final result, so it is recognized by the majority of researchers. Bayesian networks can effectively combine statistical data and subjective judgment and model in the absence of data. Bayesian network is widely used in supply chain risk management. The existing research mainly focuses on solving specific problems in supplier selection, supplier evaluation, and selection, and most of the research problems only focus on some specific problems, such as: automotive enterprise parts supplier selection and evaluation problem, Boeing Company, and supply evaluation problem [122].

By analyzing the keywords in the references, we can more accurately understand the main research contents and directions in the previous research, and also provide technical support for the follow-up research. In addition, according to the frequency of keywords in the existing relevant literature, it is of great significance to analyze the resilience of automobile supply chain disruption risk, and is also the focus of the follow-up work.

6. Research Gaps and Future Research Opportunities

The above research results provide a theoretical basis for this paper, and prove the importance of research on risk management and resilience improvement strategy of the supply chain of automotive enterprises. However, there are still some deficiencies in the existing research results, and the detailed analysis is as follows:

(1) Research on supply chains in the automotive industry is insufficient. The existing literature provides relatively comprehensive studies and discussions on manufacturing supply chains from the aspects of supply chain performance, supply chain linkages, re-source sharing games, and supply chain linkage risks. However, there is little literature specific to the automotive supply chain.

(2) Research on supply chain risk and resilience in the automotive industry is insufficient. Existing studies focus mainly on the degree of resilience of the regional spatial network structure and the factors affecting the resilience of the network structure of industry clusters. Most research on supply chain resilience has focused on exploring the factors affecting supply chain resilience. Few studies have integrated supply chain, risk, and resilience to explore supply chain resilience in the automotive industry and analysed the impact of risk factors on resilience. There is a lack of conceptual models linking supply chain risk management and performance management.

(3) Research on resilience assessment in the automotive supply chain needs to be expanded. An overview of relevant studies shows that most of the existing studies generally start from the resilience structure, select a centrality index to examine the state of the network based on the analysis of statistical characteristics of the network, such as order distribution, cluster coefficient, and mean path length, and assess the resilience of the studied object. Supply chain resilience is rarely characterized by indicators such as supply chain performance, but rather by a comprehensive examination of the degree of supply chain re-silience from holistic and multi-layered factors.

7. Conclusions

In this paper, we systematically review the literature on automotive supply chain disruption risk management research. We used Cite Space software to analyze and visualize the number of journal papers, core authors, research institutions, source journals, research hotspots, and future development directions of automotive supply chain disruption risk management research. The main findings are as follows:

First of all, in the past 20 years, the number of publications related to the topic of automotive supply chain disruption risk management increased steadily, which grew from five related publications in 2000 to a peak number of 105 in 2021. This indicates that research on automotive supply chain disruption risk management had attracted increasingly more attention from society and scholars. The influential works, their authors, and the existing and emerging research clusters/themes are identified.

Second, at present, automotive supply chain disruption risk management research has not formed a stable core author group, but has formed core research institutions and core research journals.

Third, "The effects of an integrative supply chain strategy on customer service and financial performance: an analysis of direct versus indirect relationships" published by "Journal of Operations Management" is the most cited paper among the 866 documents. Furthermore, reference co-citation analysis presented ten clustered groups and the literature of each cluster might provide potential future research opportunities regarding automotive supply chain disruption risk management. These research contents will bring useful inspiration to managers of automotive manufacturing companies, and supply chain management-related scholars and help them in their future work and study in supply chain-related management and research work.

Finally, there are some limitations that need to be addressed in future work. The data source was collected from the core collection of WOS, which may cause deviations in the results of bibliometric analysis. We could extend the data source to include more publications about the field of automotive supply chain disruption risk management such as ProQuest Dissertations and Theses. Moreover, although we could obtain objective results about the field of automotive supply chain disruption risk management based on bibliometric analysis, some underlying reasons for these results are not explained. Some social science research methods, such as expert interviews, could be employed in the future to address this limitation.

Author Contributions: Conceptualization, K.H.; Data curation, K.H.; Formal analysis, K.H.; Funding acquisition, J.W.; Methodology, J.Z.; Project administration, J.W.; Software, K.H.; Supervision, J.W.; Validation, K.H.; Visualization, K.H.; Writing—original draft, K.H.; Writing—review & editing, J.W. All authors have read and agreed to the published version of the manuscript.

Funding: This study was funded by the National Social Science Fund of China (No. 18BGL018).

Data Availability Statement: The datasets used in this study are available from the corresponding author upon reasonable request.

Conflicts of Interest: The authors declare no conflict of interest.

References

1. Ho, W.; Zheng, T.; Yildiz, H.; Talluri, S. Supply chain risk management: A literature review. *Int. J. Prod. Res.* **2015**, *53*, 5031–5069. [CrossRef]
2. Wu, T.; Blackhurst, J.; Chidambaram, V. A model for inbound supply risk analysis. *Comput. Ind.* **2006**, *57*, 350–365. [CrossRef]
3. Gaudenzi, B.; Borghesi, A. Managing risks in the supply chain using the AHP method. *Int. J. Logist. Manag.* **2006**, *17*, 114–136. [CrossRef]
4. Samvedi, A.; Jain, V.; Chan, F.T. Quantifying risks in a supply chain through integration of fuzzy AHP and fuzzy TOPSIS. *Int. J. Prod. Res.* **2013**, *51*, 2433–2442. [CrossRef]
5. Clark, G. Understanding and reducing the risk of supply chain disruptions. *J. Bus. Contin. Emerg. Plan.* **2012**, *6*, 6–12.
6. Hansen, E.N.; Nybakk, E.; Panwar, P. Firm performance, business environment, and outlook for social and environmental responsibility during the economic downturn: Findings and implications from the forest sector. *Can. J. For. Res.* **2013**, *43*, 1137–1144. [CrossRef]
7. Krause, D.; Ellram, L.M. The effects of the economic downturn on interdependent buyer-supplier relationships. *J. Bus. Logist.* **2014**, *35*, 191–212. [CrossRef]
8. Ketikidis, P.H.; Cucchiella, F.; Gastaldi, M. Risk management in supply chain: A real option approach. *J. Manuf. Technol. Manag.* **2016**, *17*, 700–720.
9. Manavazhi Mohan, R.; Dinesh, K. Material and equipment procurement delays in highway projects in Nepal. *Int. J. Proj. Manag.* **2012**, *20*, 627–632. [CrossRef]
10. Angulo, A.; Nachtmann, H.; Waller, M.A. Supply chain information sharing in a vendor managed inventory partnership. *J. Bus. Logist.* **2004**, *25*, 101–120. [CrossRef]
11. Bernanke, B.S. Bankruptcy, Liquidity, and Recession. *Am. Econ. Rev.* **1981**, *71*, 155–159.
12. Aibinu, A.A.; Jagboro, G.O. The effects of construction delays on project delivery in Nigerian construction industry. *Int. J. Proj. Manag.* **2012**, *20*, 593–599. [CrossRef]
13. Chopra, S.; Sodhi, M.M.S. Managing risk to avoid supply-chain breakdown. *Mit Sloan Manag. Rev.* **2014**, *46*, 53–61.

14. Handfifield, R.B.; Nichols, E.L. Supply chain redesign: Transforming supply chains into integrated value systems. *Natl. Assoc. Purch. Manag. Inc.* **2013**, *39*, 70–71.

15. Tummala, R.; Schoenherr, T. Assessing and managing risks using the Supply Chain Risk Management Process (SCRMP). *Supply Chain Manag. Int. J.* **2011**, *16*, 474–483. [CrossRef]

16. Lee, H.L.; Padmanabhan, V.; Whang, S. Information distortion in a supply chain: The bullwhip effect. *Manag. Sci.* **1997**, *43*, 546–558. [CrossRef]

17. Manuj, I.; Mentzer, J.T. Global supply chain risk management strategies. *Int. J. Phys. Distrib. Logist. Manag.* **2018**, *38*, 192–223. [CrossRef]

18. Trkman, P.; McCormack, K. Supply chain risk in turbulent environments—A conceptual model for managing supply chain network risk. *Int. J. Prod. Econ.* **2009**, *119*, 247–258. [CrossRef]

19. Hahn, G.J.; Kuhn, H. Value-based performance and risk management in supply chains: A robust optimization approach. *Int. J. Prod. Econ.* **2012**, *139*, 135–144. [CrossRef]

20. Bakhshi, H. Inflation and relative price variability. *Econ. Lett.* **2012**, *76*, 27–33. [CrossRef]

21. Zhi, H. Risk management for overseas construction projects. *Int. J. Proj. Manag.* **1995**, *13*, 231–237. [CrossRef]

22. Mirzaie, I.A.; Kandil, M.E. The effects of exchange rate fluctuations on output and prices: Evidence from developing countries. *J. Dev. Areas* **2013**, *38*, 345–367.

23. Yeo, K.T.; Tiong, R.L.K. Positive management of differences for risk reduction in BOT projects. *Int. J. Proj. Manag.* **2010**, *18*, 257–265. [CrossRef]

24. Zsidisin, G.A.; Ellram, L.M. An agency theory investigation of supply risk management. *J. Supply Chain. Manag.* **2013**, *39*, 15–27. [CrossRef]

25. Zsidisin, G.A.; Ellram, L.M.; Carter, J.R.; Cavinato, J.L. An analysis of supply risk assessment techniques. *Int. J. Phys. Distrib. Logist. Manag.* **2014**, *34*, 397–413. [CrossRef]

26. Wu, D.D.; Olson, D. Enterprise risk management: A DEA VAR approach in vendor selection. *Int. J. Prod. Res.* **2010**, *48*, 4919–4932. [CrossRef]

27. Tuncel, G.; Alpan, G. Risk assessment and management for supply chain networks: A case study. *Comput. Ind.* **2010**, *61*, 250–259. [CrossRef]

28. Williams, T.; Eden, C.; Tait, A.A. The effects of design changes and delays on project costs. *J. Oper. Res. Soc.* **1995**, *46*, 809–818. [CrossRef]

29. David, L. Health and safety executive. *Ind. Law J.* **2000**, 119–121.

30. Zhou, X.; Li, T.; Ma, X. A bibliometric analysis of comparative research on the evolution of international and Chinese green supply chain research hotspots and frontiers. *Environ. Sci. Pollut. Res.* **2021**, *28*, 6302–6323. [CrossRef]

31. Fan, Y.; Stevenson, M. A review of supply chain risk management: Definition, theory, and research agenda. *Int. J. Phys. Distrib. Logist. Manag.* **2018**, *48*, 205–230. [CrossRef]

32. Moosavi, J.; Naeni, L.M.; Fathollahi-Fard, A.M. Blockchain in supply chain management: A review, bibliometric, and network analysis. *Environ. Sci. Pollut. Res.* **2021**. [CrossRef]

33. Iftikhar, A.; Ali, I.; Arslan, A.; Tarba, S. Digital Innovation, Data Analytics, and Supply Chain Resiliency: A Bibliometric-based Systematic Literature Review. *Ann. Oper. Res.* **2022**, 1–24. [CrossRef]

34. Roblek, V.; Dimovski, V.; Mesko, M.; Peterlin, J. Evolution of organisational agility: A bibliometric study. *Kybernetes* **2022**, *51*, 119–137. [CrossRef]

35. Guan, G.; Jiang, Z.; Gong, Y.; Huang, Z.; Jamalnia, A. A Bibliometric Review of Two Decades' Research on Closed-Loop Supply Chain: 2001–2020. *IEEE Access* **2021**, *9*, 3679–3695. [CrossRef]

36. Zupic, I.; Čater, T. Bibliometric methods in management and organization. *Organ. Res. Methods* **2015**, *18*, 429–472. [CrossRef]

37. Wang, Q. Distribution features and intellectual structures of digital humanities: A bibliometric analysis. *J. Doc.* **2018**, *74*, 223–246. [CrossRef]

38. Waltman, L.; Van Eck, N.J. A new methodology for constructing a publication-level classification system of science. *J. Am. Soc. Inf. Sci. Technol.* **2012**, *63*, 2378–2392. [CrossRef]

39. Hen, Y.; Chen Ch, M.; Liu, Z.Y.; Hu, Z.G.; Wang, X.W. The methodology function of Cite Space mapping knowledge domains. *Stud. Sci. Sci.* **2015**, *33*, 242–253.

40. Li, X.; Qiao, H.; Wang, S. Exploring evolution and emerging trends in business model study: A co-citation analysis. *Scientometrics* **2017**, *111*, 869–887. [CrossRef]

41. Ruhanen, L.; Weiler, B.; Moyle, C.-L.; McLennan, C.-L.J. Trends and patterns in sustainable tourism research: A 25-year bibliometric analysis. *J. Sustain. Tour.* **2014**, *23*, 517–535. [CrossRef]

42. Vickery, S.K.; Jayaram, Y.; Droge, C.; Calantone, R. The effects of an integrative supply chain strategy on customer service and financial performance: An analysis of direct versus indirect relationships. *J. Oper. Manag.* **2003**, *21*, 523–539. [CrossRef]

43. Zhu, Q.H.; Sarkis, J.; Lai, K.H. Green supply chain management: Pressures, practices and performance within the Chinese automobile industry. *J. Clean. Prod.* **2007**, *15*, 1041–1052. [CrossRef]

44. Wong, C.Y.; Boon-itt, S.; Wong, C.W.Y. The contingency effects of environmental uncertainty on the relationship between supply chain integration and operational performance. *J. Oper. Manag.* **2011**, *29*, 604–615. [CrossRef]

45. Kim, Y.; Choi, T.Y.; Yan, T.T.; Dooley, K. Structural investigation of supply networks: A social network analysis approach. *J. Oper. Manag.* **2011**, *29*, 194–211. [CrossRef]
46. Ivanov, D. Predicting the impacts of epidemic outbreaks on global supply chains: A simulation-based analysis on the coronavirus outbreak (COVID-19/SARS-CoV-2) case. *Transp. Res. Part E Logist. Transp. Rev.* **2020**, *136*, 101922. [CrossRef]
47. Ivanov, D.; Dolgui, A. A position paper motivated by COVID-19 outbreak. *Int. J. Prod. Res.* **2020**, *58*, 2904–2915. [CrossRef]
48. Hosseini, S.; Ivanov, D.; Dolgui, A. Review of quantitative methods for supply chain resilience analysis. *Transp. Res. Part E Logist. Transp. Rev.* **2019**, *125*, 285–307. [CrossRef]
49. Kamalahmadi, M.; Parast, M.M. A review of the literature on the principles of enterprise and supply chain resilience: Major findings and directions for future research. *Int. J. Prod. Econ.* **2016**, *171*, 116–133. [CrossRef]
50. Zimmer, K.; Fröhling, M.; Breun, P.; Schultmann, F. Assessing social risks of global supply chains: A quantitative analytical approach and its application to supplier selection in the German automotive industry. *J. Clean. Prod.* **2017**, *149*, 96–109. [CrossRef]
51. Munir, M.; Jajja, M.S.S.; Chatha, K.A.; Farooq, S. Supply chain risk management and operational performance: The enabling role of supply chain integration. *Int. J. Prod. Econ.* **2020**, *227*, 107667. [CrossRef]
52. Govindan, K.; Khodaverdi, R.; Jafarian, A. A fuzzy multi criteria approach for measuring sustainability performance of a supplier based on triple bottom line approach. *J. Clean. Prod.* **2013**, *47*, 345–354. [CrossRef]
53. Tomlin, B. On the Value of Mitigation and Contingency Strategies for Managing Supply Chain Disruption Risks. *Manag. Sci.* **2006**, *52*, 639–657. [CrossRef]
54. Holling, C.S. Resilience and stability of ecological systems. *Annu. Rev. Ecol. Syst.* **1973**, *4*, 1–23. [CrossRef]
55. Alberti, M.; Marzluff, J.M.; Shulenberger, E.; Bradley, G.; Ryan, C.; Zumbrunnen, C. Integrating humans into ecology: Opportunities and challenges for studying urban ecosystems. *Bioscience* **2009**, *53*, 1169–1179. [CrossRef]
56. Rose, A. Defining and measuring economic resilience to disasters. *Disaster Prev. Manag. Int. J.* **2004**, *13*, 307–314. [CrossRef]
57. Kosasih, E.E.; Brintrup, A. A machine learning approach for predicting hidden links in supply chain with graph neural networks. *Int. J. Prod. Res.* **2021**, *60*, 5380–5393. [CrossRef]
58. Quang, H.T.; Hara, Y. Risks and performance in supply chain: The push effect. *Int. J. Prod. Res.* **2017**, *56*, 1–20.
59. Cox, A.; Sanderson, J.; Watson, G. Supply Chains and Power Regimes: Toward an Analytic Framework for Managing Extended Networks of Buyer and Supplier Relationships. *J. Supply Chain Manag.* **2010**, *37*, 28–35. [CrossRef]
60. Croxton, K.L.; García-Dastugue, S.J.; Lambert, D.M. The supply chain management processes. *Int. J. Logist. Manag.* **2016**, *12*, 13–36. [CrossRef]
61. Carter, C.R.; Ellram, L.M.; Tate, W. The use of social network analysis in logistics research. *J. Bus. Logist.* **2007**, *28*, 137–168. [CrossRef]
62. Kito, T.; Brintrup, A.; New, S.; Reed-Tsochas, F. The structure of the Toyota supply network: An empirical analysis. *Soc. Sci. Electron. Publ.* **2014**, *3*, 1–24. [CrossRef]
63. Brintrup, A.; Ledwoch, A.; Barros, J. Topological robustness of the global automotive industry. *Logist. Res.* **2015**, *9*, 1. [CrossRef]
64. Xiao, Y.B.; Chen, J.; Wu, P. Research on optimal inventory and production decision of ATO system under capacity and demand uncertainty. *Chin. J. Manag. Sci.* **2007**, *15*, 56–64.
65. Liu, L.; Jin, Q.; Tang, X.W. Research on supply chain repurchase contract of perishable goods considering delayed delivery risk. *Control. Decis.* **2012**, *27*, 1505–1509, 1515.
66. Cachon, G.P.; Lariviere, M.A. Supply chain coordination with revenue-sharing contracts: Strengths and limitations. *Manag. Sci.* **2005**, *51*, 30–44. [CrossRef]
67. Taylor, T.A. Supply chain under channel rebates with sales effort effects. *Manag. Sci.* **2002**, *48*, 992–1007. [CrossRef]
68. Dias, G.C.; Hernandez, C.T.; de Oliveira, U.R. Supply chain risk management and risk ranking in the automotive industry. *Gest. E Prod.* **2020**, *27*, 1–21. [CrossRef]
69. Alkahtani, M.; Omair, M.; Khalid, Q.S.; Hussain, G.; Ahmad, I.; Pruncu, C. A COVID-19 Supply Chain Management Strategy Based on Variable Production under Uncertain Environment Conditions. *Int. J. Environ. Res. Public Health* **2021**, *18*, 1662. [CrossRef]
70. Campuzano, F.; Mula, J.; Peidro, D. Fuzzy estimations and system dynamics for improving supply chains. *Fuzzy Sets Syst.* **2010**, *161*, 1530–1542. [CrossRef]
71. Khot, S.B.; Thiagarajan, S. Resilience and sustainability of supply chain management in the Indian automobile industry. *Int. J. Data Netw. Sci.* **2019**, *3*, 339–348. [CrossRef]
72. Adhikari, A.; Bisi, A.; Avittathur, B. Coordination mechanism, risk sharing, and risk aversion in a five-level textile supply chain under demand and supply uncertainty. *Eur. J. Oper. Res.* **2020**, *282*, 93–107. [CrossRef]
73. Rajesh, R. Technological capabilities and supply chain resilience of firms: A relational analysis using Total Interpretive Structural Modeling. *Technol. Forecast. Soc. Chang.* **2017**, *118*, 161–169. [CrossRef]
74. Ivanov, D.; Das, A. Coronavirus (COVID-19/SARS-CoV-2) and supply chain resilience: A research note. *Int. J. Integr. Supply Manag.* **2020**, *13*, 90–102. [CrossRef]
75. Ivanov, D.; Dolgui, A.; Sokolov, B.; Ivanova, M. Literature review on disruption recovery in the supply chain. *Int. J. Prod. Res.* **2017**, *55*, 6158–6174. [CrossRef]
76. Ivanov, D.; Dolgui, A.; Sokolov, B. The impact of digital technology and Industry 4.0 on the ripple effect and supply chain risk analytics. *Int. J. Prod. Res.* **2018**, *57*, 829–846. [CrossRef]

77. Mansour, S.; Zarei, M. A multi-period reverse logistics optimisation model for end-of-life vehicles recovery based on EU Directive. *Int. J. Comput. Integr. Manuf.* **2008**, *21*, 764–777. [CrossRef]
78. Żach, P. Plastic raw materials from end-of-life vehicles. *Environ. Prot. Eng.* **2012**, *38*, 151–156. [CrossRef]
79. Jonrinaldi; Zhang, D. An integrated production and inventory model for a whole manufacturing supply chain involving reverse logistics with finite horizon period. *Omega* **2013**, *41*, 598–620. [CrossRef]
80. Baryannis, G.; Dani, S.; Antoniou, G. Predicting supply chain risks using machine learning: The trade-off between performance and interpretability. *Futur. Gener. Comput. Syst.* **2019**, *101*, 993–1004. [CrossRef]
81. Chowdhury, N.A.; Ali, S.M.; Mahtab, Z. A structural model for investigating the driving and dependence power of supply chain risks in the readymade garment industry. *J. Retail. Consum. Serv.* **2019**, *51*, 102–113. [CrossRef]
82. Rice, J.B.; Caniato, F. Building a Secure and resilient supply network. *Supply Chain. Manag. Rev.* **2003**, *7*, 22–30.
83. Thun, J.H.; Hoenig, D. An empirical analysis of supply chain risk management in the German automotive industry. *Int. J. Prod. Econ.* **2011**, *131*, 242–249. [CrossRef]
84. Altay, N.; Ramirez, A. Impact of disasters on firms in different sectors: Implications for supply chain. *J. Supply Chain Manag.* **2010**, *46*, 59–80. [CrossRef]
85. Bergmann, A.; Stechemesser, K.; Guenther, E. Natural resource dependence theory: Impacts of extreme weather events on organizations. *J. Bus. Res.* **2016**, *69*, 1361–1366. [CrossRef]
86. Choy, K.L.; Chow HK, H.; Lee, W.B.; Chan FT, S. Development of performance measurement system in managing supplier relationship for maintenance logistics providers. *Benchmark. Int. J.* **2007**, *14*, 352–368. [CrossRef]
87. Cuthbertson, R.; Piotrowicz, W. Supply chain best practices—Identification and categorisation of measures and benefits. *Int. J. Prod. Perform. Manag.* **2008**, *57*, 389–404. [CrossRef]
88. Bueno-Solano, A.; Cedillo-Campos, M.G. Dynamic impact on global supply chains performance of disruptions propagation produced by terrorist acts. *Transp. Res. Part E Logist. Transp. Rev.* **2014**, *61*, 1–12. [CrossRef]
89. Wagner, S.M.; Silveira-Camargos, V. Managing risks in just-in-sequence supply networks: Exploratory evidence from automakers. *IEEE Trans. Eng. Manag.* **2012**, *59*, 52–64. [CrossRef]
90. Torabi, S.A.; Baghersad, M.; Mansouri, S.A. Resilient supplier selection and order allocation under operational and disruption risks. *Transp. Res-Part E* **2015**, *79*, 22–48. [CrossRef]
91. Sadghiani, N.S.; Torabi, S.; Sahebjamnia, N. Retail supply chain network design under operational and disruption risks. *Transp. Res. Part E Logist. Transp. Rev.* **2015**, *75*, 95–114. [CrossRef]
92. Ivanov, D. Disruption tails and revival policies: A simulation analysis of supply chain design and production-ordering systems in the recovery and post-disruption periods. *Comput. Ind. Eng.* **2019**, *127*, 558–570. [CrossRef]
93. Ivanov, D. Simulation-based single vs dual sourcing analysis in the supply chain with consideration of capacity disruptions, Big Data and demand patterns. *Int. J. Integr. Supply Manag.* **2017**, *11*, 24–43. [CrossRef]
94. Cui, J.; Zhao, M.; Li, X.; Parsafard, M.; An, S. Reliable design of an integrated supply chain with expedited shipments under disruption risks. *Transp. Res. Part E Logist. Transp. Rev.* **2016**, *95*, 143–163. [CrossRef]
95. Hou, Y.; Wang, X.; Wu, Y.J.; He, P. How does the trust affect the topology of supply chain network and its resilience? An agent-based approach. *Transp. Res. Part E Logist. Transp. Rev.* **2018**, *116*, 229–241. [CrossRef]
96. Zhao, K.; Zuo, Z.; Blackhurst, J.V. Modelling supply chain adaptation for disruptions: An empirically grounded complex adaptive systems approach. *J. Oper. Manag.* **2019**, *65*, 190–212. [CrossRef]
97. Hackl, J.; Dubernet, T. Epidemic Spreading in Urban Areas Using Agent-Based Transportation Models. *Future Internet* **2019**, *11*, 92. [CrossRef]
98. Klibi, W.; Martel, A. Modeling approaches for the design of resilient supply networks under disruptions. *Int. J. Prod. Econ.* **2012**, *135*, 882–898. [CrossRef]
99. Ivanov, D. *Structural Dynamics and Resilience in Supply Chain Risk Management*; Springer: New York, NY, USA, 2018.
100. Li, Y.; Zobel, C.W.; Seref, O.; Chatfield, D.C. Network characteristics and supply chain resilience under conditions of risk propagation. *Int. J. Prod. Econ.* **2020**, *223*, 107529. [CrossRef]
101. Pavlov, A.; Ivanov, D.; Pavlov, D.; Slinko, A. Optimization of network redundancy and contingency planning in sustainable and resilient supply chain resource management under conditions of structural dynamics. *Ann. Oper. Res.* **2019**. [CrossRef]
102. Ivanov, D.; Dolgui, A. A digital supply chain twin for managing the disruption risks and resilience in the era of Industry 4.0. *Prod. Plan. Control* **2021**, *32*, 775–788. [CrossRef]
103. Liu, Z.; Wang, Y.; Feng, J. Vehicle-type strategies for manufacturer's car sharing. *Kybernetes*, 2022; *ahead-of-print*. [CrossRef]
104. Chen, Y. Research on collaborative innovation of key common technologies in new energy vehicle industry based on digital twin technology. *Energy Rep.* **2022**, *8*, 15399–15407. [CrossRef]
105. Wang, L.; Zhu, S.; Evans, S.; Zhang, Z.; Xia, X.; Guo, Y. Automobile recycling for remanufacturing in China: A systematic review on recycling legislations, models and methods. *Sustain. Prod. Consum.* **2023**, *36*, 369–385. [CrossRef]
106. Spengler, T.; Püchert, H.; Penkuhn, T.; Rentz, O. Environmental integrated production and recycling management. *Eur. J. Oper. Res.* **1997**, *97*, 308–326. [CrossRef]
107. Muñoz, I.; Rieradevall, J.; Domènech, X.; Gazulla, C. Using LCA to Assess Eco-design in the Automotive Sector: Case Study of a Polyolefinic Door Panel (12 pp). *Int. J. Life Cycle Assess.* **2006**, *11*, 323–334. [CrossRef]

108. Zarei, M.; Mansour, S.; Kashan, A.H.; Karimi, B. Designing a Reverse Logistics Network for End-of-Life Vehicles Recovery. *Math. Probl. Eng.* **2010**, *2010*, 649028. [CrossRef]
109. Özceylan, E.; Paksoy, T. Fuzzy multi-objective linear programming approach for optimising a closed-loop supply chain network. *Int. J. Prod. Res.* **2013**, *51*, 2443–2461. [CrossRef]
110. Farel, R.; Yannou, B.; Bertoluci, G. Finding best practices for automotive glazing recycling: A network optimization model. *J. Clean. Prod.* **2013**, *52*, 446–461. [CrossRef]
111. Sherwin, M.D.; Medal, H.R.; MacKenzie, C.A. Identifying and mitigating supply chain risks using fault tree optimization. *IISE Trans.* **2020**, *52*, 236–254. [CrossRef]
112. Qazi, A.; Akhtar, P. Risk matrix driven supply chain risk management: Adapting risk matrix based tools to modelling interdependent risks and risk appetite. *Comput. Ind. Eng.* **2020**, *139*, 105351. [CrossRef]
113. Junaid, M.; Xue, Y.; Syed, M.W.; Zu Li, J.; Ziaullah, M. A Neutrosophic AHP and TOPSIS Framework for Supply Chain Risk Assessment in Automotive Industry of Pakistan. *Sustainability* **2020**, *12*, 154. [CrossRef]
114. Kolotzek, C.; Helbig, C.; Thorenz, A.; Reller, A.; Tuma, A. A company-oriented model for the assessment of raw material supply risks, environmental impact and social implications. *J. Clean. Prod.* **2018**, *176*, 566–580. [CrossRef]
115. Abdel-Basset, M.; Gunasekaran, M.; Mohamed, M.; Chilamkurti, N. RETRACTED: A framework for risk assessment, management and evaluation: Economic tool for quantifying risks in supply chain. *Future Gener. Comput. Syst.* **2019**, *90*, 489–502. [CrossRef]
116. Vilko, J.P.P.; Hallikas, J.M. Risk assessment in multimodal supply chains. *Int. J. Prod. Econ.* **2012**, *140*, 586–595. [CrossRef]
117. Nooraie, S.V.; Mellat Parast, M. A multi-objective approach to supply chain risk management: Integrating visibility with supply and demand risk. *Int. J. Prod. Econ.* **2015**, *161*, 192–200. [CrossRef]
118. Shin, K.S.; Shin, Y.W.; Kwon, J.H.; Kang, S.-H. Risk propagation based dynamic transportation route finding mechanism. *Ind. Manag. Data Syst.* **2012**, *112*, 102–124. [CrossRef]
119. Wu, T.; Blackhurst, J.; O'Grady, P. Methodology for supply chain disruption analysis. *Int. J. Prod. Res.* **2007**, *45*, 1665–1682. [CrossRef]
120. Wang, X.; Li, D.; Shi, X. A fuzzy model for aggregative food safety risk assessment in food supply chains. *Prod. Plan. Control* **2012**, *23*, 377–395. [CrossRef]
121. Srivastava, S.K.; Chaudhuri, A.; Srivastava, R.K. Propagation of risks and their impact on performance in fresh food retail. *Int. J. Logist. Manag.* **2015**, *26*, 568–602. [CrossRef]
122. Tang, L.; Jing, K.; He, J.; Stanley, H.E. Robustness of assembly supply chain networks by considering risk propagation and cascading failure. *Phys. A Stat. Mech. Appl.* **2016**, *459*, 129–139. [CrossRef]

Article

Bilateral Matching Decision Making of Partners of Manufacturing Enterprises Based on BMIHFIBPT Integration Methods: Evaluation Criteria of Organizational Quality-Specific Immunity

Qiang Liu [1], Hongyu Sun [1] and Yao He [2,*]

1 School of Economics and Management, Liaoning University of Technology, Jinzhou 121001, China
2 Department of Economics and Management, Weifang University of Science and Technology, Weifang 262700, China
* Correspondence: heyao@wfust.edu.cn

Abstract: This study aims to examine how the bilateral matching decision making of manufacturing enterprises that are seeking partners in the manufacturing supply chain can be improved by taking into consideration evaluation criteria for organizational quality-specific immunity. This study constructs an evaluation indicator system to measure organizational quality-specific immunity based on immune theory. The system's evaluation criteria are based on the key components of organizational quality-specific immunity. We also construct bilateral matching evaluation and decision-making models using interval-valued hesitant fuzzy information and bidirectional projection technology (BMIHFIBPT). The interval-valued bilateral fuzzy bidirectional projection technology is applied to solve a combination satisfaction and matching optimization model. Empirical analysis is carried out to assess both the supply and demand sides of representative manufacturing enterprises in the manufacturing supply chain, match the main supply and demand bodies of two subjects, and help manufacturing enterprises select the optimal cooperation partners. The empirical analysis results indicate that the bilateral matching evaluation and decision-making models based on BMIHFIBPT can overcome the lack of information to some extent and help solve interval-valued hesitant fuzzy decision-making problems. In turn, the models can provide a basis for manufacturing enterprises to effectively select the best cooperation partners and conduct bilateral matching decision making in the manufacturing supply chain area that supports organizational quality-specific immunity.

Keywords: bilateral matching decision making; interval-valued hesitant fuzzy information; bidirectional projection technology; organizational quality-specific immunity

Citation: Liu, Q.; Sun, H.; He, Y. Bilateral Matching Decision Making of Partners of Manufacturing Enterprises Based on BMIHFIBPT Integration Methods: Evaluation Criteria of Organizational Quality-Specific Immunity. *Processes* **2023**, *11*, 709. https://doi.org/10.3390/pr11030709

Academic Editors: Conghu Liu, Xiaoqian Song, Zhi Liu and Fangfang Wei

Received: 10 January 2023
Revised: 13 February 2023
Accepted: 14 February 2023
Published: 27 February 2023

1. Introduction

Today's increasingly fierce and complex competitive environment poses unprecedented challenges to manufacturing enterprises' development. Quality plays a vital role in staying competitive. Quality management thus not only offers many benefits for enterprises but also involves huge security risks [1]. In recent years, the quality management problems of manufacturing enterprises have become increasingly prominent. Many well-known manufacturing enterprises have experienced quality crises and faced criticism by the public and media. Takata Airbags had an issue with an abnormal gas generator rupture. After the incident came to light, more than 60 million vehicles globally were affected and recalled. The recall incident brought huge losses to Takada, which declared bankruptcy in June 2017. In 2016, Samsung was caught in the "Battery Gate" incident. Within a month of the release of the Galaxy Note7 mobile phone, more than 30 explosions and fires occurred due to battery defects. Samsung permanently stopped production and sales of the Galaxy Note7, and the event became one of the most shameful events in the company's history. Quality management problems cause serious economic losses for manufacturing enterprises, affecting in turn the survival and development of other manufacturing enterprises in the supply chain.

Quality is key to manufacturing enterprises' survival. There are many manufacturing enterprises in the manufacturing supply chain. These enterprises mainly select partners based on features of quality, for example, their ability to jointly complete quality tasks and performance in the manufacturing supply chain [2–5], maintain resilience [6–11], strengthen quality management practice and sustainability [12–19], achieve sustainable development [20,21], and promote green quality management and green technology innovation [22–25]. Organizational quality-specific immunity is the comprehensive embodiment of the organizational quality of a manufacturing enterprise in the manufacturing supply chain [26–35]. The stronger the organizational quality-specific immunity of a manufacturing enterprise, the easier it is for that enterprise to be selected as a partner.

Therefore, in view of the importance of evaluating organizational quality-specific immunity, this study uses immune theory as an entry point to construct relevant evaluation criteria. We introduce bidirectional projection technology to the bilateral matching decision-making method and consider both the supply and demand sides. We use data from the cooperation partners of representative manufacturing enterprises in the manufacturing supply chain as the interval-value hesitant fuzzy information. This study further carries out bilateral (two-sided) matching decision making according to the proposed evaluation criteria for organizational quality-specific immunity. The process involves applying integration methods to create bilateral matching decision-making models that incorporate interval-valued hesitant fuzzy information and bidirectional projection technology (BMIH-FIBPT). The resulting bilateral matching decision-making models and methods will provide the foundation basis for effective selection of partners in the manufacturing supply chain based on organizational quality-specific immunity, providing practical guidance as well as scientific contribution to the field.

2. Literature Review

2.1. Bilateral Matching Decision-Making Models and Methods

Bilateral matching decision making is closely related to management science, and various types of bilateral matching decision-making models and methods are widely used in the field of management science and engineering, mainly in the fields of human resources management, price and material management, e-commerce management, enterprise operation management, and risk investment management [36–43]. However, limited relevant empirical research has applied the various types of bilateral matching decision-making models and methods to the supply chain quality management and supply chain operation management fields. The models and methods available to support bilateral matching decision making mainly stem from models and methods related to emerging technology, operational optimization, statistical and artificial intelligence, simulation, and bilateral matching based on the specific preference information of the bilateral subjects [36–43].

Several representative emerging technology models and methods are as follows [36–43]: two-sided matching decision models based on advantage sequences, two-sided matching game analysis methods, two-sided matching models based on the matching efforts of a bilateral platform, two-sided matching decision-making methods based on triangular intuitionistic fuzzy number information, intuitionistic fuzzy two-sided matching methods considering regret aversion and matching aspiration, two-sided game matching methods with uncertain preference ordinals, decision-making models and methods based on two-sided matching dynamic games, and decision-making methods for stable two-sided matching based on linguistic preference information.

Several representative models and methods for operational optimization, statistical and artificial intelligence, and simulation are as follows [44–46]: intuitionistic fuzzy two-sided matching models, multi-objective stable matching methods with distributional constraints, and dynamic matching methods with unknown preferences.

Several representative models and methods for bilateral matching based on specific preference information from bilateral subjects are as follows [47–50]: network visualization of stable matching methods, dynamic two-sided matching methods based on uncoordinated preference information, multiple decision-making matching models based on the characteristics and circumstances of two-sided (bilateral) subjects, and strict two-sided matching methods based on complete preference ordinal information.

2.2. Organizational Quality-Specific Immunity

Quality accidents are not uncommon. Such issues reveal not only a regulatory mechanism with loopholes but also an imperfect quality management system. In the process of business development, quality accidents are the "symptoms" of enterprises with a quality-related disease, and enterprises infected by that disease will soon see their competitiveness drop, at the very least. The quality management departments and quality management personnel in an enterprise constitute the quality management immune system of an enterprise. Their function is to prevent disease from occurring in the enterprise, perpetually maintaining the enterprise's health, and to provide proper treatment the moment the enterprise starts to fall sick. Seeking out and eliminating "quality accident-inducing factors" is an effective means by which the quality management immune system can prevent an enterprise from getting sick.

Lv and Wang [51–53] were the first to apply immune theory to organizational management. Their work explores organizational adaptability and has led to inspiring results about general organizational immunity from the perspective of immunization [54]. They assert that organizational immunity consists of specific immunities and non-specific immunities. Specific immunities, which emerge as the organization establishes key behaviours, help to prevent "viruses" in the enterprise [55]. At the heart of general organizational immunity, i.e., perhaps the most important element in determining the rise or fall of an enterprise, is organizational quality [56]. Organizational quality-specific immunity, in turn, is fostered by specificity, adaptability, initiative attributes, and the characteristics of acquired cultivation and fostering nurture. These features enhance the efficiency and learning effectiveness of the immune system, saving immune response time if the issue arises again, essentially immunizing the enterprise. The main components of organizational quality-specific immunity are thus quality monitoring and cognition, quality defence, quality memory, and immune homeostasis. These components complement each other, allowing for coordinating development in the formation of an orderly dynamic benign cycle [57].

Existing empirical studies on organizational quality-specific immunity have drawn on relevant theoretical analysis frameworks and conceptual framework models to conduct data surveys, questionnaires, interviews, statistical investigations, and case studies. Luca et al. [58] point out that quality management is becoming an increasingly important factor in determining an enterprise's level of economic and technological development. Improving quality is an urgent need for enterprises seeking to survive and develop in the current competitive environment. Luca et al. also note the key success factors for implementing quality improvement measures and propose a performance measurement system to evaluate those factors. Kaynak and Hartley [59] and Tari et al. [60] confirm that quality management practice has a significant positive impact on the product quality and operational efficiency of the company. In biological and immune theory, each time the organism completes the immune response, its immune cells are enhanced; the organism is thus preserved by immunological memory. Considering organizational quality management from the perspective of immune theory, when an enterprise faces a threat to organizational quality due to internal abnormalities or changes in the external environment, its immune system should identify and respond to the threat and then commit the event to memory; doing so ensures that, next time, the enterprise can mobilize available resources in a timely and reasonable manner to remove similar risks or harm, ensuring healthy and sustainable development [61].

Wang et al. draw on the theory of medical immunity to study the adaptability of enterprises. They divide the concept of organizational immunity into two dimensions: non-specific immunities and specific immunities. Non-specific immunities are determined by three major components: organizational structure, institutional rules, and company culture. Specific immunities, in comparison, are determined by organizational monitoring and surveillance, organizational defence, and organizational memory [61]. Li et al. [62] propose the product quality management model for supply chains based on the characteristics and mechanisms of the biological immune response. With the biological immune system as a parallel, their supply chain quality management model has four stages: immune recognition, learning, memory, and immune effects. Based on that model, the quality identification mechanism and control methods are effective. Passing through the supply chain thus has a positive effect on ensuring product quality and safety.

Liu et al., Shi et al., and Dai and Ding [63–68] indicate that just as organizational quality-specific immunity is at the core of general organizational quality immunity, the latter is at the core of overall organizational immunity. In summary, they find that organizational quality-specific immunity has three main components: organizational quality monitoring and cognition; organizational quality defence, clearance, and repair; and organizational quality memory and immune self-stability. They propose a path towards quality performance improvement based on empirical analysis of the factors that affect quality defect management from the perspective of immune theory. Wang and Li [54], also on the basis on immune theory, consider enterprises' immune recognition ability and use it as a mediator variable to construct a multi-media model. They use the resulting AMOS model to analyse the influencing factors of and risks to immunity to enhance enterprises' ability to resist breaches of immunity. In summary, research has indicated that an enterprise's intrinsic ability to recognize risks to its own survival is equivalent to an immune system's ability to recognize threats to the body's health. Dai and Ding [68] draw on the basic ideas of immune theory in their analysis of the relationships between organizational immunity and the internal control of an enterprise. Based on their findings, they design an internal control evaluation indicator system and construct an evaluation model, which has broadened the research thoughts in the field.

Scholars have expressed concern about the application of immune theory in the field of quality management. The scope of its application remains very limited, and the existing research has not been in depth. At present, the domestic and international investment in the effort to understand quality management as an artificial immune system is insufficient. Few independent empirical analyses have focused on fusing the concept of organizational immunity with quality management. Theoretical and empirical research and evaluations of organizational quality management and organizational quality-specific immunity are relatively rare. Further, no studies have specifically used organizational quality-specific immunity to determine evaluation criteria in conjunction with integration methods and bilateral matching decision-making models based on BMIHFIBPT to solve the problem, on both the supply and demand sides, of matching manufacturing enterprises for cooperation.

3. Evaluation Indicator System Construction of Organizational Quality-Specific Immunity

Organizational quality-specific immunity reduces quality fluctuation and quality loss, addressing present and potential quality problems. Metaphorically, it can fight a stubborn disease, eliminate the virus, and generate the antigens to protect quality, in the longer term ensuring the re-activation and effectiveness of quality-protecting antibodies. Considering the function of organizational quality-specific immunity, and based on the relevant theories and previous literature review [63–68], the present study selects the following components for its evaluation indicator system: organizational quality monitoring and cognition; organizational quality defence, clearance, and repair (both hard and soft features); and organizational quality memory and immune self-stability. These components correspond to the respective scales and evaluation indicators shown in Table 1.

Table 1. Evaluation indicator system for organizational quality-specific immunity.

Construction Dimension		Scales and Evaluation Indicators
Organizational quality monitoring and cognition		External environmental monitoring of organizational quality Internal environmental monitoring of organizational quality Internal activities and behaviour monitoring of organizational quality Value judgement Cognitive motivation (intrinsic motivation, extrinsic motivation) Cognitive diversity
Organizational quality defence, clearance, and repair	Organizational quality defence, clearance, and repair of soft features	Leadership Employee participation Supplier relationship management Customer request
	Organizational quality defence, clearance, and repair of hard features	Product design Process management Statistical control and feedback
Organizational quality memory and immune self-stability		Learning Record Summary Save Spread and diffusion Communication control and supervision

4. Bilateral Matching Decision-Making Models Based on Interval-Valued Hesitant Fuzzy Information and Bidirectional Projection Technology (BMIHFIBPT)

In recent years, multi-criteria decision making has been widely used in the field of economic and management [69–79]. Bilateral matching decision making is an important branch of multi-criteria decision making. In bilateral matching (two-sided matching), two subjects (with different finite sets) are matched by a specific algorithm to each other in order to achieve the satisfaction of both parties [80–84]. At present, bilateral matching decision making is becoming an important component of management decision theory. Many economic and management activities entail matching one or more members of two groups, such as the matching of venture capital parties with start-ups [85–87]. Roth [88] first proposed the concept of bilateral matching in a study of marriage matching, in which Roth analyses actual cases of bilateral matching. Bilateral matching decision-making theory has since received extensive attention from scholars. Fan and Yue [17] consider the highest acceptability order of bilateral subjects and propose a strict bilateral matching decision-making method based on the complete preference information. Chen et al. [89] examine the dynamic matching decision-making problem based on uncertainty preference sequence information. On the basis of the theory of constructing the ordinal deviation, they apply the evidence fusion method to solve the problem with the help of uncertainty preference sequence information.

However, because the evaluation information of both subjects is uncertain, and because the decision-making environment and situation are complex, the subjects' preference information is often hesitant, fuzzy, and vague [90]. Therefore, how to overcome the lack of decision-making information is a problem worthy of attention. Interval-valued hesitant fuzzy sets have excellent expressive advantages in dealing with hesitant fuzzy information [90]. The present study therefore utilizes interval-valued hesitant fuzzy sets and bidirectional projection technology to measure fuzzy information and effectively solve the problem of bilateral matching decision making. The bilateral matching decision-making method used in this study firstly deals with the interval-valued hesitant fuzzy preference information. It then adopts the bidirectional projection technology to obtain the vector projection formed by different intervals. The TOPSIS method is also used to obtain the closeness of match, which can express the preference of both subjects. Finally, the optimization model is constructed. We then incorporate the closeness into the optimization model to solve the problem.

In our hesitant and fuzzy environment, sets $A_p(p = 2, 3, \ldots, m)$ and $B_q(q = 2, 3, \ldots, n)$ represent manufacturing enterprises in the manufacturing supply chain. Manufacturing enterprises in these sets all want partners in the supply chain. Manufacturing enterprises in set $A_p(p = 2, 3, \ldots, m)$ want to select a partner with suitable partners in the scope of set $B_q(q = 2, 3, \ldots, n)$. Set X is a non-empty set, such that $R = \{\langle x, g_R(x_i)\rangle | x_i \in X, i = 1, 2, \ldots, n\}$ refers to interval-valued hesitant fuzzy sets (IVHFSs) in X. $g_R(x_i)$ refers to interval-valued hesitant fuzzy elements (IVHFEs). The element x is from the scope of X, which belongs to the membership degree of R. l is the number and quantity of elements in interval-valued hesitant fuzzy elements. Manufacturing enterprises in set $A_p(p = 2, 3, \ldots, m)$ in the manufacturing supply chain present interval-valued hesitant fuzzy preference evaluation information for the potential partners (manufacturing enterprises in set $B_q(q = 2, 3, \ldots, n)$). That information forms interval-valued hesitant fuzzy preference matrix $G = \lfloor g_{ij} \rfloor_{m*n}$. Correspondingly, the potential partners (manufacturing enterprises in set $B_q(q = 2, 3, \ldots, n)$ present interval-valued hesitant fuzzy preference evaluation information for manufacturing enterprises in set $A_p(p = 2, 3, \ldots, m)$, and that information forms interval-valued hesitant fuzzy preference matrix $H = \lfloor h_{ij} \rfloor_{n*m}$ [84–100].

(1) The construction of bilateral matching decision-making models of partners of manufacturing enterprises based on BMIHFIBPT integration methods

The main model construction principles, procedures, and steps of the bilateral matching decision-making models of partners of manufacturing enterprises based on BMIHFIBPT integration methods are as follows [84–100]:

The interval-valued hesitant fuzzy decision matrix is normalized by the obtained interval-valued hesitant fuzzy information and assembled by the bidirectional projection method to ensure scientific and effective results.

Assuming $G' = [g'_{ij}]_{m \times n}$ as a standardized interval-valued hesitant fuzzy decision matrix, the positive and negative ideal fuzzy elements are $g^+ = (g_1^+, g_2^+, \cdots, g_l^+)$ and $g^- = (g_1^-, g_2^-, \cdots, g_l^-)$ respectively.

$$
\begin{aligned}
g^+ &= (g_1^+, g_2^+, \cdots, g_l^+) = (< [\gamma_1^{L+}, \gamma_1^{U+}], [\gamma_2^{L+}, \gamma_2^{U+}], \cdots, [\gamma_l^{L+}, \gamma_l^{U+}] >) \\
&= (< [\max_{1 \leq i \leq m}\gamma_{i1}^L, \max_{1 \leq i \leq m}\gamma_{i1}^U], [\max_{1 \leq i \leq m}\gamma_{i2}^L, \max_{1 \leq i \leq m}\gamma_{i2}^U], \cdots, [\max_{1 \leq i \leq m}\gamma_{il}^L, \max_{1 \leq i \leq m}\gamma_{il}^U]
\end{aligned}
\tag{1}
$$

$$
\begin{aligned}
g^- &= (g_1^-, g_2^-, \cdots, g_l^-) = (< [\gamma_1^{L-}, \gamma_1^{U-}], [\gamma_2^{L-}, \gamma_2^{U-}], \cdots, [\gamma_l^{L-}, \gamma_l^{U-}] >) \\
&= (< [\min_{1 \leq i \leq m}\gamma_{i1}^L, \min_{1 \leq i \leq m}\gamma_{i1}^U], [\min_{1 \leq i \leq m}\gamma_{i2}^L, \min_{1 \leq i \leq m}\gamma_{i2}^U], \cdots, [\min_{1 \leq i \leq m}\gamma_{il}^L, \min_{1 \leq i \leq m}\gamma_{il}^U]
\end{aligned}
\tag{2}
$$

Let g_p and g_q be two interval-valued hesitant fuzzy elements after standardization ($IVHFE$), so $g_p = (< [\gamma_{p1}^L, \lambda_{p1}^U], [\gamma_{p2}^L, \lambda_{p2}^U], \cdots, [\gamma_{pl}^L, \lambda_{pl}^U] >)$, and $g_q = (< [\gamma_{q1}^L, \lambda_{q1}^U], [\gamma_{q2}^L, \lambda_{q2}^U], \cdots, [\gamma_{ql}^L, \lambda_{ql}^U] >)$. Then, the correlation coefficients of g_p and g_q are:

$$
K_{IVHFE}(g_p, g_q) = \frac{C_{IVHFE}(h_p, h_q)}{\sqrt{E_{IVHFE}(h_p)} \times \sqrt{E_{IVHFE}(h_q)}}
\tag{3}
$$

where:

$$
C_{IVHFE}(h_p, h_q) = \frac{1}{2} \sum_{i=1}^{l} (\gamma_{pi}^L \gamma_{qi}^L + \gamma_{pi}^U \gamma_{qi}^U)
\tag{4}
$$

$$
E_{IVHFE}(h_p) = \frac{1}{2} \sum_{i=1}^{l} [(\gamma_{pi}^L)^2 + (\gamma_{pi}^U)^2]
\tag{5}
$$

$$E_{IVHFE}(h_q) = \frac{1}{2}\sum_{i=1}^{l}[(\gamma_{qi}^L)^2 + (\gamma_{qi}^U)^2]\tag{6}$$

The vector formed by the interval-valued hesitant fuzzy elements g_p and g_q are:

$$h_p h_q = (<[\min\gamma_1^L, \max\gamma_1^U], [\min\gamma_2^L, \max\gamma_2^U], \cdots, [\min\gamma_l^L, \max\gamma_l^U] >)\tag{7}$$

where:

$$\min\gamma_i^L = \min(\left|\gamma_{qi}^L - \gamma_{pi}^L\right|, \left|\gamma_{qi}^U - \gamma_{pi}^U\right|), \text{ and} \max\gamma_i^L = \max(\left|\gamma_{qi}^L - \gamma_{pi}^L\right|, \left|\gamma_{qi}^U - \gamma_{pi}^U\right|), i = 1, 2, \cdots, n.$$

Let the interval-valued hesitant fuzzy element be g', the positive and negative ideal interval-valued hesitant fuzzy elements be g^+, g^-, and the positive ideal interval-valued hesitant fuzzy element and negative ideal interval-valued hesitant fuzzy element be $g^+ g^-$. The positive and negative ideal interval-valued hesitant fuzzy elements and interval-valued hesitant fuzzy elements form the vector $g'g^+$, $g'g^-$. Then, the bidirectional projection formed by $g'g^-$ on $g^+ g^-$ is:

$$\begin{aligned}\Pr_{g^- g^+}^{g^- g'} &= |g^- g'|K_{IVHFE}(g^- g', g^- g^+)\\&= \frac{C_{IVHFE}(g^- g', g^- g^+)}{\sqrt{E_{IVHFE}(g^- g^+)}}\end{aligned}\tag{8}$$

The bidirectional projection formed by $g^+ g^-$ on $g'g^+$ is:

$$\begin{aligned}\Pr_{g'g^+}^{g^- g^+} &= |g^- g^+|K_{IVHFE}(g^- g^+, g'g^+)\\&= \frac{C_{IVHFE}(g^- g^+, g'g^+)}{\sqrt{E_{IVHFE}(g'g^+)}}\end{aligned}\tag{9}$$

Based on the closeness formula of TOPSIS and other methods, the interval-valued hesitant fuzzy information is further assembled effectively, and the following formula is constructed:

$$D = \frac{\Pr_{h^- h^+}^{h^- h'}}{\Pr_{h^- h^+}^{h^- h'} + \Pr_{h'h^+}^{h^- h^+}}\tag{10}$$

Thus, the principal matching closeness matrices of two-sided subjects $D_A = [a_{ij}]_{m \times n}$ and $D_B = [b_{ij}]_{m \times n}$ for matching are constructed.

(2) The optimization solution of bilateral matching decision-making models of partners of manufacturing enterprises based on BMIHFIBPT integration methods

The main principles, procedures, and steps of the optimization solution of bilateral matching decision-making models of partners of manufacturing enterprises based on BMIHFIBPT integration methods are as follows [84–100]:

The bilateral matching decision-making optimization model is constructed based on the principal matching closeness matrices of two-sided subjects $D_A = [a_{ij}]_{m \times n}$ and $D_B = [b_{ij}]_{m \times n}$.

$$\max Z = \sum_{i=1}^{n}\sum_{j=1}^{m}[\lambda(\frac{a_{ij} + b_{ij}}{2}) + (1 - \lambda)\sqrt{a_{ij} \times b_{ij}}]x_{ij}$$

$$s.t. \sum_{j=1}^{m} x_{ij} \leq 1, i = 1, 2, \cdots, n\tag{11}$$

$$\sum_{i=1}^{n} x_{ij} \leq 1, j = 1, 2, \cdots, m$$

where x_{ij} is the 0–1 variable, $x_{ij} = 0$ indicates that the two subjects do not match, and $x_{ij} = 1$ indicates that the two subjects match each other. λ is the adjustment parameter, $0 \leq \lambda \leq 1$, and the value of λ is determined by the specific needs of the actual problem. $\lambda = 0$ and $\lambda = 1$ indicate that the actual problem satisfies the preference consistency and complementarity, respectively. In this study, the combination of satisfaction matching analysis method is used to solve the optimization model.

$$c_{ij} = \lambda\left(\frac{a_{ij} + b_{ij}}{2}\right) + (1 - \lambda)\sqrt{a_{ij} \times b_{ij}}, \lambda \in [0,1] \tag{12}$$

(3) The processes of the bilateral matching decision-making model of partners of manufacturing enterprises based on BMIHFIBPT integration methods

The processes of the bilateral matching decision-making model of partners of manufacturing enterprises based on BMIHFIBPT integration methods are as follows [84–100]:

Step 1: Based on the matching subjects' preference information, obtain the interval-valued hesitant fuzzy decision matrices $G = [g_{ij}]_{m \times n}$ and $H = [h_{ij}]_{n \times m}$ respectively.

Step 2: Use the optimistic criterion or the pessimistic criterion to consistently process the interval-valued hesitant fuzzy decision matrices G and H and arrange the order of the elements to obtain the normalized interval-valued hesitant fuzzy element matrices $G' = [g'_{ij}]_{m \times n}$ and $H' = [h'_{ij}]_{n \times m}$.

Step 3: Apply Equations (1) and (2) to construct the positive and negative ideal fuzzy elements g^+, g^- and h^+, h^- according to the normalized interval-valued hesitant fuzzy element matrix.

Step 4: Use Equations (8) and (9) to calculate the bidirectional projection matrices $G_1^* = \left[\left(\Pr_{g^- g^+}^{g^- g'}\right)_{ij}\right]_{m \times n}$, $G_2^* = \left[\left(\Pr_{g' g^+}^{g^- g^+}\right)_{ij}\right]_{m \times n}$, and $H_1^* = \left[\left(\Pr_{h^- h^+}^{h^- h'}\right)_{ij}\right]_{n \times m}$, $H_2^* = \left[\left(\Pr_{h' h^+}^{h^- h^+}\right)_{ij}\right]_{n \times m}$ respectively.

Step 5: Use Equation (10) to calculate the closeness and obtain the closeness matrices $D_A = [a_{ij}]_{m \times n}$ and $D_B = [b_{ij}]_{m \times n}$.

Step 6: Use Equation (11) to obtain the matching degree matrix $C = [c_{ij}]_{m \times n}$ and further construct the optimization model.

Step 7: Use Lingo software to solve the optimization model and analyse the results to obtain bilateral matching decision-making schemes.

In summary, the bilateral matching decision-making model to help manufacturing enterprises choose partners based on BMIHFIBPT integration methods is shown in Figure 1.

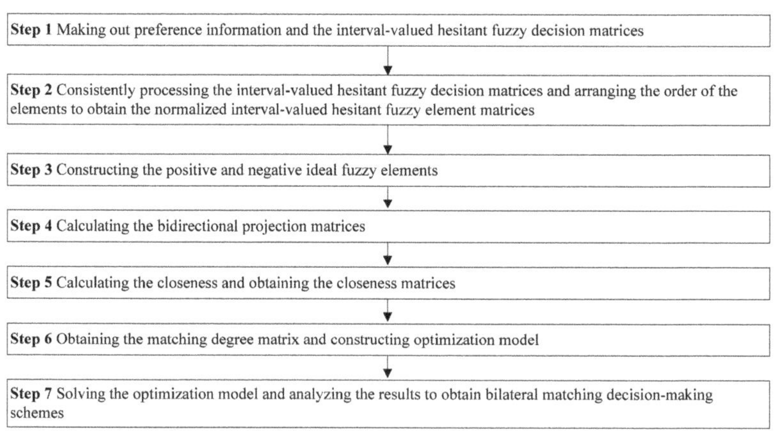

Figure 1. Process flow chart.

5. Empirical Analysis

Quality management is the key means to ensuring the manufacturing supply chain continues to develop steadily. Therefore, as the main actors in the manufacturing supply chain, manufacturing enterprises place great emphasis on quality management in the production and operation management scope. With good quality management, products will satisfy customers and core competitiveness will improve, and the cooperation and partnerships among enterprises will become closer and closer. However, to achieve satisfactory matching results and make optimal decisions, managers must be able to identify the most mature enterprises, namely, those enterprises that have established and operate an organizational quality-specific immune system. In turn, managers must also be able to achieve bilateral matching between their own enterprise and these parties. To serve this aim, the present study selects 11 large-sized manufacturing enterprises in the eastern region of China as the evaluation targets. The selected manufacturing enterprises have a quality management system certification, are good at constructing immunity mechanisms to deal with quality risk events, and display an organizational quality-specific immunity performance that has placed them in leading positions in the same industry. The reputation, product market share, assets, profit, competitiveness, finances, production and operation management performance, rank, and scale of all these enterprises are typical and representative. All 11 manufacturing enterprises are state-owned and state-holding enterprises over 20 years old. We selected and invited 80 managers of the 11 manufacturing enterprises in total. All the managers have a master's degree; have been working in a position in manufacturing enterprises for more than 20 years; and are familiar with quality management, production operation, and management of enterprises.

We collected data by extensive on-site questionnaire distribution, field interviews, and survey investigations. There are seven large-sized manufacturing enterprises (A_1, A_2, A_3, A_4, A_5, A_6, A_7) and four other large-sized manufacturing enterprises (B_1, B_2, B_3, B_4). All 11 manufacturing enterprises are seeking partners in the manufacturing supply chain with the aim of forming long-term cooperation relationships and close partnerships in the quality management scope. All the enterprises are also core actors in the same supply chain. Based on the organizational quality-specific immunity components (i.e., organizational quality monitoring and cognition; organizational quality defence, clearance, and repair, both soft and hard features; and organizational quality memory and immune self-stability), the evaluation indicator system for organizational quality-specific immunity is constructed. We use organizational quality-specific immunity to determine the evaluation criteria. Based on immune theory, organizational quality-specific immunity is the focus of the benchmarking reference system, decision-making scale, and basis. Organizational quality-specific immunity, as discussed above, is vital to both the supply and demand sides of manufacturing enterprises in the manufacturing supply chain. Therefore, it must be a central consideration in selecting optimal partners and in bilateral matching decision making.

The organizational quality-specific immunity components of the selected manufacturing enterprises are organizational quality monitoring and cognition; organizational quality defence, clearance, and repair (soft and hard features); and organizational quality memory and immune self-stability. Both parties completed a preference evaluation of the above evaluation indicators. Namely, all 80 senior managers of the manufacturing enterprises filled out the evaluation survey items and on-site evaluation questionnaires on their decision making with interval-valued hesitant fuzzy information and decision-making preferences according to the evaluation criteria of organizational quality-specific immunity. In the process of preference evaluation, the subjects' preference information was often hesitant, fuzzy, and vague. By processing the interval-valued hesitant fuzzy information, we determined the best-matched manufacturing enterprises, i.e., the enterprise pairs with optimal cooperative and coordination potential.

Step 1: According to the preference information presented by both subjects, the interval-valued hesitant fuzzy preference matrices $G = [g_{ij}]_{m \times n}$ and $H = [h_{ij}]_{n \times m}$ are obtained, as shown in Tables 2 and 3.

Table 2. Interval-valued hesitant fuzzy preference matrix 1.

	B_1	B_2	B_3	B_4
A_1	$\{[0.3,0.4],[0.6,0.8]\}$	$\{[0.1,0.3],[0.5,0.6]\}$	$\{[0.3,0.5]\}$	$\{[0.5,0.7],[0.8,0.9]\}$
A_2	$\{[0.5,0.6]\}$	$\{[0.1,0.2],[0.4,0.5]\}$	$\{[0.2,0.4],[0.5,0.6],[0.7,0.8]\}$	$\{[0.4,0.5],[0.5,0.8]\}$
A_3	$\{[0.1,0.3],[0.3,0.6]\}$	$\{[0.3,0.4],[0.5,0.6]\}$	$\{[0.2,0.3],[0.4,0.6]\}$	$\{[0.3,0.5],[0.6,0.8]\}$
A_4	$\{[0.4,0.6],[0.8,0.9]\}$	$\{[0.2,0.3],[0.5,0.7]\}$	$\{[0.2,0.3],[0.6,0.8]\}$	$\{[0.5,0.6],[0.8,0.9]\}$
A_5	$\{[0.1,0.3],[0.2,0.4]\}$	$\{[0.7,0.8]\}$	$\{[0.1,0.2],[0.5,0.7]\}$	$\{[0.3,0.5]\}$
A_6	$\{[0.3,0.4],[0.7,0.9]\}$	$\{[0.4,0.7]\}$	$\{[0.6,0.7],[0.7,0.9]\}$	$\{[0.2,0.4],[0.7,0.8]\}$
A_7	$\{[0.2,0.3],[0.5,0.6]\}$	$\{[0.3,0.5],[0.8,0.9]\}$	$\{[0.3,0.4]\}$	$\{[0.5,0.7]\}$

Table 3. Interval-valued hesitant fuzzy preference matrix 2.

	A_1	A_2	A_3	A_4
B_1	$\{[0.4,0.5],[0.7,0.8]\}$	$\{[0.2,0.6]\}$	$\{[0.6,0.7]\}$	$\{[0.1,0.3],[0.4,0.5],[0.7,0.9]\}$
B_2	$\{[0.3,0.4],[0.6,0.8]\}$	$\{[0.1,0.2],[0.5,0.9]\}$	$\{[0.3,0.8]\}$	$\{[0.1,0.4],[0.5,0.7]\}$
B_3	$\{[0.1,0.2],[0.4,0.6]\}$	$\{[0.2,0.4],[0.6,0.7]\}$	$\{[0.2,0.5]\}$	$\{[0.1,0.2],[0.4,0.5],[0.5,0.7]\}$
B_4	$\{[0.3,0.5],[0.6,0.9]\}$	$\{[0.1,0.4],[0.5,0.7]\}$	$\{[0.6,0.7]\}$	$\{[0.5,0.6],[0.8,0.9]\}$
	A_5	A_6	A_7	
B_1	$\{[0.2,0.3],[0.5,0.6]\}$	$\{[0.3,0.6]\}$	$\{[0.1,0.2],[0.3,0.4],[0.5,0.6]\}$	
B_2	$\{[0.1,0.4],[0.7,0.8]\}$	$\{[0.7,0.8]\}$	$\{[0.2,0.3],[0.5,0.7]\}$	
B_3	$\{[0.1,0.3],[0.4,0.5],[0.8,0.9]\}$	$\{[0.2,0.5]\}$	$\{[0.5,0.6],[0.7,0.8]\}$	
B_4	$\{[0.2,0.4],[0.6,0.7]\}$	$\{[0.6,0.9]\}$	$\{[0.3,0.5],[0.8,0.9]\}$	

Step 2: By consistently processing the lengths of the matrices G and H, we arranged the order of the elements (this study uses the optimistic criteria to process the length) to obtain the normalized interval-valued hesitant fuzzy element matrices $G' = [g'_{ij}]_{m \times n}$ and $H' = [h'_{ij}]_{n \times m}$, as shown in Tables 4 and 5.

Table 4. Interval-valued hesitant fuzzy preference matrix 3.

	B_1	B_2	B_3	B_4
A_1	$\{[0.3,0.4],[0.6,0.8]\}$	$\{[0.1,0.3],[0.5,0.6]\}$	$\{[0.3,0.5],[0.3,0.5],[0.3,0.5]\}$	$\{[0.5,0.7],[0.8,0.9]\}$
A_2	$\{[0.5,0.6],[0.5,0.6]\}$	$\{[0.1,0.2],[0.4,0.5]\}$	$\{[0.2,0.4],[0.5,0.6],[0.7,0.8]\}$	$\{[0.4,0.5],[0.5,0.8]\}$
A_3	$\{[0.1,0.3],[0.3,0.6]\}$	$\{[0.3,0.4],[0.5,0.6]\}$	$\{[0.2,0.3],[0.4,0.6],[0.4,0.6]\}$	$\{[0.3,0.5],[0.6,0.8]\}$
A_4	$\{[0.4,0.6],[0.8,0.9]\}$	$\{[0.2,0.3],[0.5,0.7]\}$	$\{[0.2,0.3],[0.6,0.8],[0.6,0.8]\}$	$\{[0.5,0.6],[0.8,0.9]\}$
A_5	$\{[0.1,0.3],[0.2,0.4]\}$	$\{[0.7,0.8],[0.7,0.8]\}$	$\{[0.1,0.2],[0.5,0.7],[0.5,0.7]\}$	$\{[0.3,0.5],[0.3,0.5]\}$
A_6	$\{[0.3,0.4],[0.7,0.9]\}$	$\{[0.4,0.7],[0.4,0.7]\}$	$\{[0.6,0.7],[0.7,0.9],[0.7,0.9]\}$	$\{[0.2,0.4],[0.7,0.8]\}$
A_7	$\{[0.2,0.3],[0.5,0.6]\}$	$\{[0.3,0.5],[0.8,0.9]\}$	$\{[0.3,0.4],[0.3,0.4],[0.3,0.4]\}$	$\{[0.5,0.7],[0.5,0.7]\}$

Table 5. Interval-valued hesitant fuzzy preference matrix 4.

	A_1	A_2	A_3	A_4
B_1	$\{[0.4,0.5],[0.7,0.8]\}$	$\{[0.2,0.6],[0.2,0.6]\}$	$\{[0.6,0.7]\}$	$\{[0.1,0.3],[0.4,0.5],[0.7,0.9]\}$
B_2	$\{[0.3,0.4],[0.6,0.8]\}$	$\{[0.1,0.2],[0.5,0.9]\}$	$\{[0.3,0.8]\}$	$\{[0.1,0.4],[0.5,0.7],[0.5,0.7]\}$
B_3	$\{[0.1,0.2],[0.4,0.6]\}$	$\{[0.2,0.4],[0.6,0.7]\}$	$\{[0.2,0.5]\}$	$\{[0.1,0.2],[0.4,0.5],[0.5,0.7]\}$
B_4	$\{[0.3,0.5],[0.6,0.9]\}$	$\{[0.1,0.4],[0.5,0.7]\}$	$\{[0.6,0.7]\}$	$\{[0.5,0.6],[0.8,0.9],[0.8,0.9]\}$

	A_5	A_6	A_7
B_1	$\{[0.2,0.3],[0.5,0.6],[0.5,0.6]\}$	$\{[0.3,0.6]\}$	$\{[0.1,0.2],[0.3,0.4],[0.5,0.6]\}$
B_2	$\{[0.1,0.4],[0.7,0.8],[0.7,0.8]\}$	$\{[0.7,0.8]\}$	$\{[0.2,0.3],[0.5,0.7],[0.5,0.7]\}$
B_3	$\{[0.1,0.3],[0.4,0.5],[0.8,0.9]\}$	$\{[0.2,0.5]\}$	$\{[0.5,0.6],[0.7,0.8],[0.7,0.8]\}$
B_4	$\{[0.2,0.4],[0.6,0.7],[0.6,0.7]\}$	$\{[0.6,0.9]\}$	$\{[0.3,0.5],[0.8,0.9],[0.8,0.9]\}$

Step 3: We use Equations (1)–(7) to construct positive and negative ideal fuzzy elements g^+, g^- and h^+, h^-, respectively, according to the normalized interval-valued hesitant fuzzy element matrix:

$$g^+ = (\{[0.5,0.6],[0.8,0.9]\}, \{[0.7,0.8],[0.8,0.9]\}, \{[0.6,0.7],[0.7,0.9], [0.7,0.9]\}, \{[0.5,0.7],[0.8,0.9]\})$$

$$g^- = (\{[0.1,0.3],[0.2,0.4]\}, \{[0.1,0.2],[0.4,0.5]\}, \{[0.1,0.2],[0.3,0.4], [0.3,0.4]\}, \{[0.2,0.4],[0.3,0.5]\})$$

$$h^+ = (\{[0.4,0.5],[0.7,0.9]\}, \{[0.2,0.6],[0.6,0.9]\}, \{[0.6,0.8]\}, \{[0.5,0.6],[0.8,0.9], [0.8,0.9]\}, \{[0.2,0.4],[0.7,0.8],[0.8,0.9]\}, \{[0.7,0.9]\}, \{[0.5,0.6],[0.8,0.9],[0.8,0.9]\})$$

$$h^- = (\{[0.1,0.2],[0.4,0.6]\}, \{[0.1,0.2],[0.2,0.6]\}, \{[0.2,0.5]\}, \{[0.1,0.2],[0.4,0.5], [0.5,0.7]\}, \{[0.1,0.3],[0.4,0.5],[0.5,0.6]\}, \{[0.2,0.5]\}, \{[0.1,0.2],[0.3,0.4],[0.5,0.6]\})$$

Step 4: We use Equations (8) and (9) to calculate the bidirectional projection matrices $G_1^* = [(\mathrm{Pr}_{g^-g^+}^{g^-g'})_{ij}]_{m \times n}$, $G_2^* = [(\mathrm{Pr}_{g'g^+}^{g^-g^+})_{ij}]_{m \times n}$, and $H_1^* = [(\mathrm{Pr}_{h^-h^+}^{h^-h'})_{ij}]_{n \times m}$, $H_2^* = [(\mathrm{Pr}_{h'h^+}^{h^-h^+})_{ij}]_{n \times m}$ respectively.

$$G_1^* = \begin{bmatrix} 0.419 & 0.097 & 0.215 & 0.543 \\ 0.404 & 0.071 & 0.425 & 0.295 \\ 0.130 & 0.223 & 0.234 & 0.304 \\ 0.625 & 0.166 & 0.455 & 0.516 \\ 0.016 & 0.665 & 0.283 & 0.083 \\ 0.503 & 0.388 & 0.812 & 0.295 \\ 0.244 & 0.430 & 0.123 & 0.331 \end{bmatrix}$$

$$G_2^* = \begin{bmatrix} 0.608 & 0.716 & 0.782 & 0.063 \\ 0.550 & 0.721 & 0.713 & 0.511 \\ 0.644 & 0.720 & 0.806 & 0.510 \\ 0.283 & 0.695 & 0.684 & 0.212 \\ 0.656 & 0.400 & 0.749 & 0.535 \\ 0.589 & 0.620 & 0.081 & 0.427 \\ 0.636 & 0.594 & 0.792 & 0.451 \end{bmatrix}$$

$$H_1^* = \begin{bmatrix} 0.389 & 0.185 & 0.519 & 0.113 & 0.080 & 0.105 & 0.023 \\ 0.283 & 0.230 & 0.354 & 0.161 & 0.356 & 0.409 & 0.255 \\ 0.179 & 0.306 & 0.267 & 0.124 & 0.206 & 0.283 & 0.594 \\ 0.354 & 0.251 & 0.519 & 0.620 & 0.229 & 0.398 & 0.622 \end{bmatrix}$$

$$H_2^* = \begin{bmatrix} 0.212 & 0.354 & 0.283 & 0.585 & 0.422 & 0.453 & 0.707 \\ 0.424 & 0.350 & 0.285 & 0.594 & 0.286 & 0.354 & 0.682 \\ 0.426 & 0.400 & 0.354 & 0.620 & 0.316 & 0.453 & 0.567 \\ 0.300 & 0.447 & 0.705 & 0.319 & 0.402 & 0.353 & 0.392 \end{bmatrix}$$

Step 5: We use Equation (10) to calculate the closeness and obtain the closeness matrices $D_A = [a_{ij}]_{m \times n}$ and $D_B = [b_{ij}]_{m \times n}$.

$$D_A = \begin{bmatrix} 0.408 & 0.119 & 0.216 & 0.896 \\ 0.423 & 0.090 & 0.373 & 0.366 \\ 0.168 & 0.236 & 0.225 & 0.370 \\ 0.688 & 0.193 & 0.400 & 0.709 \\ 0.024 & 0.624 & 0.274 & 0.134 \\ 0.461 & 0.385 & 0.910 & 0.409 \\ 0.277 & 0.420 & 0.135 & 0.424 \end{bmatrix} \quad D_B = \begin{bmatrix} 0.647 & 0.403 & 0.296 & 0.541 \\ 0.343 & 0.397 & 0.433 & 0.360 \\ 0.657 & 0.554 & 0.430 & 0.424 \\ 0.162 & 0.213 & 0.167 & 0.660 \\ 0.159 & 0.565 & 0.395 & 0.363 \\ 0.188 & 0.536 & 0.385 & 0.530 \\ 0.032 & 0.272 & 0.512 & 0.613 \end{bmatrix}$$

Step 6: We use Equation (11) to obtain the matching degree matrix $C = [c_{ij}]_{m \times n}$ and further construct the optimization model.

$$\eta = \begin{bmatrix} 0.514 & 0.219 & 0.253 & 0.696 \\ 0.381 & 0.189 & 0.402 & 0.363 \\ 0.332 & 0.362 & 0.311 & 0.396 \\ 0.334 & 0.203 & 0.258 & 0.684 \\ 0.062 & 0.594 & 0.329 & 0.221 \\ 0.294 & 0.454 & 0.592 & 0.466 \\ 0.094 & 0.338 & 0.263 & 0.510 \end{bmatrix}$$

Step 7: We bring the matching degree matrix η into the optimization model (11), further solving and obtaining the optimization model with the help of Lingo software:

$$X_{11} = 1, X_{12} = 0, X_{13} = 0, X_{14} = 0, X_{21} = 0, X_{22} = 0, X_{23} = 0, X_{24} = 0, X_{31} = 0, X_{32} = 0, X_{33} = 0, X_{34} = 0, X_{41} = 0$$
$$X_{42} = 0, X_{43} = 0, X_{44} = 1, X_{51} = 0, X_{52} = 1, X_{53} = 0, X_{54} = 0, X_{61} = 0, X_{62} = 0, X_{63} = 1, X_{64} = 0, X_{71} = 0, X_{72} = 0,$$
$$X_{73} = 0, X_{74} = 0.$$

The results are as follows: $A_1 \leftrightarrow B_1$, $A_4 \leftrightarrow B_4$, $A_5 \leftrightarrow B_2$, $A_6 \leftrightarrow B_3$, and for the manufacturing enterprises A_2, A_3, and A_7, there are no suitable manufacturing enterprises to match.

6. Conclusions and Discussion

6.1. Conclusions

Decision-making science has undergone continuous development and improvement and is now widely used in many different fields. This study, taking immune theory as a starting point, has applied decision-making science to the organizational quality management field in the context of manufacturing enterprises. The study constructs an evaluation indicator system for organizational quality-specific immunity. The components of organizational quality-specific immunity act as the evaluation criteria. This study further integrates interval-valued hesitant fuzzy information and bidirectional projection technology into bilateral matching decision making, constructing the bilateral matching evaluation and decision-making models based on interval-valued hesitant fuzzy information and bidirectional projection technology (BMIHFIBPT). To use the interval-valued hesitant fuzzy evaluation information to solve the combination satisfaction and matching optimization model, we apply the score formation of left interval value and right interval value, the interval-valued hesitant fuzzy preference, the interval-valued hesitant fuzzy decision-making matrices, the interval-valued hesitant fuzzy elements and closeness, and the interval-valued bilateral fuzzy bidirectional projection technology. We conduct empirical analysis to reflect the supply and demand sides of representative manufacturing enterprises in the manufacturing supply chain, match the main bodies of two parties and subjects, and help manufacturing enterprises achieve optimal bilateral matching. The empirical analysis results indicate that bilateral matching decision-making models that incorporate interval-valued hesitant fuzzy information and bidirectional projection technology (BMIHFIBPT) via integration methods can offer the following: a bilateral matching evaluation and decision-making process for both supply- and demand-side partners; interval-valued hesitant fuzzy evaluation and decision information; and bidirectional projection technology, which possesses the consistency, feasibility, operability, and rationality to solve the interval-valued hesitant fuzzy decision-making problems. Thus, this study provides a basis for manufacturing enterprises in the manufacturing supply chain area to effectively select the best partners based on organizational quality-specific immunity. Bilateral matching evaluation and decision-making models and methods are embedded to provide a reference for manufacturing enterprises to make optimal decisions. These models and methods offer effectiveness, accuracy, robustness, and convenience in selecting the optimal supply–demand matching relationships, determining coordination and cooperation partners on the basis of organizational quality-specific immunity, and achieving satisfactory matching evaluation and optimal coupling decisions by operating the organizational quality-specific immune system.

6.2. Discussion

Bilateral subjects usually give vague, fuzzy, and hesitant preference information [90]. This vague, fuzzy, and hesitant ambiguity emerges in the process of bilateral matching decision making as well [89–94]. The ability to deal with vague, fuzzy, and hesitant ambiguity in bilateral matching decision making has important theoretical value and practical significance. Using the methods and models of the relevant literature [36–43], consistent empirical research results and bilateral matching decision-making results were obtained. Compared with existing studies [44–50], the present study offers the following advantages due to its research methods and models: First, the bilateral matching decision-making models and the methods based on interval-valued hesitant fuzzy information and bidirectional projection technology are different from the bilateral matching decision-making models and methods based on complete preference order information, which effectively help determine the evaluation criteria of organizational quality-specific immunity. The bilateral matching decision-making methods based on interval-valued hesitant fuzzy information and bidirectional projection are determined by the interval-valued hesitant fuzzy information given by the subjects; these methods build on existing methods [44–46]. After processing the length of the interval-valued hesitant fuzzy decision matrix and arranging the order of elements, the normalized interval-valued hesitant fuzzy element matrix is obtained, and the positive and negative ideal fuzzy elements are constructed. The bidirectional projection value matrix is calculated using bidirectional projection technology, and reference is made to TOPSIS. The TOPSIS method is used to calculate the closeness and obtain the closeness matrix. The satisfaction and matching optimization model is thus solved, and the best solution is obtained. Through empirical analysis, this study indicates that the bi-level and bidirectional projection decision-making method with interval-valued hesitant fuzzy information and bidirectional projection technology can solve the problem of matching enterprises on both the demand and supply sides based on organizational quality-specific immunity components. The method, in other words, can select optimal and sustainable partners for representative manufacturing enterprises in the manufacturing supply chain on the basis of organizational quality-specific immunity. The research results show that bilateral matching evaluation models can be achieved using integration methods and bilateral matching decision-making methods that incorporate interval-valued hesitant fuzzy information and bidirectional projection technology based on a biological perspective and an empirical point of view.

With the help of analogy, we can use immune theory to map the operations of organizational quality management and, as a result, generate an understanding of how to achieve organizational quality-specific immunity. There is potential for further application of interval-valued hesitant fuzzy information in decision-making problems in the field of organizational quality management. The results of the present empirical analysis in the field of organizational quality management drew on immune theory, which offers new ideas and methodological systems. The organizational quality-specific immunity evaluation criteria, based on the results, can also inform theoretical frameworks to help manufacturing enterprises effectively select partners based on organizational quality-specific immunity with a view to forming close and long-term relationships in the manufacturing supply chain.

6.3. Research Limitations and Prospects

This study is only a preliminary exploration of the application of interval-valued hesitant fuzzy information and bidirectional projection technology to find optimal partners for manufacturing enterprises in the manufacturing supply chain based on organizational quality-specific immunity. The proposed bilateral (two-sided) matching decision-making models that incorporate interval-valued hesitant fuzzy information and bidirectional projection technology via integration methods are only evaluation instruments; they are only part of an approach for the selection of optimal partners of manufacturing enterprises. While the models will ensure the suitability of two partners, the models do not offer an explanation in terms of other partners and subjects in the manufacturing supply chain. The original data on which the models are based were mainly derived from the subjective responses of manufacturing enterprises' managers gathered via on-site questionnaire distribution, field interviews, and survey investigations. In future research, we will carry out longitudinal tracking in order to obtain longitudinal tracking time series data with the aim of combining objective data with subjective data for the models and methods. Incorporating objective data, as well as introducing more partners and subjects of manufacturing enterprises in the manufacturing supply chain, will further improve the proposed bilateral matching decision-making models based on the evaluation criteria of organizational quality-specific immunity.

Author Contributions: Q.L. contributed to the motivation, the interpretation of the methods, the formal data analysis, the validation of the results, and provided the draft versions and references. H.S. contributed to the data analysis and results, the software and resources, the literature review and references. Y.H. contributed to the related concepts and major recommendations, investigation, supervision, the extraction of the conclusions and discussion. All authors have read and agreed to the published version of the manuscript.

Funding: This research is funded by the National Natural Science Foundation of China (72001055), Research Base of Science and Technology Innovation Think Tank of Liaoning Province (Research Base of High Quality Development of Equipment Manufacturing Industry, NO. 09), and 2022 Scientific Research Project of Department of Education of Liaoning Province (LJKMR20220986).

Data Availability Statement: The data presented in this study are available on request from the corresponding author.

Conflicts of Interest: The authors declare no conflict of interest.

References

1. Samson, D.; Terziovski, M. The relationship between total quality management practices and operational performance. *J. Oper. Manag.* **1999**, *17*, 393–409. [CrossRef]
2. Muango, C.O.; Abrokwah, E.; Shaojian, Q. Revisiting the link between information technology and supply chain management practices among manufacturing firms. *Eur. J. Int. Manag.* **2021**, *16*, 647–667. [CrossRef]
3. Lu, L.J.; Navas, J. Advertising and quality improving strategies in a supply chain when facing potential crises. *Eur. J. Oper. Res.* **2021**, *288*, 839–851. [CrossRef]
4. Groot-Kormelinck, A.; Trienekens, J.; Bijman, J. Coordinating food quality: How do quality standards influence contract arrangements? A study on Uruguayan food supply chains. *Supply Chain. Manag. Int. J.* **2021**, *26*, 449–466. [CrossRef]
5. Yang, X.; Guo, Y.; Liu, Q.; Zhang, D.M. Empirical evaluation on the effect of enterprise quality immune response based on EMIBSGTD-TAS. *J. Intell. Fuzzy Syst.* **2021**, *40*, 11587–11606. [CrossRef]
6. Birkinshaw, J.; Ridderstråle, J. The diffusion of management ideas within the MNC: Under the sway of the corporate immune system. *Rev. Int. Bus. Strategy* **2021**, *31*, 576–595.

7. Sajko, M.; Boone, C.; Buyl, T. CEO greed, corporate social responsibility, and organizational resilience to systemic shocks. *J. Manag.* **2021**, *47*, 957–992. [CrossRef]
8. Santoro, G.; Messeni, P.A.; Del, G.M. Searching for resilience: The impact of employee-level and entrepreneur-level resilience on firm performance in small family firms. *Small Bus. Econ.* **2021**, *57*, 455–471. [CrossRef]
9. Ramezani, J.; Camarinha, M.L.M. Approaches for resilience and antifragility in collaborative business ecosystems. *Technol. Forecast. Soc. Chang.* **2020**, *151*, 119846. [CrossRef]
10. Conz, E.; Magnani, G. A dynamic perspective on the resilience of firms: A systematic literature review and a framework for future research. *Eur. Manag. J.* **2020**, *38*, 400–412. [CrossRef]
11. Schweizer, R.; Lagerström, K. "Wag the dog" initiatives and the corporate immune system. *Multinatl. Bus. Rev.* **2020**, *28*, 109–127. [CrossRef]
12. Zhou, H.G.; Li, L. The impact of supply chain practices and quality management on firm performance: Evidence from China's small and medium manufacturing enterprises. *Int. J. Prod. Econ.* **2020**, *230*, 1–13. [CrossRef]
13. Yaw, A.M.; Esther, A.; Ebenezer, A.; Dallas, O. The influence of lean management and environmental practices on relative competitive quality advantage and performance. *J. Manuf. Technol. Manag.* **2020**, *3*, 1–22.
14. Kharub, M.; Sharma, R. An integrated structural model of QMPs, QMS and firm's performance for competitive positioning in MSMEs. *Total Qual. Manag. Bus. Excell.* **2020**, *31*, 312–341. [CrossRef]
15. Hong, J.T.; Zhou, Z.H.; Li, X.; Lau, K.H. Supply chain quality management and firm performance in China's food industry-The moderating role of social co-regulation. *Int. J. Logist. Manag.* **2020**, *31*, 99–122. [CrossRef]
16. Soltani, E.; Wilkinson, A. TQM and performance appraisal: Complementary or incompatible? *Eur. Manag. Rev.* **2020**, *17*, 57–82. [CrossRef]
17. Parast, M.M. A learning perspective of supply chain quality management: Empirical evidence from US supply chains. *Supply Chain. Manag. Int. J.* **2020**, *25*, 17–34. [CrossRef]
18. Salimian, H.; Rashidirad, M.; Soltani, E. Supplier quality management and performance: The effect of supply chain oriented culture. *Prod. Plan. Control.* **2020**, *32*, 942–958. [CrossRef]
19. Heydari, J.; Govindan, K.; Basiri, Z. Balancing price and green quality in presence of consumer environmental awareness: A green supply chain coordination approach. *Int. J. Prod. Res.* **2021**, *59*, 1957–1975. [CrossRef]
20. Zhang, Y.; Liu, X.M.; Hu, R.; Zhang, L.L.; Zhou, W. Research on the mediating effect of organizational cultural intelligence on the relationships between innovation and total quality management. *Manag. Rev.* **2021**, *33*, 116–127.
21. Chen, Y.Y.; Wu, L. Does ISO9000 certification improve financial performance of the producer service firms?—The crowding-out effect of signaling on efficiency enhancement of the certification. *Manag. Rev.* **2021**, *33*, 271–282.
22. Khorshidvand, B.; Soleimani, H.; Sibdari, S.; Mehdi, S.E.M. A hybrid modeling approach for green and sustainable closed-loop supply chain considering price, advertisement and uncertain demands. *Comput. Ind. Eng.* **2021**, *157*, 107326. [CrossRef]
23. Singh, S.K.; Giudice, M.D.; Chierici, R.; Graziano, D. Green innovation and environmental performance: The role of green transformational leadership and green human resource management. *Technol. Forecast. Soc. Chang.* **2020**, *150*, 119762. [CrossRef]
24. Jayashree, S.; Reza, M.N.H.; Malarvizhi, C.A.N.; Gunasekaran, A.; Rauf, M.A. Testing an adoption model for Industry 4.0 and sustainability: A Malaysian scenario. *Sustain. Prod. Consum.* **2022**, *31*, 313–330. [CrossRef]
25. Yang, M.; Wang, E.Z.; Ye, C.S. Environmental management system certification and Chinese manufacturing enterprises' export. *Chin. Ind. Econ.* **2022**, *6*, 155–173.
26. Liu, Q.; Lian, Z.F.; Guo, Y. Empirical analysis of organizational quality defect management enabling factors identification based on SMT, interval-valued hesitant fuzzy set ELECTRE and QRA methods. *J. Intell. Fuzzy Syst.* **2020**, *38*, 7595–7608. [CrossRef]
27. Liu, Q.; Hu, H.Y.; Guo, Y. Organizational quality specific immune maturity evaluation based on continuous interval number medium operator. *Symmetry.* **2020**, *12*, 918. [CrossRef]
28. Liu, Q.; Lian, Z.F.; Guo, Y.; Tang, S.L.; Yang, F.X. Heuristic decision of planned shop visit products based on similar reasoning method: From the perspective of organizational quality-specific immune. *Open Phys.* **2020**, *18*, 126–138. [CrossRef]
29. Liu, Q.; Lian, Z.F.; Guo, Y.; Yang, F.X. Intuitively fuzzy multi-attribute group decision making of organizational quality specificity immune evolution ability based on evidential reasoning. *J. Intell. Fuzzy Syst.* **2020**, *6*, 7609–7622. [CrossRef]
30. Liu, Q.; Hu, H.Y.; Guo, Y.; Yang, F.X. Evaluation and decision making of organizational quality specific immune based on MGDM-IPLAO method. *Open Phys.* **2019**, *17*, 863–870. [CrossRef]
31. Liu, Q.; Qu, X.L.; Zhao, D.Y.; Guo, Y. Qualitative simulation of organization quality specific immune decision-making of manufacturing enterprises based on QSIM algorithm simulation. *J. Comput. Methods Sci. Eng.* **2021**, *21*, 2059–2076. [CrossRef]
32. Kang, X.; Zhao, D.N.; Liu, Q. Transmission mechanism of simmelian ties on the knowledge spiral-the conducive combination of a high-performance work practice and knowledge fermentation. *J. Organ. Chang. Manag.* **2021**, *34*, 1003–1017. [CrossRef]
33. Yang, X.; Guo, Y.; Liu, Q.; Zhang, D.M. Dynamic co-evolution analysis of low-carbon technology innovation compound system of new energy enterprise based on the perspective of sustainable development. *J. Clean. Prod.* **2022**, *349*, 131330. [CrossRef]
34. Yang, X.; Guo, Y.; Liu, Q. Research on innovation factors, innovation radiation and product quality frontier: The transmission mechanism of organizational quality specific immune under innovation oriented logic. *Nankai Bus. Rev.* **2021**, *24*, 38–52.
35. Yang, X.; Guo, Y.; Liu, Q. Deconvoluting the mechanism of the continuous quality improvement technological discontinuities on the continuous quality improvement on-track innovation: The mediating effect of the organizational quality specific immune and the two-stage moderating effect of the technological perception. *J. Ind. Eng. Eng. Manag.* **2022**, *36*, 37–52.

36. Zhang, L.L.; Ma, D.N.; Lou, Y. A two-sided matching decision model based on advantage sequences. *Oper. Res. Manag. Sci.* **2021**, *30*, 50–56.
37. Yu, Q. Two-sided matching analysis between the government and the social capital in public private partnership. *Fisc. Sci.* **2021**, *70*, 57–71.
38. Xu, Q.; Yang, Y.Y. Dynamic pricing strategy of shared supply chain based on matching efforts of bilateral platform. *Oper. Res. Manag. Sci.* **2022**, *31*, 82–90.
39. Yue, Q.; Zhu, J.G. Two-sided matching decision based on triangular intuitionistic fuzzy number information. *Oper. Res. Manag. Sci.* **2022**, *30*, 57–62.
40. Zhang, D.; Dai, H.J.; Liu, X.R. Intuitionistic fuzzy two-sided matching method considering regret aversion and matching aspiration. *Oper. Res. Manag. Sci.* **2020**, *29*, 132–139.
41. Lin, Y.; Wang, Y.M. Two-sided game matching with uncertain preference ordinal. *J. Oper. Res.* **2020**, *24*, 155–162.
42. Liu, L.; Zhang, Z.S.; Wang, Z. Carbon emission reduction technology investment decision based on two-sided matching-dynamic game. *Oper. Res. Manag. Sci.* **2020**, *29*, 20–26.
43. Zhang, D.; Sun, T.; Chen, Y.; Wan, L.Q. Decision making method for stable two-sided matching under linguistic preference information. *Oper. Res. Manag. Sci.* **2019**, *28*, 60–66.
44. Yu, D.; Xu, Z. Intuitionistic fuzzy two-sided matching model and its application to personnel-position matching problems. *J. Oper. Res. Soc.* **2019**, *2*, 1–10. [CrossRef]
45. Gharote, M.; Phuke, N.; Patil, R.; Lodha, S. Multi-objective stable matching and distributional constraints. *Soft Comput.* **2019**, *23*, 2995–3011. [CrossRef]
46. Fershtman, D.; Pavan, A. Pandora's auctions: Dynamic matching with unknown preferences. *Am. Econ. Rev.* **2017**, *107*, 186–190. [CrossRef]
47. Morizumi, Y.; Hayashi, T.; Ishida, Y. A network visualization of stable matching in the stable marriage problem. *Artif. Life Robot.* **2011**, *16*, 40–43. [CrossRef]
48. Ackermann, H.; Goldberg, P.W.; Mirrokni, V.S.; Glin, H.; Cking, B. Uncoordinated two-sided matching markets. *Siam J. Comput.* **2011**, *40*, 92–106. [CrossRef]
49. Alpern, S.; Katrantzi, I. Equilibria of two-sided matching games with common preferences. *Eur. J. Oper. Res.* **2009**, *196*, 1214–1222. [CrossRef]
50. Fan, Z.P.; Yue, Q. Strict two-sided matching method based on complete preference ordinal information. *J. Manag. Sci.* **2014**, *17*, 21–34.
51. Lv, P.; Wang, Y.H. Research on enterprise adaptability based on organizational immune perspective. *Res. Manag.* **2008**, *29*, 164–171.
52. Lv, P.; Wang, Y.H. Research on organizational immunization behavior and mechanism. *J. Manag.* **2009**, *6*, 607–614.
53. Lv, P.; Wang, Y.H. Construction and operation mechanism of organizational immune system—A case study of Daya Bay Nuclear Power Station. *Sci. Technol. Manag.* **2007**, *28*, 151–156+173.
54. Wang, N.; Li, Q.X. Empirical search on influence factors of enterprise risk immune ability under the mechanism of immune identification. *Soc. Sci. Front.* **2015**, *2*, 267–270.
55. Shi, L.P.; Liu, Q.; Tang, S.L. Research on quality performance upgrading paths based on organizational specific immune-Empirical analysis of projection pursuit and enter methods. *Nankai Bus. Rev.* **2012**, *15*, 123–134.
56. Walsh, J.P.; Ungson, G.R. Organizational memory. *Acad. Manag. Rev.* **1991**, *16*, 57–91. [CrossRef]
57. Shi, L.P.; Liu, Q.; Teng, Y. Quality performance upgrading paths based on OSI-PP-Enter: Theoretical framework and empirical analysis. *J. Manag. Eng.* **2015**, *29*, 152–163.
58. Luca, C.; Paolo, T.; Alessandro, B. The role of performance measurement systems to support quality improvement initiatives at supply chain level. *Int. J. Product. Perform. Manag.* **2010**, *59*, 163–185.
59. Kaynak, H.; Hartley, J.L. A replication and extension of quality management into the supply chain. *J. Oper. Manag.* **2008**, *26*, 468–489. [CrossRef]
60. Tari, J.J.; Molina, J.F.; Castejon, J.L. The relationship between quality management practices and their effects on quality outcomes. *Eur. J. Oper. Res.* **2007**, *183*, 483–501. [CrossRef]
61. Wang, Y.H.; Lu, P.; Xu, B.; Yang, Z.N.; Su, X.Y.; Du, D.B.; Song, Z.C.; Duan, X.G. Preliminary study on organizational immune. *Sci. Technol. Manag.* **2006**, *27*, 133–139.
62. Li, Q.X.; Sun, P.S.; Jin, F.H. Research on supply chain quality management model based on the perspective of immune. *Commer. Res.* **2010**, *7*, 72–76.
63. Liu, Q.; Shi, L.P.; Su, Y. Influence factors of quality defect management and function mechanism of influence factors on quality performance: Research status and tendency. *Technol. Econ.* **2015**, *34*, 92–99+107.
64. Liu, Q.; Shi, L.P.; Su, Y. Team quality defect management influence factors based on medical analogy metaphor: An empirical analysis of sensemaking-fuzzy extraction-pair. *Mod. Financ.* **2015**, *35*, 71–84.
65. Liu, Q.; Shi, L.P.; Chen, W.; Su, Y. Team quality defect management influence factors identification and fuzzy rules extraction based on SYT-FR-CA: Theoretical framework of main logic of medical immune and evidence. *Ind. Eng. Manag.* **2016**, *21*, 110–117.

66. Shi, L.P.; Liu, Q.; Jia, Y.N.; Yu, X.Q. Quality performance upgrading paths optimization based on Projection Pursuit-RAGA-NK-GERT-Explanatory framework of setting organizational quality specific immune and product life cycle as main logic. *Oper. Res. Manag. Sci.* **2015**, *24*, 188–197.
67. Shi, L.P.; Liu, Q.; Wu, K.J.; Du, Z.W. Function mechanism of organizational quality specific immune elements consistency on quality performance—Empirical analysis based on PP, fit and hierarchy regression analysis method. *Ind. Eng. Manag.* **2013**, *18*, 84–91.
68. Dai, C.L.; Ding, X.S. Enterprise internal control evaluation system construction based on organizational immune theory. *Financ. Account. Mon.* **2013**, *10*, 30–33.
69. Hayat, K.; Tariq, Z.; Lughofer, E.; Aslam, M.F. New aggregation operators on group-based generalized intuitionistic fuzzy soft sets. *Soft Comput.* **2021**, *25*, 13353–13364. [CrossRef]
70. Hayat, K.; Raja, M.S.; Lughofer, E.; Yaqoob, N. New group-based generalized interval-valued q-rung orthopair fuzzy soft aggregation operators and their applications in sports decision-making problems. *Comput. Appl. Math.* **2023**, *42*, 4. [CrossRef]
71. Hayat, K.; Ali, M.I.; Karaaslan, F.; Cao, B.Y.; Shah, M.H. Design concept evaluation using soft sets based on acceptable and satisfactory levels: An integrated TOPSIS and Shannon entropy. *Soft Comput.* **2020**, *24*, 2229–2263. [CrossRef]
72. Hayat, K.; Ali, M.I.; Alcantud, J.C.R.; Cao, B.Y.; Tariq, K.U. Best concept selection in design process: An application of generalized intuitionistic fuzzy soft sets. *J. Intell. Fuzzy Syst.* **2018**, *35*, 5707–5720. [CrossRef]
73. Chan, J.; Du, Y.X.; Liu, W.F. Pythagorean hesitant fuzzy risky multi-attribute decision making method based on cumulative prospect theory and VIKOR. *Oper. Res. Manag. Sci.* **2022**, *31*, 50–56.
74. Zhang, F.M.; Zhu, S.Q. Probabilistic language multi-attribute large group decision-making method based on group consistency in social network analysis. *J. Syst. Manag.* **2022**, *31*, 679–688.
75. Qu, G.H.; Wang, B.Y.; Qu, W.H.; Xu, Z.S.; Zhang, Q. A multiple-attribute decision-making method based on the dual hesitant fuzzy geometric heronian means operator and its application. *Chin. J. Manag. Sci.* **2022**, *30*, 216–228.
76. Yin, X.; Liu, Q.S.; Ding, Z.W.; Zhang, Q.T.; Wang, X.Y.; Huang, X. Toward intelligent early-warning for rockburst in underground engineering: An improved multi-criteria group decision-making approach based on fuzzy theory. *J. Basic Sci. Eng.* **2022**, *30*, 374–395.
77. Wang, W.M.; Xu, H.Y.; Zhu, J.J. Large-scale group DEMATEL decision making method from the perspective of complex network. *Syst. Eng.-Theory Pract.* **2021**, *41*, 200–212.
78. Di, P.; Ni, Z.C.; Yin, D.L. A multi-attribute decision making optimization algorithm based on cloud model and evidence theory. *Syst. Eng.-Theory Pract.* **2021**, *41*, 1061–1070.
79. Lin, P.P.; Li, D.F.; Jiang, B.Q.; Yu, G.F.; Wei, A.P. Bipolar capacities multi-attribute decision making VIKOR method with interaction attributes. *Syst. Eng.-Theory Pract.* **2021**, *41*, 2147–2156.
80. Amanda, R.; Atangana, A. Derivation of a groundwater flow model within leaky and self-similar aquifers: Beyond Hantush model. *Chaos Soliton Fract.* **2018**, *116*, 414–423. [CrossRef]
81. Temidayo, J.O.; Emmanuel, A.; Sulemain, A.S. Differential transformation method for solving Malaria-Hygiene mathematical model. *Matrix Sci. Math.* **2021**, *5*, 1–5.
82. Gomez-Aguilar, J.F.; Atangana, A. Time-fractional variable-order telegraph equation involving operators with Mittag-Leffler kernel. *J. Electromagnet Wave* **2019**, *33*, 165–177. [CrossRef]
83. Zhao, C.H.; Li, J.Y. Equilibrium selection under the Bayes-based strategy updating rules. *Symmetry* **2020**, *12*, 739. [CrossRef]
84. Bloch, F.; Ryder, H. Two-sided search, marriage and matchmakers. *Int. Econ. Rev.* **2000**, *41*, 93–115. [CrossRef]
85. Casamatta, C. Financing and advising: Optimal financial contracts with venture capitalists. *J. Financ.* **2002**, *58*, 2059–2086. [CrossRef]
86. Sorensen, M. How smart is smart money? A two-sided matching model of venture capital. *J. Financ.* **2007**, *62*, 2725–2762. [CrossRef]
87. Silveira, R.; Wright, R. Venture capital: A model of search and bargaining. *Rev. Econ. Dyn.* **2015**, *19*, 232–246. [CrossRef]
88. Roth, A.E. Common and conflicting interests in two-sided matching markets. *Eur. Econ. Rev.* **1985**, *27*, 75–96. [CrossRef]
89. Chen, S.Q.; Wang, Y.M.; Shi, H.L. Dynamic matching decision method based on ordinal deviation fusion degree. *Oper. Res. Manag. Sci.* **2014**, *23*, 59–65.
90. Zhao, X.D.; Zang, Y.Q.; Sun, W. Two-sided matching decision method with interval-valued hesitant fuzzy information based on bidirectional projection method. *Oper. Res. Manag. Sci.* **2017**, *26*, 104–109.
91. Park, D.G.; Kwun, Y.C.; Park, J.H.; Park, Y. Correlation coefficient of interval-valued intuitionistic fuzzy sets and its application to multiple attribute group decision making problems. *Math. Comput. Model.* **2009**, *50*, 1279–1293. [CrossRef]
92. Wang, Z.X.; Wang, S.; Liu, F. Two-sided matching decision making method based on optimization model. *Pract. Cogn. Math.* **2006**, *44*, 177–183.
93. Wu, W.Y.; Jin, F.F.; Guo, S.; Chen, H.Y.; Zhou, L.G. Correlation coefficient and its application of dual hesitation fuzzy sets. *Comput. Eng. Appl.* **2015**, *51*, 38–42+61.
94. Xu, Z.S.; Hu, H. Projection models for intuitionistic fuzzy multiple attribute decision making. *Int. J. Inf. Technol. Decis. Mak.* **2010**, *9*, 267–280. [CrossRef]
95. Malviya, P.S.; Yadav, N.; Ghosh, S. Acousto-optic modulation in ion implanted semiconductor plasmas having SDDC. *Appl. Math. Nonlinear Sci.* **2018**, *3*, 303–310. [CrossRef]

96. Nizami, A.R.; Pervee, A.; Nazeer, W.; Baqir, M. Walk polynomial: A new graph invariant. *Appl. Math. Nonlinear Sci.* **2018**, *3*, 321–330. [CrossRef]
97. Pandey, P.K.; Jaboob, S.S.A. A finite difference method for a numerical solution of elliptic boundary value problems. *Appl. Math. Nonlinear Sci.* **2018**, *3*, 311–320. [CrossRef]
98. Long, Q.; Wu, C.; Wang, X. A system of nonsmooth equations solver based upon subgradient method. *Appl. Math. Comput.* **2015**, *251*, 284–299. [CrossRef]
99. Chen, N.; Xu, Z.S.; Xia, M.M. Interval-valued hesitant preference relations and their applications to group decision making. *Knowl.-Based Syst.* **2013**, *37*, 528–540. [CrossRef]
100. Wang, Z.X.; Huang, S.; Liu, F. Decision making method for two-sided matching based on optimization model. *Math. Pract. Theory* **2014**, *44*, 178–183.

Article

Control-Centric Data Classification Technique for Emission Control in Industrial Manufacturing

Zihao Chen [1,*] and Jian Chen [2]

[1]   International School, Beijing University of Posts and Telecommunications, Beijing 100876, China
[2]   Jiangxi University of Technology, Nanchang 330098, China
*   Correspondence: z1haoc@bupt.edu.cn

Abstract: Artificial intelligence-based hardware devices are deployed in manufacturing units and industries for emission gas monitoring and control. The data obtained from the intelligent hardware are analyzed at different stages for standard emissions and carbon control. This research article proposes a control-centric data classification technique (CDCT) for analyzing as well as controlling pollution-causing emissions from manufacturing units. The gas and emission monitoring AI hardware observe the intensity, emission rate, and composition in different manufacturing intervals. The observed data are used for classifying its adverse impact on the environment, and as a result industry-adhered control regulations are recommended. The classifications are performed using deep neural network analysis over the observed data. The deep learning network classifies the data according to the environmental effect and harmful intensity factor. The learning process is segregated into classifications and analysis, where the analysis is performed using previous emission regulations and manufacturing guidelines. The intensity and hazardous components levels in the emissions are updated after the learning process for recommending severe lookups over the varying manufacturing intervals.

Keywords: artificial intelligence hardware; data classification; deep learning; emission control; industrial manufacturing

Citation: Chen, Z.; Chen, J. Control-Centric Data Classification Technique for Emission Control in Industrial Manufacturing. *Processes* **2023**, *11*, 615. https://doi.org/10.3390/pr11020615

Academic Editors: Conghu Liu, Xiaoqian Song, Zhi Liu and Fangfang Wei

Received: 10 January 2023
Revised: 9 February 2023
Accepted: 10 February 2023
Published: 17 February 2023

1. Introduction

Artificial intelligence (AI) hardware is the most used in various fields to enhance the efficiency and reliability of systems. AI hardware is also used in industries that reduce latency and workload in production. Industries require various AI hardware to improve performance [1–3]. AI-based hardware is utilized for emission monitoring and reducing hardware computation costs and maintenance charges. Industries emit certain gases and components during production. Emission monitoring is a complicated task to perform in industry management systems [4]. AI hardware is commonly used in monitoring systems so as to decrease the energy consumption range in the computation process. The emission monitoring system identifies the exact emission ranges from industries to provide feasible information for environmental protection [5–7]. Important factors, components, patterns, and principles of industries are detected based on AI hardware. AI technology-based hardware is commonly used in monitoring systems that detect industries' sources and range of emissions [8]. Artificial neural network (ANN) and support vector machine (SVM) algorithms are generally used in AI hardware that identifies the emission level from both companies and industries [9,10].

Data analysis is crucial in every field that provides necessary information for further processes. Logistics and strategic techniques are implemented in management systems to analyze data [11]. Information and details are stored in a database that provides feasible data for analysis and detection processes. AI hardware is employed in data analysis to discover new patterns and features of data [12–14]. AI techniques and algorithms are used in AI hardware to reduce complexity and latency of the analysis process. AI-based hardware detects necessary factors and data principles, decreasing the energy consumption range in the computation process [15]. Big data analytics (BDA) is widely used for analysis processes in various industries and hospitals. AI hardware-based data analysis finds the actual data required to perform a certain task in an application [16]. Important patterns and features contain details about data that enhance the efficiency and performance of analysis systems. Behavioral- and activities-based data are also identified by AI hardware that handles a huge amount of data using the analysis process. Data analytics tools based on AI are also used in data analysis systems. As a result, both significance and reliability are increased in data management systems [17,18].

Machine learning (ML) algorithms and techniques are used for emission data analysis in industries. Real-time emission data analysis is a difficult process to perform in industries. ML techniques reduce computation time and also the range of energy consumption [19]. In addition, ML techniques maximize accuracy in detection and prediction processes, improving the systems' efficiency. The random forest (RF) algorithm is normally used in emission data analysis systems [20]. RF detects the exact intensity level of carbon emissions of industries. The RF algorithm extracts the patterns of emission ranges to gather necessary information for data analysis systems. Both renewable and non-renewable emissions are analyzed by RF, which reduces complexity in the further detection process [21]. The artificial neural network (ANN) algorithm is also used for emission data analysis. ANN scrutinizes the datasets of the database based on certain features and patterns. ANN increases the reliability and mobility of data analysis systems [22]. The deep reinforcement learning (DRL) algorithm is used for emission control measures in the industries. DRL trains the data which are required for control policies. DRL also detects the impact of emission on the environment, which produces relevant data for the environment management process [23].

2. Related Works

Tao et al. [24] introduced a channel-enhanced spatiotemporal network (CENet) for industry smoke emission recognition. Supervision information and patterns are required to detect the exact smoke emission level of industries. The loss function is used here for detection of essential characteristics and features. It also reduces the latency and energy consumption range in computation. The introduced CENet achieves high emission detection accuracy, enhancing the production efficiency in industries.

Fiscante et al. [25] proposed a new detection method that determines the atmospheric trace gases using hyperspectral satellite data. The main aim of the proposed method is to measure the actual gases that are emitted by industries. Unsupervised sparse mixing gases cause various environmental problems. Temperature and pressure of environmental data are collected, which provide feasible information for the detection process. The proposed method increases detection accuracy, thereby, improving industrial systems' performance.

Guo et al. [26] designed a global meta-analysis for greenhouse gas (GHG) emissions in nitrogen fertilizer (NF) applications. The NF increases crops' cultivation and product range as well as it maximizes overall cultivation. Crop production emits a huge amount of GHG into the environment. NF-based crop production reduces the GHG emission range in the environment. Experimental results show that the proposed meta-analysis identifies the actual GHG emission ratio, which provides necessary data for emission control policies.

Sikdar et al. [27] have developed a deep learning approach for classification and damage-resource detection. A convolutional neural network (CNN) algorithm is employed to classify damages which are based on certain patterns and principles. The feature extraction technique in CNN brings out the important features from scalogram images. Acoustic emission (AE) is also detected by CNN, which reduces false alarm rates in industries. The developed approach maximizes accuracy in damage resource detection, increasing the systems' efficiency and reliability.

Choi et al. [28] presented a machine learning (ML)-based classification model in urban areas. The proposed model mainly aims to recognize odor sources and content in urban areas. Odor-causing substances and materials emit gases and smoke that cause certain environmental problems. A decision tree (DT) is used in the classification model that identifies the source of the odor. The proposed DT model achieves high accuracy in detection and classification, enhancing the application's significance and effectiveness.

Tacchino et al. [29] developed a multi-scale model for steam methane reforming reactors in industries. The finite element method (FEM) is used here to detect the gas and pressure range of gas emissions from industries. FEM divides the resultant gases into their types based on features and patterns collected by reactors. The multi-scale model validates the actual gas range which is emitted by industries. Compared with other models, the proposed model increases accuracy in emission detection.

Tuttle et al. [30] proposed a nonlinear support vector machine (SVM)-based NOx emission prediction model. Both spatial and temporal features are detected from the database, which leads to the production of optimal information for further prediction. Furthermore, an artificial neural network (ANN) is also used to identify features about NOx emission details. As a result, the proposed SVM model achieves high accuracy in NOx emission prediction, enhancing the industries' efficiency and product range.

Sun et al. [31] introduced a VIIRS thermal anomaly data-based detection method for heavy industries. The main aim of the proposed method is to detect the air pollution emission range of the industries. Industrial management systems gather air quality, gas emission, energy charges, and spatiotemporal features. Spatiotemporal patterns provide relevant data which are required for the emission detection process. The introduced method increases detection accuracy, reducing the computation and further processing complexity.

Ju et al. [32] proposed a new atmospheric pollutant emission prediction method for industries. Quantification results of pollutant emission standards (QRPES) are used in the paper to produce the necessary information for the proposed prediction method. In addition, machine learning (ML) techniques such as random forest regression (RFR) and support vector regression (SVR) are used in the emission prediction process. As a result, the proposed method maximizes accuracy, improving industries' performance and efficiency levels.

Sun et al. [33] designed a mechanism that results in reduction and verification, validation, and accreditation (VV&A) for NO emission prediction. Computational fluid dynamics (CFD) is availed to predict the structure and environment of polluted areas. CFD reduces both time and energy consumption range of computation, thereby, enhancing the systems' efficiency. CFD also reduces the error ratio in prediction, maximizing the industries' production. Experimental results show that the proposed method predicts the accurate level of NO emission and atmospheric temperature of the industries.

Milkevych et al. [34] developed a matched filter for gas emission measurement in a dairy cattle field. Data synchronization was performed to identify the exact emission ratio of gases in cattle fields. Cattle field emits a huge amount of greenhouse gases into the environment that causes various problems. The main aim of the proposed approach is to detect the methane emission range from the cattle fields. The proposed approach maximizes prediction accuracy, reducing cost and latency in the computation process.

Martinez et al. [35] proposed a new prediction method for quantity surfactants in the environment using fluorescent spectroscopy measurements. The actual goal of the proposed method is to detect the textile wastewater range. Excitation–emission second-order data are used to reduce the latency in classification and identification processes. The proposed method decreases the wastewater content by providing feasible information to control policies. The proposed method improves both the efficiency and reliability of industries.

Lee et al. [36] introduced an industrial energy system model for industries. The proposed model is mainly used to reduce the emission range into the environment. Technology learning is developed to identify the relationship among spillovers that produce optimal data for the prediction process. Characteristics, features, and patterns of data are analyzed for the emission detection process. The proposed energy system model increases the industry's effectiveness, robustness, and production levels.

Ren et al. [37] discussed the probability density function control to investigate the controller design methods where the random variable for the stochastic processes was adjusted to follow the desirable distributions. Once the relationship between control inputs and outputs PDF is expressed, the control aim can be defined as determining the control input signals which would modify the system output PDFs to trail the pre-specified target PDFs.

Zhang et al. [38] proposed the non-Gaussian stochastic distribution control (NGSDC). Through the influence of data science, the performance has been elevated leading to improved industrial artificial intelligence. Stochastic distribution control has been further established by recently concentrating on the data-driven design and multi-agent system. This article summarizes the most recent published outcomes in the last 5 years of stochastic distribution control work in modelling, controls, fault diagnosis, filtering, and industrial applications.

3. Proposed Control-Centric Data Classification Technique

In the proposed technique, intelligent hardware devices were used to compute a data analysis for monitoring and controlling the emission gas at different stages using artificial intelligence (AI). The data are requested from the AI-based intelligent hardware and analysis is performed to control emission as well as carbon control through the CDC technique. According to controlled emission, gas refers to emissions produced by the industry at various stages from the intelligent hardware within the devices that is frequently monitored and analyzed. In addition, there are some causes for the occurrence of emission gas identified in the industry based on natural disasters. Therefore, the AI hardware simultaneously observes the emission rate, intensity, and composition of the gas. The emission monitoring is also performed at various manufacturing intervals. In Figure 1, the CDCT is illustrated.

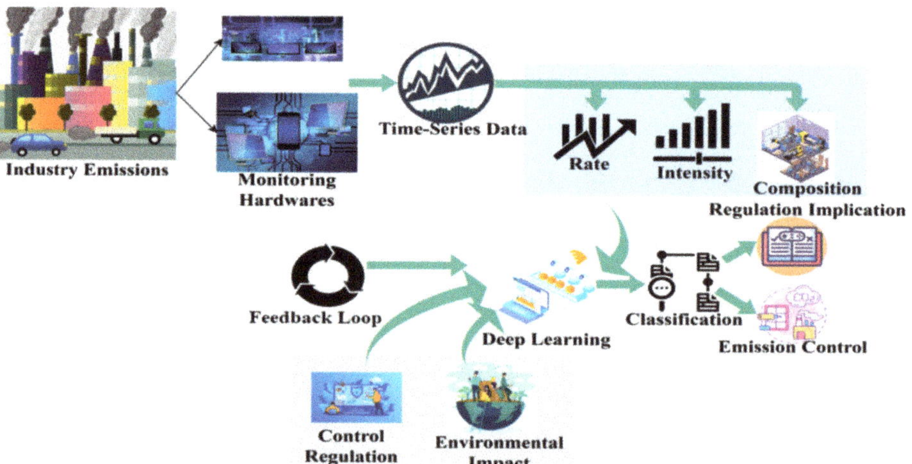

Figure 1. Working Process of Control-centric Data Classification Process.

This data analysis of previous emission regulations and manufacturing guidelines has been modified to current improvable regulations and guidelines for controlling emission gas produced by the industry. In addition, the AI-assisted intelligent hardware analyzed for controlling pollution caused by emission gas and carbon release from the manufacturing units at various time series are also analyzed. The important factors in this technique, namely, intensity, emission rate, and composition of the gas and emission, are observed continuously from the manufacturing units in a sequential manner. The gas and emission occurrences are identified in the AI hardware at different manufacturing intervals depending upon emission and carbon control for each production at various time series. After identifying emission gas from the industry, it is analyzed based on intensity and composition. The observed data are utilized for segregating the adverse impact of emission gas on that environment, and the industry-adhered control regulations are recommended for further processing. The data from day-to-day functions, activities, and production are performed by the AI hardware. For the benefit of the industry, data are recurrently analyzed through deep neural network learning for reducing emissions and gas. Therefore, the learning process is responsible for data classification through harmful intensity and adverse environmental effects identified from the industry emissions using previous emission regulations and manufacturing guidelines for confining gas and emission occurrence. The learning process is classified for data classification, and where the analysis takes place based on control regulations and adverse environmental impact. The observed data are analyzed by a reliable system employed in several industries and manufacturing units. The proposed data classification technique aims to improve control regulations and manufacturing guidelines for identifying high-emission intensity and hazardous components in the gas. Emissions are updated after the performance of deep learning leading to different recommendations as well as lookups over various time series. A feedback loop is set up as the part of the system in which the system's output is utilized as input for future operations in industrial manufacturing.

4. Emission Control Recommendations

Emission and gas sensors are used for sensing information from the AI hardware setup in the manufacturing units for monitoring and controlling adverse impact as well as high intensity. The data classification was performed to analyze the factors as observed from the current instance by the AI hardware monitoring. During emission occurrence, the considerable features in this research article correlate with the AI hardware process. The study focuses on the hardware monitoring of the industries and manufacturing units. It analyzes if the emission takes place or not in that environment and identifies the industry emission as $Industry_{Em}$. The probability of industrial gas and emission occurrence is identified using CDCT and it can be expressed as:

$$\rho(Industry_{Em}) = \frac{Em^{-1}(1 - Em)^{-1}}{t_s(R,\ I,\ C)} \tag{1}$$

where the condition $t_s(R,\ I,\ C)$ can be expressed as:

$$t_s(R,\ I,\ C) = \int_0^1 Em^{-1}(1 - Em)^{-1} d \cdot Em \tag{2}$$

The considerable factors in industry emission, R, I, and C are computed as:

$$R = \frac{R}{H_{int}} \left[E_{imp} - \left(E_{imp}\right)^2 - \left(H_{int}\right)^2 \right] \tag{3}$$

$$I = \frac{(1 - E_{imp})}{H_{int}} \left[E_{imp} - \left(E_{imp}\right)^2 - \left(H_{int}\right)^2 \right] \tag{4}$$

$$C = \frac{\left(R + \left(E_{imp}\right)^2\right)}{\left(H_{int}\right)^2} \left[E_{imp} - \left(E_{imp}\right)^2 - \left(H_{int}\right)^2 \right] \tag{5}$$

In the above Equations (1) to (5), the variables R, I, and C represent emission rate, intensity, and composition observed from the individual manufacturing units at different time intervals, t_s, respectively, where H_{int} and E_{imp} denote the high emission volume (intensity) and adverse environmental impact addressed by the industries and manufacturing units. H_{int} indicates the highest level of gas and emission intensity based on its mean values and E_{imp} is the hazardous components (chemical) level in the emission. The values of H_{int} and E_{imp} are required as explained in Equations (6) and (7).

$$H_{int} = \frac{2(R-I+C)}{R+I+C+2}\left[\frac{R+I+C+2}{R*I*C}\right]^{\frac{1}{2}} \quad (6)$$

$$E_{imp} = \frac{4\left\{(R-I)^2(R+I+C) - C(R+I+2)\right\}}{C(R+I+2)(R+I+3)} \quad (7)$$

Here, the observed data from the manufacturing units are analyzed at different emission stages, and carbon occurrences are identified using placed sensors and AI hardware. AI hardware can monitor these data to reduce pollution-causing hazardous gas and emissions. The values of H_{int} and E_{imp} were estimated for different time series. The AI hardware is reliable in providing precise data about the actual source of energy from the environment within the devices/machinery. Several data analysis techniques were used to analyze the machinery's intelligence hardware at different time series. The time-series data observation process is portrayed in Figure 2.

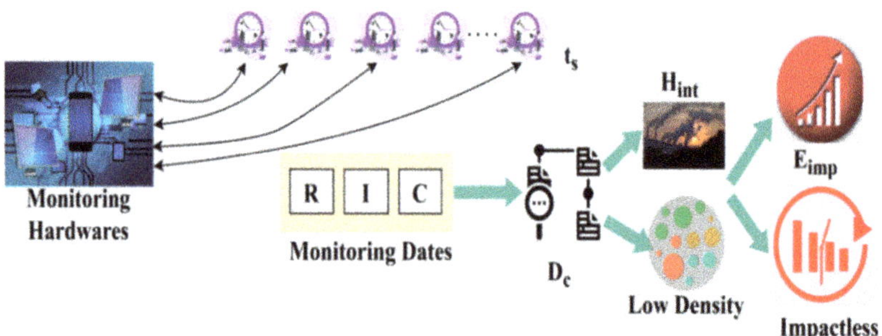

Figure 2. Time-Series Data Observation Process.

The time series data observation is instigated from the monitoring process to the classification. The $R, I,$ and C in t_s are classified as H_{int} and low intensity for which the impact is validated. In this t_s, the D_c performs classification $\forall\, R$ such that E_{imp} for all H_{int} is identified. This is required for $Industry_{Em}$ regulation and emission control (Figure 2). The performance of artificial intelligence hardware is monitored and analyzed through the CDCT technique in the particular industrial environment. The proposed technique is classified into two processes, namely, control regulation and environmental impact. Based on the control regulations, the data are observed to analyze and identify harmful intensity and adverse environmental effects after prolonged control regulations and manufacturing guidelines for time series. Instead, the observed data are analyzed and the resultant pollution is measured. Similar time emission occurrence in manufacturing units is identified through AI hardware. After the analysis, the adverse impact on the environment and industry-adhered control regulations is recommended and classified for emission rate, intensity, and composition measure. This classification process is performed through a deep neural network to reduce the chance of causing emissions and pollutions in industries and manufacturing units. The data classification is performed on any specific artificial intelligence hardware for recognizing harmful intensity and environmental impact in that machinery. The proposed technique uses deep neural network learning to focus on such factors in various manufacturing intervals.

D_C represents the continuous data classification analysis at different manufacturing intervals. The actual energy observed from sources A_E is also analyzed for successive big data classification which is expressed as:

$$A_E = D_C - H_{int} * E_{imp} \tag{8}$$

$$= arg \sum H_{int}(t_s) + E_{imp}(t_s) \ \forall \ D_C \tag{9}$$

As per Equations (8) and (9), the harmful emission intensity and adverse environmental impact are controlled through some control regulations and manufacturing guidelines. The objective of controlling harmful intensity in all $D_C \in A_E$ is defined in the equation. In this technique, the time series is divided into three instances based on control regulations and guidelines, i.e., emission rate (E_R), emission intensity (E_I), and emission composition (E_C) from the pursuing manufacturing instance. The final estimation $t_s = E_R + E_I + E_C$ is performed for measuring emission rate and intensity in the industry for gas and emission monitoring and control using AI hardware. If i represents the number of machinery in that manufacturing unit, then $E_C = (i \times t_s) - E_I$ is the discrete instance for identifying emissions in this industry, and the required data is to be classified and recommended. Through deep learning, let $C_r(E_R)$, $C_r(E_I)$, and $C_r(E_C)$ represent the control regulation-based data classification processed at different t_s intervals. Therefore, H_{int} and E_{imp} are identified in all AI hardware-assisted industries and manufacturing such that:

$$C_r(E_R) = \frac{(i \times t_s)}{E_{imp}} : A_E \forall \ H_{int} = 0 \tag{10}$$

Such that,

$$C_r(E_I) = \frac{(R + I \times t_s)}{E_{imp}} C : A_E, \ \forall \ H_{int} \neq 0 \tag{11}$$

and,

$$C_r(E_C) = \frac{(R + I + C)}{E_{imp}} t_s : D_C + A_E, \ \forall \ H_{int} = 1 \tag{12}$$

Equations (10)–(12) compute the actual gas and emission observed from the industries and manufacturing units in the current instances and are recommended with data classification. Now, based on the control regulations as in the above equations, Equation (8) is re-written as:

$$D_C(t_s) = [C_r(E_R) - C_r(E_I) + C_r(E_C)] = (i \times t_s) : A_E - \frac{H_{int}}{i} : E_{imp} * A_E \tag{13}$$

In Equation (13), the continuous data classification of $E_R + E_I + E_C \in t_s$ is to be again estimated for identifying the first H_{int} and E_{imp} in specific industries or manufacturing units. This is computed to identify high-volume emissions that occur in industry which is based on control regulations using deep neural network analysis. The correlating time series, control regulation, and environmental impact analysis using the available observed data from the AI hardware is processed through a deep learning paradigm. For this instance, the sequence of $i \in C_r$ is expressed as:

$$i(C_r) = \left(1 - \frac{E_R + E_C - E_I}{i}\right) + \sum_{i=1}^{t_s} \frac{\left(1 - \frac{A_E}{D_C}\right)^{i-1} E_{imp-i}}{H_{int}} \tag{14}$$

Equation (14) compares the current control regulations with the previous emission regulations and guidelines for precise data analysis. Therefore, based on the data classification, the deep learning process is performed to gain the final output for the $H_{int} \neq 0$ case.

The regulation implications $\left(\mu_{E_R}\right)$, $\left(\mu_{E_I}\right)$, and $\left(\mu_{E_C}\right)$ for sequential data classification and analysis at the first level are given as:

$$\mu_{E_R} = \frac{\mathsf{C}_r(E_R)}{\sum_{i \in t_s}[i + A_E(t)]_{t_s}} \tag{15}$$

$$\mu_{E_I} = \frac{\mathsf{C}_r(E_R).\,\mathsf{C}_r(E_I)}{\sum_{i \in t_s}[i + A_E(t)]_{t_s}[1 - H_{int}]_{t_s}} \tag{16}$$

$$\mu_{E_C} = \frac{\mathsf{C}_r(E_R).\,\mathsf{C}_r(E_I).\mathsf{C}_r(E_C).t_s}{\sum_{i \in t_s}[i + A_E(t)]_{t_s}\{[1 - i(t_s)] - E_{imp}\}_{t_s}} \tag{17}$$

Equations (15)–(17) estimate the modified control regulations observed and are recommended for updating the regulations and guidelines based on the current tools and hardware process with the previous emission regulations. In this first level, data classification is the serving input for the regulation implication for reducing emissions in the industry. Figure 3 presents the learning for classifying A_E.

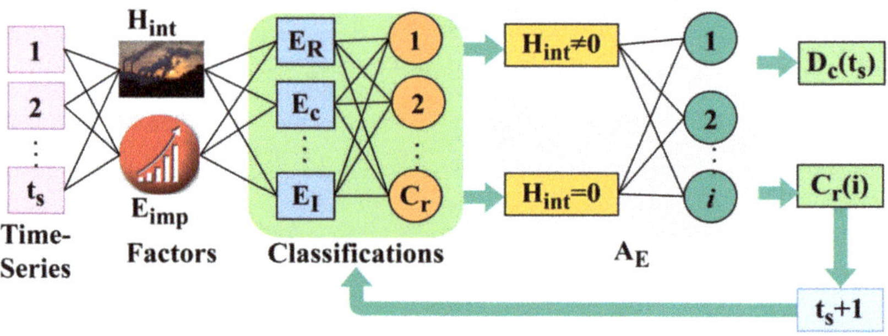

Figure 3. Learning for Classifying Actual Energy Observed from Sources A_E.

The learning for A_E relies on two different classifications, i.e., C_r and H_{int} as presented in Figure 3. The pre-classification for H_{int} and E_{imp} are verified for E_R, E_C, and E_I over the t_s. This classification extracts $H_{int} = 0$ (or) $H_{int} \neq 0$ across i; the i is validated for A_E. In this output, the $D_c(t_s)$ or $\mathsf{C}_r(i)$ is the extracting process. If the system gives $D_c(t_s)$ as the output then regulations are performed; otherwise, classification is performed. The consecutive deep learning for analyzing the harmful intensity of industrial gas and emission helps to identify the adverse impact on the environment through control regulations and manufacturing guidelines. This deep neural network analysis is discussed in the following section.

5. Control Regulation Recommendation Using Learning

In the deep neural network analysis, the control regulations are segregated into classifications, and further analysis is performed to update the regulations and guidelines for the current instance. The deep learning identifies high harmful intensity of the industry leakage through data segregation or classification. In the article, important factors are measured for reliable system processing. The emission rate can be measured for different stages through conventional procedures. However, the government protects these emission occurrence regions if high emission composition and intensity lead to severe hazards in those surroundings. The emission rate intensity and composition are sequentially analyzed and monitored using deep learning to reduce emissions from the industries. The proposed technique considers the severe lookups over the varying manufacturing intervals; thus, the control regulation and environmental impact are analyzed. The computation of the intensity and hazardous components levels in the emission is represented as E^L; the hardware processor is computed in Equation (18) between actual energy and AI.

$$E^L = \sqrt{\frac{I * C}{3H_{int}}} \tag{18}$$

where the harmful intensity value and high components level are identified in the manufacturing unit, the industrial process and manufacturing will be halted, and the government will protect the surrounding with effective control regulation. The previous emission regulations and manufacturing guidelines were also analyzed to modify the current regulations for controlling emissions from the manufacturing units. Industries' harmful intensity and chemicals are monitored, carefully handled, and processed. The control regulation recommendation process is illustrated in Figure 4.

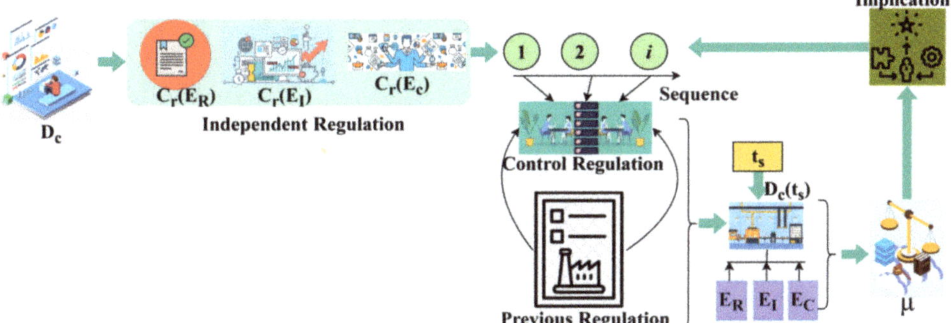

Figure 4. Emission Control Regulation Recommendation.

The D_c is used for independent regulation for i with new implications. This is used for $C_r \, \forall \, E_R$, E_I and E_C such that D_c is further instigated. Therefore, the $D_c(t_s)$ as in Equation (13) is required for μ; the μ is independent for E_R, E_I, and E_C. The $(t_S + 1)$ is required for A_E classification for preventing $i(C_r)$ mismatch and hence E^L is validated. Therefore, the new control regulations are implied for which i is validated using the intelligent hardware (Figure 4). In particular, the contrary process is analyzed using deep neural network learning to control emissions from the industry. If an occurrence is identified for any instance of emission, then the harmful intensity and environmental impact of that manufacturing unit is predicted so that it can be protected with the control regulations. Hence, the data classification and high emission intensity occurrence in the manufacturing units leads to dangers and proper recommendation and control regulation helps to control emissions and pollution produced by the industries. The gas and emission rate and level of the industry and manufacturing units are monitored sequentially to maximize control regulations and manufacturing guidelines. Based on this learning, less emission intensity and composition increase the product manufacturing and also improve recommendations through data classification. Hence, the pollution and emission-causing damages are reduced. Therefore, control regulations and manufacturing guidelines are used to maximize industrial performance.

6. Performance Assessment

This section presents the analysis of industrial emission-based environmental impact using the data from [39]. A series of emission information has been observed in a specific power plant industry under 1 h variation. The information from 11 artificial hardwares are obtained at 36,733 instances for 5 years. The data classification is represented in Figure 5.

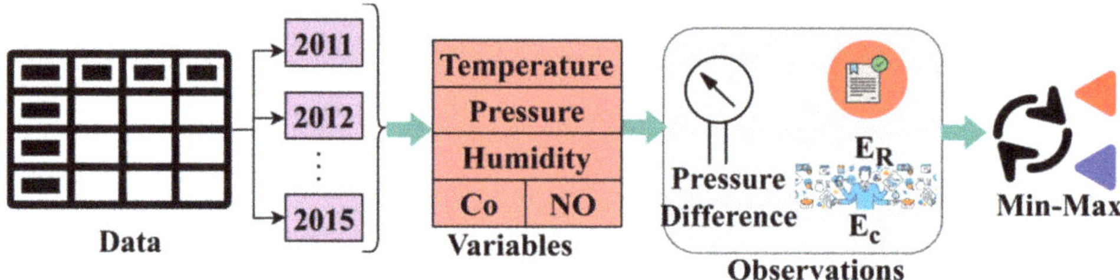

Figure 5. Emission Data Classification.

From this data set, the CO and NO_x emissions are jointly analyzed for their impact on the environment. The variations such as pressure difference (5), emission rate (1 h), and the $E_c(No_x$ or CO or both) are extracted for analysis. The first analysis is presented for E_C, E_I and E_R for $E_C = NO_X$ and $E_C = CO$ between 2011 and 2015 (Figure 6).

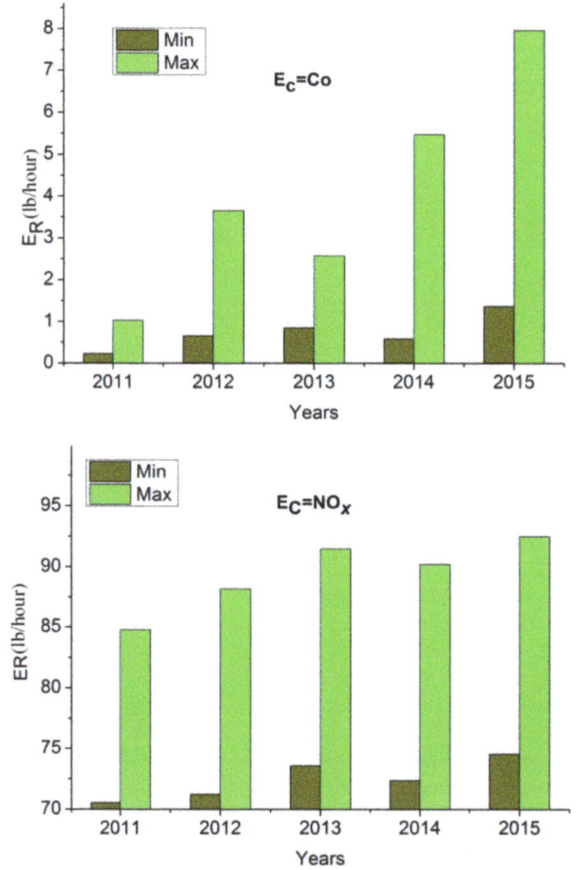

Figure 6. Emission rate E_R for carbon monoxide C_o and nitric oxide NO_X.

The emission rate is classified as the minimum data, including the maximum value in a 24 h observation from an electric power plant. The (lb/hour) value varies for different years (2011 to 2015), so they are classified using pressure, humidity, and intensity. The production increases the intensity and generates emission across different demands. Based on this intensity, the E_x is estimated using the pressure difference (mbar), as presented in Figure 7.

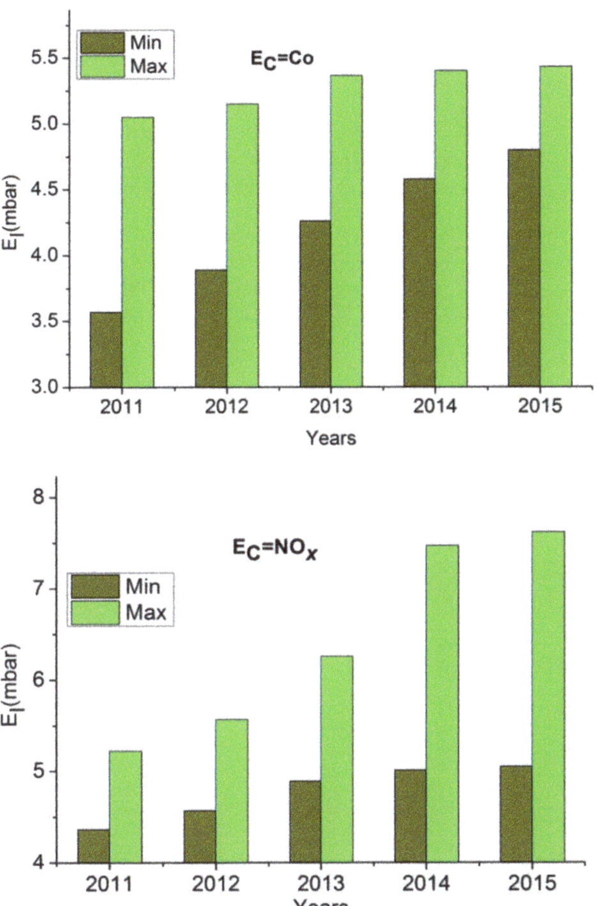

Figure 7. Emission Intensity E_I for Carbon Monoxide Co and Nitric Oxide NO_x.

The variables considered in Figure 5 are used for identifying the *min* and *max* intensities. Compared with E_R, the E_x increases the variation when E_R is higher (uneven/less) when compared with the previous years. Therefore, as the years increase, the production increases as the E_R and E_I for $E_C = C_O$ and NO_X. The E_{imp} for the different years (2011 to 2015) of Co and NO_X is presented in Table 1.

Table 1 values are computed according to the observed and correlated values. For the distinguishable (mbar) and $\left(\text{mg/m}^3\right)$ the E_c determines the E_{imp} over the t_S. As t_s is continuous, then the E_C for detecting No_X or Co or both is consistent. The E_{imp} is the joint detection of Co and NO_X over the impact estimated as the range exceeding the actual level (Table 1). Based on these features, the actual C_r is presented in Figure 8.

The analysis for the recommendation and implications that is different from the previous regulations is presented in Figure 8. The implied regulations are optimal for confining the emissions across various industrial processes. The process implications are performed for confining H_{int} over A_E. In this confining, classifications are prominent over the available data in which the adverse impact is measured. Depending on the environmental impact and regulation policies, control measures are provided. Therefore, the distinguishable sequences provide further recommendations over the μ_{E_R}, μ_{E_I}, and μ_{E_C} independently. These implications are regular for controlling E^L (Figure 8). Table 2 presents the sequence from H_{int} to E^L (high) for which regulations are required.

Table 1. Adverse Environmental Impact E_{imp} for Carbon Monoxide Co and Nitric Oxide NO_X.

E_c	Years	R	Variation	I	Variation	E_{imp} (%)
	2011	0.231	−0.058	3.57	−1.04	12.63
	2012	1.89	−0.064	4.12	−0.95	15.47
Co	2013	2.36	+1.25	4.69	1.58	21.36
	2014	4.59	+1.36	5.23	2.12	19.47
	2015	7.96	+1.47	5.41	2.62	28.31
	2011	70.558	−0.1	4.365	−0.51	17.64
	2012	75.25	−0.095	5.46	−0.31	28.63
NOx	2013	81.25	1.56	6.53	0.15	32.54
	2014	90.47	2.56	6.85	0.46	38.25
	2015	92.498	3.04	7.62	0.8	41.63
	2011	13.61	−1.34	3.061	−0.23	11.28
	2012	25.14	−0.58	4.63	−0.15	15.36
$Co + NOx$	2013	36.14	−0.12	6.98	1.69	21.58
	2014	52.98	1.58	7.47	2.58	32.56
	2015	70.28	2.72	8.74	3.1	46.25

The recommendation using H_{int} and E^L is analyzed as given in Table 2. The H_{int} over the different i is presented as dots in which the green denotes the lesser impact and red denotes the higher impact. Based on the available C_r (recommended), the E^L is confined. However, this is verified using further D_c and, therefore, the recommendations are strong over the available sequences (Table 2). Unlike the above discussion, the following section briefs about the comparative analysis using classifications, data analysis, recommendation rate, effect identification, and analysis time. The methods OC-SVM [30], CTRP [31], and QRPES [32] are added in this comparative analysis study along with the proposed CDCT.

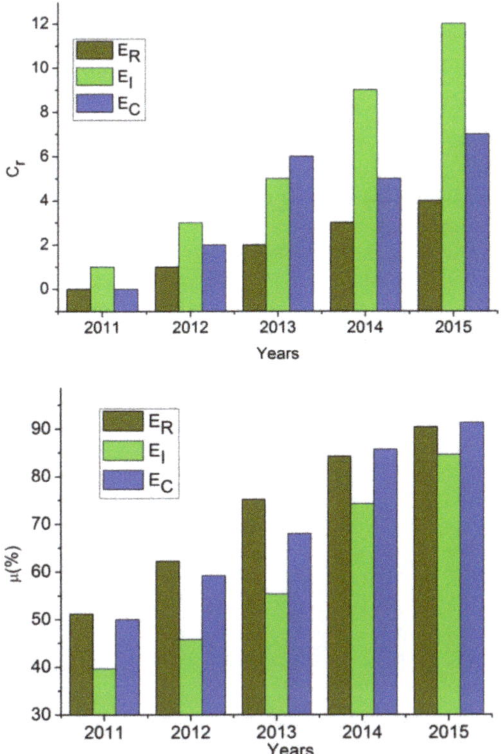

Figure 8. Deep Learning based Data Classification C_r and Regulation Implication μ Analysis.

Table 2. Analysis of Sequence from H_{int} to H_{int} with Recommendation.

H_{int} %	Sequence	E^L	C_r	Classifications	Recommendation Rate
5	{ ● ● ● ● ● ● }	37.76	6	85	0.295
10	{ ● ● ● ● ● ● }	38.25	3	25	0.254
15	{ ● ● ● ● ● ● }	40.25	4	69	0.241
20	{ ● ● ● ● ● ● }	39.65	7	75	0.33
25	{ ● ● ● ● ● ● }	41.25	8	98	0.348
30	{ ● ● ● ● ● ● }	45.21	12	162	0.472
35	{ ● ● ● ● ● ● }	43.64	1	9	0.193
40	{ ● ● ● ● ● ● }	47.58	3	25	0.201
45	{ ● ● ● ● ● ● }	46.89	8	98	0.348
50	{ ● ● ● ● ● ● }	48.92	11	136	0.385

7. Classification

In Figure 9, the emission gas leakage from the industries and manufacturing units is identified through considerable factor values of AI-based hardware processing. It is analyzed for improving the recommendation rate. The emission intensity and composition value is continuously monitored to reduce environmental effects. The industry-adhered control regulations are created to protect humans from the harmful intensity and environmental impact. Depending upon the control regulations and manufacturing guidelines using deep neural network learning, the data classification is performed to segregate the adverse impact on the environment at different manufacturing intervals. The learning process updates the control regulations with the current data observation condition and H_{int} and E_{imp} are analyzed with previous emission control regulations to enhance the data analysis and the recommendation rate. The emission rate modified due to high intensity and composition over the varying manufacturing intervals can be observed in this data analysis for industry emission occurrence in identification and monitoring. This emission occurrence is addressed using deep neural network analysis and regulation implication for achieving successive data classification, preventing harmful intensity. Therefore, the emission rate from the industry is analyzed for any complexity occurrences, preventing high data classification due to regulation implications.

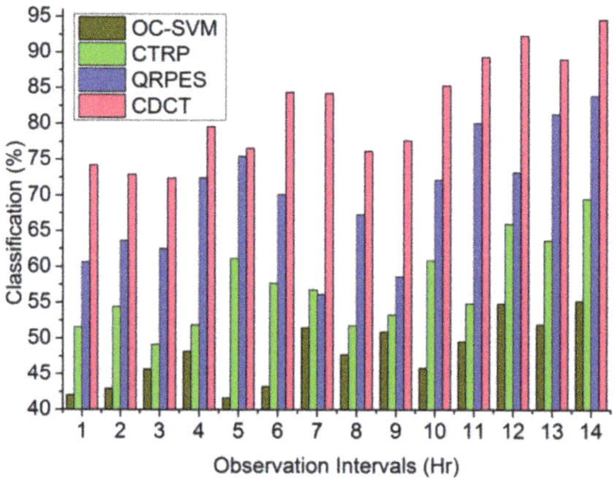

Figure 9. Emission Data Classifications.

8. Data Analysis

The control regulations and manufacturing guidelines are recommended for the industry based on the harmful intensity and composition used in those manufacturing units. The recommendation ensures the emission control can be classified based on the adverse impact on the environment as represented in Figure 10. The data analysis is performed with observed industrial information for analyzing and controlling emission and pollution caused by the industry. The emission rate is analyzed with some control regulations at different intervals for the first input data. The observed data are analyzed to provide precise recommendations for that manufacturing unit and then $E_C = (i \times t_s) - E_I$ is computed for individual industries. This proposed technique satisfies high classification and environmental impact identification by measuring the specific industry's emission rate, intensity, and composition. In this analysis, continuous monitoring and observation are performed in manufacturing units and industries to reduce the harmful intensity and adverse environmental impact on those surroundings. This impact can be addressed through deep neural network learning until new control regulations are updated for maintaining an accurate measure of emission rate, intensity, and composition used in manufacturing units, preventing harmful intensity. Therefore, the data analysis is high in this proposed technique with the recommended precision.

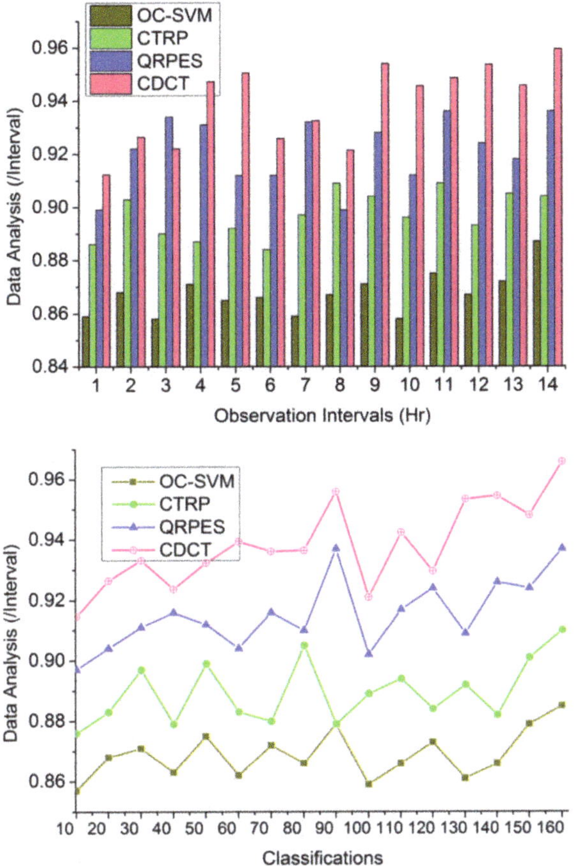

Figure 10. Emission Control Data Analysis.

9. Recommendation Rate

This proposed control-centric data classification process achieves a high recommendation rate for gas and emission monitoring. The analysis relies on AI hardware observation with control regulations (refer to Figure 11). Based on the harmful intensity and environmental impact of industry emissions identified at different manufacturing intervals, time series is performed for classifying its adverse impact. The data classification is processed for monitoring industry and manufacturing units wherein the industry-adhered control regulations are recommended. The observed data from industry and classifications are analyzed to identify the environmental impact due to high harmful intensity. The high emission intensity is identified from the manufacturing units using the accumulated data and calculated emission rate using the deep neural network at different time intervals. The adverse environmental impact is identified through control regulations for controlling the gas and emissions in the industry to enhance the recommendation and classification. The control regulation and manufacturing guidelines are updated with previous emission regulations depending upon other factors in the proposed technique. Therefore, the recommendation rate is high, and the effect identification also increases.

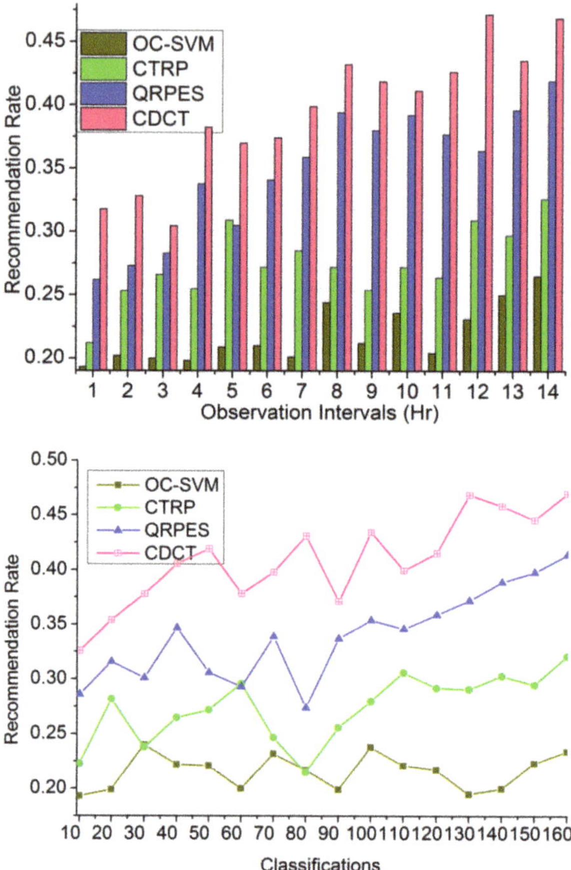

Figure 11. Recommendation Rate for Gas and Emission Monitoring.

10. Effect Identification

This proposed technique satisfies high-effect identification of individual manufacturing units and industrial information working under control regulations and guidelines that aid in monitoring AI hardware for providing reliable recommendations (refer to Figure 12). The harmful intensity and environmental impact is mitigated to classify the data for analyzing and controlling emissions due to high intensity and hazardous components used in the particular industry. This impact is addressed through deep learning and control regulation implications for reduced emissions of gas and carbon output. The data classification is processed between the time series. The control regulations are performed to identify the adverse environmental impact through the condition $t_s = E_R + E_I + E_C$. It is performed to measure emission rate and intensity of the industry. The data classification and recommendation are performed within control regulations and manufacturing guidelines of the specific industry resulting in some implications. From the different manufacturing intervals, the AI hardware performance data are observed for measuring the considerable factors in that industry so as to achieve high-effect identification.

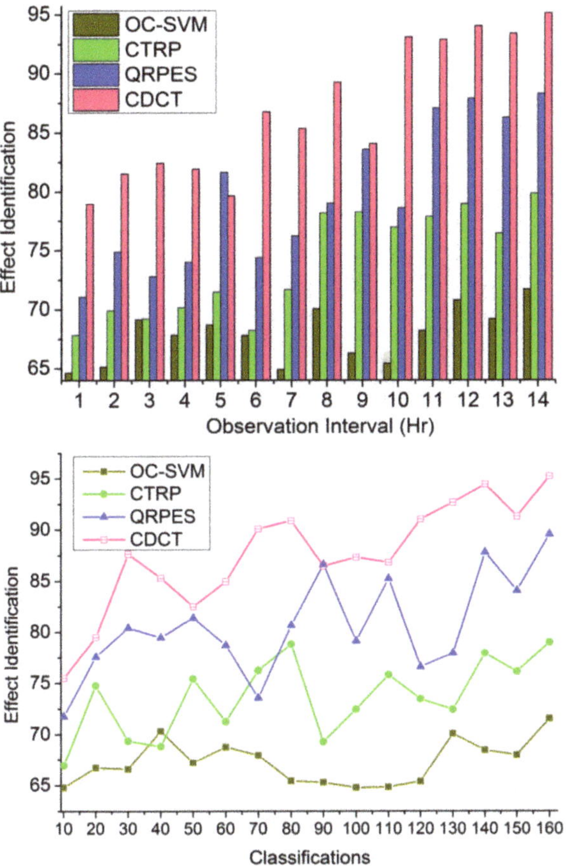

Figure 12. Effect Identification for Individual Manufacturing.

11. Analysis Time

In this proposed technique, the data analysis time is less than other factors for monitoring and analyzing the industrial information for controlled emission and carbon control. Reliable AI hardware processing is maintained for better standardization in a specific industry using previous control regulations and manufacturing guidelines. The computation of emission rate, intensity, and composition for preventing harmful intensity and environmental impact from the industry $D_C \in A_E$ is determined. The data analysis and monitoring of the machinery or devices are sequentially performed using control regulations for time series emission control. The regulation implication is validated for the current instances. Based on the regulation implication for classified data along with previous regulations and guidelines, deep learning is employed to prevent complexity in identifying emission gas leakage. The proposed technique analyzes the intensity and hazardous component levels of the emissions. They are updated after the deep learning process for recommending severe lookups in manufacturing units for data analysis achieving less analysis time as represented in Figure 13. Tables 3 and 4 present the summary of the above discussion.

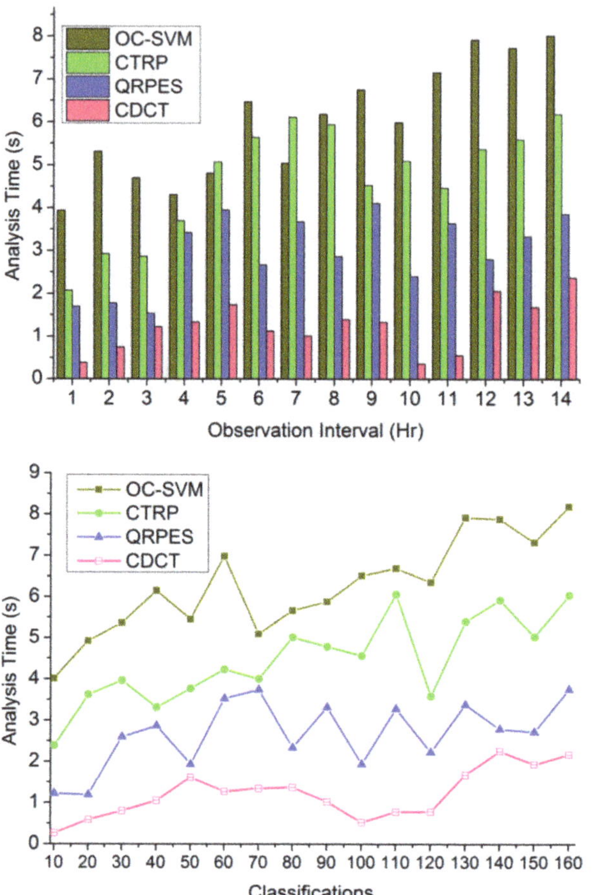

Figure 13. Data Analysis Time of Industrial Monitoring and Analysis Process.

Table 3. Overall Comparative Summary (Observation Intervals).

Metrics	OC-SVM	CTRP	QRPES	CDCT
Classification (%)	55.18	69.47	83.88	94.597
Data Analysis (/Interval)	0.887	0.904	0.936	0.9594
Recommendation Rate	0.265	0.326	0.419	0.4685
Effect Identification	71.7	79.86	88.29	95.122
Analysis Time (s)	8.02	6.18	3.85	2.376

Observations: The proposed technique maximizes classification, data analysis, recommendations, and effect identification by 12.54%, 10.28%, 13.18%, and 15.17%, respectively. It reduces the analysis time by 10.08%.

Table 4. Overall Comparative Summary (Classifications).

Metrics	OC-SVM	CTRP	QRPES	CDCT
Data Analysis (/Interval)	0.885	0.910	0.937	0.9658
Recommendation Rate	0.234	0.321	0.414	0.4698
Effect Identification	71.55	79.02	89.58	95.246
Analysis Time (s)	8.18	6.04	3.75	2.179

Observations: The proposed technique maximizes data analysis, recommendations, and effect identification by 11.03%, 14.68%, and 15.2%, respectively. It reduces the analysis time by 10.6%.

12. Conclusions

This article discussed the proposed control-centric data classification technique for emissions control and regulation implications for industrial productions. This technique focuses on cases where regulations are implied for controlling environmentally impacting gases. The emission rate, intensity, and composition are segregated using artificial intelligence hardware-based data measured at different intervals. Depending on the classification, the adverse impact is estimated in correlation with the actual regulations. In this process, deep learning classification is deployed to identify the high intensity and adverse impact of different gases. The classification at different levels is performed to improve the regulation implications and the modified rule adherence of the industry in pollution control and harmful emissions. The recommendations for industrial operations and data analysis are identified from the AI hardware-sensed information for which the regulation implications and monitoring are pursued. The proposed technique maximizes classification, data analysis, recommendation, and effect identification by 12.54%, 10.28%, 13.18%, and 15.17%, respectively. Furthermore, it reduces the analysis time by 10.08%.

Author Contributions: Conceptualization, Methodology, Visualization, Z.C.; Writing—Original Draft, Writing—Review & Editing, J.C. All authors have read and agreed to the published version of the manuscript.

Funding: This research received no external funding.

Data Availability Statement: The data used in the research is in the manuscript.

References

1. Fort, A.; Landi, E.; Mugnaini, M.; Parri, L.; Pozzebon, A.; Vignoli, V. A LoRaWAN carbon monoxide measurement system with low-power sensor triggering for the monitoring of domestic and industrial boilers. *IEEE Trans. Instrum. Meas.* **2020**, *70*, 5500609. [CrossRef]
2. Lv, Z.; Chen, D.; Lou, R.; Song, H. Industrial security solution for virtual reality. *IEEE Internet Things J.* **2020**, *8*, 6273–6281. [CrossRef]
3. Li, D.; Yu, H.; Tee, K.P.; Wu, Y.; Ge, S.S.; Lee, T.H. On time-synchronized stability and control. *IEEE Trans. Syst. Man Cybern.-Syst.* **2021**, *52*, 2450–2463. [CrossRef]
4. Oliva, G.; Zarra, T.; Pittoni, G.; Senatore, V.; Galang, M.G.; Castellani, M.; Belgiorno, V.; Naddeo, V. Next-generation instrumental odour monitoring system (IOMS) for the gaseous emissions control in complex industrial plants. *Chemosphere* **2021**, *271*, 129768. [CrossRef]
5. Timofeev, A.V.; Groznov, D.I. Classification of seismoacoustic emission sources in fiber optic systems for monitoring extended objects. *Optoelectron. Instrum. Data Process.* **2020**, *56*, 50–60. [CrossRef]
6. Guo, B.; Wang, Y.; Zhou, H.; Hu, F. Can environmental tax reform promote carbon abatement of resource-based cities? Evidence from a quasi-natural experiment in China. *Environ. Sci. Pollut. Res.* **2022**. [CrossRef]
7. Lu, C.; Zhou, H.; Li, L.; Yang, A.; Xu, C.; Ou, Z.; Wang, J.; Wang, X.; Tian, F. Split-core magnetoelectric current sensor and wireless current measurement application. *Meas. J. Int. Meas. Confed.* **2022**, *188*, 110527. [CrossRef]
8. Wu, X.; Yu, X.; Xu, R.; Cao, M.; Sun, K. Nonlinear dynamic soft-sensing modeling of NOx emission of a selective catalytic reduction denitration system. *IEEE Trans. Instrum. Meas.* **2022**, *71*, 2504911. [CrossRef]
9. Dang, W.; Guo, J.; Liu, M.; Liu, S.; Yang, B.; Yin, L.; Zheng, W. A semi-supervised extreme learning machine algorithm based on the new weighted kernel for machine smell. *Appl. Sci.* **2022**, *12*, 9213. [CrossRef]

10. Parente, A.P.; Valdman, A.; Folly, R.O.M.; de Souza, M.B.; Fileti, A.M.F. An online calibration tool for soft sensors: Development and experimental tests in a semi-industrial boiler plant. *Braz. J. Chem. Eng.* **2020**, *37*, 189–199. [CrossRef]

11. Liu, S.; Xiao, Q. An empirical analysis on spatial correlation investigation of industrial carbon emissions using SNA-ICE model. *Energy* **2021**, *224*, 120183. [CrossRef]

12. Liu, S.; Peng, G.; Sun, C.; Balezentis, T.; Guo, A. Comparison of improving energy use and mitigating pollutant emissions from industrial and non-industrial activities: Evidence from a variable-specific productivity analysis framework. *Sci. Total Environ.* **2022**, *806*, 151279. [CrossRef] [PubMed]

13. Xu, K.; Guo, Y.; Liu, Y.; Deng, X.; Chen, Q.; Ma, Z. 60-GHz compact dual-mode on-chip bandpass filter using GaAs technology. *IEEE Electron Device Lett.* **2021**, *42*, 1120–1123. [CrossRef]

14. Shen, Y.; Ding, N.; Zheng, H.T.; Li, Y.; Yang, M. Modeling relation paths for knowledge graph completion. *IEEE Trans. Knowl. Data Eng.* **2021**, *33*, 3607–3617. [CrossRef]

15. Cárdenas-Mamani, Ú.; Kahhat, R.; Vázquez-Rowe, I. District-level analysis for household-related energy consumption and greenhouse gas emissions: A case study in Lima, Peru. *Sustain. Cities Soc.* **2022**, *77*, 103572. [CrossRef]

16. Taranina, O.A.; Burkat, V.S.; Volkodaeva, M.V. Analysis of the concentration of gas-phase and solid-phase polyaromatic hydrocarbons in industrial emissions from aluminum production. *Metallurgist* **2020**, *63*, 1227–1236. [CrossRef]

17. Lilley, A.; Coleman, M.D.; Goddard, S.L.; Mills, E.G.; Brown, R.J.; Robinson, R.A.; Clack, M.J. Inter-comparability of analytical laboratories in quantifying polycyclic aromatic hydrocarbons collected from industrial emission sources. *Accredit. Qual. Assur.* **2022**, *27*, 155–163. [CrossRef]

18. De Oliveira Gabriel, R.; Leal Braga, S.; Pradelle, F.; Torres Serra, E.; Coutinho Sobral Vieira, C.L. Numerical simulation of an on-grid natural gas PEMFC-solar photovoltaic micro CHP unit: Analysis of the energy, economic and environmental impacts for residential and industrial applications. *Technol. Econ. Smart Grids Sustain. Energy* **2022**, *7*, 5. [CrossRef]

19. Qin, L.; Lu, G.; Hossain, M.M.; Morris, A.; Yan, Y. A flame imaging-based online deep learning model for predicting NO_x emissions from an oxy-biomass combustion process. *IEEE Trans. Instrum. Meas.* **2021**, *71*, 2501811. [CrossRef]

20. Chen, S.; Zhu, X.; Chen, K.; Liu, Z.; Li, P.; Liang, X.; Jin, X.; Du, Z. Applying deep learning-based regional feature recognition from macro-scale image to assist energy saving and emission reduction in industrial energy systems. *J. Adv. Res.* **2022**, *in press*. [CrossRef]

21. Friedel, J.E.; Holzer, T.H.; Sarkani, S. Development, optimization, and validation of unintended radiated emissions processing system for threat identification. *IEEE Trans. Syst. Man Cybern. Syst.* **2018**, *50*, 2208–2219. [CrossRef]

22. Zhang, X.; Dong, F. Determinants and regional contributions of industrial CO_2 emissions inequality: A consumption-based perspective. *Sustain. Energy Technol. Assess.* **2022**, *52*, 102270. [CrossRef]

23. Huang, K.; Eckelman, M.J. Estimating future industrial emissions of hazardous air pollutants in the United States using the National Energy Modeling System (NEMS). *Resour. Conserv. Recycl.* **2021**, *169*, 105465. [CrossRef]

24. Tao, H.; Xie, C.; Wang, J.; Xin, Z. CENet: A channel-enhanced spatiotemporal network with sufficient supervision information for recognizing industrial smoke emissions. *IEEE Internet Things J.* **2022**, *9*, 18749–18759. [CrossRef]

25. Fiscante, N.; Addabbo, P.; Biondi, F.; Giunta, G.; Orlando, D. Unsupervised sparse unmixing of atmospheric trace gases from hyperspectral satellite data. *IEEE Geosci. Remote Sens. Lett.* **2022**, *19*, 6006405. [CrossRef]

26. Guo, C.; Liu, X.; He, X. A global meta-analysis of crop yield and agricultural greenhouse gas emissions under nitrogen fertilizer application. *Sci. Total Environ.* **2022**, *831*, 154982. [CrossRef]

27. Sikdar, S.; Liu, D.; Kundu, A. Acoustic emission data based deep learning approach for classification and detecting damage-sources in a composite panel. *Compos. Part B Eng.* **2022**, *228*, 109450. [CrossRef]

28. Choi, Y.; Kim, K.; Kim, S.; Kim, D. Identification of odor emission sources in urban areas using machine learning-based classification models. *Atmos. Environ. X* **2022**, *13*, 100156. [CrossRef]

29. Tacchino, V.; Costamagna, P.; Rosellini, S.; Mantelli, V.; Servida, A. Multi-scale model of a top-fired steam methane reforming reactor and validation with industrial, experimental data. *Chem. Eng. J.* **2022**, *428*, 131492. [CrossRef]

30. Tuttle, J.F.; Blackburn, L.D.; Powell, K.M. Online classification of coal combustion quality using nonlinear SVM for improved neural network NOx emission rate prediction. *Comput. Chem. Eng.* **2020**, *141*, 106990. [CrossRef]

31. Sun, S.; Li, L.; Wu, Z.; Gautam, A.; Li, J.; Zhao, W. Variation of industrial air pollution emissions based on VIIRS thermal anomaly data. *Atmos. Res.* **2020**, *244*, 105021. [CrossRef]

32. Ju, T.; Lei, M.; Guo, G.; Xi, J.; Zhang, Y.; Xu, Y.; Lou, Q. A new prediction method of industrial atmospheric pollutant emission intensity based on pollutant emission standard quantification. *Front. Environ. Sci. Eng.* **2023**, *17*, 8. [CrossRef] [PubMed]

33. Sun, J.; Zhang, Z.; Liu, X.; Zheng, H. Reduced methane combustion mechanism and verification, validation, and accreditation (VV&A) in CFD for NO emission prediction. *J. Therm. Sci.* **2021**, *30*, 610–623.

34. Milkevych, V.; Villumsen, T.M.; Løvendahl, P.; Sahana, G. Data synchronization for gas emission measurements from dairy cattle: A matched filter approach. *Comput. Electron. Agric.* **2022**, *201*, 107299. [CrossRef]

35. Martínez, R.A.; Fechner, D.C.; Delfino, M.R.; Pellerano, R.G.; Goicoechea, H.C. Rapidly determined three textile surfactants in environmental samples by modeling excitation-emission second-order data with multi-way calibration methods. *Environ. Sci. Pollut. Res.* **2022**, *29*, 25869–25880. [CrossRef] [PubMed]

36. Lee, H.; Kim, H.; Choi, D.G.; Koo, Y. The impact of technology learning and spillovers between emission-intensive industries on climate policy performance based on an industrial energy system model. *Energy Strategy Rev.* **2022**, *43*, 100898. [CrossRef]

37. Ren, M.; Zhang, Q.; Zhang, J. An introductory survey of probability density function control. *Syst. Sci. Control. Eng.* **2019**, *7*, 158–170. [CrossRef]

38. Zhang, Q.; Zhou, Y. Recent advances in non-Gaussian stochastic systems control theory and its applications. *Int. J. Netw. Dyn. Intell.* **2022**, *1*, 111–119. [CrossRef]

39. Available online: https://archive.ics.uci.edu/ml/datasets/Gas+Turbine+CO+and+NOx+Emission+Data+Set (accessed on 23 February 2022).

Article

Employment Effect of Structural Change in Strategic Emerging Industries

Li Liu [1,2], Cisheng Wu [1,*] and Yiyan Zhu [1]

[1] School of Management, Hefei University of Technology, Hefei 230009, China
[2] School of Business, Fuyang Normal University, Fuyang 236037, China
* Correspondence: cswu@hfut.edu.cn

Abstract: Stable development of strategic emerging industries promotes its industrial transformation and upgrading, which has affected the development of not only the society and the economy but also other fields, thereby having a great impact on employment. To measure the impact of structural change of strategic emerging industries on employment in China, this paper constructs a regression equation, in which the employment of strategic emerging industries is the dependent variable, while the change direction of strategic emerging industry structure, the employment elasticity of strategic emerging industries and the change speed of industrial structure are the independent variables. The research results are as follows: (i) The change direction of strategic emerging industries is positively correlated with employment. (ii) The employment elasticity of strategic emerging industries is on the rise, and is positively correlated with employment. (iii) The speed of change of strategic emerging industries is unstable, and is negatively correlated with employment. As a result, the structural change in strategic emerging industries has played a role in promoting employment. The government should recognize the impact of structural changes in strategic emerging industries on China's employment. By implementing the existing strategic emerging industry policies and improving the external environment for the development of strategic emerging industries, the strategic emerging industries will play the role of "innovation, growth and leadership" in economic and social development.

Keywords: strategic emerging industries; industrial structure; employment effect

Citation: Liu, L.; Wu, C.; Zhu, Y. Employment Effect of Structural Change in Strategic Emerging Industries. *Processes* **2023**, *11*, 599. https://doi.org/10.3390/pr11020599

Academic Editors: Conghu Liu, Xiaoqian Song, Zhi Liu and Fangfang Wei

Received: 31 January 2023
Revised: 14 February 2023
Accepted: 14 February 2023
Published: 16 February 2023

1. Introduction

In recent years, emerging industries have gradually become the main driving force of global economic recovery and growth. Strategic emerging industries can promote the progress and development of science, technology and society [1,2]. With the rapid development of a new round of global scientific and technological revolution and industrial transformation, strategic emerging industries have been highly valued by countries all over the world. The United States and other countries have sought to accelerate the development of strategic emerging industries in order to maintain their position at the forefront of science and technology and maintain their advantages in global competition in the future. Strategic emerging industries are gradually becoming the main driving force promoting the development of the national economy [3]. Strategic emerging industries can drive social development, change the environment and space in which humans live, and create a considerable number of related industries and supporting facilities [4]. This strategy has become a global trend of industrial development and an inevitable requirement of optimal economic growth [5].

In October 2010, the State Council of China issued the Decision of the State Council on Accelerating the Cultivation and Development of Strategic Emerging Industries (http://www.gov.cn/zwgk/2010-10/18/content_1724848.htm (accessed on 18 October

2010)). The decision regards seven industries, namely energy conservation and environmental protection, new-generation information technology, biology, high-end equipment manufacturing, new energy, new materials and new-energy vehicles, as China's strategic emerging industries. In November 2018, the Chinese government had identified nine major industries as strategic emerging industries according to its national conditions and global development status, with the high-end equipment manufacturing industry as the core, the new generation of information technology industry as the support, and the energy conservation and environmental protection industry, new materials industry and new energy vehicle industry as the pilot (http://www.gov.cn/zhengce/zhengceku/2018-12/31/content_5433037.htm (accessed on 26 Novezmber 2018)). In November 2019, the guide catalogue for industrial restructuring was issued by the national development and reform commission of China, which clarified the goal and direction of the development of strategic emerging industries (https://www.ndrc.gov.cn/fgsj/tjsj/cyfz/zzyfz/201911/t20191107_1201849.html (accessed on 7 November 2019)). Therefore, it is necessary to actively cultivate and vigorously develop strategic emerging industries, and promote the transformation of industries for digitalization, networking and intelligence through the effective integration of traditional industries and strategic emerging industries [6]. Strategic emerging industries based on major technological breakthroughs and development needs have become China's new competitive advantage and the decisive force to achieve leapfrog development [7,8].

Industrial structural change is one of the most robust features of economic development [9]. The process of national or regional economic growth is the dynamic evolution process of industrial structure from uncoordinated to coordinated, from lower-level coordination to higher-level coordination, that is, the process of continuous optimization of industrial structures [10]. With industrial structure adjustment, the share of agriculture, manufacturing and service industries in employment has changed significantly. Employment has been a major concern in the growth and economic development of societies and countries [11]. As an important parameter determining employment scale, industry structure has the most significant influence on growing job employment. The strategic emerging industrial structure refers to the development level of each specific strategic emerging industry in a specific region and the proportional relationship among those specific industries. The development and expansion of strategic emerging industries can expand the scope of the economy, increase employment opportunities and create more jobs. Compared with the change of traditional industry structure, the change of strategic emerging industry structure has a greater impact on employment.

Strategic emerging industries paly a very import role in economic and social development and are the main direction in the future [12]. Strategic emerging industries can promote industrial transformation and upgrading as well as increase employment. In recent years, changes in the structure of strategic emerging industries have had an increasing impact on employment. However, most of the strategic emerging industries belong to the high-tech industry, and changes in the structure of strategic emerging industries will have uncertain effects on the growth of labor and employment. Some strategic emerging industries may have positive employment elasticity as a result of increased demand for their products and services, resulting in increased employment. However, due to technological advancements and automation, some strategic emerging industries may have negative employment elasticity, reducing the need for labor. Therefore, it is of great theoretical and practical significance to study the structural changes of strategic emerging industries and increase the number of labor employment. This is conducive for promoting industrial integration, development and optimizing the allocation of labor resources.

The main contributions of our research are as follows. Firstly, using real data, we calculate and analyze the direction of the structural change, the employment absorption capacity and the speed of change of the industrial structure in China's strategic emerging industries from 2009 to 2020. Secondly, we build a model to empirically analyze the employment effect of the structural changes in strategic emerging industries. Thirdly, we empirically analyze the promoting effect of structural change of strategic emerging industries on employment.

The structural arrangement of this paper is as follows. In Section 2, we provide a literature review. In Section 3, we present research design that includes the mechanism of action, puts forward the hypotheses, and the regression equation. The variables and Data Sources are presented in Section 4. Section 5 gives the empirical results. Section 6 presents the important conclusions of this paper.

2. Literature Review

The relationship between changes in industrial structure and employment has attracted increasing attention. To explain why industry structural change impacts employment, many empirical research papers and models have been used to investigate this problem. This research examines the characteristics of industry structural change and the relationship between industrial structure upgrading and employment.

2.1. The Characteristics of Industrial Structure Change

Industrial structure refers to the composition of various industries and their proportional relationship. The structural transformation of the process of workers transferring from agriculture to other sectors is a significant feature of development [13]. The change of industrial structure in a region is usually related to consumer demand, factor endowment and technological progress. The upgrading of industrial structure depends on the upgrading of factor endowment structure. The factor endowment structure mainly includes natural resources, material capital, labor force, technology, infrastructure and system. Kang and Lei [14] investigated the space-time evolution characteristics of industrial economy in Central Asia from the perspective of industrial scale, structural rationality, industrial competitiveness and industrial isomorphism, and pointed out that the industrial structure of Central Asian countries is becoming increasingly advanced, but there are certain differences in this evolution characteristics.

Most scholars study the change of traditional industrial structures. However, Gabardo et al. [15] underline that structural change cannot be restricted to the three broad sectors but instead it covers the change in the structure of production and employment between and within sectors as well. Lauri Hetemäki et al. [16] discussed structural changes of the forest sector and the realistic contribution of the forest-based sector to the global sustainability challenges. Duan [17] proposed a modeling method for evaluating the influence of industrial structure change on promoting coastal forestry economic growth based on big data and piecewise sample autoregression feature decomposition models and analyzed the influencing factors of industrial structure change and promoting the economic growth of coastal forestry. Xie et al. [18] studied the change of China's Marine industry structure through two indicators: industrial structure rationalization and industrial structure promotion. Jin et al. [19] employed the added values of three land-sea industries in the three marine economic circles of northern, eastern, and southern China to clarify the evolutionary behavior of the industrial structure of these three circles on the land and sea. Studying the characteristics of industry structural change can grasp the direction of structural changes in strategic emerging industries and provide a basis for the analysis of employment effects.

2.2. The Relationship between Industrial Structure Change and Employment

Scholars have examined the interplay between industrial structural change and employment from various perspectives. Although scholars' views vary, they all agree that to some extent, changes in industrial structure affect employment. One view is that upgrading the industrial structure based on technological innovation can promote employment. The change of industrial structure can promote employment in two ways: first, industrial development needs to absorb labor force; second, the development of this industry can promote the development of other industries, and then promote employment. Basile

Rosanna et al. [20] used Italian labor data from 1981 to 2008 to study the impact of changes in industrial structure on employment growth and concluded that the relationship between industrial structure and employment structure is nonlinear. Chivu et al. [21] analyzed the number and scale of industrial structure adjustment, industrial output structural change, employment number and employee compensation in Romania. Zhang et al. [22] established a spreadsheet-based analytical model to estimate the employment effects of the solar PV industry among China's strategic emerging industries during the period 2009–2015. Monteforte [13] presented a dynamic dual economy model of structural transformation and found that labor market institutions influence structural transformation because they determine how quickly the urban sector can absorb labor from agriculture.

However, another view is that upgrading the industrial structure based on technological progress will harm employment. Hong et al. [23] taken the Korean ICT industry from 1995 to 2009 as the research object to analyze the relationship between structural changes and employment, and pointed out that unemployment growth occurred in the field of information and communication technology manufacturing. Although the output of the information and communication technology manufacturing industry increased significantly, the employment rate continued to decline. From the three dimensions of employment elasticity, structural deviation and labor productivity ratio, Ding et al. [24] investigated the degree of synergy between the three sectors, and found that the employment elasticity of the three sectors presents different characteristics, and the third sector has the most significant impact on employment.

Through a review of the literature on industrial structural change and employment, we found the following three research shortcomings. Firstly, most theories developed regarding the impact of industrial structural change on labor and employment are based on the early stage of industrial development or on mature industries. Although these research results have reference values and do provide some theoretical support for the research focus of this paper, they are not fully applicable to the situation in China. Secondly, although Chinese scholars have demonstrated the impact of industrial structural improvements on labor employment, mainly focusing on economic growth, the wide range of industrial structures, and the econometric calculations, their conclusions are disparate. Thirdly, the scope of the existing literature is limited to primary industry, secondary industry and tertiary industry, whereas scant attention is given to the effects of changes in strategic emerging industrial structure on employment and on its industrial employment flexibility.

Strategic emerging industries are new, rapidly growing industries that are deemed strategic for a country's economic development and future competitiveness. These industries typically involve cutting-edge technologies, goods, and services, and they have the potential to drive significant economic growth and job creation. According to our review of the literature, there are few researches on the employment effect of structural changes in strategic emerging industries. Strategic emerging industries are key areas to promote industrial restructuring in China [25]. Stable employment is the top priority for us. The research object of this paper is more targeted and time-sensitive, providing theoretical support for future research in related fields.

3. Research Design

3.1. Mechanism of Action

The change of industrial structure is believed to be formed by the interaction of different technological developments on the supply side and factors on the demand side [26]. The two classic components of industrial structural change are industrial structural rationalization and upgrading [27,28]. This involves the reallocation of production factors (labor, capital and environment) among different sectors. Among those production factors, labor resource is the most dynamic [29,30]. The development of strategic emerging industries can help transform the economic development model and realize the optimization and upgrading of industrial organizations [31]. With the development of technology and economics, the structure of strategic emerging industries is adjusting through dynamic development, and in the process, the forces of "creation and destruction" coexist, which produces the mechanism of destruction and compensation in labor employment.

The employment destruction mechanism of structural changes in strategic emerging industries involves the upgrading of the structure of emerging industries, in which labor-intensive industries are gradually replaced by technology- and capital-intensive industries, which reduces the demand for labor and causes structural unemployment. These changes will reduce the demand for labor, which in turn will have a crowding-out effect on employment, namely, a destructive effect. These changes are reflected in the following four aspects. First, in the process of changing the structure of strategic emerging industries, the technological level will be improved, machines will replace the labor force, and production will be increasingly automated and intelligently improved, which will reduce the demand for labor and the number of jobs [32]. Second, the change of strategic industrial structure changes the demand for labor force, which results in the lag effect and the decrease in employment. Third, the structural adjustment of strategic emerging industries will lead to the continuous optimization and upgrading of industries and the improvement of management levels, which will lead to a reduction of redundant personnel and a decline in employment. Fourth, in the process of restructuring strategic emerging industries, institutional and policy obstacles will appear, hindering the flow of labor between different industries, which will also lead to a decline in employment.

To sum up, the process of the employment destruction mechanism involved in structural changes in strategic emerging industries is as follows (see Figure 1). The changes in strategic emerging industrial structure will improve production technology, change job demand, and improve the level of management so as to make production automatic and intelligent; machines will replace the labor force redundant personnel will be cut; and eventually, with the reduced demand and supply of labor, employment will decline.

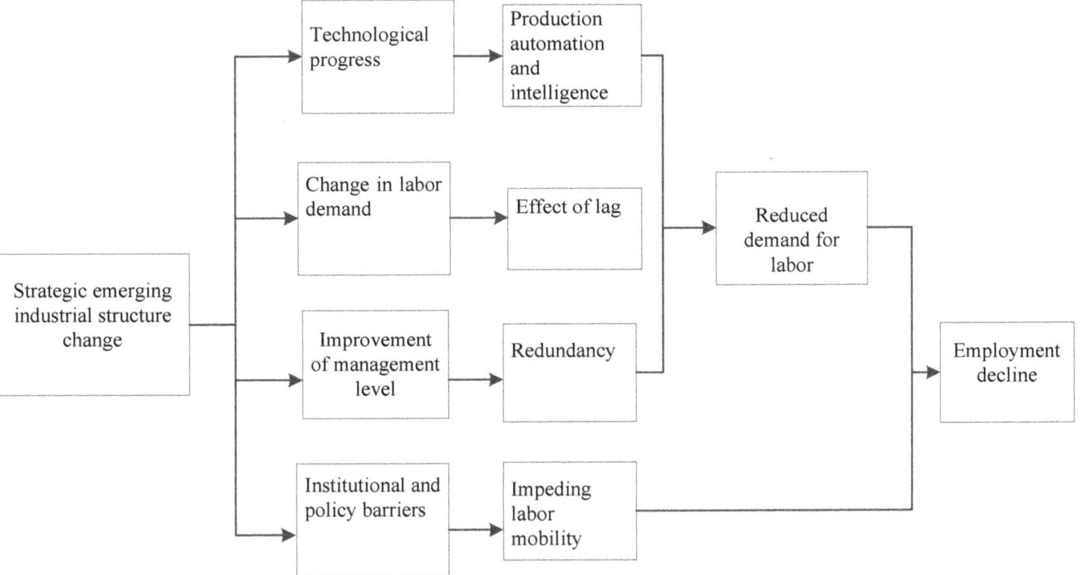

Figure 1. The formation process of the employment destruction mechanism of the changes in strategic emerging industrial structure.

The employment compensation mechanism [33] underlying the change in strategic emerging industries involves the creation of new products and emerging industries via technological innovation, leading to changes in the number of strategic emerging industries and the composition of the industrial structure. These changes, in turn, lead to the flow of production factors between different industrial sectors and to the transfer of labor and drive the development of related industries. All these processes will increase the demand for labor force and have a promoting effect on employment, that is, a compensation effect. These changes are reflected in the following five aspects. We can summarize the formation process of the employment compensation mechanism caused by the structural changes in strategic emerging industries as follows (see Figure 2).

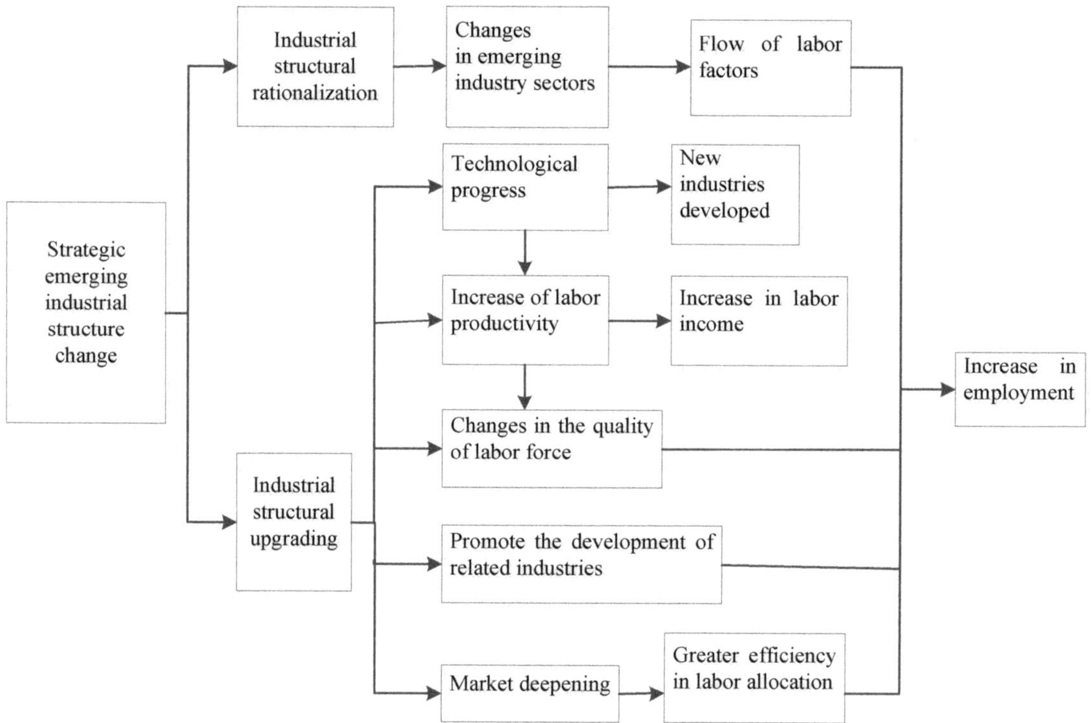

Figure 2. The formation process of the employment compensation mechanism of the change in strategic emerging industrial structure.

Firstly, the change in strategic emerging industries is accompanied by technological progress and increased labor productivity, which promotes the flow of labor factors within industries and among different industries. Thus, the reallocation of labor resources is realized, and the amount of employment is increased. Second, in the process of upgrading strategic emerging industrial structure, enterprises begin to develop new technologies in order to increase their market share and generate advantages in the market, resulting in the continuous improvement of the technological level of each industry [34]. This advancement will bring new products and new industries, create new jobs, and increase the number of people employed. Third, in the process of changing strategic emerging industries, the newly generated industries have good development prospects and relatively high labor productivity. Fourth, in the process of changing strategic emerging industries, the quality of the labor force will change and human capital investment will increase, easing the pressure of employment. This effect occurs because the rise of new industries results in higher requirements for the quality of the labor force, requiring more highly educated and skilled workers [35]. The demand for higher qualifications requires workers to extend their education time and improve their basic skills in order to adapt to the changes in strategic emerging industrial structure. This requirement for a more highly skilled labor force slows the entry of workers into the labor market, thus reducing the pressure on employment.

In addition, technological innovation brought about by the changes in the structure of strategic emerging industries leads to higher income levels, enabling people to continue to develop their education and increasing investment in human capital. Human capital investment also improves the quality and competitiveness of the labor force in emerging industries, which is conducive to labor mobility. Fifth, changes in the structure of strategic emerging industries directly or indirectly promote the expansion of relevant industries, deepen the market, transform the mode of economic growth [36] and improve the efficiency of labor resource allocation, thus increasing employment.

Changes in the structure of strategic emerging industries will promote the flow of factors of production among different departments and encourage the innovative allocation of labor resources. It will promote technological innovation, new products and new industries as well as create new jobs. The labor productivity of emerging industries increases, the enterprise income increases and the labor force income increases, which attracts more workers. This process will change the quality of the labor force, increase the investment in human capital and reduce the pressure of employment. It will promote the expansion of related industries, promote the deepening of the market and improve the efficiency of labor resource allocation. All of these factors will create jobs.

The effects of changes in strategic emerging industry on employment are complex and systematic, and the impact of industrial structural changes on employment depends on the combined effect of the employment compensation mechanism and the destructive mechanism. When the destructive mechanism of structural change in strategic emerging industry is greater than the compensation mechanism, total employment decreases, and when the compensation mechanism of the emerging industrial structure is greater than the destructive mechanism, total employment increases.

3.2. Hypotheses

Based on the above theoretical analysis, we can find that the level of employment in strategic emerging industries is not only related to the expansion of the industrial scale but also to the adjustment of the industrial structure. There are many indicators to measure the change in industrial structure, but the main indicators are the direction and speed of the change. In addition, employment elasticity mainly emphasizes the absorptive capacity of labor resources for economic growth. Employment elasticity is an indicator to measure industrial employment absorptive capacity. Based on the above theoretical analysis, this paper examines the effect of the change in strategic emerging industrial structure on employment from the three perspectives of the direction of change in the industrial structure change, employment elasticity and change speed. It also puts forward the following three assumptions.

The scale of industrial structural change can be measured by the direction of this change. The value added of the industry is the value added by the production activities of enterprises in a certain period of time, which can accurately reflect the scale and speed of industrial production. Therefore, this paper uses the proportion of Gross Domestic Product contributed by the added value of strategic emerging industries to express the direction of industrial structural change and proposes the first hypothesis.

Hypothesis 1. *The change of direction in strategic emerging industrial structure is positively correlated with the level of employment.*

Employment elasticity refers to the percentage change of employment corresponding to the change of economic growth [37]. Analogously, industrial employment elasticity refers to the proportion of change caused by the change in employment quantity with a 1% growth in an industry's output. The industrial employment elasticity formula is as follows.

$$\varepsilon = \frac{\Delta L / L_i}{\Delta Y / Y_i} \tag{1}$$

In Formula (1), ε represents employment elasticity, $\Delta L/L_i$ refers to the rate of employment growth in industry i, and $\Delta Y/Y_i$ is the output growth rate of industry i. When the elasticity coefficient ε is positive, a higher value means that the pull effect of economic growth on employment is stronger; when the elasticity coefficient ε is negative, it means that the higher its absolute value is, the greater the "crowding out" effect on employment. Generally, as the industry matures, its employment elasticity will gradually decrease. Therefore, this paper adopts industrial employment elasticity as an index to measure the employment absorption capacity of strategic emerging industries and proposes the second hypothesis.

Hypothesis 2. *The employment elasticity of strategic emerging industries is in the positive direction of the level of employment.*

The speed of change in industrial structure is an indicator of the degree of volatility (or change) in various industries. The change value of industrial structure is an index used to measure the change amplitude of industrial structure. The faster the change in industrial structure, the greater the adjustment of industrial dynamics and the more obvious the impact on employment. The value of industrial structural change was first proposed by the American economist Simon Kuznets and was improved by Gan Chunhui and others. The formula is as follows.

$$K = \frac{\sum |X_{it} - X_{io}|}{N} \tag{2}$$

In Formula (2), K is the value of industrial structure change, which describes the speed of change of the structure of strategic emerging industries. X_{io} is the proportion of the beginning industry; X_{it} is the proportion of the end-of-term industry; N is the number of industries. The greater the K value, the greater the rate of change in the structure of strategic emerging industries. On the other hand, the smaller the K value, the smaller the rate of change in the structure of strategic emerging industries. Generally, the faster the change in industrial structure, the more violent the fluctuations in various industries and the more serious the structural unemployment. Therefore, this paper uses the value of industrial structural change as an index to measure the speed of change of strategic emerging industrial structure and proposes the third hypothesis.

Hypothesis 3. *The speed of change of the structure of strategic emerging industries is oppositely correlated with the level of employment.*

In summary, assuming that other conditions such as technological progress remain unchanged, the employment effect of the change in strategic emerging industries is the result of the combined effect of the direction of industrial structural change, the elasticity of industrial employment and the speed of change of industrial structure.

3.3. The Regression Equation

Based on the above theoretical analysis and assumptions, assuming that other conditions such as technological progress remain unchanged, this paper uses the three indicators of change direction, employment elasticity and speed of change in the strategic emerging industrial structure to construct a model and empirically study the employment effect of the change in China's strategic emerging industrial structure. This paper draws on the Cobb-Douglas production function to construct a nonlinear function relationship that includes the employment level of strategic emerging industries, the direction of industrial structure change, the elasticity of employment and the speed of change of industrial structure, that is,

$$L = f(ISCD, IEE, ISCS) = \eta (ISCD)^{\alpha}(IEE)^{\beta}(ISCS)^{\gamma} \tag{3}$$

In Formula (3), L represents the number of people employed in strategic emerging industries; $ISCD$ represents the change direction of strategic emerging industries; IEE represents the employment elasticity of strategic emerging industries; $ISCS$ represents the change speed in the structure of strategic emerging industries; α, β and γ represent the coefficients of the change direction, employment elasticity and change speed of the structure of strategic emerging industries, respectively, and η represents a constant.

By taking logarithms from both sides of Formula (3) simultaneously, Formula (4) is obtained.

$$\ln L = \ln \eta + \alpha \ln(ISCD) + \beta \ln(IEE) + \gamma \ln(ISCS) \tag{4}$$

The regression Formula (5) is established on the basis of Formula (4).

$$\ln L = \rho + a \ln(ISCD) + b \ln(IEE) + c \ln(ISCS) + \mu \tag{5}$$

In Formula (5), ρ is the constant item; a, b, and c represent the respective coefficients; and μ represents the random error.

4. Variables and Data Sources

4.1. Explained Variables: The Level of Employment in Strategic Emerging Industries

Table 1 shows that the number of people employed in China's strategic emerging industries increased from 2009 to 2020, from 47.97 million in 2009 to 97.72 million in 2020, showing different degrees of this upward trend year by year, indicating that China's strategic emerging industries are developing well.

Table 1. Employment in strategic emerging industries from 2009 to 2020 (Unit: 10,000 people).

Year	2009	2010	2011	2012	2013	2014	2015	2016	2017	2018	2019	2020
Number of people in business at the end of the year	4797	5174	5398	5767	6052	6452	6803	7172	7595	8251	8877	9772

4.2. Explaining Variables

4.2.1. The Changing Direction of Strategic Emerging Industrial Structure

In Table 2, we can see that from 2009 to 2020, the proportion of GDP contributed by China's strategic emerging industries increased, from 3.7% in 2009 to 11.7% in 2020, showing a steady upward trend year by year.

Table 2. The proportion of added value of China's strategic emerging industries in GDP in 2009–2020.

Year	2009	2010	2011	2012	2013	2014	2015	2016	2017	2018	2019	2020
Proportion of industry added value in GDP	3.7%	4.0%	4.8%	5.9%	7.3%	7.6%	8.0%	8.4%	10%	11%	11.5%	11.7%

4.2.2. The Employment Elasticity of Strategic Emerging Industries

Using Formula (1), this paper measures the employment elasticity of China's strategic emerging industries from 2009 to 2020, as shown in Table 3. Table 3 reflects the employment elasticity and total employment elasticity of the seven subsectors of China's strategic emerging industries in 2009–2020.

Table 3. Total employment elasticity and employment elasticity of China's seven subsectors of strategic emerging industries in 2009–2020.

Industry / Year	Energy-Saving and Environmental Protection Industry	A New Generation of the Information Technology Industry	Bio-industry	High-End Equipment Manufacturing	New Energy Industry	New Materials Industry	New Energy Vehicle Industry	Total Employment Elasticity
2009	0.332	1.236	0.589	0.732	0.023	0.544	0.407	0.500
2010	0.357	0.567	0.740	0.711	0.040	0.532	0.759	0.529
2011	0.135	0.411	0.460	0.942	0.340	0.458	0.367	0.445
2012	0.119	0.919	0.722	0.405	0.199	0.455	0.978	0.542
2013	0.200	0.459	0.553	0.273	0.139	0.459	0.521	0.372
2014	0.236	0.668	0.939	0.393	0.179	0.661	1.222	0.614
2015	0.134	0.924	1.164	0.324	0.435	0.255	1.252	0.641
2016	0.131	0.746	1.109	0.226	0.338	0.283	1.087	0.560
2017	0.184	0.749	0.849	1.251	0.486	0.635	0.485	0.663
2018	0.330	1.011	0.801	1.301	0.876	0.669	0.517	0.786
2019	0.303	1.033	1.375	1.425	1.285	0.454	0.114	0.855
2020	0.660	1.078	0.995	1.877	0.541	0.256	0.523	0.847
Mean value	0.260	0.817	0.858	0.792	0.407	0.472	0.686	0.613

In terms of subindustries, the employment elasticity of each subsector industry is greater than 0, which indicates that China's seven strategic emerging industries have a pull effect on employment. However, the employment elasticity of each subsector of strategic emerging industries shows an unstable trend. Regarding the average employment elasticity of each subsector, the bioindustry had the greatest pull effect on employment between 2009 and 2020, with an average employment elasticity value of 0.858. Second, the new generation of the information technology industry and high-end equipment manufacturing industry have average employment elasticities of 0.817 and 0.792, respectively. In addition, the average employment elasticities of the new materials industry and new energy industry are also relatively close, with values of 0.472 and 0.407, respectively, hovering at approximately 0.5, indicating that the pull effect on employment is not obvious. The energy-saving and environmental protection industry has the lowest average employment elasticity value, at only 0.260, indicating that the employment absorption capacity of energy-saving and environmental protection is weak. On the whole, the employment elasticity of China's strategic emerging industries fluctuates between 0.855 and 0.372 from 2009 to 2020, with an average of 0.613, and all of the values are positive, which indicates that strategic emerging industries have a pull effect on employment. The ability of strategic emerging industries to absorb employment is strong, and the role of strategic emerging industries in absorbing employment is enhancing.

4.2.3. The Change Speed in the Structure of Strategic Emerging Industries

Using Formula (2), this paper measures the change value K of the structure of strategic emerging industries for 2009 to 2020, as shown in Tables 4 and 5. In order to make the data easier to understand, we added the percentage after the data.

Table 4. The speed of change in the structure of China's seven subsectors strategic emerging industries in 2009–2020.

Year	Energy-Saving and Environmental Protection Industry	A New Generation of the Information Technology Industry	Bioindustry	High-End Equipment Manufacturing	New Energy Industry	New Materials Industry	New Energy Vehicle Industry
2009–2010	0.923%	3.046%	1.334%	0.675%	0.126%	0.276%	0.077%
2010–2011	0.325%	1.432%	0.593%	0.038%	0.093%	0.016%	0.271%
2011–2012	0.267%	0.709%	0.730%	1.202%	0.112%	0.299%	0.177%
2012–2013	0.218%	2.213%	0.467%	2.836%	0.013%	0.209%	0.367%
2013–2014	0.566%	2.472%	0.007%	3.186%	0.057%	0.213%	0.261%
2014–2015	0.573%	2.288%	0.184%	1.221%	0.743%	0.163%	0.113%
2015–2016	0.133%	2.697%	0.435%	1.560%	0.285%	1.254%	0.005%
2016–2017	0.185%	2.742%	0.029%	1.327%	0.661%	0.239%	0.445%
2017–2018	0.477%	0.450%	0.020%	0.722%	0.110%	0.470%	0.206%
2018–2019	0.967%	1.162%	0.274%	0.084%	0.084%	0.753%	0.358%
2019–2020	0.711%	1.226%	0.833%	0.654%	0.255%	0.313%	0.435%

Table 5. China's strategic emerging industry structure: changes in R value in 2009–2020.

Year	$\sum |X_{it}-X_{i0}|$	K
2009	6.457%	0.922%
2010	2.737%	0.391%
2011	3.496%	0.499%
2012	6.296%	0.899%
2013	6.763%	0.966%
2014	5.285%	0.755%
2015	6.359%	0.908%
2016	5.571%	0.796%
2017	2.455%	0.351%
2018	3.682%	0.526%
2019	4.427%	0.632%
2020	7.110%	1.016%

Table 4 reflects the speed of change in the structure of the seven subsectors of strategic emerging industries from 2009 to 2020. As shown in Table 4, the industrial structure of each subsector of emerging industry changed to varying degrees between 2009 and 2020. Taking the energy-saving and environmental protection industry as an example, the speed of industrial structure change in 2009–2010 was 0.923%, the change in industrial structure decreased to 0.218% in 2012–2013, and then in 2014–2015, there was an upward trend reaching 0.573%. Between 2015–2016, it fell to 0.133%, but then increased to 0.711% in 2019–2020. It can be seen that the change in the energy-saving and environmental protection industrial structure shows an unstable fluctuation trend. The structures of other subsectors of strategic emerging industry also show different degrees of change, which shows that there are obvious differences in the strategic emerging industrial structures of each subdivision.

Table 5 reflects the speed of change in the structure of strategic emerging industries from 2009 to 2020. From 2009 to 2020, the change in *K* value of the industrial structure of strategic emerging industries increased in volatility from 0.922% in 2009 to 0.351% in 2017 and to 1.016% in 2020, showing a declining trend and then an upward trend.

4.3. Data Sources

In 2012, China's Ministry of Industry and Information Technology announced the Strategic Emerging Industry Classification (2012). In 2018, China's National Bureau of Statistics published the Strategic Emerging Industries Classification (2018). In view of the availability and reliability of the data, this paper takes the 2012 edition of China's strategic emerging industry classification, defines seven industries, and selects 2009–2020 as the research interval. The raw data are taken from the China Statistical Yearbook, China High-tech Industry Statistical Yearbook, China Science and Technology Statistical Yearbook, China Torch Statistical Yearbook, and China Strategic Emerging Industry Development Report (2022). Some of the data come from China's National Bureau of Statistics, strategic emerging industries database and so on. Data that are difficult to obtain directly, such as the number of people employed in the new energy vehicle industry, are calculated using the statistics and data of the relevant industries.

5. Results

This paper takes strategic emerging employment as the dependent variable and the direction of industrial structure change, the elasticity of industrial employment and the speed of industrial structure change as the independent variables. Based on regression Equation (5), by using Eviews 9.0 analysis software, this study further analyzes the employment effect of changes in the strategic emerging industrial structure.

The presence of non-smooth time series in regression modeling analysis will lead to pseudo-regression analysis. To avoid this phenomenon, it is generally necessary to perform a unit root test of individual variables before the econometric modeling analysis.

According to the results of unit root test in Table 6, it can be found that the p value of the ADF of the original series of the four variables is greater than 0.05. The hypothesis of no unit root is rejected. Therefore, the four variables are not stable sequences. However, the ADF values of the difference series of the four variables are all less than the critical value of 5% and the probability p is all less than 0.05. It is considered that there is no unit root in all difference sequences. Therefore, the four variables meet the conditions for further regression analysis. Using Eviews 9.0 to test the regression estimate and test Equation (5), the following regression results are obtained:

Table 6. Test of all enterprises.

Variable	Original Sequence			First-Order Difference			Second-Order Difference		
	ADF	p Value	Conclusion	ADF	p Value	Conclusion	ADF	p Value	Conclusion
$\ln(L)$	−0.3796	0.5183	Not smooth	−3.6825	0.0761	Not Smooth	−7.4192	0.0000	Smooth
$\ln(ISCD)$	−1.7733	0.6486	Not smooth	−2.3360	0.3822	Not smooth	−3.0793	0.0065	Smooth
$\ln(IEE)$	−3.3602	0.1089	Not smooth	−2.3873	0.3598	Not Smooth	−3.2505	0.0060	Smooth
$\ln(ISCS)$	−2.3238	0.3912	Not smooth	−3.2885	0.1255	Not Smooth	−4.1978	0.0009	Smooth

According to the regression results, we can see that $R^2 = 0.9839$, the F value = 163.6847, and the p value = 0.000, indicating that regression Equation (5) is well fitted. There is a stable relationship between the number of employments in strategic emerging industries and the direction of industrial structure change, the elasticity of employment and the speed of industrial structure change.

From the regression coefficient of Table 7, the following regression model is obtained,

$$\ln L = 9.9491 + 0.4264 \ln(ISCD) + 0.1917 \ln(IEE) - 0.0133 \ln(ISCS) \tag{6}$$

Table 7. Regression coefficient table.

	Coefficient	Standard Error	Trial	Prob
C	9.9491	0.1434	69.3411	0.0000
$\ln(ISCD)$	0.4264	0.0357	11.9205	0.0000
$\ln(IEE)$	0.1917	0.0600	3.1909	0.6499
$\ln(ISCS)$	−0.0133	0.0282	−0.4715	0.0128

From regression Equation (6) and Table 7, we can see the following:

(1) The relationship among the three explanatory variables is different: the strategic emerging industrial structure change direction ($ISCD$), industrial employment elasticity (IEE) and the speed of change ($ISCS$), and the number of people employed (L) of the interpreted variables. The directions of change of the strategic emerging industrial structure ($ISCD$) and the number of people employed (L) are positively correlated, and the correlation is significant. The industrial employment elasticity (IEE) and the number of people employed (L) are positively correlated, but the correlation is not significant. The speed of change ($ISCS$) and the number of people employed (L) show an inverse relationship, and they are significantly correlated.

(2) The hypothesis proposed in this paper is verified: Firstly, it can be seen that hypothesis 1 is verified. The direction of structural change in strategic emerging industries ($ISCD$) and the change in employment (L) are changing in the same direction, indicating that the greater the proportion of the value added of strategic emerging industries in GDP, the greater the number of people employed. This finding shows that the development direction of strategic emerging industries determines the level of employment as well as the growth of employment in these industries. Therefore, expanding the scale of the development of strategic emerging industries and giving full play to their leading and expanding roles in economic and social development is an important measure to promote the high-quality development of China's economy and expand employment. Secondly, it can be seen that hypothesis 2 is verified. The employment elasticity of strategic emerging industries and the number of people employed (L) are changing in positive directions, indicating that with the development of strategic emerging industries, the employment elasticity of each industry is gradually creasing and the level of employment absorption is gradually creasing. Thirdly, it can be seen that hypothesis 3 is verified. The speed of change in the structure of strategic emerging industries ($ISCS$) and the number of people employed (L) are changing in oppositive directions, indicating that the faster the structural change in strategic emerging industries, the fewer people will be employed. Thus, we can see that the three hypotheses in this paper are supported.

(3) Three indicators, that is, the direction of change in the structure of strategic emerging industries, the elasticity of industrial employment, and the speed of change, have different effects on labor employment. First, the change in the structure of strategic emerging industry has the greatest impact on employment, and its coefficient is 0.4246, $p < 0.001$. Second, the employment elasticity of strategic emerging industries has a lesser impact on employment, and its coefficient is 0.1917, $p > 0.05$. Third, the rate of change of the structure of strategic emerging industries has the least impact on employment, and its coefficient is −0.0133, $p < 0.05$. It can be seen that the direction of structural change of strategic emerging industries and their employment elasticity has a greater positive effect on employment than the negative effect of the speed of structural change of strategic emerging industries on employment.

In total, the effect of changes in strategic emerging industrial structure on employment is positive. This indicates that the structural adjustment of strategic emerging industries has a positive effect on employment growth. That is to say, the structural change in strategic emerging industries have played a role in promoting employment. The change in China's strategic emerging industrial structure provides strong support for the expansion of China's employment. The results show that in the development of strategic emerging industries, the change in industrial structure has played a role in promoting employment. Therefore, we should rationally adjust the structure of strategic emerging industries, accelerate the development of strategic emerging industries and enhance the proportion of their industrial added value to the GDP by expanding the boundaries of these industries, accelerating the transformation and application of high-tech research results and cultivating new industries and new employment growth points, so as to improve the employment absorption capacity of strategic emerging industries.

6. Conclusions and Implications

This paper constructs a regression equation, in which the employment of strategic emerging industries is the dependent variable, while the change direction of strategic emerging industry structure, the employment elasticity of strategic emerging industries and the change speed of industrial structure are the independent variables. Using the relevant data of China's strategic emerging industries from 2009 to 2020, this paper uses the regression equation to analyze to the employment effect of structural change in strategic emerging industries. The main research conclusions are as follows:

(1) The direction of the change in the structure of China's strategic emerging industries is basically the same as its GDP and moves in a positive direction with the employment of strategic emerging industries. This shows that in the process of the development and growth of strategic emerging industries, the direction of the change in the structure of strategic emerging plays a role in promoting its employment. (2) The employment elasticity of China's strategic emerging industries shows a creasing trend year by year in its fluctuation, and changes positively with the number of its employees. This means that with the development of strategic emerging industries, its industrial employment absorption capacity is enhancing year by year and its employment is increasing. (3) The speed of change of the structure of China's strategic emerging industries is not stable at present, and this change is negatively correlated with the number of employments in strategic emerging industries. This trend shows that China's strategic emerging industrial structure is still in a state of dynamic adjustment, and further clarification of industrial boundaries is required.

The research results of this paper have certain theoretical significance and practical value. From the perspective of theoretical significance, the extension of industrial structure change to strategic emerging industry structure change has enriched the theoretical research on the employment effect of industrial structure change. From the perspective of practical value, based on the main research conclusions, this paper proposes the following countermeasures and suggestions for the formulation of strategic emerging industrial policies: (1) It is clear that changes in the structure of strategic emerging industries have a positive impact on employment. They have the potential not only to drive significant economic growth, but also increase the number of employments. The government should frequently provide assistance and incentives to these industries in order to promote their development and global competitiveness. (2) It is necessary to recognize that employment elasticity of strategic emerging industries is fluctuating. More research and analysis are required to determine the employment elasticity of strategic emerging industries. The government has further explored new channels for improve employment flexibility of strategic emerging industries to promote employment promoting by building effective links between emerging industries and traditional industries. (3) Although the structural change in strategic emerging industries has a positive effect on employment, the destructive effect of the speed of change of strategic emerging industries on employment cannot be ignored. The government should establish a sound labor market mobility mechanism, enhance the education and training of strategic emerging industries professionals, and as much as possible curb the destruction of employment caused by the rapid change of strategic emerging industry.

Although this paper has some contributions in theory and practice, it still has some shortcomings. On the one hand, this paper only constructs a regression model to analyze the employment effect of structural change in strategic emerging industries and does not aim at the specific employment effect of strategic emerging industries such as high-end equipment manufacturing and the new-generation information technology industry for further analysis. On the other hand, this paper does not consider the influence of those factors such as economic growth, technological advancements, and government policies preference on employment levels of strategic emerging industries.

Stepping into the stage of high-quality and innovation-driven development, the 14th Five-Year Plan for China places a high priority on advancing the strategic emerging industry (SEI) [38]. These problems need to be further discussed in the follow-up research. In the future, with the improvement of statistical data, more theoretical methods [39] and empirical studies are needed to enrich our understanding of the direction of change, the employment absorption capacity and the speed of change of the industrial structure in strategic emerging industries. We also can continue to study and understand the employment effect of structural changes in strategic emerging industries among different regions and provinces in China.

Author Contributions: Conceptualization, C.W. and L.L.; methodology, C.W.; software, L.L.; validation, L.L.; formal analysis, L.L.; investigation, Y.Z. and L.L.; resources, Y.Z.; data curation, Y.Z.; writing—original draft preparation, L.L.; writing—review and editing, L.L.; visualization, C.W.; supervision, C.W.; project administration, C.W.; funding acquisition, L.L. All authors have read and agreed to the published version of the manuscript.

Funding: Key program of humanities and social sciences in Anhui universities of China (Grant No. SK2021A0415 and SK2021A0387).

Data Availability Statement: The data used to support the findings of this study are available from the corresponding author upon request.

Conflicts of Interest: The authors declare that there are no conflict of interest regarding the publication of this paper.

References

1. Wang, X.; Li, B.; Yin, S. The convergence management of strategic emerging industries: Sustainable design analysis for facilitating the improvement of innovation networks. *Sustainability* **2020**, *12*, 900. [CrossRef]
2. Chen, X.T.; Li, Z.H.; Wang, J.J. Impact of subsidy policy on remanufacturing industry's donation strategy. *Processes* **2023**, *11*, 118. [CrossRef]
3. Yin, K.D.; Zhang, K.; Huang, C. Institutional supply, market cultivation and the development of marine strategic emerging industries. *Mar. Policy* **2022**, *139*, 105045. [CrossRef]
4. Zuo, W.C.; Li, Y.Q.; Wang, Y.H. Research on the optimization of new energy vehicle industry research and development subsidy about generic technology based on the three-way decisions. *J. Clean. Prod.* **2019**, *212*, 46–55. [CrossRef]
5. Wang, X.; Li, B.Z.; Yin, S.; Zeng, J.W. Formation mechanism for integrated innovation network among strategic emerging industries: Analytical and simulation approaches. *Comput. Ind. Eng.* **2021**, *162*, 107705. [CrossRef]
6. Zeng, G.; Guo, H.X.; Geng, C.X. Evaluation of financing efficiency of strategic emerging industries in the context of green development: Evidence from China. *Environ. Sci. Pollut. Res.* **2022**, *29*, 63472–63493. [CrossRef]
7. Lv, Z. Review of Tracking research on the cultivation and development of strategic emerging industries in China. *J. Manag. World* **2019**, *35*, 189.
8. Tang, X.; Gao, S.T.; Zhao, T.Q. Can digital finance help increase the value of strategic emerging enterprises? *Sustain. Cities Soc.* **2022**, *81*, 103829. [CrossRef]
9. Święcki, T. Determinants of structural change. *Rev. Econ. Dyn.* **2017**, *24*, 95–131. [CrossRef]
10. Liu, S.F.; Deng, J. On the model of industrial structure coordination degree and optimization planning of industrial structure in Jiangsu Province and China. *J. Grey Syst.* **2021**, *33*, 29–38.
11. Yang, F. Resource collection algorithm for entrepreneurship and employment education in universities based on data mining. *Mob. Inf. Syst.* **2022**, *2022*, 6038255. [CrossRef]
12. Liu, W.; Yang, J. The evolutionary game theoretic analysis for sustainable cooperation relationship of collaborative innovation network in strategic emerging industries. *Sustainability* **2018**, *10*, 4585. [CrossRef]
13. Monteforte, F. Structural change, the push-pull hypothesis and the Spanish labour market. *Econ. Model.* **2020**, *86*, 148–169. [CrossRef]
14. Kang, L.; Liu, Y. Characteristics of industrial structure evolution and isomorphism in Central Asia. *J. Geogr. Sci.* **2020**, *30*, 1781–1801. [CrossRef]
15. Gabardo, F.A.; Pereima, J.B.; Einloft, P. The incorporation of structural change into growth theory: A historical appraisal. *Economia* **2017**, *18*, 392–410. [CrossRef]

16. Hetemäki, L.; Hurmekoski, E. Forest products markets under change: Review and research implications. *Curr. For. Rep.* **2016**, *2*, 177–188. [CrossRef]
17. Duan, Q.R. Influence of industrial structure change on coastal forestry economic growth. *J. Coast. Res.* **2019**, *93*, 895–900. [CrossRef]
18. Xie, B.; Zhang, R.; Sun, S. Impacts of marine industrial structure changes on marine economic growth in China. *J. Coast. Res.* **2019**, *98*, 314–319. [CrossRef]
19. Jin, X.; Zhou, S.; Sumaila, U.R.; Yin, K.; Lv, X. Coevolution of economic and industrial linkages within the Land-Sea industrial structure of China. *Water* **2021**, *13*, 3452. [CrossRef]
20. Basile, R.; Donati, C.; Pittiglio, R. Industry structure and employment growth: Evidence from semiparametric geoadditive models. *Reg. Dev.* **2013**, *10*, 121–160.
21. Chivu, L.; Ciutacu, C. About industrial structures decomposition and recomposition. *Procedia Econ. Financ.* **2014**, *8*, 157–166. [CrossRef]
22. Zhang, S.; Chen, Y.; Liu, X.; Yang, M.; Xu, L. Employment effects of solar PV industry in China: A spreadsheet-based analytical model. *Energy Policy* **2017**, *109*, 59–65. [CrossRef]
23. Hong, J.; Byun, J.; Kim, P. Structural changes and growth factors of the ICT industry in Korea: 1995–2009. *Telecommun. Policy* **2016**, *40*, 502–513. [CrossRef]
24. Ding, Y.Y.; Li, Z.; Ge, X.; Hu, Y. Empirical analysis of the synergy of the three sectors' development and labor employment. *Technol. Forecast. Soc. Chang.* **2020**, *160*, 120–130. [CrossRef]
25. Zeng, G.; Geng, C.X.; Guo, H.X. Spatial spillover effect of strategic emerging industry agglomeration and green economic efficiency in China. *Pol. J. Environ. Stud.* **2020**, *29*, 3901–3914. [CrossRef]
26. Samaniego, R.M.; Sun, J.Y. Productivity growth and structural transformation. *Rev. Econ. Dyn.* **2016**, *21*, 266–285. [CrossRef]
27. Peneder, M. Industrial structure and aggregate growth. *Struct. Chang. Econ. Dyn.* **2003**, *14*, 427–448. [CrossRef]
28. Hartwig, J. Testing the growth effects of structural change. *Struct. Chang. Econ. Dyn.* **2012**, *23*, 11–24. [CrossRef]
29. Audretsch, D.B.; Lehmann, E.E.; Menter, M.; Wirsching, K.; Linton, J. Intrapreneurship and absorptive capacities: The dynamic effect of labor mobility. *Technovation* **2021**, *99*, 102129. [CrossRef]
30. Li, S.A.; Gong, L.; Pan, S.; Luo, F. Wage and price differences, technology gap and labor flow dynamics. *Econ. Model.* **2020**, *88*, 211–222. [CrossRef]
31. Su, J. Economic efficiency measurement algorithm of strategic emerging industries based on multifeature Fusion. *Mob. Inf. Syst.* **2022**, *2022*, 1118941.
32. Novakova, L. The impact of technology development on the future of the labour market in the Slovak Republic. *Technol. Soc.* **2020**, *62*, 101256. [CrossRef]
33. Vivarelli, M. Innovation, employment and skills in advanced and developing countries: A survey of economic literature. *J. Econ. Issues* **2014**, *14*, 449–474. [CrossRef]
34. Hou, J.; Teo, T.S.H.; Zhou, F.; Lim, M.K.; Chen, H. Does industrial green transformation successfully facilitate a decrease in carbon intensity in China? An environmental regulation perspective. *J. Clean. Prod.* **2018**, *184*, 1060–1071. [CrossRef]
35. Balsmeier, B.; Woerter, M. Is this time different? How digitalization influences job creation and destruction. *Res. Policy* **2019**, *48*, 103765. [CrossRef]
36. Vu, K.M. Structural change and economic growth: Empirical evidence and policy insights from Asian economies. *Struct. Chang. Econ. Dyn.* **2017**, *41*, 64–77. [CrossRef]
37. Xue, X.X.; Wang, X.H.; Li, L.W. Employment absorption capacity of e-commerce service industry. *J. Coast. Res.* **2019**, *93*, 879–882. [CrossRef]
38. Ding, D.; Li, R.J.; Guo, J.H. An entropy-based TOPSIS and optimized grey prediction model for spatiotemporal analysis in strategic emerging industry. *Expert Syst. Appl.* **2023**, *213*, 119169. [CrossRef]
39. Guo, Y.X.; Ding, H.P. Coupled and coordinated development of the data-driven logistics industry and digital economy: A case study of Anhui province. *Processes* **2022**, *10*, 2036. [CrossRef]

Article

Dynamic Optimal Decision Making of Innovative Products' Remanufacturing Supply Chain

Lang Liu [1,2], Zhenwei Liu [2,*], Yutao Pu [2] and Nan Wang [1]

[1]   School of Business Administration, Guizhou University of Finance and Economics, Guiyang 550025, China
[2]   School of Economic and Management, East China Jiaotong University, Nanchang 330013, China
*   Correspondence: 2020048025200008@ecjtu.edu.cn; Tel.: +86-189-4235-0433

Abstract: In order to realize the recyclability of innovative product resources, we explored the optimal dynamic path of each decision variable in the remanufacturing supply chain and analyzed the impact of each decision variable on supply chain performance. Based on the Bass innovation diffusion model, we established a remanufacturing supply chain model in which a single manufacturer leads and a single retailer follows, and the retailer is responsible for recycling. The optimal wholesale price, retail price, and recovery effort path were obtained through optimal control theory. We also discussed the influence of different innovation coefficients and imitation coefficients on the overall long-term profit of each member in the supply chain, and at the same time, found the optimal market share of the product. The research results show that the larger the market innovation coefficient and the imitation coefficient are, the larger the overall long-term profit of the manufacturer and the greater the market share of the product, while the overall long-term profit of the retailer and the entire supply chain will increase first and then decrease; when the innovation coefficient and imitation coefficient are above a certain level, retailers will not enter the market. In a market with a small innovation coefficient and a large imitation coefficient, the overall long-term profits of retailers and supply chains will be higher. This study provides a theoretical basis for the decision making of the remanufacturing supply chain of innovative products in a dynamic environment, and also provides guidance for the practice of nodal enterprises in the supply chain.

Keywords: innovative products; remanufacturing; supply chain management; differential game; bass model

Citation: Liu, L.; Liu, Z.; Pu, Y.; Wang, N. Dynamic Optimal Decision Making of Innovative Products' Remanufacturing Supply Chain. *Processes* **2023**, *11*, 295. https://doi.org/10.3390/pr11010295

Academic Editors: Conghu Liu, Xiaoqian Song, Zhi Liu and Fangfang Wei

Received: 16 December 2022
Revised: 2 January 2023
Accepted: 4 January 2023
Published: 16 January 2023

1. Introduction

In recent years, with the rapid development of the economy, the speed of innovation product update iteration is accelerating. While consuming a lot of resources, it also produces a large number of waste products and causes serious environmental pollution. If these waste products are not recycled, it will not only cause environmental pollution, but also leads to a great waste of resources. Therefore, environmental issues have received more and more attention [1], and awareness of environmental issues and social responsibility is increasing [2]. The White Paper 2018 on intelligent remanufacturing products edited by Tsinghua University pointed out that the implementation of waste product recycling and remanufacturing can save 70% of materials, reduce 60% of energy consumption, and reduce 80% of pollutant emissions [3]. At present, governments and enterprises are vigorously promoting waste product recycling and remanufacturing industry to achieve sustainable economic development. European countries have the world's most advanced waste information and communication equipment recycling management systems [4]. The Chinese government has also continuously strengthened the conservation and utilization of resources, strengthened ecological construction and governance, and resolutely taken a new path of green, low-carbon, and environmental protection. Therefore, the remanufacturing supply chain has become an important way to achieve the goals of 'carbon peaking and carbon neutrality'. Recycling and remanufacturing end-of-life products is a better way for the closed-loop supply chain, which can reduce waste discharge and resource consumption and achieve the purpose of sustainable development [5]. Some large international companies, such as Xerox, Kodak, Hewlett-Packard, Volvo, and BMW, have begun to adopt remanufacturing strategies [6,7]. Many Chinese companies are also following up, such as the recycling and remanufacturing of old mobile phones by Chinese domestic mobile phone manufacturers OPPO and VIVO. Companies adopting remanufacturing strategies are widely used in industries such as computers, printers, ink transportation, furniture, auto parts, and medical equipment [8]. SpaceX's recovery of the Falcon rocket is the pinnacle of the current remanufacturing supply chain. With the vigorous development of the global economy and the widespread use of advanced technology, many diversified products are flooding consumers' lives like an unstoppable wave. The life cycle of products has become shorter and shorter, and the speed of updating products has become faster and faster, making the total amount and growth rate of waste continue to rise. The remanufacturing supply chain is considered one of the most effective ways to deal with waste economically and environmentally, and it has become the consensus of industry and academia on the recycling and reuse of end-of-life products [9,10]. Therefore, it is especially important to analyze the game behavior among the members of the remanufacturing supply chain, coordinate the relationship between them, and maximize the profit of each member.

The remanufacturing supply chain is the supply chain that includes the remanufacturing industry. Remanufacturing refers to the process of product disassembly and recovery in the reverse or closed-loop supply chain; this disassembly and recovery process entails the repair or replacement of outdated product components and/or of expired products [11]. Through the implementation of repair and transformation of these products, their quality and performance are brought back to the standards and requirements of new products. Scholars around the world have conducted a lot of research on the remanufacturing supply chain. Masoudipour et al. [10] first established a reverse-channel design framework from the supply chain perspective and proposed three recycling methods for manufacturers. Jena et al. [12] discussed the impact of whether or not to share advertising costs on second-hand product recycling activities. Tsao et al. [13] proposed a remanufacturing supply chain network adopted by RFID. Bhattacharya et al. [14] explored the problem of second-hand product price optimization in a remanufacturing supply chain system. Ma et al. [15] studied a three-level supply chain consisting of a manufacturer, a retailer, and two recyclers, and considered various cooperation models. Xiong et al. [16] discussed and compared the remanufacturing supply chain in two different situations: manufacturer remanufacturing and retailer remanufacturing. Xie et al. [17] developed a remanufacturing supply chain with online and offline dual-channel sales. They obtained manufacturer–retailer advertising cooperation strategies in two different situations: centralized decision making and decentralized decision making. Zhang et al. [9] analyzed and compared three remanufacturing supply chain models of manufacturer recycling, retailer recycling, and third-party platform recycling from the perspectives of the environment, the economy, and social welfare. Zhang et al. [18] explored the pricing and collection strategies of different closed-loop supply chain models in regard to a quality and marketing effort-dependent demand in a fuzzy environment. He et al. [19] proposed a closed-loop supply chain model consisting of one manufacturer and one third-party collector to investigate competitive collection and channel convenience.

The above studies have greatly enriched the theoretical results of the remanufacturing supply chain, but they are all static models. They do not take into account the dynamic characteristics of the recycling process. With the advent of the information age and the popularity of the Internet, information transmission is becoming more and more rapid, the response of the supply chain to market changes is becoming more and more rapid, and the decision of each member of the supply chain will be able to change in real time; in this circumstance, the production, sales, recycling, and remanufacturing between manufacturers and retailers is a long-term differential game process, so the static decision-making assumption will no longer apply. Lee [20] first proposed and studied the remanufacturing supply chain model under decentralized decision making with dynamic features. In the dynamic remanufacturing supply chain model, De Giovanni [21] considered a remanufacturing supply chain composed of a manufacturer and a retailer investing in green advertising based on the goodwill model. Xiang and Xu [22] researched a remanufacturing supply chain consisting of a manufacturer, a retailer, and an Internet service platform that invests in research and development advertising and extensive data marketing. He et al. [23] studied a low-carbon service supply chain composed of a service provider and a service integrator and found that a two-way contract can benefit the entire service supply chain and its members. Song et al. [24] constructed a remanufacturing supply chain model of shared bicycles that considers the impact of recycling efforts on the riding experience. Yang and Xu [25] developed a remanufacturing supply chain model with multiple manufacturing and remanufacturing plants and multiple recycling and supply centers in a low-carbon context and derived the optimal decisions of supply chain participants.

The above studies have discussed dynamic remanufacturing supply chains in various situations. Still, all assume that the product market is relatively mature, and use classic demand models or goodwill models to describe changes in demand. However, there is a gradual promotion process for a new product entering the market. Few studies describe this process in remanufacturing supply chain coordination. According to the China Household Electrical Appliance Research Institute 2022 report data, 2021 China's computer population comprised 320 million units, 1.5 billion mobile phone units, 2421.8 million units of theoretical scrap of computers, and more than 400 million units of theoretical scrap of mobile phones [26]. The remanufacturing of these used innovative products can effectively protect the environment and save a lot of resources, so it is important for us to study the remanufacturing of innovative products. In the face of the growing market for innovative products, we can not only use the goodwill model and the classic demand model; the diffusion of innovations model to study the remanufacturing supply chain is also particularly important. Therefore, this paper introduces the diffusion of innovations model to characterize the market's changes in demand for new products. The diffusion of innovations model (Bass model) was proposed by Frank Bass [27] in 1969 to describe how durable new products expand the market among consumers. Subsequently, Bass et al. [28] proposed a diffusion model of the duopoly market under competitive conditions and studied the advertising strategy of oligopoly. Since the application of this model was quite successful, it has been widely used in various fields. Quan et al. [29] introduced the Bass model into the open-loop supply chain to characterize the immediate market demand. They studied the master–slave dynamic game process led by a supplier and followed by a retailer.

Based on the above literature, we find the following deficiencies in the existing remanufacturing supply chain studies.

1. In the growing market of innovative products such as mobile phones, computers, automobiles, and other smart devices, scholars only consider the goodwill model and the classical demand model. Few scholars have studied the supply chain of innovative products, and the research on the supply chain of innovative products needs to be deepened.
2. The existing research on the remanufacturing supply chain mainly focuses on the static supply chain, without considering the game between manufacturers and retailers in continuous time. The static model cannot adapt to the continuous and real-time modeling requirements, so it is urgent to increase research on the dynamic supply chain.

Therefore, this paper introduces the Bass model to establish a remanufacturing supply chain model based on the above research. Retailers are responsible for product sales and recycling, and manufacturers are responsible for new product manufacturing and recycling product remanufacturing, to find the best decision path for the best retail price, recycling effort, and wholesale price. The main contributions of this thesis are (1) the introduction of the Bass model into the remanufacturing supply chain; (2) exploring the optimal dynamic decision path of manufacturers and retailers in the process of new product diffusion; (3) analyzing the overall long-term profits of manufacturers, retailers, and the entire supply chain under different market conditions.

The structure of this paper is as follows: Section 2 is Model Background Description, which introduces the Bass model, model-related parameters, and the remanufacturing supply chain model, respectively. In Section 3, Model Building and Solving, we solve the differential game model of remanufacturing supply chain and obtain the optimal pricing strategy for each member of the supply chain. Section 4 is optimal decision analysis, and Section 5 is the numerical example, where we analyze the impact of different innovation coefficients and imitation coefficients on the supply chain. Finally, Section 6 is conclusions.

2. Model Background Description

2.1. Introduction to the Bass Model

The core idea of the Bass model is that the group initially purchasing a new durable product is an innovative group, and their purchasing decisions are independent of other members of the social system. The group that buys the durable product later is used as an imitation group. The time for this group to buy a new product is affected by the innovation group, and this influence increases with the increase in the number of purchasers. In the context of achieving the "carbon peaking and carbon neutrality goals", customers who are the first to use remanufactured products can be regarded as innovative groups, and the other group can be regarded as imitating groups. The mathematical source of this theory is the reliability function model, which is widely used in reliability theory. The total market demand is equal to the "total number of experiments" in the reliability function. The number of products sold at time t is equivalent to the "accumulated number of failed products" in the reliability function, and the number of products that have not been sold at time t. is equivalent to the "number of products still available" in the reliability function. The new product market share $x(t)$ at time t is equivalent to the probability distribution function $F(\cdot)$ in the reliability function. Therefore, the Bass model can well describe the dynamic game evolution process of the remanufacturing supply chain of new durable products.

When a manufacturer develops a new durable product and puts it on the market, the retailer is responsible for selling the latest product and recycling it. The manufacturer is responsible for producing new products and remanufacturing waste products. The product will be replaced by next-generation products after being sold and will no longer be sold after exiting the market. As the leader of the remanufacturing supply chain, the manufacturer first determines the product's wholesale price. Then, the retailer determines its retail price and recycling strategy based on the wholesale price selected by the manufacturer.

2.2. Symbol Description

(1)　Main parameters and their meanings

c_m: the marginal production cost of new products; c_r: marginal production cost of remanufactured products; Δ: the difference between the marginal production cost of new products and the marginal production cost of reproducts, namely $c_m - c_r$; N: the largest demand in the market; a: the influence coefficient of price on the conversion of innovators (first-time consumers) to imitators (potential consumers); j: innovation coefficient, which indicates the proportion of new products that can be sold without any sales effort after being launched; k: imitation coefficient, which indicates the influence coefficient of innovators affected by sales efforts on imitators; γ: the manufacturer's incentive coefficient for the retailer's recycling; k_r: the retailer's product recycling activity cost coefficient; α: the recycling effort's influence coefficient on the recycling rate; β: the natural recession coefficient of the recovery rate; T: the total time the new product is put on the market; $x(t)$: the market share of the new product at time t; $D(t)$: the demand for new products at time t; $\tau(t)$: retailer's product recovery rate at time t.

(2) Variables (including objective function) and their meanings

$w(t)$: the wholesale price of the new product at time t, which represents the manufacturer's decision variable; $p(t)$: the retail price of the new product at time t, which represents the retailer's decision variable; $R(t)$: the retailer's recycling effort at time t, which represents the retailer's decision variable; π_M: the manufacturer's overall long-term profit; π_R: the overall long-term profit of the retailer; π_h: the overall long-term profit of the supply chain, namely $\pi_M + \pi_R$.

2.3. Optimal Control Theory

Optimal control theory is a branch of mathematical optimization that aims to find an optimal control of a dynamical system over a specific period of time, which can lead to a maximum (or minimum) value of a specific system performance index. The basic optimal control problem is described as follows:

$$
\begin{aligned}
&\max V = \int_0^T F(t, y, u)\,dt \\
&s.t. \\
&\dot{y}(t) = f(t, y, u) \\
&y(0) = A \,,\; y(T) \text{ free } (A, T \text{ given}) \\
&u(T) \in U \text{ for all } t \in [0, T]
\end{aligned}
\tag{1}
$$

where V is the performance metric to be maximized, F is the immediate performance metric, t is the time variable, y is the state variable, and u is the control variable to find an optimal control path u^* that maximizes the performance metric in time $[0, T]$. Another class of variables, costate variables $\lambda(t)$, is introduced in the solution process. Costate variables enter the optimal control problem through Hamilton functions. Hamilton functions are defined as follows:

$$
H(t, y, u, \lambda) \equiv F(t, y, u) + \lambda(t) f(t, y, u)
\tag{2}
$$

The most important problem for solving optimal control is the first-order condition, also known as the maximum principle. After defining the Hamilton function, the maximum principle conditions are as follows:

$$
\begin{aligned}
&\underset{u}{Max}\, H(t, y, u, \lambda) \text{ for all } t \in [0, T] \\
&\dot{y}(t) = \frac{\partial H}{\partial \lambda} \quad \text{(equation of motion for } y\text{)} \\
&\dot{\lambda}(T) = -\frac{\partial H}{\partial y} \quad \text{(equation of motion for } \lambda\text{)} \\
&\lambda(T) = 0 \quad \text{(transversality condition)}
\end{aligned}
\tag{3}
$$

Symbol $\underset{u}{\max} H$ indicates that the Hamilton function is maximized, and this equivalent condition is expressed as

$$
H(t, y, u^*, \lambda) \geq H(t, y, u, \lambda) \text{ for all } t \in [0, T]
\tag{4}
$$

where u^* is the optimal control variable. If $\underset{u}{Max}\, H(t, y, u, \lambda)$ is differentiable with respect to u, it can be represented by the first-order condition $\frac{\partial H}{\partial u} = 0$. The optimal control variables, state variables, and costate variables of the optimal control problem can be solved by the maximum principle.

2.4. Model Function

(1) Consumer market

An innovative product is a product with product innovation. Product innovation is defined as the development of new products, changes in design of established products, or use of new materials or components in the manufacture of established products [30], e.g., smart phones (iPhones, Samsung smart phones, etc.), smartwatches (apple watch, etc.), computers (Dell, HP, Lenovo, etc.), graphics cards (Geforce RTX graphics cards, Radeon

RX graphics cards, etc.), etc. Once sold, the innovative product will spread in the consumer market, gradually accumulating sales through publicity, promotion, and verbal communication among consumers. Assuming that the target market demand for the innovative product is N, according to the assumptions of the Bass model, we divide consumers into two categories. One category is consumers who have accepted and purchased new products, also known as innovators (first-time consumers). Its market share at time t is $x(t)$, then the purchase quantity of the innovator at time t is $Nx(t)$; The other category is consumers who have not yet accepted—but may be influenced by—innovators in the future and will accept new products in the future, also known as imitators (potential consumers). Its proportion at time t is $1 - x(t)$, and the potential consumption amounts of potential consumers at time t is $N(1 - x(t))$. According to Reference [29], we also assume that the change in market share is

$$\dot{x}(t) = e^{-ap(t)}(j + kx(t))(1 - x(t)) \tag{5}$$

where $e^{-ap(t)}$ represents the influence coefficient of price on the conversion of imitators to innovators. The higher the price, the slower the conversion. $kx(t)$ represents the degree of influence that innovators affected by sales efforts have on potential consumers. j represents the degree of influence that innovators who are not affected by sales efforts have on potential consumers. Suppose $x(0) = 0$ indicates that the new product has just entered the market and has not yet spread among consumers. Assuming that the innovator only buys one copy of the product and does not repeat the purchase, the demand at time t is equal to the amount of change in existing consumers at time t, namely:

$$D(t) = N\dot{x} = Ne^{-ap(t)}(j + kx(t))(1 - x(t)) \tag{6}$$

(2) Manufacturer

As the remanufacturing supply chain leader, the manufacturer is responsible for the production and remanufacturing of new products and rewards retailers for product recycling. Assuming that there is no essential difference between a newly produced product and a remanufactured product, because through the remanufacturing of these used products, their quality and performance are brought back to the standards and requirements of new products. The marginal production cost of a manufacturer's new product is c_m, and the marginal production cost of a remanufactured product is c_r. Let $\Delta = c_m - c_r$. The average cost of producing a marginal new product is $c_m - \Delta \tau(t)$. The manufacturer sells the new product to the retailer at the wholesale price of $w(t)$, and provides the retailer with a reward of $\gamma \tau(t)$ according to the recycling ratio, where γ is the reward coefficient and $\tau(t)$ is the recycling rate at t time. Electronic durable goods such as mobile phones, TVs, and computers are updated very quickly. Manufacturers often let these products withdraw from the market after a period of time and introduce new products at the same time. Mark the total time of the new product on the market as T. Manufacturers pursue the maximization of their own interests by choosing the optimal path of the optimal wholesale price $w(t)$.

(3) Retailer

As a follower of a remanufacturing supply chain, retailers are responsible for the sales of new products and recycling the products. The retailer sells new products to consumers at time t at price $p(t)$. The recycling effort of the retailer at time t is $R(t)$, the recycling rate of waste products at time t is $\tau(t)$, and $\tau(0) = 0$ when the new product just enters the market. Assuming that the recovery effort has a positive effect on the change in the recovery rate, and considering the natural decline of the recovery rate, the change in the recovery rate is:

$$\dot{\tau}(t) = \alpha R(t) - \beta \tau(t) \tag{7}$$

where $\alpha > 0$ is the influence coefficient of the recovery effort on the change of the recovery rate, and β is the recovery rate decline coefficient. Assuming that the marginal return of recovery costs decreases with the increase in input, the recovery effort is assumed to be a quadratic function $\frac{1}{2}k_r R^2(t)$, $k_r > 0$. After the manufacturer chooses the optimal wholesale price $w(t)$, the retailer pursues the maximization of its own interests by choosing the corresponding optimal retail price $p(t)$ and the optimal recycling effort $R(t)$.

3. Model Building and Solving

In this remanufacturing supply chain, both the manufacturer and the retailer constitute a Stackelberg game in which the manufacturer leads the retailer to follow. The specific design is shown in Figure 1 below. The manufacturer first decides its wholesale price $w(t)$, and then the retailer decides the retail price $p(t)$ and the recycling effort $R(t)$. The objective functions of the manufacturer and retailer can be obtained as:

$$\pi_M = \int_0^T \{[w(t) - c_m + \Delta\tau(t)]N\dot{x}(t) - \gamma\tau(t)\}dt \tag{8}$$

$$\pi_R = \int_0^T \{[p(t) - w(t)]N\dot{x}(t) - \frac{1}{2}k_r R^2(t) + \gamma\tau(t)\}dt \tag{9}$$

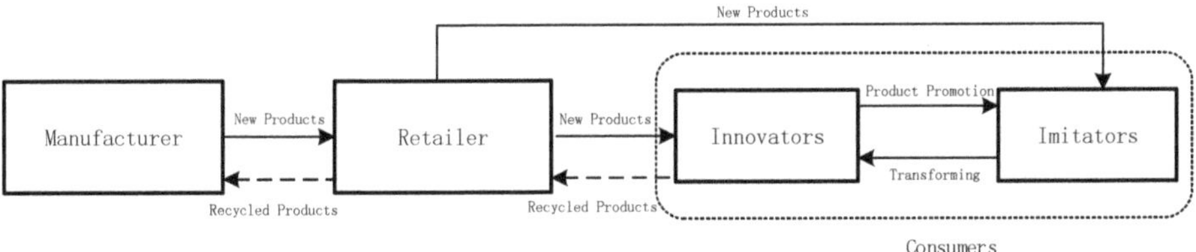

Figure 1. Supply chain design diagram.

When the manufacturer determines the optimal wholesale price $w(t)$, the retailer's optimal retail price $p(t)$ and the optimal recovery effort $R(t)$ are sought to maximize its profit π_R. That is, to solve the following optimal control problem:

$$max\ \pi_R = \max_{p,R} \int_0^T \{[p(t) - w(t)]N\dot{x}(t) - \frac{1}{2}k_r R^2(t) + \gamma\tau(t)\}dt \tag{10}$$

$$s.t. \begin{cases} \dot{x}(t) = e^{-ap(t)}(j + kx(t))(1 - x(t)) \\ \dot{\tau}(t) = \alpha R(t) - \beta\tau(t) \\ x(0) = 0 \\ \tau(0) = 0 \end{cases} \tag{11}$$

Proposition 1. *When the manufacturer's optimal wholesale price is* $w^*(t)$ *, the retailer's optimal retail price* $p^*(t)$ *is:*

$$p^*(t) = \frac{1}{a} + w(t) - \frac{1}{a}\ln\frac{(j + kx(T))(1 - x(T))}{(j + kx(t))(1 - x(t))} \tag{12}$$

The retailer's optimal recycling effort $R^*(t)$ is:

$$R^*(t) = \frac{\alpha\gamma}{\beta k_r(1 - e^{\beta(T-t)})} \tag{13}$$

Proof. According to [31]. The Hamilton function of the system is:

$$H_R = [Np(t) - Nw(t) + \lambda_{R1}(t)]e^{-ap(t)}[j + kx(t)](1 - x(t)) \\ - \tfrac{1}{2}k_r R^2(t) + \gamma\tau(t) + \lambda_{R2}(t)[\alpha R(t) - \beta\tau(t)] \tag{14}$$

Necessary conditions for the retailer's optimal control problem are:

$$\begin{cases} \frac{\partial H_R}{\partial p} = 0 \\ \frac{\partial H_R}{\partial R} = 0 \\ \dot{\lambda}_{R1}(t) = -\frac{\partial H_R}{\partial x} \\ \dot{\lambda}_{R2}(t) = -\frac{\partial H_R}{\partial \tau} \\ \dot{x}(t) = e^{-ap(t)}(j + kx(t))(1 - x(t)) \\ \dot{\tau}(t) = \alpha R(t) - \beta\tau(t) \end{cases} \tag{15}$$

Transverse conditions are: $\lambda_{R1}(T) = 0$, $\lambda_{R2}(T) = 0$, $x(0) = 0$, $\tau(0) = 0$.

From $\frac{\partial H_R}{\partial p} = (N - aNp(t) + aNw(t) - a\lambda_{R1}(t))e^{-ap(t)}(j + kx(t))(1 - x(t)) = 0$, we can obtain

$$p(t) = \frac{1}{a} + w(t) - \frac{\lambda_{R1}}{N} \tag{16}$$

From

$$\begin{aligned} \dot{\lambda}_{R1}(t) &= -\frac{\partial H_R}{\partial x} = -(Np(t) - Nw(t) + \lambda_{R1}(t))e^{-ap(t)}(k - j - 2kx(t)) \\ &= -(Np(t) - Nw(t) + \lambda_{R1}(t))\dot{x}(t)\frac{(k-j-2kx(t))}{(j+kx(t))(1-x(t))} \\ &= -\dot{x}(t)\frac{N(k-j-2kx(t))}{a(j+kx(t))(1-x(t))} \end{aligned} \tag{17}$$

then

$$\begin{aligned} \int_t^T \dot{\lambda}_{R1}(t)dt &= -\int_t^T \dot{x}(t)\frac{N(k - j - 2kx(t))}{a(j + kx(t))(1 - x(t))}dt \\ \lambda_{R1}(T) - \lambda_{R1}(t) &= -\int_{x(t)}^{x(T)} \frac{N(k - j - 2kx(t))}{a(j + kx(t))(1 - x(t))}dx \\ \lambda_{R1}(t) &= \frac{N}{a}ln\frac{(j + kx(T))(1 - x(T))}{(j + kx(t))(1 - x(t))} \end{aligned} \tag{18}$$

Bring it into Equation (12) to obtain

$$p(t) = \frac{1}{a} + w(t) - \frac{1}{a}ln\frac{(j + kx(T))(1 - x(T))}{(j + kx(t))(1 - x(t))} \tag{19}$$

From $\frac{\partial H_R}{\partial R} = -k_r R(t) + \alpha\lambda_{R2}(t) = 0$, $\dot{\lambda}_{R2}(t) = -\frac{\partial H_R}{\partial \tau} = \beta\lambda_{R2}(t) - \gamma$, we can obtain

$$R(t) = \frac{\alpha\lambda_{R2}(t)}{k_r} \tag{20}$$

$$\lambda_{R2}(t) = C_1 e^{\beta t} + \frac{\gamma}{\beta} \tag{21}$$

From $\lambda_{R2}(T) = 0$, we obtain $C_1 = -\frac{\gamma}{\beta}e^{-\beta T}$. Putting it into Equation (21), we obtain $\lambda_{R2}(t) = \frac{\gamma}{\beta}(1 - e^{\beta(t-T)})$, and then

$$R^*(t) = \frac{\alpha\gamma}{\beta k_r}(1 - e^{\beta(t-T)}) \tag{22}$$

Substituting Equation (22) into $\dot{\tau}(t) = \alpha R(t) - \beta\tau(t)$, and $\tau(0) = 0$, we obtain

$$\tau^*(t) = \frac{\alpha^2\gamma}{\beta^2 k_r}(1 - e^{-\beta t} + \frac{1}{2}e^{-\beta T}(e^{-\beta t} - e^{\beta t})) \tag{23}$$

The proof is complete. □

Proposition 2. *The manufacturer's optimal retail price $w^*(t)$ is*

$$w^*(t) = \frac{1}{a} + c_m - \Delta\tau(t) = \frac{1}{a} + c_m - \Delta(\frac{\alpha^2\gamma}{\beta^2 k_r}(1 - e^{-\beta t} + \frac{1}{2}e^{-\beta T}(e^{-\beta t} - e^{\beta t}))) \tag{24}$$

Proof. The Hamilton function of the system is

$$H_M = (Nw(t) - Nc_m + N\Delta\tau(t) + \lambda_M(t))e^{-(1+aw(t))}(j + kx(T))(1 - x(T)) - \gamma\tau(t) \tag{25}$$

The necessary conditions for the manufacturer's optimal control problem are

$$\begin{cases} \frac{\partial H_M}{\partial w} = 0 \\ \dot{\lambda}_M(t) = -\frac{\partial H_M}{\partial x} \\ \dot{x}(t) = e^{-(1+aw(t))}(j + kx(T))(1 - x(T)) \end{cases} \tag{26}$$

The transverse conditions are $\lambda_M(T) = 0, x(0) = 0$.
From

$$\frac{\partial H_M}{\partial w} = (N - aNw(t) + aNc_m - aN\Delta\tau(t) - a\lambda_M(t))e^{-(1+aw(t))}(j + kx(T))(1 - x(T)) = 0 \tag{27}$$

we can obtain

$$w(t) = \frac{1}{a} + c_m - \Delta\tau(t) - \frac{\lambda_M(t)}{N} \tag{28}$$

and $\dot{\lambda}_M(t) = -\frac{\partial H_M}{\partial x} = 0, \lambda_M(T) = 0$, so $\lambda_M(t) \equiv 0$, then

$$w^*(t) = \frac{1}{a} + c_m - \Delta\tau(t) = \frac{1}{a} + c_m - \Delta\frac{\alpha^2\gamma}{\beta^2 k_r}(1 - e^{-\beta t} + \frac{1}{2}e^{-\beta T}(e^{-\beta t} - e^{\beta t})) \tag{29}$$

The proof is complete. □

Proposition 3. *For the above system, the optimal paths for manufacturers and retailers in a remanufacturing supply chain dominated by manufacturers are:*

(1) *The manufacturer's optimal pricing path strategy is*

$$w^*(t) = \frac{1}{a} + c_m - \Delta(\frac{\alpha^2\gamma}{\beta^2 k_r}(1 - e^{-\beta t} + \frac{1}{2}e^{-\beta T}(e^{-\beta t} - e^{\beta t}))) \tag{30}$$

(2) *The retailer's optimal pricing path strategy and recycling effort decision are*

$$p^*(t) = \frac{1}{a} + w(t) - \frac{1}{a}ln\frac{(j + kx(T))(1 - x(T))}{(j + kx(t))(1 - x(t))} \tag{31}$$

$$R^*(t) = \frac{\alpha\gamma}{\beta k_r}(1 - e^{\beta(t-T)}) \tag{32}$$

See Propositions 1 and 2 for specific proofs.

Proposition 4. *The path of new product market share is*

$$x(t) = B(t)(j + kx(T))(1 - x(T)) \tag{33}$$

where

$$B(t) = \int_0^t e^{-(2+ac_m-a\Delta(\frac{\alpha^2\gamma}{\beta^2 k_r}(1-e^{-\beta t}+\frac{1}{2}e^{-\beta T}(e^{-\beta t}-e^{\beta t}))))}\, dt \tag{34}$$

and

$$x(T) = \frac{-1-B(T)j+B(T)k+\sqrt{4B^2(T)jk+(-1-B(T)j+B(T)k)^2}}{2B(T)k} \tag{35}$$

Proof. From Proposition 2:

$$\begin{aligned}
x(t) &= (j+kx(T))(1-x(T))\int_0^t e^{-(1+aw(t))}\,dt \\
&= (j+kx(T))(1-x(T))\int_0^t e^{-(2+ac_m-a\Delta(\frac{\alpha^2\gamma}{\beta^2 k_r}(1-e^{-\beta t}+\frac{1}{2}e^{-\beta T}(e^{-\beta t}-e^{\beta t}))))}\,dt
\end{aligned} \tag{36}$$

Let $B(t) = \int_0^t e^{-(2+ac_m-a\Delta(\frac{\alpha^2\gamma}{\beta^2 k_r}(1-e^{-\beta t}+\frac{1}{2}e^{-\beta T}(e^{-\beta t}-e^{\beta t}))))}\,dt$, then

$$x(T) = \frac{-1-B(T)j+B(T)k+\sqrt{4B^2(T)jk+(-1-B(T)j+B(T)k)^2}}{2B(T)k}$$

When $t = T$, $x(T) = B(t)(j+kx(T))(1-x(T))$, then

$$x(T) = \frac{-1-B(T)j+B(T)k+\sqrt{4B^2(T)jk+(-1-B(T)j+B(T)k)^2}}{2B(T)k},$$

or $x(T) = \frac{-1-B(T)j+B(T)k-\sqrt{4B^2(T)jk+(-1-B(T)j+B(T)k)^2}}{2B(T)k} < 0$ (drop out).

The proof is complete. $\square$

Proposition 5. *The long-term profit of the manufacturer is*

$$\pi_M = \frac{N}{a}x(T) - \frac{\alpha^2\gamma^2}{2\beta^3 k_r}(2\beta T - 3 + 4e^{-\beta T} - e^{-2\beta T}) \tag{37}$$

The long-term benefits of the retailer are

$$\pi_R = \frac{N}{a}\left(x(T) + \ln\frac{(j+kx(T))(1-x(T))}{j}\right) + \frac{\alpha^2\gamma^2}{4\beta^3 k_r}(2\beta T - 3 + 4e^{-\beta T} - e^{-2\beta T}) \tag{38}$$

Proof. The long-term benefits of the manufacturer are

$$\begin{aligned}
\pi_M &= \int_0^T (w(t) - c_m + \Delta\tau(t))N\dot{x}(t) - \gamma\tau(t)\,dt \\
&= \int_0^T \frac{N}{a}\dot{x}(t)\,dt - \int_0^T \gamma\tau(t)\,dt \\
&= \frac{N}{a}x(T) - \frac{\alpha^2\gamma^2}{2\beta^3 k_r}(2\beta T - 3 + 4e^{-\beta T} - e^{-2\beta T})
\end{aligned} \tag{39}$$

The long-term benefits of the retailer are

$$\begin{aligned}
\pi_R &= \int_0^T (p(t) - w(t))N\dot{x}(t) - \frac{1}{2}k_r R^2(t) + \gamma\tau(t)\,dt \\
&= \int_0^T \left(\frac{1}{a} - \frac{1}{a}\ln\frac{(j+kx(T))(1-x(T))}{(j+kx(t))(1-x(t))}\right)N\dot{x}(t) - \frac{1}{2}k_r R^2(t) + \gamma\tau(t)\,dt \\
&= \frac{1}{a}\int_{x(0)}^{x(T)} \left(1 - \ln\frac{(j+kx(T))(1-x(T))}{(j+kx(t))(1-x(t))}\right)N\,dx - \int_0^T \frac{\alpha^2\gamma^2}{2\beta^2 k_r}(1-e^{\beta(t-T)})^2\,dt + \int_0^T \gamma\tau(t)\,dt \\
&= \frac{1}{a}\int_{x(0)}^{x(T)} \left(1 - \ln\frac{(j+kx(T))(1-x(T))}{(j+kx(t))(1-x(t))}\right)N\,dx + \frac{\alpha^2\gamma^2}{4\beta^3 k_r}(2\beta T - 3 + 4e^{-\beta T} - e^{-2\beta T}) \\
&= \frac{N}{a}\left(x(T) - \int_{x(0)}^{x(T)} \ln\frac{(j+kx(T))(1-x(T))}{(j+kx(t))(1-x(t))}\,dx\right) + \frac{\alpha^2\gamma^2}{4\beta^3 k_r}(2\beta T - 3 + 4e^{-\beta T} - e^{-2\beta T})
\end{aligned} \tag{40}$$

The proof is complete. □

4. Optimal Decision Analysis

Result 1. *Although the manufacturer's optimal wholesale price changes over time, the marginal contribution of each new product is $\frac{1}{a}$. That is to say, the optimal wholesale price strategy for the manufacturer to obtain the maximum profit must make the marginal contribution of each new product $\frac{1}{a}$.*

Result 2. *According to Equation (31), $\frac{\partial(p-w)}{\partial x} = \frac{k-j-2kx(t)}{a(j+kx(t))(1-x(t))}$ is obtained. When $k - j > 0$, that is, when the influence of the innovator on the purchase of imitators is greater than the influence of the imitator's innovative purchase effect, the retailer sets prices so that the marginal contribution of each product increases first and then decreases as the proportion of innovators increases; when $k - j \leq 0$, that is, when the influence of the imitator's innovative purchase effect is greater than the influence of the innovator on the imitator's purchase, the retailer's pricing makes the marginal contribution of each product decrease as the proportion of adopters increases.*

Result 3. *From Equation (32), we can see that $\frac{dR(t)}{dt} = -\frac{\alpha\gamma}{k_r}e^{\beta(t-T)} < 0$. The recycling effort of retailers decreases with time. The recycling effort is greatest when the new product is first launched. With the continuation of the launch time, the recycling effort gradually decreases. When the product is withdrawn from the market, the recycling effort is reduced to 0.*

Result 4. *From Equation (35), we know that $\frac{\partial x(T)}{\partial j} = \frac{B(T)(1+B(T)j+B(t)k-\sqrt{(1+B(T)j+B(t)k)^2-4B(T)j})}{2B(T)k\sqrt{4B^2(T)jk+(-1-B(T)j+B(T)k)^2}} >$ 0. That is, the greater the innovation coefficient of a product, the greater the final sales volume $Nx(T)$ of the product, and it can be seen from Equation (37) that the overall long-term profit of the manufacturer will also be greater.*

Result 5. *From Equation (35), we can see that $\frac{\partial x(T)}{\partial T} = \frac{B'(T)[\sqrt{4B^2(T)jk+(-1-B(T)j+B(T)k)^2}-(1+B(T)j-B(T)k)]}{2kB^2(T)\sqrt{4B^2(T)jk+(-1-B(T)j+B(T)k)^2}}.$*

From $B'(T) > 0$, we know that $\frac{\partial x(T)}{\partial T} > 0$. That is, the longer the selected sales period, the higher the market share $x(T)$ when the product exits the market, and the greater the final sales volume $Nx(T)$ will be.

5. Numerical Example

The iPhone is a typical representative of innovative products; according to TechInsights' teardown analysis, the estimated cost of the 256 GB version of the iPhone 13 Pro is around USD 500, and its sell price is USD 1099, i.e., $c_m = 500$. Apple Inc., Cupertino, CA, USA, which uses recycling technology to disassemble and recycle old iPhones, has set up a special department to study how to improve the technology of recycling old iPhones. They have developed three devices, named Daisy, Taz, and Dave, to take apart old iPhones and collect the valuable raw materials for making new products. Nearly 20% of Apple products shipped in fiscal year 2021 were made from recycled materials [32], so we can obtain $c_r = 400$, $\Delta = 100$. Based on the above references [20–25,29], we can obtain $a = 0.002$, $k_r = 2000$, $\gamma = 10000$, $\alpha = 0.2$, $\beta = 0.5$, $T = 30$, $N = 100,000$. The numerical analysis is performed by bringing these values into the model as follows.

The overall long-term profit of the manufacturer, the overall long-term profit of the retailer, the overall long-term profit of the supply chain, and the proportion of innovators at time T vary with j and k, as shown in the figures below.

Result 6. Table 1 and Figures 2–5 show that when the innovation coefficient j and the imitation coefficient k are larger, the proportion of innovators at time T is larger, and the overall long-term profit of the manufacturer and retailer is higher.

Table 1. The influence of changes in parameters j and k on $X(T)$, π_m, π_r, and π_h.

	$k = 0.1$	$k = 0.3$	$k = 0.5$	$k = 0.7$	$k = 0.9$
$j = 0.1$	$x(T) = 16.77\%$ $\pi_m = 8{,}169{,}477$ $\pi_r = 8{,}653{,}421$ $\pi_h = 16{,}822{,}898$	$x(T) = 22.39\%$ $\pi_m = 10{,}978{,}126$ $\pi_r = 10{,}149{,}939$ $\pi_h = 21{,}128{,}065$	$x(T) = 30.24\%$ $\pi_m = 14{,}903{,}680$ $\pi_r = 12{,}203{,}997$ $\pi_h = 27{,}103{,}677$	$x(T) = 39.29\%$ $\pi_m = 19{,}428{,}616$ $\pi_r = 14{,}857{,}422$ $\pi_h = 34{,}286{,}038$	$x(T) = 47.76\%$ $\pi_m = 23{,}665{,}577$ $\pi_r = 17{,}959{,}185$ $\pi_h = 41{,}624{,}762$
$j = 0.3$	$x(T) = 36.75\%$ $\pi_m = 18{,}160{,}691$ $\pi_r = 21{,}972{,}993$ $\pi_h = 40{,}133{,}684$	$x(T) = 42.44\%$ $\pi_m = 21{,}005{,}584$ $\pi_r = 24{,}194{,}929$ $\pi_h = 45{,}200{,}513$	$x(T) = 48.31\%$ $\pi_m = 23{,}937{,}359$ $\pi_r = 26{,}665{,}438$ $\pi_h = 50{,}602{,}797$	$x(T) = 53.89\%$ $\pi_m = 26{,}727{,}890$ $\pi_r = 29{,}315{,}981$ $\pi_h = 56{,}043{,}871$	$x(T) = 58.89\%$ $\pi_m = 29{,}226{,}630$ $\pi_r = 32{,}065{,}608$ $\pi_h = 61{,}292{,}238$
$j = 0.5$	$x(T) = 48.63\%$ $\pi_m = 24{,}099{,}991$ $\pi_r = 32{,}304{,}674$ $\pi_h = 56{,}404{,}665$	$x(T) = 53.24\%$ $\pi_m = 26{,}402{,}195$ $\pi_r = 34{,}592{,}296$ $\pi_h = 60{,}994{,}491$	$x(T) = 57.63\%$ $\pi_m = 28{,}597{,}894$ $\pi_r = 36{,}981{,}292$ $\pi_h = 65{,}579{,}186$	$x(T) = 61.65\%$ $\pi_m = 30{,}608{,}553$ $\pi_r = 39{,}425{,}744$ $\pi_h = 70{,}034{,}297$	$x(T) = 65.23\%$ $\pi_m = 32{,}397{,}806$ $\pi_r = 41{,}884{,}117$ $\pi_h = 74{,}281{,}923$
$j = 0.7$	$x(T) = 56.63\%$ $\pi_m = 28{,}098{,}253$ $\pi_r = 40{,}789{,}355$ $\pi_h = 68{,}889{,}608$	$x(T) = 60.32\%$ $\pi_m = 29{,}944{,}437$ $\pi_r = 429{,}90{,}543$ $\pi_h = 72{,}934{,}980$	$x(T) = 63.74\%$ $\pi_m = 31{,}654{,}302$ $\pi_r = 45{,}225{,}646$ $\pi_h = 76{,}879{,}948$	$x(T) = 66.84\%$ $\pi_m = 33{,}201{,}773$ $\pi_r = 47{,}466{,}725$ $\pi_h = 80{,}668{,}498$	$x(T) = 69.59\%$ $\pi_m = 34{,}580{,}783$ $\pi_r = 49{,}690{,}895$ $\pi_h = 84{,}271{,}678$
$j = 0.9$	$x(T) = 62.42\%$ $\pi_m = 30{,}992{,}496$ $\pi_r = 48{,}004{,}289$ $\pi_h = 78{,}996{,}785$	$x(T) = 65.42\%$ $\pi_m = 32{,}493{,}335$ $\pi_r = 50{,}080{,}667$ $\pi_h = 82{,}574{,}002$	$x(T) = 68.17\%$ $\pi_m = 33{,}866{,}554$ $\pi_r = 52{,}158{,}179$ $\pi_h = 86{,}024{,}733$	$x(T) = 70.64\%$ $\pi_m = 35{,}105{,}751$ $\pi_r = 54{,}219{,}755$ $\pi_h = 89{,}325{,}506$	$x(T) = 72.86\%$ $\pi_m = 36{,}214{,}147$ $\pi_r = 56{,}252{,}149$ $\pi_h = 92{,}466{,}296$

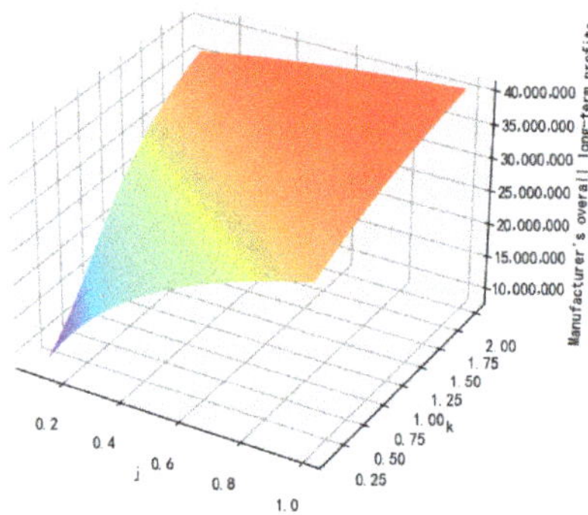

Figure 2. The relationship between parameters j, k and the overall long-term profit of the manufacturer.

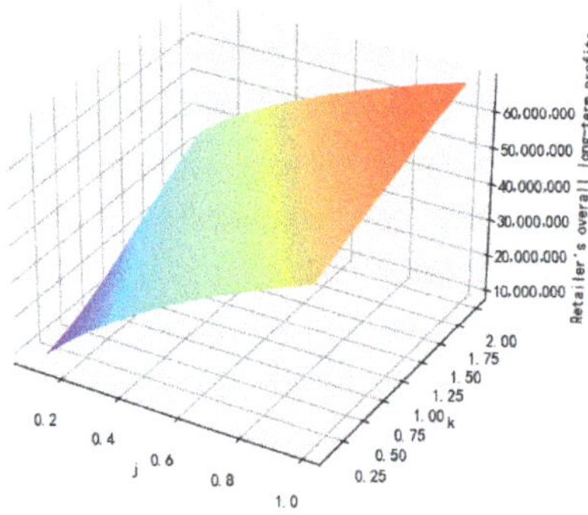

Figure 3. The relationship between parameters j, k and the overall long-term profit of the retailer.

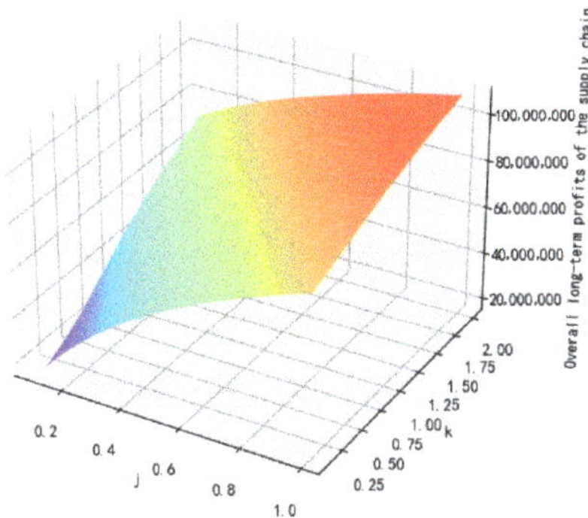

Figure 4. The relationship between parameters j, k and the overall long-term profit of the supply chain.

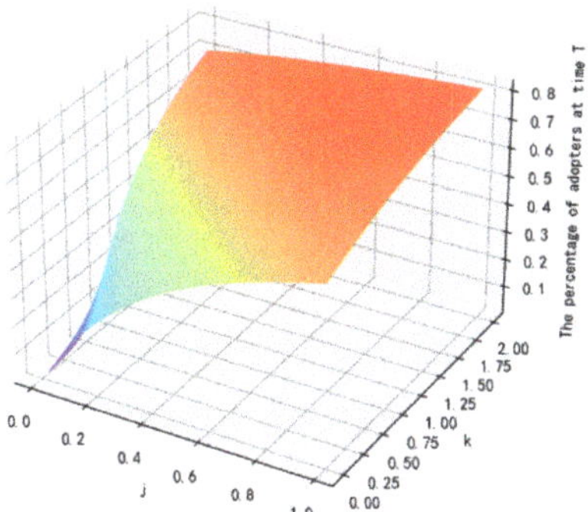

Figure 5. The relationship between parameters j, k and the proportion of innovators at time T.

Observing Figures 2 and 4, we can find that when the innovation coefficient j and the imitation coefficient k change, the overall long-term profit of the manufacturer is similar to the change graph of the proportion of innovators at time T. Because the amount of change in $\frac{\alpha^2\gamma^2}{2\beta^3 k_r}(2\beta T - 3 + 4e^{-\beta T} - e^{-2\beta T})$ in $\pi_M = \frac{N}{a}x(T) - \frac{\alpha^2\gamma^2}{2\beta^3 k_r}(2\beta T - 3 + 4e^{-\beta T} - e^{-2\beta T})$ is much smaller than the amount of change in $\frac{N}{a}x(T)$, when j, k, α, and β change, the overall long-term profit of the manufacturer and the proportion of innovators at time T are roughly linear. Similarly, the image of the retailer's overall long-term profit is similar to the image of the total supply chain profit because the degree of change in the overall long-term profit of the retailer is greater than the degree of change in the overall long-term profit of the manufacturer.

From Proposition 3, it can be seen that the parameters j and k do not affect the recovery effort and recovery rate. The change path of the recovery effort and recovery rate in the above situation is shown in the figure below.

Result 7. *Observing Figure 6, it is found that it is consistent with Result 3, and the degree of recovery effort decreases with time. When $t < 25$, it decreases slowly; when $t > 25$, it decreases sharply; when the product is withdrawn from the market ($t = 30$), the recovery intensity decreases to 0. Figure 7 shows that the recovery first gradually increases, maintains a high level of recovery from around $t = 5$ to around $t = 22$, and then begins to gradually decrease after reaching the highest recovery.*

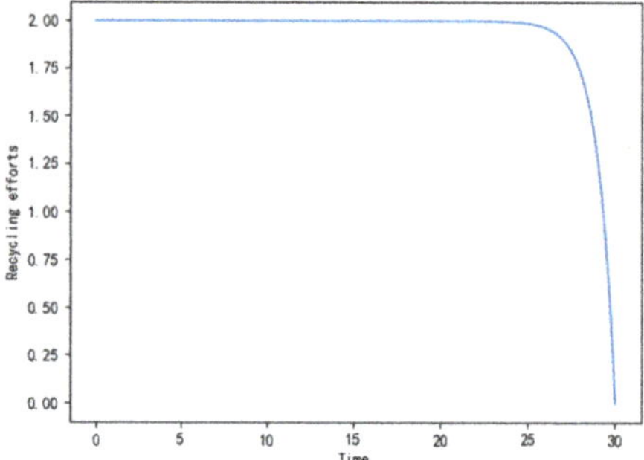

Figure 6. Change path of recycling effort.

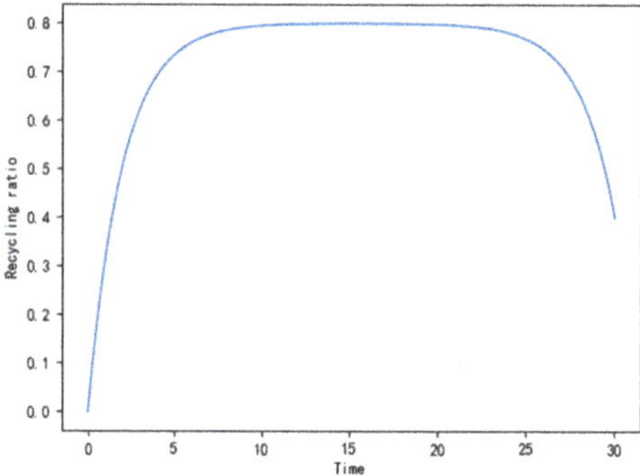

Figure 7. Change path of recovery rate.

6. Conclusions

This paper establishes a remanufacturing supply chain modeled by a single manufacturer and followed by a single retailer based on the Bass model. The optimal control method obtains the best path for the manufacturer's wholesale price, the retailer's retail price, and the best path for recycling efforts, and further obtains the optimal market share, recovery rate, and overall long-term profits of manufacturers, retailers, and supply chains. It also compares the supply chain's overall long-term profit and market share with different innovation coefficients and imitation coefficients. The main conclusions are as follows:

Conclusion (1): From Result 1, when the marginal profit of each product is $\frac{1}{a}$, the manufacturer can obtain the optimal path for the wholesale price. The retailer's pricing strategy depends on different imitation coefficients and innovation coefficients. It can be seen from Result 2 that when the imitation coefficient is greater than the innovation coefficient, retailer pricing makes the marginal profit of each product increase first and then decrease as the proportion of innovators increases. This shows that in real life, imitators limit the curiosity of new products. When a new product was first launched, the imitators became more curious. Over time, this curiosity will gradually disappear. This also reminds the participants in the supply chain that when making sales efforts (such as advertising) for a new product, it is better to do so in the early stages of product launch than in the middle and late stages. When the imitation coefficient is less than or equal to the Innovation coefficient, retailer pricing will make the marginal profit of each product decrease as the proportion of innovators increases. This shows that when more and more people use a new product in real life, the price reduction will become more and more apparent. This conclusion is consistent with objective reality. This also shows that there is no need to increase sales efforts for a new product in the middle and late stages. When the initial innovation population (innovation coefficient j) is small, the product market is mainly expanded by increasing sales efforts (increasing the imitation coefficient k).

Conclusion (2): Results 3 and 7 show that retailers' recycling efforts maintain a high level when new products enter the market and then slowly reduce the degree of their recycling efforts. At the end of the product sales period, the attempt to recycle the product was reduced to a greater extent until the product was no longer recycled. The corresponding recovery rate gradually increased as the retailer's recovery efforts strengthened. Then, the recovery efforts of the retailer continued to weaken and then began to decline after the recovery rate reached the maximum. When many products were first launched in real life, the product recycling promotion was powerful and maintained for an extended period, and the recycling rate was also increasing. However, as the product launch time increased, the recycling effort gradually weakened, and the recycling rate reached its peak and began to decline. When the product was close to exiting the market, the promotion of product recycling would drop significantly. When the product was withdrawn from the market, the product would no longer be recycled. From Result 5, the longer the product sales time, the greater the sales volume under its optimal strategy. These conclusions are consistent with reality.

Conclusion (3): Results 4 and 6 show that in a market with a higher innovation coefficient j and imitation coefficient k, manufacturers and retailers can obtain higher profits, so the overall long-term profits of the supply chain will be higher as well. The larger the innovation coefficient, the more attractive the innovative product is to consumers; the larger imitation coefficient means the larger the customer group that imitates, thus leading to the faster diffusion of the innovative product in the market, i.e., the larger the sales quantity will be at a fixed sales time. As a result, the manufacturer and the retailer can all earn higher profits, as can the entire supply chain.

Conclusion (4): It can be seen from Result 6 that in a market with a high imitation coefficient and innovation coefficient, the proportion of innovators at the end of product sales is also relatively high. That is, more products can be sold in such a market. Fewer products are sold in markets where both the imitation coefficient and the innovation coefficient are small. The innovation coefficient is formed naturally, and the human factor is relatively small. However, the imitation coefficient can be improved by increasing sales efforts. Therefore, when the innovation coefficient is not significant, the participants in the supply chain must increase their sales efforts in the early stage of sales and strive to improve the imitation coefficient to expand sales.

Conclusion (5): Compared with Reference [29], the decision making of supply chain participants is more complicated in this paper. The manufacturer's wholesale price strategy is no longer a fixed value but changes over time. Retailers' retail prices also change with changes in wholesale prices. The proportion of new product innovators is no longer a linear function of time.

Our findings have following managerial implications:

1. To maximize profits, the manufacturer's optimal pricing strategy is to maintain a fixed value of marginal profit for each product, and in the early stage, the innovation factor is increased through advertising and other sales efforts to speed up the product diffusion, which can increase the final sales of the product.
2. The specific pricing strategy of the retailer should be determined according to different markets, and the recycling strategy of the retailer should be to make a lot of recycling publicity and preparation in the early stage to raise consumers' recycling awareness as soon as possible, so that the recycling rate of innovative products can quickly reach a high level, and then the retailer can maintain this high recycling rate with lower recycling costs.
3. In markets with higher innovation and imitation coefficients, the sales of the innovative product are higher and the profits of the closed-loop supply chain are higher, so manufacturers and retailers should choose to put the innovative product into markets with higher innovation and imitation coefficients, while in markets with lower

innovation and imitation coefficients, manufacturers and retailers should consider sales efforts such as advertising and better customer services to improve consumer satisfaction and increase the speed of product diffusion, thus increasing the profits of manufacturers and retailers.

4. The closed-loop innovative product supply chain is much more complex than the open-loop one. In the closed-loop innovative product supply chain, the dynamic demand changes in the market become more complex, and thus the optimal decisions of manufacturers and retailers are also more complex, so the members of the closed-loop supply chain need to have stronger market information acquisition ability and faster response to market changes.

The research of this paper is based on the premise of a complete market monopoly. It only considers the situation of a secondary remanufacturing supply chain when a dominant supplier and the following retailer are selling new products. And it assumes that the time when the product is no longer sold is specific. In the future, we can also consider an oligopoly situation, such as one supplier and two retailers, or two suppliers and one retailer. The time for new product sales to exit the market can be used as a control variable to find the optimal time for product replacement.

Author Contributions: Conceptualization, L.L. and Z.L.; methodology, Y.P.; software, N.W.; validation, L.L., Y.P. and N.W.; formal analysis, Z.L.; investigation, Z.L.; resources, L.L.; data curation, N.W.; writing—original draft preparation, Z.L.; writing—review and editing, L.L.; visualization, Y.P.; supervision, Y.P.; project administration, L.L.; funding acquisition, L.L. All authors have read and agreed to the published version of the manuscript.

Funding: This research was funded by National Natural Science Foundation of China (No. 72162015), Social Science Foundation of Jiangxi Province (21GL17).

Data Availability Statement: Not applicable.

Acknowledgments: This work was supported by National Natural Science Foundation of China (No. 72162015), Social Science Foundation of Jiangxi Province (21GL17). This support is gratefully appreciated.

Conflicts of Interest: The authors declare no conflict of interest.

References

1. Deng, D.S.; Long, W.; Li, Y.Y.; Shi, X.Q. Building Robust Closed-Loop Supply Networks against Malicious Attacks. *Processes* **2021**, *9*, 39. [CrossRef]
2. Guo, Y.; Shi, Q.; Guo, C. A Fuzzy Robust Programming Model for Sustainable Closed-Loop Supply Chain Network Design with Efficiency-Oriented Multi-Objective Optimization. *Processes* **2022**, *10*, 1963. [CrossRef]
3. Institute of Internet Industry; Tsinghua University. *White Paper 2018 on Intelligent Remanufacturing*; Institute of Internet Industry: Beijing, China, 2018.
4. Ding, Y. Game Analysis and Decision Research of Remanufacturing Supply Chain System for Information and Communication Equipment. Master's Thesis, Zhejiang University of Technology, Hangzhou, China, 2020.
5. Mota, B.; Gomes, M.I.; Carvalho, A.; Barbosa-Povoa, A.P. Sustainable supply chains: An integrated modeling approach under uncertainty. *Omega* **2017**, *77*, 32–57. [CrossRef]
6. Savaskan, R.C.; Bhattacharya, S.; Wassenhove, L. Closed-Loop Supply Chain Models with Product Remanufacturing. *Manag. Sci.* **2004**, *50*, 239–252. [CrossRef]
7. Savaskan, R.C.; Wassenhove, L. Reverse channel design: The case of competing retailers. *Oper. Res.* **2006**, *46*, 621–624. [CrossRef]
8. Ferguson, M.E.; Toktay, L.B. The effect of competition on recovery strategies. *Prod. Oper. Manag.* **2006**, *15*, 351–368. [CrossRef]
9. Jian, J.; Li, B.; Zhang, N.; Su, J. Decision-making and coordination of green closed-loop supply chain with fairness concern. *J. Clean. Prod.* **2021**, *298*, 126779. [CrossRef]
10. Masoudipour, E.; Amirian, H.; Sahraeian, R. A novel closed-loop supply chain based on the quality of returned products. *J. Clean. Prod.* **2017**, *151*, 344–355. [CrossRef]
11. Diaz, R.; Marsillac, E. Evaluating strategic remanufacturing supply chain decisions. *Int. J. Prod. Res.* **2016**, *55*, 2522–2539. [CrossRef]
12. Jena, S.K.; Sarmah, S.P.; Sarin, S.C. Joint-advertising for collection of returned products in a closed-loop supply chain under uncertain environment. *Comput. Ind. Eng.* **2017**, *113*, 305–322. [CrossRef]

13. Tsao, Y.C.; Linh, V.T.; Lu, J.C. Closed-loop supply chain network designs considering RFID adoption. *Comput. Ind. Eng.* **2016**, *113*, 716–726. [CrossRef]
14. Bhattacharya, R.; Kaur, A.; Amit, R.K. Price optimization of multi-stage remanufacturing in a closed loop supply chain. *J. Clean. Prod.* **2018**, *186*, 943–962. [CrossRef]
15. Ma, Z.J.; Zhang, N.; Dai, Y.; Hu, S. Managing channel profits of different cooperative models in closed-loop supply chains. *Omega* **2016**, *59*, 251–262.
16. Xiong, Y.; Zhao, Q.; Zhou, Y. Manufacturer-remanufacturing vs. supplier-remanufacturing in a closed-loop supply chain. *Int. J. Prod. Econ.* **2016**, *176*, 21–28. [CrossRef]
17. Xie, J.P.; Liang, L.; Liu, L.H.; Ieromonachou, P. Coordination contracts of dual-channel with cooperation advertising in closed-loop supply chains. *Int. J. Prod. Econ.* **2016**, *183*, 528–538. [CrossRef]
18. Zhang, H.; Wang, Z.; Hong, X.; Zhong, Q. Fuzzy closed-loop supply chain models with quality and marketing effort-dependent demand. *Expert Syst. Appl.* **2022**, *207*, 118081. [CrossRef]
19. He, Q.; Wang, N.; Browning, T.R.; Jiang, B. Competitive collection with convenience-perceived customers. *Eur. J. Oper. Res.* **2022**, *303*, 239–254. [CrossRef]
20. Lee, C.; Realff, M.; Ammons, J. Integration of channel decisions in a decentralized reverse production system with retailer collection under deterministic non-stationary demands. *Adv. Eng. Inform.* **2011**, *25*, 88–102. [CrossRef]
21. Giovanni, P.D. Environmental collaboration in a closed-loop supply chain with a reverse revenue sharing contract. *Ann. Oper. Res.* **2014**, *220*, 135–157. [CrossRef]
22. Xiang, Z.; Xu, M. Dynamic cooperation strategies of the closed-loop supply chain involving the Internet service platform. *J. Clean. Prod.* **2019**, *220*, 1180–1193. [CrossRef]
23. He, P.; He, Y.; Shi, C.V.; Xu, H.; Zhou, L. Cost-sharing contract design in a low-carbon service supply chain. *Comput. Ind. Eng.* **2019**, *139*, 106160. [CrossRef]
24. Song, J.; Bian, Y.; Liu, G. Decisions of Closed-Loop Supply Chain Based on Recycling Effort and Differential Game. *Discret. Dyn. Nat. Soc.* **2020**, *2020*, 7493942. [CrossRef]
25. Yang, Y.; Xu, X. A differential game model for closed-loop supply chain participants under carbon emission permits. *Comput. Ind. Eng.* **2019**, *135*, 1077–1090. [CrossRef]
26. China Household Electric Appliance Research Institute. *White Paper on WEEE Recycling Industry in China (2021)*; China Household Electric Appliance Research Institute: Beijing, China, 2022.
27. Bass, F.M. A New Product Growth Model for Consumer Durables. *Manag. Sci.* **1976**, *15*, 215–227. [CrossRef]
28. Bass, F.M.; Krishnamoorthy, A.; Sethi, P. Generic and Brand Advertising Strategies in a Dynamic Duopoly. *Mark. Sci.* **2005**, *24*, 556–568. [CrossRef]
29. Quan, X.; Tu, F.; Wei, J. A Manufacturer-Retailer Stackelberg Game of Innovation Pricing. *J. Syst. Eng.—Theory Pract.* **2007**, *8*, 111–117.
30. White, M.; Braczyk, J.; Ghobadian, A.; Niebuhr, J. *Small Firms' Innovation: Why Regions Differ*; Policy Studies Institute: Addis Ababa, Ethiopia, 1988.
31. Chiang, A.C. *Elements of Dynamic Optimization*; McGraw-Hill: New York, NY, USA, 2000.
32. Apple Inc. *Environmental Progress Report*; Apple Inc.: Cupertino, CA, USA, 2022.

Article

Joint Economic–Environmental Benefit Optimization by Carbon-Abatement Cost Sharing in a Capital-Constrained Green Supply Chain

Jinzhao Shi [1,2], Wenxin Jiao [1,2,*], Kewen Jing [1,2], Qi Yang [1] and Kin Keung Lai [3,*]

[1] School of Economics and Management, Chang'an University, Middle Section of South Second Ring Road, Xi'an 710064, China
[2] Research Center for Green Engineering and Sustainable Development, Chang'an University, Middle Section of South Second Ring Road, Xi'an 710064, China
[3] College of Economics, Shenzhen University, Nanhai Ave 3688, Shenzhen 518060, China
* Correspondence: jwxjacky@chd.edu.cn (W.J.); mskklai@outlook.com (K.K.L.)

Abstract: This paper studies the potential of carbon-abatement cost-sharing contracts in optimizing the joint economic–environmental benefit of a green supply chain. One-way and two-way cost-sharing contracts were investigated, respectively, in scenarios in which a capital-constrained manufacturer has a dominant downstream retailer or a dominant upstream supplier. The manufacturer obtains financing from a competitively priced bank to fulfill its production, carbon-abatement investment, and even insufficient emission permit purchase given the fact that the cap-and-trade regulation exists. Results show that in both one-way and two-way cost-sharing cases, cost sharing of carbon abatement has no effect on the manufacturer's output or its counterparty's wholesale price decisions; however, it improves the carbon abatement level of the supply chain. As a result, such cost-sharing of carbon abatement is proven to hamper the profit of the overall supply chain, but it improves the joint "economic-environmental" benefit of the supply chain if the cost-sharing coefficient is properly chosen. Furthermore, this problem is studied in the case of consumers' green preferences, and carbon-abatement cost sharing is also verified to have the potential to optimize joint economic–environmental benefits.

Keywords: carbon abatement; cost sharing; capital constraint; environmental externality

Citation: Shi, J.; Jiao, W.; Jing, K.; Yang, Q.; Lai, K.K. Joint Economic–Environmental Benefit Optimization by Carbon-Abatement Cost Sharing in a Capital-Constrained Green Supply Chain. *Processes* **2023**, *11*, 226. https://doi.org/10.3390/pr11010226

Academic Editors: Conghu Liu, Xiaoqian Song, Zhi Liu and Fangfang Wei

Received: 22 November 2022
Revised: 1 January 2023
Accepted: 3 January 2023
Published: 10 January 2023

1. Introduction

With the rapid development of the manufacturing industry, the carbon dioxide emissions problem becomes a worldwide issue that has brought a variety of climate anomalies and disasters. Since the agreement on reducing greenhouse gas emissions has already been reached in the late 20th century by the Kyoto Protocol, many countries have been sparing no efforts in lowering carbon emissions [1–4]. When it comes to China, at the 75th United Nations General Assembly held in September 2020, President Xi Jinping solemnly promised the world that China would strive to peak carbon dioxide emissions by 2030 and achieve carbon neutrality by 2060 (accessed on 22 September 2021, http://www.qstheory.cn/yaowen/2020-09/22/c_1126523612.htm).

In order to effectively implement the task of energy saving and emission reduction, carbon cap, cap and trade, carbon tax, and other policies have been widely used, among which, the cap-and-trade regulation has been proven to be efficient [5–8]. Apart from government policies, consumers also play a role in stimulating enterprises to make carbon reduction efforts due to their green preferences [9,10]. Therefore, the task of building green supply chains is imminent. As an innovative measure to build a green supply chain, collaborative emission abatement in supply chains is becoming a global consensus and widely practiced, e.g., as a retail giant, Walmart plans to make a joint carbon abatement with suppliers for 1 billion tons by 2030 through "Project Gigaton" (accessed on 6 April 2022, https://corporate.walmart.com/esgreport/esg-issues/climate-change), and BMW announced to cooperate with suppliers such as CATL and Shougang Group to reduce emissions by 20% in 2030 (accessed on 27 September, 2021, https://www.shougang.com.cn/sgweb/html/sgyw/20190927/3856.html).

The above shows that joint emission abatement has become an essential action for more and more supply chains while it may bring huge financial pressure for the small- and medium-sized enterprises (SMEs) among these chains [11,12]. This paper aims to study how cost sharing of carbon abatement affects the operational decisions as well as the performances of supply chains with a capital-constrained manufacturer. One-way and two-way cost-sharing contracts were investigated in scenarios in which the manufacturer has a dominant downstream retailer and upstream supplier, respectively. The following questions were addressed: (1) What are the equilibriums of supply chain financing systems consisting of supply members and a bank in the cases of one-way and two-way cost sharing? (2) How does cost sharing of carbon abatement affect the operational decisions as well as the benefit of the supply chain in each case? (3) Can and in what circumstance can cost-sharing mechanisms bring satisfactory joint economic–environmental benefits to capital-constrained supply chains? (4) When considering consumers' green preferences, how are the above problems influenced? Our main results show that in each of the one-way and two-way cost-sharing cases, cost-sharing of carbon abatement will hamper the profit of the overall supply chain but enhance the joint "economic-environmental" benefit of the supply chain if the parameter of the contract is properly chosen. Consumers' green preferences along with cost-sharing will further enhance the carbon abatement of the supply chain, but a similar role of cost-sharing on the joint "economic-environmental" benefit of the supply chain is concluded.

The rest of this article is organized as follows: Section 2 reviews the literature to clarify research gaps. Sections 3 and 4 present the one-way and two-way cost-sharing cases, respectively. In Section 5, the problem is extended to consider consumers' green preferences. Section 6 concludes the paper and provides discussions.

2. Literature Status

This paper examines the role of carbon-abatement cost-sharing contracts in low-carbon supply chains with a capital-constrained manufacturer. Two streams of research, i.e., "operations and financing of the low-carbon supply chain" and "cost-sharing in supply chain operations", are most related to the current paper. Below, we give a summary of the literature and highlight our innovations.

2.1. Operations and Financing of Low-Carbon Supply Chain

Aside from various research on traditional supply chain operations, i.e., ordering, pricing, production, etc., more and more scholars have been paying attention to the operations of low-carbon supply chains where the carbon abatement decision of the chain members is further emphasized [13–15]. Impacts of different low-carbon policies, e.g., carbon trading [8,16–18], carbon tax [19–21], etc., on the emission reduction efficiency of low-carbon supply chains are widely studied. Xu et al. [22] examined the impact of carbon trading price on manufacturer carbon abatement policies in a make-to-order (MTO) system while Bai et al. [23] studied manufacturers' integrated decisions on selling prices and carbon reduction level in the context of considering two kinds of products. Bian and Guo [20] investigated the roles of different environmental policies on emissions reduction in manufacturing and indicated that compared to a carbon tax, a carbon abatement subsidy is more effective in abating carbon emissions, but it may lead to lower social welfare.

In recent years, the problem of financing is becoming a new issue for sustainable operations of supply chains since low-carbon activities, such as carbon abatement investment, may increase the financing requirements of supply chain members. Then, many studies began to explore issues such as integrated operations (e.g., ordering, pricing, etc.), financing, and carbon abatement decisions in supply chains [12,24,25]. At the same time, various motivational factors for low-carbon investment, e.g., consumers' green preferences [26,27], etc., have also been widely characterized. Representatively, Wu et al. [26] studied downstream retailers' ordering and financing decisions as well as their suppliers' pricing and carbon abatement decisions while Xu and Fang [28] and Zou et al. [12] examined manufacturers' integrated ordering, carbon abatement, and financing decisions. For the scenario of upstream financing, Qin et al. [24] researched green suppliers' choices between bank financing and advance payment while Huang et al. [29] explored how government subsidy strategies affect low-carbon supply chain operations and financing.

Different from the above studies, this paper mainly focuses on examining how cost sharing of carbon abatement affects the operations as well as the benefit of low-carbon supply chains given the manufacturer in the chain faces capital constraints.

2.2. Cost-Sharing in Supply Chain Operations

Cost-sharing is a common contract or mechanism in supply chain management. When it comes to low-carbon supply chains, Ghosh and Shah [30] studied the coordination effect of a cost-sharing contract for a supply chain consisting of a manufacturer and a retailer. Wang et al. [9] combined "cost-sharing" and "altruistic preference" to improve the profit of a retailer-led, low-carbon supply chain. Under the cap-and-trade regulation, Wang et al. [31] implemented a two-way cost-sharing contract to enhance the carbon abatement level as well as the profit of a supply chain. In the context of capital constraint, Wu et al. [26] proposed a contract portfolio of revenue sharing, cost sharing, and transfer payment to coordinate a low-carbon supply chain in which a capital-constrained retailer may apply bank credit financing or trade credit financing. Lai et al. [32] explored the production, financing, and green investment decisions of a manufacturer who lacks money and can obtain financing from a bank or its supplier's investment. They found that a portfolio of cost-sharing and quantity discount contracts can achieve supply chain coordination.

Different from the above studies, this paper focuses on exploring how cost sharing of carbon abatement affects the operational decisions of a supply chain where the capital-constrained manufacturer has a dominant downstream retailer or upstream supplier. One-way and two-way cost-sharing contracts were investigated in these two supply chain settings, respectively, and the joint "economic-environmental" benefit of the contracts is confirmed in this research.

3. The One-Way Cost-Sharing Case

3.1. Problem Statement

In this case, we consider the supply chain to consist of a capital-constrained manufacturer and a creditworthy downstream retailer. The retailer is a retail giant who acts as the leader in the chain, and the manufacturer who encounters capital constraint acts as the follower. In response to the retailer's request, the manufacturer develops a single kind of green product and sells them to the end market through the retailer's supermarkets. This follows the common practice in retailer-led, low-carbon supply chains, e.g., Walmart has developed Project Gigaton to engage its suppliers to reduce carbon emissions by one billion metric tons by 2030. Due to its capital constraint, the manufacturer is assumed to obtain funds from a competitively priced bank to fulfill its needs of production and a low-carbon investment. The carbon abatement occurs as a cost for the manufacturer, which follows a quadratic form to the abatement level e, i.e., $\frac{1}{2}k(e)^2$, where k is a cost coefficient [12,26]. There also exists a carbon trading market wherein the manufacturer can obtain a free permit quota from the government and buy (sell) insufficient (excess) carbon emission permits at a certain price [2,3,12,28]. Assuming G is the free quota, e_0 is the initial carbon emissions for unit production, and p_e is the carbon trading price, we have $(e_0q - e - G)p_e$ denoting the carbon trading cost or revenue of the manufacturer in the cases of over emitting or under emitting, where q is the output of the manufacturer which is a decision variable.

Given the manufacturer jointly deciding its output q and carbon abatement level e, the retailer decides the wholesale price of the product that it would like to purchase, i.e., w. After that, the bank in the competitive banking market sets a suitable interest rate r to obtain an expected zero profit after examining the manufacturer's repayment abilities in the random market, where X is the random demand whose PDF, CDF, complementary CDF, and failure rate are $f(x)$, $F(x)$, $\overline{F}(x)$ and $h(x) = f(x)/\overline{F}(x)$, respectively. In addition, without loss of generality, the manufacturer's self-capital is normalized to zero [28,33]. The time value of funds and the salvage value of leftovers are not considered [34,35].

To incentivize the manufacturer to invest in carbon emission reduction, the retailer may promise to share a proportion of the carbon-abatement cost for the manufacturer, and our intention is to investigate whether such a cost-sharing mechanism can benefit the whole supply chain. We summarize the main notations in Notations section.

3.2. The Case without Cost-Sharing

In the Stackelberg game between the manufacturer and the retailer, by backward induction, we start with solving the problem of the manufacturer to identify its response function and then derive the optimal strategy of the retailer to find the equilibrium of the game. Given the loan interest rate announced by the bank and the wholesale price quoted by the retailer, the supplier decides on its production quantity and carbon abatement level to maximize its expected profit. The total amount of capital that the manufacturer needs to borrow from the bank for production, carbon abatement investment, and carbon trading activities would be $cq + \frac{1}{2}k(e)^2 + (e_0q - e - G)p_e$. In a competitive banking market, by considering the bank's competitively-priced equation as a constraint, the manufacturer's decision-making problem can be formulated as follows:

$$Max\pi_m(e,q;w,r) = E\left\{ w\min[X,q] - \left[cq + \tfrac{1}{2}ke^2 + (e_0q - e - G)p_e\right](1+r)\right\}^+$$
$$s.t.\,E\min\left\{ w\min[X,q], \left[cq + \tfrac{1}{2}ke^2 + (e_0q - e - G)p_e\right](1+r)\right\} = cq + \tfrac{1}{2}ke^2 + (e_0q - e - G)p_e \tag{1}$$

Lemma 1. *In the case without cost-sharing of carbon abatement, the manufacturer's optimal carbon abatement and production decisions are*

$$\begin{cases} e^* = \frac{p_e}{k} \\ q^* = \overline{F}^{-1}\left(\frac{c+e_0 p_e}{w}\right) \end{cases} \tag{2}$$

Proof. See Appendix A. $\square$

Next, we discuss the retailer's problem. By manipulating the manufacturer's best response, the retailer determines the optimal wholesale price to maximize its profit:

$$Max\pi_r(w;q^*) = (p - w)\min[X,q^*] \tag{3}$$

Lemma 2. *In the case without cost-sharing of carbon abatement, the retailer's optimal wholesale price quotation is*

$$w^* = \frac{p\overline{F}(q^*) - c - e_0 p_e}{h(q^*)\left[q^* - \int_0^{q^*} F(x)dx\right]} \tag{4}$$

Proof. See Appendix A. $\square$

As for the bank, if q^*, e^*, and w^* are obtained, the competitively priced bank's interest rate can be determined with the constraint condition in Equation (1). Then, we can obtain the equilibrium of the supply chain financing game.

Proposition 1. *In the case without cost-sharing of carbon abatement, the equilibrium of the supply chain financing system, i.e., (e^*, q^*, w^*, r^*), solves the following:*

$$\begin{cases} e^* = \frac{p_e}{k} \\ q^* = \overline{F}^{-1}\left(\frac{c+e_0p_e}{w^*}\right) \\ w^*h(q^*)\left[q^* - \int_0^{q^*} F(x)dx\right] = p\overline{F}(q^*) - c - e_0p_e \\ \text{Emin}\left\{w^*\min[X,q^*], \left[cq^* + \frac{1}{2}k(e^*)^2 + (e_0q^* - e^* - G)p_e\right](1+r^*)\right\} = cq^* + \frac{1}{2}k(e^*)^2 + (e_0q^* - e^* - G)p_e \end{cases} \tag{5}$$

Proof. See Appendix A. $\square$

From Equation (5), it is obvious that compared to the centralized supply chain where $p\overline{F}(q^*) - c - e_0p_e = 0$ holds, $w^*h(q^*)\left[q^* - \int_0^{q^*} F(x)dx\right]$ is the efficiency loss of the supply chain attributed to the double marginalization.

3.3. The Case with Cost-Sharing

In this part, we consider the case where the retailer shares a proportion of carbon abatement investment, i.e., β, for the manufacturer. The manufacturer's decision-making problem is thus transformed into:

$$Max\pi_m(e,q;w,r) = E\left\{w\min[X,q] - \left[cq + \frac{1}{2}k(1-\beta)e^2 + (e_0q - e - G)p_e\right](1+r)\right\}^+$$
$$s.t.\text{Emin}\left\{w\min[X,q], \left[cq + \frac{1}{2}k(1-\beta)e^2 + (e_0q - e - G)p_e\right](1+r)\right\} = cq + \frac{1}{2}k(1-\beta)e^2 + (e_0q - e - G)p_e \tag{6}$$

Lemma 3. *In the case of cost-sharing of carbon abatement, the manufacturer's optimal carbon abatement and production decisions are*

$$\begin{cases} e^{**} = \frac{p_e}{k(1-\beta)} \\ q^{**} = \overline{F}^{-1}\left(\frac{c+e_0p_e}{w}\right) \end{cases} \tag{7}$$

Proof. See Appendix A. $\square$

Next, we discuss the retailer's problem. By manipulating the manufacturer's best response, the retailer determines the optimal wholesale price to maximize its profit:

$$Max\pi_r(w;e^{**},q^{**}) = (p-w)\min[X,q^{**}] - \frac{1}{2}\beta k(e^{**})^2 \tag{8}$$

Lemma 4. *In the case of cost-sharing of carbon abatement, the retailer's optimal wholesale price quotation is*

$$w^{**} = \frac{p\overline{F}(q^{**}) - c - e_0p_e}{h(q^{**})\left[q^{**} - \int_0^{q^{**}} F(x)dx\right]} \tag{9}$$

Proof. See Appendix A. $\square$

As for the bank, if q^{**}, e^{**}, and w^{**} are obtained, the competitively priced bank's interest rate can be determined by the constraint condition in Equation (6). Then, we can obtain the equilibrium of the supply chain financing game.

Proposition 2. *In the case of cost-sharing of carbon abatement, the equilibrium of the supply chain financing system, i.e., $(e^{**},q^{**},w^{**},r^{**})$, solves the following:*

$$\begin{cases} e^{**} = \frac{p_e}{k(1-\beta)} \\ q^{**} = \overline{F}^{-1}\left(\frac{c+e_0 p_e}{w^{**}}\right) \\ w^{**}h(q^{**})\left[q^{**} - \int_0^{q^{**}} F(x)dx\right] = p\overline{F}(q^{**}) - c - e_0 p_e \\ \mathrm{Emin}\left\{w^{**}\min[X, q^{**}], \left[cq^{**} + \frac{1}{2}k(1-\beta)(e^{**})^2 + (e_0 q^{**} - e^{**} - G)p_e\right](1+r^{**})\right\} = cq^{**} + \frac{1}{2}k(1-\beta)(e^{**})^2 + (e_0 q^{**} - e^{**} - G)p_e \end{cases}$$

(10)

Proof. See Appendix A. □

In the case of cost-sharing of carbon abatement, the efficiency loss attributed to the double marginalization still exists and remains the same as that in the case without the cost-sharing contract.

3.4. The Value of Cost-Sharing

By comparing the equilibriums in Propositions 1 and 2, we obtain the following corollary:

Corollary 1. *The one-way cost-sharing contract on carbon abatement provided by the retailer has no influence on the manufacturer's output or the retailer's wholesale price quotation decisions; however, it will undoubtedly improve the carbon abatement level of the manufacturer, i.e., $q^{**} = q^*$, $w^{**} = w^*$, $e^{**} > e^*$*

Proof. See Appendix A. □

We are more interested in how the cost-sharing contract of the retailer affects the profits of the chain members. As Corollary 1 shows, the contract has positive environmental externalities. Then, we further examine the joint "economic-environmental" benefit of the contract. The following Proposition 3 summarizes all these results.

Proposition 3. *The one-way cost-sharing contract for carbon abatement provided by the retailer can:*
(a) *Increase the profit of the manufacturer while reducing the profits of the retailer as well as the overall supply chain;*
(b) *Enhance the joint "economic-environmental" benefit only when the proportion of cost-sharing is smaller than a threshold, i.e., $0 < \beta < \frac{2}{3}$.*

Proof. See Appendix A. □

From Proposition 3, the one-way cost-sharing contract for carbon abatement cannot benefit all members in the supply chain, but it undoubtedly reduces the carbon emissions of the manufacturer, which in turn may bring positive environmental externalities and enhance the joint "economic-environmental" benefit, showing the potential value of the contract.

4. The Two-Way Cost-Sharing Case

4.1. Problem Statement

Different from that in Section 3, in this section, we consider a supply chain consisting of a capital-constrained manufacturer and a creditworthy upstream supplier. The supplier (acting as the leader) and the manufacturer (acting as the follower) jointly complete the production process of green products. Both parties are assumed to put efforts into carbon abatement, and their respective levels of carbon abatement are e_s and e_m. Following the similar quadratic form as in Section 3, the corresponding carbon-abatement costs for both parties are $\frac{1}{2}k_s(e_s)^2$ and $\frac{1}{2}k_m(e_m)^2$, where k_s and k_m are cost coefficients. In order to incentivize the joint emissions reduction among the supply chain members, we assume the supplier and manufacturer promise to share a proportion of the carbon-abatement costs with each other, where β (α) is the proportion of the cost that the supplier (manufacturer) shares with the manufacturer (supplier). What we are interested in is whether the two-way cost-sharing mechanism could benefit the supply chain.

4.2. The Case without Cost-Sharing

In this case, the supplier and the manufacturer invest in their carbon abatement, respectively. We first solve the problem of the follower, i.e., the manufacturer, who makes joint decisions on an output q (production quantity) and carbon abatement level e_m. Given that it is capital-constrained, the manufacturer should raise money from the bank to cover its cost of production, carbon abatement, and even insufficient carbon emission permit purchase. Its total financing amount can be formulated as $(c_m + w)q + \frac{1}{2}k_m(e_m)^2 + (e_0^m q - e_m - G_m)p_e$. Considering the bank's competitively priced equation as a constraint, the manufacturer's decision-making problem is as follows:

$$
\begin{aligned}
Max\pi_m(q, e_m; w, r) &= E\left\{ p\min[q, X] - \left[(c_m + w)q + \tfrac{1}{2}k_m(e_m)^2 + (e_0^m q - e_m - G_m)p_e\right](1+r)\right\}^+ \\
s.t.\,E\min&\left\{ p\min[q, X], \left[(c_m + w)q + \tfrac{1}{2}k_m(e_m)^2 + (e_0^m q - e_m - G_m)p_e\right](1+r)\right\} \\
&= (c_m + w)q + \tfrac{1}{2}k_m(e_m)^2 + (e_0^m q - e_m - G_m)p_e
\end{aligned}
\tag{11}
$$

Lemma 5. *In the case without cost-sharing of carbon abatement, the manufacturer's optimal carbon abatement and production decisions are*

$$
\begin{cases}
e_m^* = \frac{p_e}{k_m} \\
q^* = \bar{F}^{-1}\left(\frac{c_m + w + e_0^m p_e}{p}\right)
\end{cases}
\tag{12}
$$

Proof. See Appendix A. $\square$

By manipulating the manufacturer's best response in Equation (12), as the leader, the supplier decides its optimal wholesale price and carbon abatement level for producing the semifinished products to maximize its profit:

$$
Max\pi_s(w, e_s; q^*, e_m^*) = (w - c_s)q^* - \frac{1}{2}k_s(e_s)^2 - (e_0^s q^* - e_s - G_s)p_e
\tag{13}
$$

Lemma 6. *In the case without cost-sharing of carbon abatement, the supplier's optimal wholesale price quotation and carbon abatement level are*

$$
\begin{cases}
w^* = pq^* f(q^*) + c_s + e_0^s p_e \\
e_s^* = \frac{p_e}{k_s}
\end{cases}
\tag{14}
$$

Proof. See Appendix A. $\square$

If q^*, e_m^*, w^*, and e_s^* are obtained, the competitively priced bank interest rate can be determined by the constraint condition in Equation (11). Then, we obtain the equilibrium of the supply chain financing game.

Proposition 4. *In the case without cost-sharing of carbon abatement, the equilibrium of the supply chain financing system, i.e.,* $(q^*, e_m^*, w^*, e_s^*, r^*)$ *, solves the following.*

$$
\begin{cases}
q^* = \overline{F}^{-1}\left(\frac{c_m + w^* + e_0^m p_e}{p}\right) \\
e_m^* = \frac{p_e}{k_m} \\
w^* = pq^* f(q^*) + c_s + e_0^s p_e \\
e_s^* = \frac{p_e}{k_s} \\
\text{Emin}\left\{p\min[q^*, X], \left[(c_m + w^*)q^* + \frac{1}{2}k_m(e_m^*)^2 + (e_0^m q^* - e_m^* - G_m)p_e\right](1 + r^*)\right\} \\
= (c_m + w^*)q^* + \frac{1}{2}k_m(e_m^*)^2 + (e_0^m q^* - e_m^* - G_m)p_e
\end{cases}
\tag{15}
$$

Proof. See Appendix A. □

From Equation (15), it is obvious that both the carbon abatement decisions of the supplier and manufacturer are decoupled from their prospective operational decisions (output and pricing). By some transformations, we have $p\overline{F}(q^*) = c_m + c_s + e_0^s p_e + e_0^m p_e + pq^* f(q^*)$, where the term $pq^* f(q^*)$ corresponds to the efficiency loss of the supply chain attributing to the double marginalization.

4.3. The Case with Cost-Sharing

In this part, we consider the case where the supplier (manufacturer) shares a proportion of carbon abatement investment, i.e., β (α), for the manufacturer (supplier). Under such a two-way cost-sharing contract, the manufacturer's decision-making problem becomes

$$
\begin{aligned}
&Max\pi_m(q, e_m; w, e_s, r) = E\left\{p\min[q, X] - \left[(c_m + w)q + \frac{1}{2}k_m(1 - \beta)(e_m)^2 + \frac{1}{2}k_s\alpha(e_s)^2 + (e_0^m q - e_m - G_m)p_e\right](1 + r)\right\}^+ \\
&s.t.\text{Emin}\left\{p\min[q, X], \left[(c_m + w)q + \frac{1}{2}k_m(1 - \beta)(e_m)^2 + \frac{1}{2}k_s\alpha(e_s)^2 + (e_0^m q - e_m - G_m)p_e\right](1 + r)\right\} \\
&= (c_m + w)q + \frac{1}{2}k_m(1 - \beta)(e_m)^2 + \frac{1}{2}k_s\alpha(e_s)^2 + (e_0^m q - e_m - G_m)p_e
\end{aligned}
\tag{16}
$$

Lemma 7. *In the case of two-way cost-sharing of carbon abatement, the manufacturer's optimal carbon abatement and production decisions are*

$$
\begin{cases}
e_m^{**} = \frac{p_e}{(1 - \beta)k_m} \\
q^{**} = \overline{F}^{-1}\left(\frac{c_m + w + e_0^m p_e}{p}\right)
\end{cases}
\tag{17}
$$

Proof. See Appendix A. □

By manipulating the manufacturer's best response, as the leader, the supplier decides its optimal wholesale price and carbon abatement level for producing the semifinished products to maximize its expected profit:

$$
Max\pi_s(w, e_s; q^*, e_m^*) = (w - c_s)q^* - \frac{1}{2}(1 - \alpha)k_s(e_s)^2 - \frac{1}{2}\beta k_m(e_m^*)^2 - (e_0^s q^* - e_s - G_s)p_e
\tag{18}
$$

Lemma 8. *In the case of two-way cost-sharing of carbon abatement, the supplier's optimal wholesale price quotation and carbon abatement level are*

$$
\begin{cases}
w^{**} = pq^{**} f(q^{**}) + c_s + e_0^s p_e \\
e_s^{**} = \frac{p_e}{(1 - \alpha)k_s}
\end{cases}
\tag{19}
$$

Proof. See Appendix A. □

Then, the competitively priced bank chooses an interest rate based on the constraint condition in Equation (16). The equilibrium of the supply chain financing game is yielded as follows:

Proposition 5. *In the case of two-way cost-sharing of carbon abatement, the equilibrium of the supply chain financing system, i.e.,* $(q^{**}, e_m^{**}, w^{**}, e_s^{**}, r^{**})$ *solves the following:*

$$\begin{cases} q^{**} = \overline{F}^{-1}\left(\frac{c_m + w^{**} + e_0^m p_e}{p}\right) \\ e_m^{**} = \frac{p_e}{(1-\beta)k_m} \\ w^{**} = pq^{**}f(q^{**}) + c_s + e_0^s p_e \\ e_s^{**} = \frac{p_e}{(1-\alpha)k_s} \\ \mathrm{Emin}\left[p\min[q^{**}, X], \left[(c_m + w^{**})q^{**} + \frac{1}{2}k_m(1-\beta)(e_m^{**})^2 + \frac{1}{2}k_s\alpha(e_s^{**})^2 + (e_0^m q^{**} - e_m^{**} - G_m)p_e\right](1 + r^{**})\right] \\ = (c_m + w^{**})q^{**} + \frac{1}{2}k_m(1-\beta)(e_m^{**})^2 + \frac{1}{2}k_s\alpha(e_s^{**})^2 + (e_0^m q^{**} - e_m^{**} - G_m)p_e \end{cases} \quad (20)$$

Proof. See Appendix A. □

4.4. The Value of Cost-Sharing

Comparing the two equilibriums in Equations (15) and (20), it can be easily found that the two-way cost-sharing contract does not affect the quantity or pricing decisions of the manufacturer or the supplier but undoubtedly increases the carbon abatement levels of both parties, as summarized in Corollary 2.

Corollary 2. *The two-way cost-sharing contract for carbon abatement has no influence on the manufacturer's output or the supplier's wholesale price decisions; however, it will undoubtedly improve the carbon abatement levels of both parties, i.e.,* $q^{**} = q^*$, $w^{**} = w^*$, $e_m^{**} > e_m^*$, $e_s^{**} > e_s^*$.

Proof. See Appendix A. □

We then examine the joint "economic-environmental" benefit of the contract. The following Proposition 6 summarizes the results:

Proposition 6. *The two-way cost-sharing contract for carbon abatement can*

(a)   *Reduce the profits of the overall supply chain;*
(b)   *Enhance the joint "economic-environmental" benefit only when the proportions of the cost-sharing meet* $(1-\alpha)^2(3\beta - 2\beta^2)k_s < (1-\beta)^2(2\alpha - 3\alpha^2)k_m$.

Proof. See Appendix A. □

Based on Proposition 6, the two-way cost-sharing contract for carbon abatement will hamper the profit of the overall supply chain, but it can bring positive environmental externalities, and the joint "economic-environmental" benefit will be enhanced if the parameters of the contract are properly chosen.

5. Extension for the Case with Green Consumers

5.1. Game Equilibrium

Since consumers have low-carbon preferences in certain markets, in this part, we extend the problem to meet an emission-dependent demand scenario. Taking the more complicated two-way cost-sharing case in Section 4 as an example, according to the conventions in previous studies [26,28,33], the emission-dependent demand can be described as $\hat{X} = X + a_s e_s + a_m e_m$, where X is the initial demand, e_s and e_m are the carbon abatement levels of the supplier and the capital-constrained manufacturer, and a_s and a_m are two coefficients measuring the elasticity of demand for carbon abatement. In this case, the decision-making problems of the manufacturer and supplier without the two-way cost-sharing contract in Equations (11) and (13) can be rewritten as Equations (21) and (22). To distinguish the optimal results, we added a "-" to the top of the new solutions.

$$
\begin{aligned}
Max\pi_m(q, e_m; w, e_s, r) &= E\left\{ pmin[q, \hat{X}] - \left[(c_m + w)q + \tfrac{1}{2}k_m(e_m)^2 + (e_0^m q - e_m - G_m)p_e \right](1 + r) \right\}^+ \\
s.t. Emin&\left\{ pmin[q, \hat{X}], \left[(c_m + w)q + \tfrac{1}{2}k_m(e_m)^2 + (e_0^m q - e_m - G_m)p_e \right](1 + r) \right\} \\
&= (c_m + w)q + \tfrac{1}{2}k_m(e_m)^2 + (e_0^m q - e_m - G_m)p_e
\end{aligned}
\tag{21}
$$

$$
Max\pi_s(w, e_s; \overline{q}^*) = (w - c_s)\overline{q}^* - \frac{1}{2}k_s(e_s)^2 - (e_0^s \overline{q}^* - e_s - G_s)p_e
\tag{22}
$$

Proposition 7. *In the case of green consumers, the equilibrium of the supply chain financing system, i.e.,* $(\overline{q}^*, \overline{e}_m^*, \overline{w}^*, \overline{e}_s^*, \overline{r}^*)$ *solves the following:*

$$
\begin{cases}
\overline{q}^* = \overline{F}^{-1}\left(\frac{c_m + \overline{w}^* + e_0^m p_e}{p} \right) + a_m \overline{e}_m^* + a_s \overline{e}_s^* \\
\overline{e}_m^* = \frac{a_m(p - c_m - \overline{w}^* - e_0^m p_e) + p_e}{k_m} \\
\overline{w}^* = \frac{p\overline{q}^* k_m f(\overline{q}^* - a_m \overline{e}_m^* - a_s \overline{e}_s^*)}{a_m^2 p f(\overline{q}^* - a_m \overline{e}_m^* - a_s \overline{e}_s^*) + k_m} + c_s + e_0^s p_e \\
\overline{e}_s^* = \frac{a_s(\overline{w}^* - c_s - e_0^s p_e) + p_e}{k_s} \\
Emin\left[pmin[\overline{q}^*, X + a_m \overline{e}_m^* + a_s \overline{e}_s^*], \left[(c_m + \overline{w}^*)\overline{q}^* + \tfrac{1}{2}k_m(\overline{e}_m^*)^2 + (e_0^m \overline{q}^* - \overline{e}_m^* - G_m)p_e \right](1 + \overline{r}^*) \right] \\
\quad = (c_m + \overline{w}^*)\overline{q}^* + \tfrac{1}{2}k_m(\overline{e}_m^*)^2 + (e_0^m \overline{q}^* - \overline{e}_m^* - G_m)p_e
\end{cases}
\tag{23}
$$

Proof. See Appendix A. □

Comparing the results in Equations (23) and (15), it can be found that $\overline{e}_m^* > e_m^*$ and $\overline{e}_s^* > e_s^*$. This means that the green consumers in the market can undoubtedly stimulate the carbon abatement investments of both the manufacturer and supplier. When further considering two-way cost-sharing of carbon abatement, the decision-making problems of both parties are changed into:

$$
\begin{aligned}
Max\pi_m(q, e_m; w, e_s, r) &= E\left\{ pmin[q, \hat{X}] - \left[(c_m + w)q + \tfrac{1}{2}k_m(1 - \beta)(e_m)^2 + \tfrac{1}{2}k_s\alpha(e_s)^2 + (e_0^m q - e_m - G_m)p_e \right](1 + r) \right\}^+ \\
s.t. Emin&\left\{ pmin[q, \hat{X}], \left[(c_m + w)q + \tfrac{1}{2}k_m(1 - \beta)(e_m)^2 + \tfrac{1}{2}k_s\alpha(e_s)^2 + (e_0^m q - e_m - G_m)p_e \right](1 + r) \right\} \\
&= (c_m + w)q + \tfrac{1}{2}k_m(1 - \beta)(e_m)^2 + \tfrac{1}{2}k_s\alpha(e_s)^2 + (e_0^m q - e_m - G_m)p_e
\end{aligned}
\tag{24}
$$

$$
Max\pi_s(w, e_s; \overline{q}^{**}, \overline{e}_m^{**}) = (w - c_s)\overline{q}^{**} - \frac{1}{2}(1 - \alpha)k_s(e_s)^2 - \frac{1}{2}\beta k_m(\overline{e}_m^{**})^2 - (e_0^s \overline{q}^{**} - e_s - G_s)p_e
\tag{25}
$$

Proposition 8. *In the case of green consumers and two-way cost-sharing, the equilibrium of the supply chain financing system, i.e.,* $(\overline{q}^{**}, \overline{e}_m^{**}, \overline{w}^{**}, \overline{e}_s^{**}, \overline{r}^{**})$ *solves the following:*

$$
\begin{cases}
\overline{q}^{**} = \overline{F}^{-1}\left(\frac{c_m + \overline{w}^{**} + e_0^m p_e}{p} \right) + a_m \overline{e}_m^{**} + a_s \overline{e}_s^{**} \\
\overline{e}_m^{**} = \frac{a_m(p - c_m - \overline{w}^{**} - e_0^m p_e) + p_e}{k_m(1 - \beta)} \\
\overline{w}^{**} = \frac{pk_m[\overline{q}^{**}(1 - \beta) + a_m \beta \overline{e}_m^{**}]f(\overline{q}^{**} - a_m \overline{e}_m^{**} - a_s \overline{e}_s^{**})}{k_m(1 - \beta) + a_m^2 p f(\overline{q}^{**} - a_m \overline{e}_m^{**} - a_s \overline{e}_s^{**})} + c_s + e_0^s p_e \\
\overline{e}_s^{**} = \frac{a_s(\overline{w}^{**} - c_s - e_0^s p_e) + p_e}{k_s(1 - \alpha)} \\
Emin\left[pmin[\overline{q}^{**}, X + a_m \overline{e}_m^{**} + a_s \overline{e}_s^{**}], \left[(c_m + \overline{w}^{**})\overline{q}^{**} + \tfrac{1}{2}k_m(1 - \beta)(\overline{e}_m^{**})^2 + \tfrac{1}{2}k_s\alpha(\overline{e}_s^{**})^2 + (e_0^m \overline{q}^{**} - \overline{e}_m^{**} - G_m)p_e \right](1 + \overline{r}^{**}) \right] \\
\quad = (c_m + \overline{w}^{**})\overline{q}^{**} + \tfrac{1}{2}k_m(1 - \beta)(\overline{e}_m^{**})^2 + \tfrac{1}{2}k_s\alpha(\overline{e}_s^{**})^2 + (e_0^m \overline{q}^{**} - \overline{e}_m^{**} - G_m)p_e
\end{cases}
\tag{26}
$$

Proof. See Appendix A. □

Comparing Equations (26) and (23), there is no doubt that when considering consumers' green preferences, introducing the two-way cost-sharing contract can also enhance the carbon abatement levels of both the manufacturer and the supplier.

5.2. Numerical Analysis

In this part, we use the method of numerical analysis to intuitively show the properties of the equilibrium obtained in Proposition 8 as well as to explore how the two-way cost-sharing contract affects the benefit of the whole supply chain. Without loss of generality, we assume $p = 1.0$, $c_m = c_s = e_0^m = e_0^s = 0.1$, $k_m = k_s = a_s = a_m = 1.0$, $p_e = 0.5$, $G_s = G_m = 5.0$, and the initial market demand follows $X \sim \exp(0.01)$ [34–36]. When two-way cost-sharing of carbon abatement does not exist, the initial equilibrium of the game is $\overline{q}^* = 51.13$, $\overline{e}_m^* = 0.89$, $\overline{w}^* = 0.46$, $\overline{e}_s^* = 0.81$. When cost-sharing exists, how the equilibrium changes according to the two sharing coefficients, i.e., α and β, is shown in Figure 1. It can be observed that as the supplier increases the proportion of cost-sharing for the manufacturer, i.e., β rises, the manufacturer would like to abate more emissions, i.e., $\overline{e}_m^*$ increases. The influence of α on the supplier's carbon abatement level has the same rule. However, offering cost-sharing for the manufacturer will increase the cost of the supplier, which leads to a rise in the supplier's wholesale price $\overline{w}^*$. The change in the wholesale price finally decreases the output of the manufacturer $\overline{q}^*$.

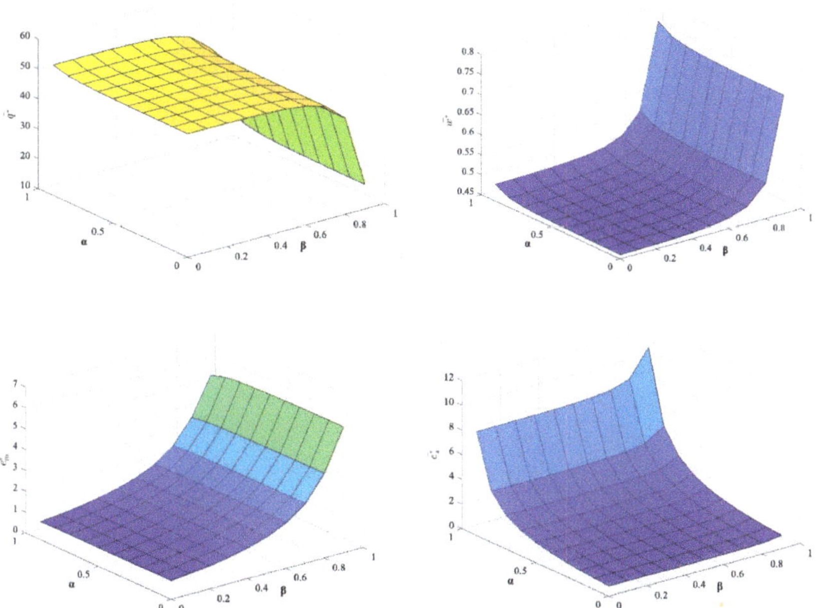

Figure 1. Changes of the equilibrium with two sharing coefficients.

We further explored the effects of two-way cost-sharing on carbon abatement on the profits of the supplier and the manufacturer, as shown in Figure 2. It can be observed that as the two sharing coefficients, i.e., α and β, rise, the profits of both parties will decline. Specifically, the decline of one party's profit is more sensitive to the cost-sharing proportion that it offers to the other party. Overall, the introduction of the two-way cost-sharing contract on carbon abatement will hamper the overall profit of the supply chain, as shown in Figure 3a. However, it undoubtedly promotes the carbon abatement of both parties, and when further considering the value of positive environmental externalities, the contract can enhance the joint "economic-environmental" benefit of the supply chain as long as the two sharing coefficients fall within some favorable intervals, i.e., $\alpha \in (0, 0.46)$ and $\beta \in (0, 0.07)$ in the example, as shown in Figure 3b.

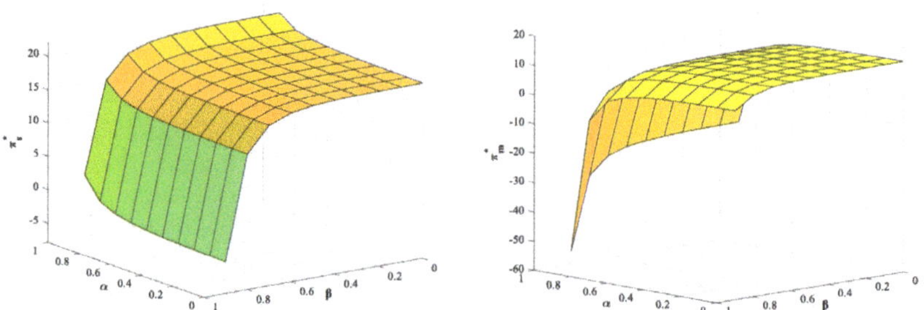

Figure 2. Profit changes of the supply chain members as two sharing coefficients varying.

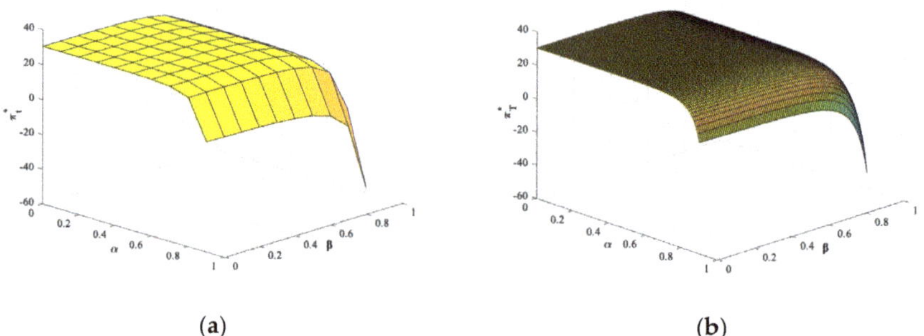

(a)   (b)

Figure 3. Benefit of the two-way cost-sharing contract on the supply chain. (a) Economic benefit and (b) joint "economic-environmental" benefit.

6. Conclusions

This paper studies the role of carbon-abatement cost-sharing contracts in supply chains with a capital-constrained manufacturer. The manufacturer obtains financing from a competitively priced bank to fulfill its production, carbon-abatement investment, and even insufficient emission permits purchase given that the cap-and-trade regulation exists. Firstly, a one-way cost-sharing contract was investigated in the scenario where the manufacturer has a dominant downstream retailer, and the retailer shares a proportion of the carbon-abatement cost for the manufacturer. The results show that the one-way cost-sharing contract has no effect on the manufacturer's output or the retailer's wholesale price quotation decisions; however, it improves the carbon abatement level of the manufacturer. As a result, the contract hampers the profit of the retailer as well as the overall supply chain but enhances the joint "economic-environmental" benefit of the supply chain in some circumstances. Then, a two-way cost-sharing contract was investigated in the scenario where the manufacturer has a dominant supplier, and they share a proportion of the carbon-abatement cost with each other. The results also show the potential "economic-environmental" benefit of the two-way cost-sharing contract if the parameter of the contract is properly chosen. Finally, the problem was studied in the case of consumers' green preferences, and numerical analysis also indicates similar roles as described above.

The managerial insights of this research mainly lie in how upstream and downstream larger firms should share carbon abatement costs with capital-constrained manufacturers since, from the perspective of low-carbon supply chain management, such approaches help reduce emissions of the supply chain and achieve satisfactory joint "economic-environmental" benefits. This is also a manifestation of the social responsibility of large enterprises in the supply chain. However, such cost-sharing measures may be disadvantageous to upstream and downstream enterprises in terms of purely economic benefits, which requires further discussions of multiple contracts within the supply chain and external government incentives in future research.

Author Contributions: Conceptualization, J.S.; methodology, J.S.; writing—original draft preparation, W.J.; supervision, Q.Y., K.K.L. and K.J. All authors have read and agreed to the published version of the manuscript.

Funding: This work was supported by the National Natural Science Foundation of China [grant number 72202021], the Postdoctoral Science Foundation of China [grant number 2019M663605], the MOE (Ministry of Education in China) Project of Humanities and Social Sciences for Young Scholars [grant number 20YJC790115], and the Fundamental Research Funds for the Central Universities, CHD [grant number 300102231674].

Data Availability Statement: Not applicable.

Conflicts of Interest: The authors have no relevant financial or nonfinancial interest to disclose.

Notations

Summary of notations in the one-way cost-sharing case:

e (Decision Variable)	Manufacturer's carbon abatement level
q (Decision Variable)	Manufacturer's production quantity
w (Decision Variable)	Retailer's unit wholesale price quotation
c	Manufacturer's unit production cost
p	Retailer's unit selling price
r (Decision Variable)	Bank's loan interest rate ($0 \leq r \leq 1$)
X	Random market demand
$f(x)$, $F(x)$	PDF and CDF of X respectively
$h(x)$, $H(x)$	Failure rate and the generalized failure rate of X
k	Coefficient of carbon abatement investment cost
e_0	Initial carbon emission of the manufacturer for unit production
p_e	Unit carbon trading price
G	Free carbon emission quota from the government
β	Cost-sharing rate of carbon abatement investment shared by the retailer for the manufacturer ($0 < \beta < 1$)
The variables with superscript "*" and "**"	Optimal solutions of the decision variables in the cases without and with the cost-sharing contract, respectively
π_m, π_r	Profits of the manufacturer and the retailer, respectively

Appendix A

Proof of Lemma 1. For the case without cost-sharing, based on $(a - b)^+ = a - \min[a, b]$, Equation (1) can be converted into

$$Max\,\pi_m(e, q; w) = wE\min[X, q] - cq - \tfrac{1}{2}k(e)^2 - (e_0q - e - G)p_e$$
$$= \int_0^q wxf(x)dx + wq\overline{F}(q) - cq - \tfrac{1}{2}k(e)^2 - (e_0q - e - G)p_e \tag{A1}$$

Then, the first- and second-order derivatives of π_m with respect to q and e are, respectively, as follows:

$$\frac{\partial \pi_m}{\partial q} = w\overline{F}(q) - c - e_0p_e, \quad \frac{\partial \pi_m}{\partial e} = -ke + p_e$$
$$\frac{\partial^2 \pi_m}{\partial(q)^2} = -wf(q) < 0, \quad \frac{\partial^2 \pi_m}{\partial(e)^2} = -k < 0 \tag{A2}$$

Obviously, Equation (A1) is concave. Then, we can derive the optimal solution of the problem from the first-order conditions, as shown in Equation (2). $\square$

Proof of Lemma 2. For the retailer's problem, Equation (3) can be converted into

$$Max\pi_r(w;q^*) = (p-w)\left[\int_0^{q^*} xf(x)dx + q^*\overline{F}(q^*)\right] \tag{A3}$$

Taking the total derivative of π_r with respect to w, we have

$$\begin{aligned}
\frac{d\pi_r}{dw} &= \frac{\partial \pi_r}{\partial w} + \frac{d\pi_r}{dq^*}\frac{dq^*}{dw} \\
&= \int_0^{q^*} F(x)dx - q^* + (p-w)\overline{F}(q^*)\frac{1}{wh(q^*)} \\
&= \int_0^{q^*} F(x)dx - q^* + \frac{p\overline{F}(q^*)-c-e_0 p_e}{wh(q^*)}
\end{aligned} \tag{A4}$$

Letting Equation (A4) be zero, we have w^*, as shown in Equation (4). $\square$

Proof of Proposition 1. The conclusions are obvious, so we omit the proof. $\square$

Proof of Lemma 3. Similar to the method in the proof of Lemma 1, Equation (6) can be converted into

$$Max\pi_m(e,q;w) = \int_0^q wxf(x)dx + wq\overline{F}(q) - cq - \frac{1}{2}k(1-\beta)e^2 - (e_0 q - e - G)p_e \tag{A5}$$

Then, the first- and second-order derivatives of π_m with respect to q and e are, respectively, as follows:

$$\begin{aligned}
&\frac{\partial \pi_m}{\partial q} = w\overline{F}(q) - c - e_0 p_e, \quad \frac{\partial \pi_m}{\partial e} = -k(1-\beta)e + p_e \\
&\frac{\partial^2 \pi_m}{\partial q^2} = -wf(q) < 0, \quad \frac{\partial^2 \pi_m}{\partial e^2} = -k(1-\beta) < 0
\end{aligned} \tag{A6}$$

It is obvious that Equation (A5) is concave. Then, we can derive the optimal solutions, as shown in Equation (7). $\square$

Proof of Lemma 4. This part of the proof is similar to Lemma 2, so we omit the proof. $\square$

Proof of Proposition 2. The conclusions are obvious, so we omit the proof. $\square$

Proof of Corollary 1. From the conclusions in Propositions 1 and 2, the results in Corollary 1 can be directly drawn. $\square$

Proof of Proposition 3. For part (a): Based on the conclusions in Corollary 1, recall the equilibriums in both cases. The differences in expected profits of the manufacturer and retailer after introducing the contract are

$$\begin{aligned}
\triangle \pi_m &= \frac{p_e^2}{2k(1-\beta)} - \frac{p_e^2}{2k} > 0 \\
\triangle \pi_r &= -\frac{\beta p_e^2}{2k(1-\beta)^2} < 0
\end{aligned} \tag{A7}$$

Then, the difference in the profit of the overall supply chain is

$$\triangle \pi_{SC} = \triangle \pi_m + \triangle \pi_r = -\frac{\beta^2 p_e^2}{2k(1-\beta)^2} < 0 \tag{A8}$$

For part (b): From Corollary 1, the one-way cost-sharing contract has a positive environmental externality. By assuming the environmental benefit equals the value of the reduced carbon emissions in the cap-and-trade market, we have

$$\triangle \pi_e = (e^{**} - e^*)p_e = \frac{\beta p_e^2}{k(1-\beta)} \tag{A9}$$

To ensure that $\triangle \pi_{SC} + \triangle \pi_e > 0$, we have $0 < \beta < \frac{2}{3}$, under which, the cost-sharing contract can enhance the joint "economic-environmental" benefit. $\square$

Proof of Lemma 5. For the case without two-way cost-sharing, based on $(a - b)^+ = a - \min[a, b]$, Equation (11) can be converted into

$$Max\pi_m(q, e_m; w) = p \int_0^q x f(x) dx + pq\overline{F}(q) - (c_m + w)q - \frac{1}{2}k_m(e_m)^2 - (e_0^m q - e_m - G_m)p_e \quad (A10)$$

Then, the first- and second-order derivatives of π_m with respect to q and e_m are, respectively, as follows:

$$\begin{aligned} \frac{\partial \pi_m}{\partial q} &= p\overline{F}(q) - (c_m + w) - e_0^m p_e, \frac{\partial \pi_m}{\partial e_m} = -k_m e_m + p_e \\ \frac{\partial^2 \pi_m}{\partial (q)^2} &= -pf(q) < 0, \frac{\partial^2 \pi_m}{\partial (e_m)^2} = -k_m < 0 \end{aligned} \quad (A11)$$

Obviously, Equation (11) is concave. Then, we can derive the optimal solution for the problem from the first-order conditions, as shown in Equation (12). $\square$

Proof of Lemma 6. Taking the total derivative of π_s with respect to w and e_s, we have

$$\begin{aligned} \frac{d\pi_s}{dw} &= \frac{\partial \pi_s}{\partial w} + \frac{d\pi_s}{dq^*}\frac{dq^*}{dw} + \frac{d\pi_s}{de_m^*}\frac{de_m^*}{dw} \\ &= q - (w - c_s - e_0^s p_e)\frac{1}{pf(q)} \end{aligned} \quad (A12)$$

$$\begin{aligned} \frac{d\pi_s}{de_s} &= \frac{\partial \pi_s}{\partial e_s} + \frac{d\pi_s}{dq^*}\frac{dq^*}{de_s} + \frac{d\pi_s}{de_m^*}\frac{de_m^*}{de_s} \\ &= -k_s e_s + p_e \end{aligned} \quad (A13)$$

Letting Equations (A12) and (A13) be zero, we have w^* and e_s^*, as shown in Equation (14). $\square$

Proof of Proposition 4. The conclusions are obvious, so we omit the proof. $\square$

Proof of Lemma 7. Similar to the Proof of Lemma 5, the first- and second-order derivatives of π_m in Equation (16) with respect to q and e_m are, respectively, as follows:

$$\begin{aligned} \frac{\partial \pi_m}{\partial q} &= p\overline{F}(q) - (c_m + w_2) - e_0^m p_e, \frac{\partial \pi_m}{\partial e_m} = -(1 - \beta)k_m e_m + p_e \\ \frac{\partial^2 \pi_m}{\partial q^2} &= -pf(q) < 0, \frac{\partial^2 \pi_m}{\partial (e_m)^2} = -k_m < 0 \end{aligned} \quad (A14)$$

The problem is concave, and we can derive the optimal solutions from the first-order conditions, as shown in Equation (17). $\square$

Proof of Lemma 8. This part of the proof is similar to Lemma 6, so we omit the proof. $\square$

Proof of Proposition 5. The conclusions are obvious, so we omit the proof. $\square$

Proof of Corollary 2. From the conclusions in Propositions 4 and 5, the results in Corollary 2 can be directly drawn. $\square$

Proof of Proposition 6. For part (a): Based on the conclusions in Corollary 2, recall the equilibriums in both cases, the differences in the expected profits of the manufacturer and supplier after introducing the contract are

$$\begin{aligned} \triangle \pi_m &= \frac{\beta p_e^2}{2k_m(1-\beta)} - \frac{\alpha p_e^2}{2k_s(1-\alpha)^2} \\ \triangle \pi_s &= \frac{\alpha p_e^2}{2k_s(1-\alpha)} - \frac{\beta p_e^2}{2k_m(1-\beta)^2} \end{aligned} \quad (A15)$$

Thus, the difference in the expected profit of the overall supply chain is

$$\triangle \pi_{SC} = \triangle \pi_m + \triangle \pi_s = -\frac{\beta^2 p_e^2 (1-\alpha)^2 k_s + \alpha^2 p_e^2 (1-\beta)^2 k_m}{2k_m k_s (1-\beta)^2 (1-\alpha)^2} < 0 \qquad (A16)$$

For part (b): From Corollary 2, since the two-way cost-sharing contract has a positive environmental externality, by assuming the environmental benefit equals the value of the reduced carbon emissions in the cap-and-trade market, we have

$$\triangle \pi_e = (\triangle e_m + \triangle e_s) p_e = \frac{2\alpha(1-\alpha)(1-\beta)^2 k_m p_e^2 + 2\beta(1-\beta)(1-\alpha)^2 k_s p_e^2}{2k_s k_m (1-\alpha)^2 (1-\beta)^2} \qquad (A17)$$

Letting $\triangle \pi_{SC} + \triangle \pi_e > 0$, we have $(1-\alpha)^2 (3\beta - 2\beta^2) k_s < (1-\beta)^2 (2\alpha - 3\alpha^2) k_m$, under which, the two-way cost-sharing contract can enhance the joint "economic-environmental" benefit. $\square$

Proof of Proposition 7. For the case without two-way cost-sharing, based on $(a-b)^+ = a - \min[a, b]$, Equation (21) can be converted into

$$Max\pi_m(q, e_m; w, e_s, r) = p[q - \int_0^{q-a_m e_m - a_s e_s} F(x)dx] - (c_m + w)q - \frac{1}{2}k_m(e_m)^2 - (e_0^m q - e_m - G_m)p_e \qquad (A18)$$

Then, the first- and second-order derivatives of π_m with respect to q and e_m are, respectively, as follows:

$$\frac{\partial \pi_m}{\partial q} = p\overline{F}(q - a_m e_m - a_s e_s) - (c_m + w) - e_0^m p_e, \frac{\partial \pi_m}{\partial e_m} = pF(q - a_m e_m - a_s e_s) - k_m e_m + p_e$$
$$\frac{\partial^2 \pi_m}{\partial (q)^2} = -pf(q - a_m e_m - a_s e_s) < 0, \frac{\partial^2 \pi_m}{\partial (e_m)^2} = -a^2 pf(q - a_m e_m - a_s e_s) - k_m < 0 \qquad (A19)$$

Then, the first- and second-order derivatives of π_s with respect to w and e_s are, respectively, as follows:

$$\frac{d\pi_s}{dw} = \frac{\partial \pi_s}{\partial w} + \frac{d\pi_s}{dq^*}\frac{dq^*}{dw} + \frac{d\pi_s}{de_m^*}\frac{de_m^*}{dw}$$
$$= q - (w - c_s - e_0^s p_e)\left(\frac{1}{pf(q-a_m e_m^* - a_s e_s)} + \frac{a^2}{k_m}\right) \qquad (A20)$$

$$\frac{d\pi_s}{de_s} = \frac{\partial \pi_s}{\partial e_s} + \frac{d\pi_s}{dq^*}\frac{dq^*}{de_s} + \frac{d\pi_s}{de_m^*}\frac{de_m^*}{de_s}$$
$$= -k_s e_s + p_e + (w - c_s - e_s^0 p_e)a_s \qquad (A21)$$

Letting Equations (A20) and (A21) be zero, we have $\overline{q}^*, \overline{e}_m^*, \overline{w}^*, \overline{e}_s^*, \overline{r}^*$, as shown in Equation (23). $\square$

Proof of Proposition 8. This part of the proof is similar to Proposition 7, so we omit the proof. $\square$

References

1. Yang, H.; Chen, W. Retailer-driven carbon emission abatement with consumer environmental awareness and carbon tax: Revenue-sharing versus cost-sharing. *Omega* **2018**, *78*, 179–191. [CrossRef]
2. Cao, E.; Yu, M. The bright side of carbon emission permits on supply chain financing and performance. *Omega* **2019**, *88*, 24–39. [CrossRef]
3. An, S.; Li, B.; Song, D.; Chen, X. Green credit financing versus trade credit financing in a supply chain with carbon emission limits. *Eur. J. Oper. Res.* **2021**, *292*, 125–142. [CrossRef]
4. De, A.; Gorton, M.; Hubbard, C.; Aditjandra, P. Optimization model for sustainable food supply chains: An application to Norwegian salmon. *Transp. Res. Part E Logist. Transp. Rev.* **2022**, *161*, 102723. [CrossRef]
5. Mirzaee, H.; Samarghandi, H.; Willoughby, K. A three-player game theory model for carbon cap-and-trade mechanism with stochastic parameters. *Comput. Ind. Eng.* **2022**, *169*, 108285. [CrossRef]

6.   Zeng, Y.; Cai, X.; Feng, H. Cost and emission allocation for joint replenishment systems subject to carbon constraints. *Comput. Ind. Eng.* **2022**, *168*, 108074. [CrossRef]

7.   Ji, T.; Xu, X.; Yan, X.; Yu, Y. The production decisions and cap setting with wholesale price and revenue sharing contracts under cap-and-trade regulation. *Int. J. Prod. Res.* **2020**, *58*, 128–147. [CrossRef]

8.   Yang, Y.; Goodarzi, S.; Bozorgi, A.; Fahimnia, B. Carbon cap-and-trade schemes in closed-loop supply chains: Why firms do not comply? *Transp. Res. Part E Logist. Transp. Rev.* **2021**, *156*, 102486. [CrossRef]

9.   Wang, Y.; Yu, Z.; Jin, M.; Mao, J. Decisions and coordination of retailer-led low-carbon supply chain under altruistic preference. *Eur. J. Oper. Res.* **2021**, *293*, 910–925. [CrossRef]

10.   Zhang, Q.; Zheng, Y. Pricing strategies for bundled products considering consumers' green preferences. *J. Clean. Prod.* **2022**, *344*, 130962. [CrossRef]

11.   Reza-Gharehbagh, R.; Arisian, S.; Hafezalkotob, A.; Makui, A. Sustainable supply chain finance through digital platforms: A pathway to green entrepreneurship. *Ann. Oper. Res.* **2022**, 1–35. [CrossRef]

12.   Zou, T.; Zou, Q.; Hu, L. Joint decision of financing and ordering in an emission-dependent supply chain with yield uncertainty. *Comput. Ind. Eng.* **2021**, *152*, 106994. [CrossRef]

13.   Agi, M.A.N.; Faramarzi-Oghani, S.; Hazır, Ö. Game theory-based models in green supply chain management: A review of the literature. *Int. J. Prod. Res.* **2021**, *59*, 4736–4755. [CrossRef]

14.   Wang, X.; Sethi, S.P.; Chang, S. Pollution abatement using cap-and-trade in a dynamic supply chain and its coordination. *Transp. Res. Part E Logist. Transp. Rev.* **2022**, *158*, 102592. [CrossRef]

15.   Cheng, F.; Chen, T.; Chen, Q. Cost-reducing strategy or emission-reducing strategy? The choice of low-carbon decisions under price threshold subsidy. *Transp. Res. Part E Logist. Transp. Rev.* **2022**, *157*, 102560. [CrossRef]

16.   Guo, Y.; Wang, M.; Yang, F. Joint emission reduction strategy considering channel inconvenience under different recycling structures. *Comput. Ind. Eng.* **2022**, *169*, 108159. [CrossRef]

17.   Xu, X.; Choi, T.M. Supply chain operations with online platforms under the cap-and-trade regulation: Impacts of using blockchain technology. *Transp. Res. Part E Logist. Transp. Rev.* **2021**, *155*, 102491. [CrossRef]

18.   Entezaminia, A.; Gharbi, A.; Ouhimmou, M. A joint production and carbon trading policy for unreliable manufacturing systems under cap-and-trade regulation. *J. Clean. Prod.* **2021**, *293*, 125973. [CrossRef]

19.   Dou, G.W.; Guo, H.N.; Zhang, Q.Y.; Li, X.D. A two-period carbon tax regulation for manufacturing and remanufacturing production planning. *Comput. Ind. Eng.* **2019**, *128*, 502–513. [CrossRef]

20.   Bian, J.; Guo, X. Policy analysis for emission-reduction with green technology investment in manufacturing. *Ann. Oper. Res.* **2021**, *316*, 5–32. [CrossRef]

21.   Yu, W.; Wang, Y.; Feng, W.; Bao, L.; Han, R. Low carbon strategy analysis with two competing supply chain considering carbon taxation. *Comput. Ind. Eng.* **2022**, *169*, 108203. [CrossRef]

22.   Xu, X.; Zhang, W.; He, P.; Xu, X. Production and pricing problems in make-to-order supply chain with cap-and-trade regulation. *Omega* **2017**, *66*, 248–257. [CrossRef]

23.   Bai, Q.; Xu, J.; Zhang, Y. Emission reduction decision and coordination of a make-to-order supply chain with two products under cap-and-trade regulation. *Comput. Ind. Eng.* **2018**, *119*, 131–145. [CrossRef]

24.   Qin, J.; Han, Y.; Wei, G.; Xia, L. The value of advance payment financing to carbon emission reduction and production in a supply chain with game theory analysis. *Int. J. Prod. Res.* **2020**, *58*, 200–219. [CrossRef]

25.   Moon, I.; Jeong, Y.J.; Saha, S. Investment and coordination decisions in a supply chain of fresh agricultural products. *Oper. Res.* **2020**, *20*, 2307–2331. [CrossRef]

26.   Wu, D.D.; Yang, L.; Olson, D.L. Green supply chain management under capital constraint. *Int. J. Prod. Econ.* **2019**, *215*, 3–10.

27.   Shi, J.; Liu, D.; Du, Q.; Cheng, T.C.E. The role of the procurement commitment contract in a low-carbon supply chain with a capital-constrained supplier. *Int. J. Prod. Econ.* **2023**, *255*, 108681. [CrossRef]

28.   Xu, S.; Fang, L. Partial credit guarantee and trade credit in an emission-dependent supply chain with capital constraint. *Transp. Res. Part E Logist. Transp. Rev.* **2020**, *135*, 101859. [CrossRef]

29.   Huang, S.; Fan, Z.; Wang, N. Green subsidy modes and pricing strategy in a capital-constrained supply chain. *Transp. Res. Part E Logist. Transp. Rev.* **2020**, *136*, 101885. [CrossRef]

30.   Ghosh, D.; Shah, J. Supply chain analysis under green sensitive consumer demand and cost sharing contract. *Int. J. Prod. Econ.* **2015**, *164*, 319–329. [CrossRef]

31.   Wang, Z.; Brownlee, A.E.I.; Wu, Q. Production and joint emission reduction decisions based on two-way cost-sharing contract under cap-and-trade regulation. *Comput. Ind. Eng.* **2020**, *146*, 106549. [CrossRef]

32.   Lai, Z.; Lou, G.; Zhang, T.; Fan, T. Financing and coordination strategies for a manufacturer with limited operating and green innovation capital: Bank credit financing versus supplier green investment. *Ann. Oper. Res.* **2021**, 1–35. [CrossRef]

33.   Cao, E.; Du, L.; Ruan, J. Financing preferences and performance for an emission-dependent supply chain: Supplier vs. *bank. Int. J. Prod. Econ.* **2019**, *208*, 383–399. [CrossRef]

34.   Shi, J.; Du, Q.; Lin, F.; Li, Y.; Bai, L.; Fung, R.Y.K.; Lai, K.K. Coordinating the supply chain finance system with buyback contract: A capital-constrained newsvendor problem. *Comput. Ind. Eng.* **2020**, *146*, 106587. [CrossRef]

35. Shi, J.; Du, Q.; Lin, F.; Lai, K.K.; Cheng, T. Optimal financing mode selection for a capital-constrained retailer under an implicit bankruptcy cost. *Int. J. Prod. Econ.* **2020**, *228*, 107657. [CrossRef]
36. Buzacott, J.A.; Zhang, R.Q. Inventory management with asset-based financing. *Manag. Sci.* **2004**, *50*, 1274–1292. [CrossRef]

Article

Quality Control of Water-Efficient Products Based on DMAIC Improved Mode—A Case Study of Smart Water Closets

Yan Bai [1], Jialin Liu [1,2], Rui Zhang [1] and Xue Bai [1,2,*]

[1]   China National Institute of Standardization, Beijing 100191, China
[2]   Key Laboratory of Energy Efficiency, Water Efficiency and Greenization for State Market Regulation, Beijing 102200, China
*   Correspondence: bai_xue2022@163.com

Abstract: Water-efficient products, a key component of water-saving technology, are widely installed and utilized in all sectors of society. Due to China's extensive and varied use of this product, advancements in effectiveness and quality will significantly enhance people's standard of living. In recent years, manufacturers, corporate purchasers, and individual customers have given more attention to the quality of these items due to the spike in local market and export demands for water-efficient products in China. It has been a pressing problem to find a practical solution for increasing product quality in a reasonable and scientific manner. In order to build a DECIA quality improvement model for water-efficient product quality that is quantifiable and technically practical, this paper investigates how to improve the quality of smart water closets based on six-sigma management. Thus, the development of a water-efficient industry can be green and sustainable.

Keywords: water-efficient products; quality improvement; DMAIC; smart water closets

1. Introduction

Water-efficient products refer to a type of daily-used device that meets certain water conservation requirements, including regular water closets, smart water closets, squat toilets, water nozzles, showers, and water purifiers. China has been the largest producer and consumer of water-efficient products in the world. Since 2010, the industry of water-efficient products has experienced rapid development, with an average annual growth rate of 10%. In 2021, the sales volume of water-efficient products in China reached 500 million units. Water-efficient products are daily necessities, so their quality directly affects water use efficiency, environmental protection, and safety. The Chinese people are better off due to the rapid economic development, which stimulates the demand for water-efficient products. As traditional production methods cannot meet the needs of energy and resource conservation and environmental protection, there is an urgent need for high-quality water-efficient products that are green and low-carbon. There will be opportunities and difficulties for the water-conservation device market as a result of the trend toward utilizing high-quality, water-efficient products and a healthy lifestyle. However, there is little research conducted in this area, especially concerning the theory and technology of water-efficient upgrades, and therefore they are unable to influence the process of product development.

Smart water closets were introduced in China in the 1990s, and their popularity has exploded over the past decade. China is currently the world's largest manufacturer and distributor of smart water closets. As living standards continue to rise, so do consumer demands and requirements for items that use less water. Consumers readily accept and buy smart water closets with humanized and personalized features, such as warm water cleansing, warm air drying, automatic deodorization, music, and health monitoring. These features have made smart water closets an integral part of home furnishing. Roughly 10 million sets of smart water closets were sold in China in 2021, a nearly 130% increase from 2016. Along with the fast spread of smart water closets, quality has been widely

Citation: Bai, Y.; Liu, J.; Zhang, R.; Bai, X. Quality Control of Water-Efficient Products Based on DMAIC Improved Mode—A Case Study of Smart Water Closets. *Processes* **2023**, *11*, 131. https://doi.org/10.3390/pr11010131

Academic Editors: Conghu Liu, Xiaoqian Song, Zhi Liu and Fangfang Wei

Received: 5 December 2022
Revised: 27 December 2022
Accepted: 28 December 2022
Published: 1 January 2023

concerned by consumers. A smart water closet is a high-power wading electrical product that is powered by highly complicated technology that spans a variety of industries and specialties, including water, electricity, heating, machinery, sensing, injection molding, welding, etc. When using, it is common to have issues with controlling the temperature of warm air, water efficiency, input power and current, and grounding procedures. For the steady and quick development of the industry, it is imperative to create a quality model that analyzes the influencing and evaluating variables of the quality of smart water closets over the entire life cycle.

The first section of this paper examines the current research findings on Six Sigma management. The DECIA model of water-efficient product quality improvement, which is based on the DMAIC model in the Six Sigma framework, is studied in the second section. The five dimensions of the model's construction are definition, assessment, inspection, improvement, and application. On the basis of a case study involving smart water closets, the contents of each dimension were then introduced. Finally, a few ideas and actions were put out to raise the standard of water-efficient products in accordance with the DECIA model.

2. Research Status

Motorola was the first company to embrace Six Sigma, and it has since gained widespread acceptance throughout the world, particularly in the manufacturing, service, government, and other institutions or enterprise management sectors. It has developed into a sophisticated management style that offers businesses significant financial advantages [1]. Through business process re-engineering and procedure optimization, Six Sigma management is intended to increase the effectiveness of enterprise management in a scientific and efficient manner. It is also known as the DMAIC model for short since it comprises the five stages of Define, Measure, Analyze, Improve, and Control in its application to quality improvement [2]. DMAIC is a cycle of continuous improvement and promotion that combines project management, schedule management, and statistical analysis. Table 1 shows the processes, techniques, and technologies under the DMAIC process, which provides a solution to intricate and systematic issues [3].

Table 1. Key points and tools of DMAIC process activities.

Stages	Key Points of Activities	Commonly Used Tools and Techniques
D (Define)	Identifying problems Determining Y(CPQ/CTP)	Balanced scorecard, SIPOC, Quality Function Deployment (QFD), cost of poor quality, etc.
M (Measure)	Determining a baseline Measuring Y, Xs	Relationship matrix, flow chart, sampling technique, measurement system analysis (MSA), value flow chart, process capability analysis, etc.
A (Analyze)	Determining major factors Determining Y = f(X)	Causality diagram, hypothesis testing, level comparison, analysis of variance, experimental design, regression analysis, process analysis, etc.
I (Improve)	Eliminating major factors Optimizing Y = f(X)	Experimental design, response surface methodology, tuning operations, lean improvement technique, etc.
C (Control)	Maintaining results Updating Y = f(X)	Control charts, standard operating procedures, error prevention measures, process capability index, process document control, etc.

Six Sigma management starts with D for Define, which stands for the definition of a problem and boundaries. The definition of the issues to be resolved and the goals to be attained is necessary for this demand analysis step, setting the fundamental groundwork for project implementation [4–6]. The purpose of M for Measure, also known as the assessment of project improvement, is to identify the measurement indicators that demand improvement with a focus on detecting issues. To assist the reflection of shortcomings and process obstacles, it is necessary to gather a sizable amount of data from the project's operational phases [7]. In order to accurately reveal the project's causality, A for Analysis is the key to identifying problems that require transformation and examining their internal logical relationships in light of feedback returns from the measured data. I for Improve refers to the process of problem-solving through iteration and bridging the gap between the measurement indicator and the goal by adhering to the technical path and solution designed for project improvement based on the analysis results and the thorough consideration of factors like cost. The final stage of the model, C for Control, which stands for controlling the project, is crucial to sustaining the project's ability to improve in order to support continuous improvement. The entire process of creating improvements needs to be monitored to ensure that they are resolved permanently [8,9].

The DMAIC approach has been widely employed in industrial manufacturing [10–12] to effectively decrease defects or errors resulting from manual handling, materials, processes, machinery, or environment [13,14]. In addition, the DMAIC approach can be implemented in the urban service sector, government agencies, and other regulatory departments [4] in order to minimize handling errors caused by people or systems. Successful DMAIC applications have been made in service sectors including banking [15,16], education [17,18], hospitals [19,20], etc. The following are benefits of the DMAIC model. First, it is a data-driven model. As the project is completed, statistics are gathered to identify difficulties with an emphasis on a data-driven strategy. Secondly, it focuses on continuous improvement. The DMAIC is a model that is continuously optimized and iterated until the measurement index reaches the target [18,21]. Thirdly, it emphasizes the shift in thinking style. Since DMAIC optimizes company culture beyond a mere instrument, it adds fresh significance to the initial project for improvement by altering employees' thinking and behavior [22]. However, DMAIC has inherited drawbacks. First, the reasonability of the boundary and problem definition in the DMAIC model will directly affect the subsequent measurement, analysis, improvement, and control links. Definition errors could lead to project delays or even project failure. Second, it is unclear how the improvement cycle works. The DMAIC model places an emphasis on studying the process of quality improvement, but disregards the completion of the improvement cycle, leading to unmanageable expenses and service quality. Third, it discourages innovation. The DMAIC model streamlines the procedure, emphasizes its uniformity, and lowers the dynamic of innovation.

The boundary issues for items that conserve water are defined by the green, leader, and green indicator systems. The evaluation system for water-efficient products is formed by national, industrial, or association standards with distinct cycles for change and review. Such standards also provide technical support for policies that have been issued relating to water-efficient products. In addition, innovative procedures and product lines will drive the development of water-efficient products in order to meet assessment requirements. In conclusion, the DMAIC model's advantages can be utilized while avoiding its drawbacks by taking into account the industry's current features when improving water-efficient products.

3. Construction of Water-Efficient Product Quality Improvement Model

Taking into consideration the features of water-efficient products and based on the internal quality control of water-efficient products and external policy market promotion, we developed a DECIA improvement model (Define, Evaluate, Check, Improve, and Application) based on the six-sigma DMAIC quality management model for performance evaluation and quality improvement under the novel use scenario of water-saving appliances. Define: compiling a database by screening performance indicators. Evaluate: building an evaluation system on the criteria of water efficiency, green energy, and leader.

Check: innovate inspection method and research relevant devices. Improve: On-going technology improvement or upgrade for water-saving facilities to develop new products. Application: Researching evaluation techniques for the use of water efficiency standards and product marketing advancement to tap the water-saving potential of the product, form marketing policies to publicize and promote products. The DECIA model's details are shown in Figure 1, which can determine the quality evaluation index system of water-efficient products and guide the collection of corresponding data, through objective index data analysis to find the existence of quality problems. Moreover, the evaluation system established will effectively promote the quality of water-efficient products via iterative upgrade, clear quality improvement goals, and innovative water-saving technologies, thus solving the problem of unclear transformation cycle and insufficient innovation momentum in the DMAIC model. In summary, the DECIA model provides theoretical and technical support for the quality evaluation and improvement of water-efficient products, respectively, and enriches the application scenarios of the DMAIC quality management model.

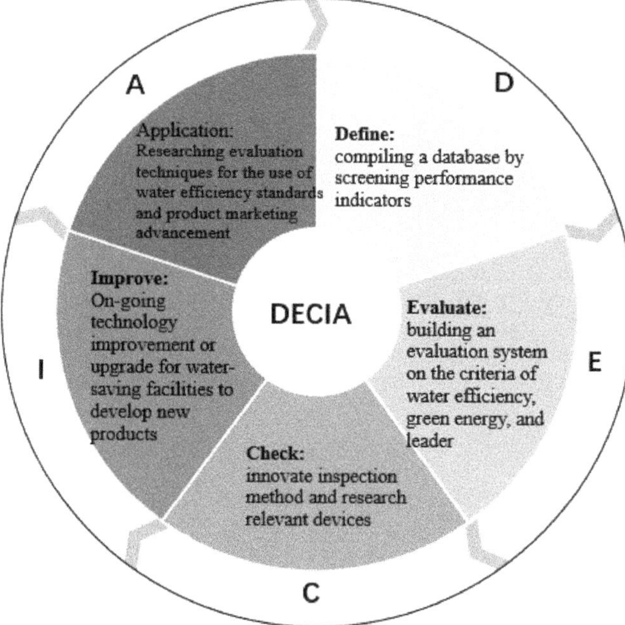

Figure 1. Schematic diagram of DECIA model.

3.1. Definition

For water-efficient products, e.g., water nozzle, toilet, and water purifier, the Definition represents the boundary featuring the quality of water-saving appliances and the indicators reflecting the quality attributes over the product life cycle [23]. The water efficiency indicator is a necessary performance criterion in order to reflect the characteristics of water conservation. The entire process, including the acquisition of raw materials, manufacturing, circulation, usage, and recovery, should be covered in the selection of indicators, along with economy, feasibility, and operability.

The defined indicator database of smart water closet is shown in Table 2.

Table 2. Indicator database of smart water closets.

Type of Indicator	Indicators	Type of Indicator	Indicators
Water efficiency indicators [24]	Water consumption for full flushing of double flush toilets	Leader indicators [25,26]	Safety performance
	Average water consumption for cleaning		Warm-water characteristic
	Average water consumption for flushing		Self-cleaning nozzle
	. . .		Seat heating
Green indicators [27]	Waste porcelain utilization rate		Water seal recovery function
			Sewage replacement function
	Utilization rate of waste base (including glazed body)		Waterproof grade of the whole product
			Cleaning function
	Utilization rate of waste glaze slurry		Ball discharge
	Waste sludge recovery rate		Particulate emission
	Product packaging		Mixed media discharge
	Individual quality		Drainage pipeline transport characteristics
	Comprehensive energy consumption per unit product		Antimicrobial property
	Service life of pressure flush valve		Antimicrobial durability
	Service life of seat and cover		Artificial test and paper ball test
	. . .		. . .

3.2. Evaluation

Indicators are classified into three groups: water efficiency evaluation, green evaluation, and leader to build a quality assessment system for water conservation evaluation, as shown in Figure 2. Three tiers and three dimensions make up the system. The three-level evaluation framework combines leading, comprehensive, and independent indicators. The independent indicators specifically refer to the ratings for water efficiency for appliances applied throughout the entire evaluation system. Comprehensive evaluation incorporates green evaluation indicators, as shown in Figure 3, including resource attributes (water intake per unit product, recycling, and utilization of production wastes), energy attributes, environmental attributes, quality attributes, etc. Leader evaluations refer to a series of evaluation indicators for standard enterprise leaders. For the three dimensions, the first dimension consists of top-down evaluation indicators that enrich assessment criteria and technical requirements to gradually improve the overall quality and production level. The second dimension includes bottom-up evaluation indicators that allocate from the basic to the premier level based on water-saving capability, green, and leader, highlighting the core function of water efficiency evaluation. A horizontal evaluation model covering the complete life cycle of appliances makes up the third dimension.

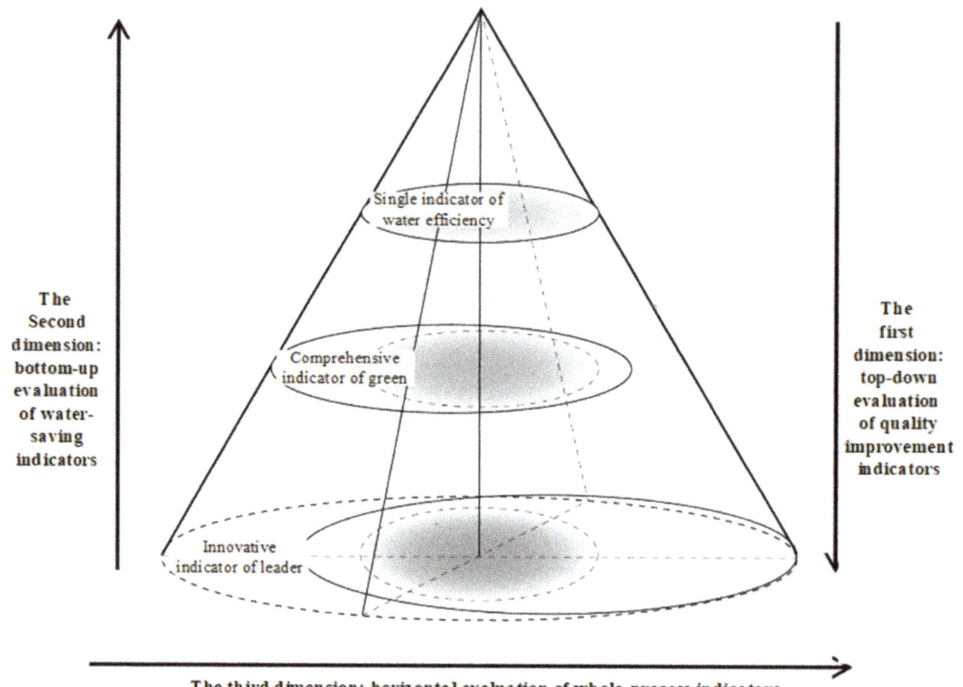

Figure 2. "3 + 3" performance evaluation system.

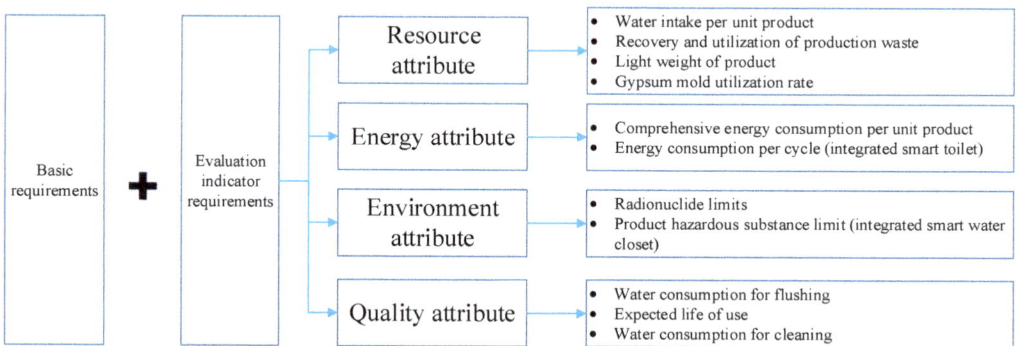

Figure 3. Construction of green evaluation system.

Take smart water closet for example, as shown in Figure 4. Its water efficiency evaluation dimension includes the average water consumption for cleaning and flushing. The green evaluation dimension incorporates water intake per unit, recycling rate of waste water, unit quality, a carbon footprint report, and the expected life of use. Leader evaluation dimensions cover safety performance, cleaning performance, antibacterial activity, and other indicators.

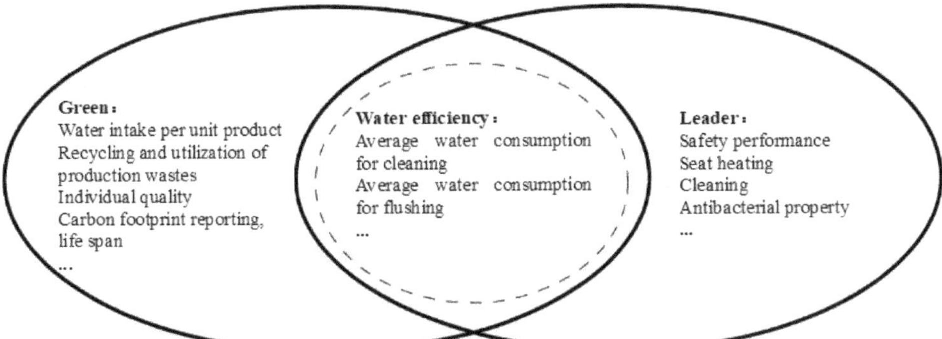

Figure 4. Example of evaluation system of smart water closets.

3.3. Checking

The checking procedures and facilities are improved based on the evaluation system that satisfies market expectations and achieves the indication value. When testing the smart water closet, leader evaluation indicators for water efficiency, environmental friendliness, and other factors were taken into consideration.

Improving detection methods: Targeting existing smart water closets' problems of technical weakness, low precision, and big testing errors from the aspects of average flushing water consumption and heating temperature, we modified the checking method and redesigned the checking devices (Figure 5). In view of the rapid change of the water temperature, the detection device with a fast thermocouple as the main element was developed to realize fast, accurate, and dynamic temperature checking. At the same time, the new method also satisfies data statistical analysis by tracing the changing pattern of test parameters, such as water temperature and power. The improved checking method and device meet the indicator requirements in the evaluation system, provide technical support, and improve the checking efficiency.

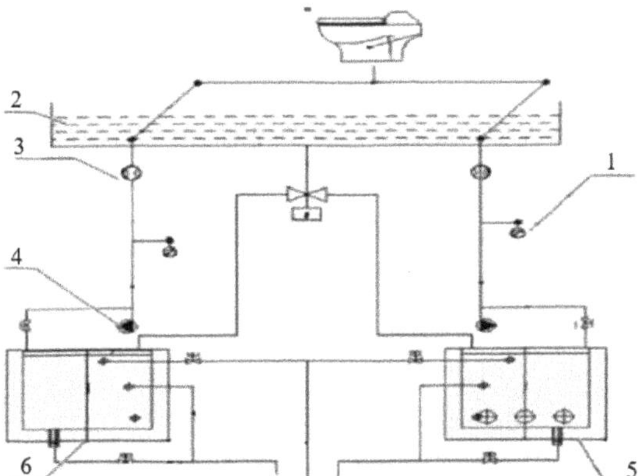

Figure 5. Schematic diagram of detection for a smart water closet. Note: 1—Pressure gauge; 2—Weight scale; 3—Flow meter; 4—Water pump; 5—Cold water tank; 6—Hot water tank.

Developing smart monitor: A comprehensive smart water closet monitoring platform is built for the safety attribute in the leader indicators by integrating the smart water supply control system, the electrical control system, the data acquisition and analysis system, and a test bench. The smart monitoring platform can also effectively acquire and process data such as pressure and temperature. The platform is an embedded system based on the

Internet of Things. It adopts high-power switching technology and realizes stable water source temperature and pressure testing through a constant temperature and pressure water supply system controlled by an adaptive tuning controller driven by DAQmx. As a result, the smart water closet is guaranteed to operate safely under various conditions. Meanwhile, the pipeline water supply system and the platform of the testing equipment were scientifically designed based on fluid mechanics and ergonomics to achieve the functions of human-computer interface interaction parameter setting. Other functions, such as automatic storage of test data and remote data transmission, also standardize the test conditions for flushing and cleaning water volume and provide technical support for performance parameter detection (Figure 6).

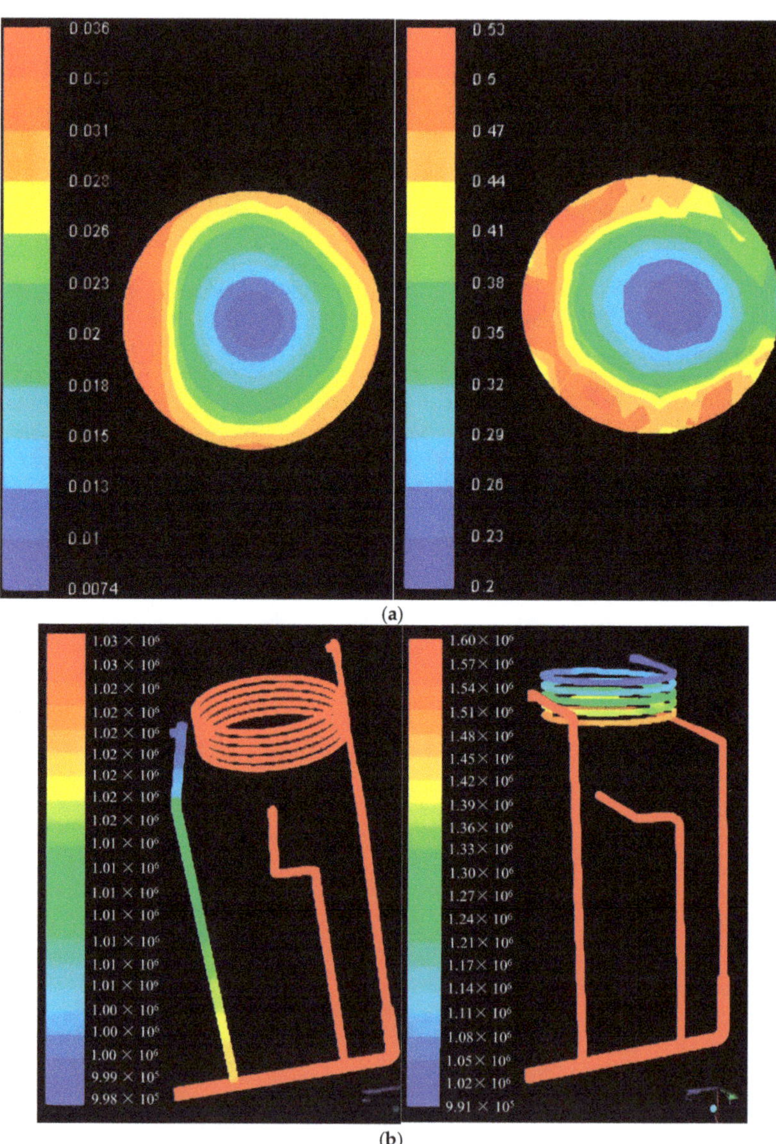

(a)

(b)

Figure 6. *Cont.*

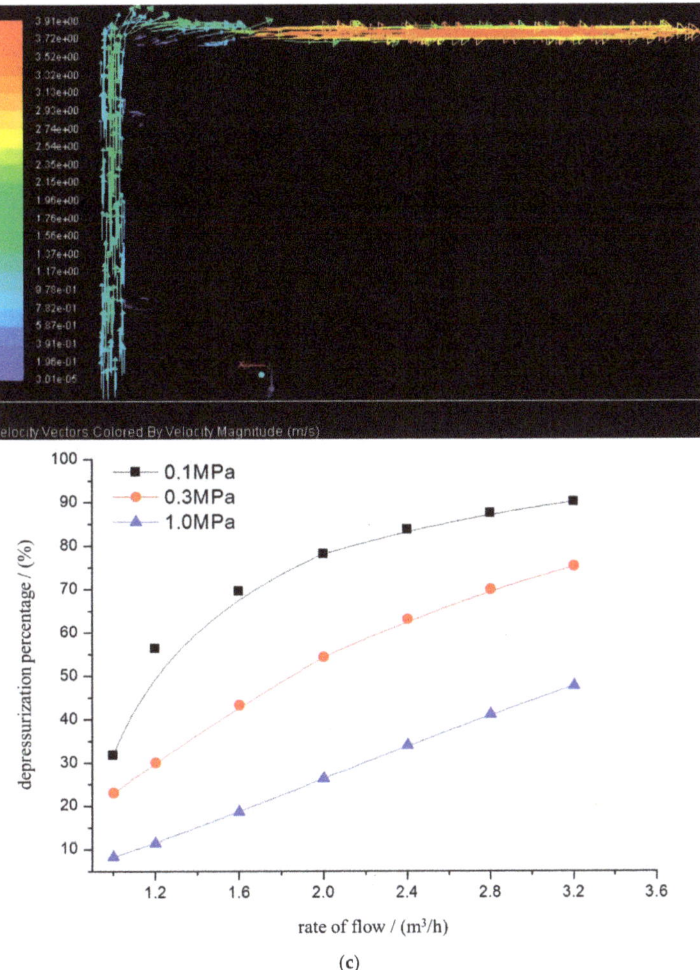

(c)

Figure 6. Determination of the standardized piping system through test verification. (**a**) Analysis of influence of pipe diameter on turbulence intensity; (**b**) Nephogram of overall pressure distribution in pipeline; (**c**) Analysis of pipe diameter velocity and pressure.

3.4. Improvement

This stage is designed to encourage enterprises to invest in R&D, technology innovation, and market competitiveness under the guidance of the evaluation indicators and testing results, so as to continuously improve/upgrade water-efficient technology and create new products [28].

For example, the evaluation system of the smart water closets will specify the value of the indicators, including water efficiency indicators of the average cleaning water consumption level 1 indicator value of 0.3 L, level 2 for 0.5 L, level 3 for 0.7 L; with seat heating function of energy consumption indicators level 1 for 0.03 kW·h, level 2 for 0.04 kW·h, level 3 for 0.06 kW·h, which are mandatory in China, and products that do not meet the relevant standards will no longer be allowed to be sold. Therefore, in order to meet the water efficiency index and the requirements of the frontrunner, manufacturers will increase investment in new technology research and development and develop new products. In particular, energy-saving water efficiency of more than two levels of products in the market is very popular with consumers. The entry of new products will drive positive growth in product sales, feeding the investment in technology research and development. In order to meet the requirements of water efficiency and leader indicators, manufacturers have been optimizing their technologies to strengthen the producing process as well as innovating new products. In particular, China's water efficiency level 2 and above products are very popular with consumers in the market, and the entry of new products will drive positive growth in product sales and promote investment in technology research and development.

Optimizing the process: Due to the limitations of the process, many manufacturers of smart water closet products cannot meet the requirements of the leader indicators. In light of the leader indicator that requires top water efficiency, a certain Chinese smart water closet design represents a first-of-its-kind inner water retaining structure, breaking through the traditional water diversion structure. The new design solves the water wasting problem caused by the baffle plate of the flush toilet with a waterless ring structure, and the water efficiency reaches the first class level. By fixing the baffle plate and the flow division plate behind it but at the front end of the main water outlet of the toilet body, the new product installs a water blocking bump in the interior of the baffle plate, so the flushing water will only flow down in the middle of the plate and save water. The water retaining design utilizes the flushing water more efficiently and prevents water from flowing on both sides of the water baffle.

New product R&D: In order to occupy the highly smart water closet market, a manufacturer has developed a height adjusting component for the defects in the original water seal, so that the new smart water closet can reach the limit of first-class water efficiency. The height of the water seal point in the sewage discharge pipeline, which will lower when flushing. The height difference between the water surface in the toilet and the water seal point increases the pressure difference, pushing the sewage up into the upper pipeline faster to trigger the siphon effect. As a result, when maintaining the water seal height unchanged, the water volume used for replacing the storage water is rather low, thus enhancing its flushing and water-saving capabilities. At the same time, the siphon pipe structure, water tank structure, water spray hole, and water distribution hole of the smart water closets were optimized. A double-pump toilet flushing structure was designed and developed to solve problems such as backwater or reverse suction by installing an air isolation valve on the spray water outlet pipe. Meanwhile, the structure of pump flushing + air isolation valve was adopted to reduce the average water consumption to less than 4 L, reaching the requirement of the water efficiency indicator. The design is shown in Figure 7. Due to the greatly improved water-saving performance and better product quality, the product became the main brand of the enterprise after entering the market, with annual sales volume greater than 100,000 sets, and soon occupied the high-end market and achieved greater economic benefits.

In addition, the water purifier products, for example, in order to achieve the water efficiency requirements of the water purifier, a manufacturer developed a new product to achieve the water purifier rinse water does not need to be discharged, reducing water consumption and achieving water efficiency level 2 requirements. This product solves the problem of large discharge of rinse water from water purifiers and the inability to utilize rinse water by developing a small flow throttle device. The inlet of the low-flow throttle valve is connected to the inlet pipe of the reverse osmosis processor or the outlet pipe of the concentrated water to realize the recycling of concentrated water and limit the concentration of water in the circulating pipeline of the reverse osmosis system through the regulation technology to reduce the blockage of the inlet pipe, and to extend the service life of the membrane, improve the water efficiency, and enhance the product quality.

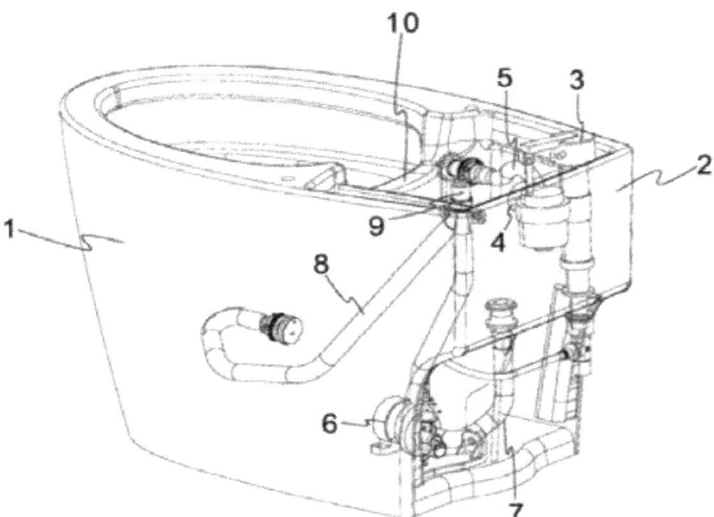

Figure 7. Schematic diagram of the R&D process of smart water closet. Note: 1—Toilet body; 2—Water tank; 3—Water inlet valve; 4—Swipe circle pump; 5—Arch bridge-shaped bend; 6—Jet pump; 7—Jet inlet pipe; 8—Injection water pipe; 9—Air isolation valve; 10—Overflow hole.

3.5. Application

Marketing strategies help promote the application of water-efficient products while the product evaluation model will tap water-saving potential and continuously stimulate the updating of Definition indicators. Such effort enables continuous water-efficient products' quality iteration and improvement.

For instance, the water efficiency labeling system, water efficiency pioneering system, green product system and leader system have been formulated in the field of smart water closet for wider promotion and application. The pioneering system water efficiency is based on China's National Development and Reform Commission, the Ministry of Water Resources, the Ministry of Housing and Urban-Rural Development, and other departments issuing Measures for the Administration of Water Efficiency Labels and Implementation Rules for Water Efficiency Labeling of smart water closets and other policies, which clearly put forward the average water consumption of cleaning, average water consumption of flushing, flushing full flush water consumption, and unit cycle energy consumption should be based on GB 38,448 intelligent toilet energy efficiency water efficiency limit value and level for the determination of evaluation indicators. The system has become compulsory in China where products without these labels are not allowed to be sold in the market. The leader system is based on the Implementation Plan for Leader Action by Water Efficiency Leaders issued by China's National Development and Reform Commission, Ministry of Water Resources, State Administration of Market Regulation and other departments, and the Implementation Plan for Energy Efficiency Leader Systems (Trial), which clearly proposed to meet the requirements of the relevant indicators to be named "frontrunner" products, such as water efficiency level to meet the average flush water consumption ≤4.0 L, wash function without residue, mixed media emissions for the first flush number of 28. The system of selection of the best, to establish the benchmark in the field of intelligent toilets, the formation of benchmarking effect, to promote new product updates and technical upgrades in the field of intelligent toilets. The system selects the best among the best, sets up a benchmark in the smart water closets industry, and promotes the new product and technology upgrade. The green product system is based on Certification and Accreditation Administration of the People's Republic of China issue of the Announcement of Certification and Accreditation Administration on Issuing the Qualification Conditions of Green Product Certification Organizations and the First Batch of Certification Implementation Rules, which proposed that water closets' resource attributes, energy attributes, environmental attributes, and quality attributes need to meet the relevant indicators to be recognized as green products. The system promotes the green certification of smart water closet products and promotes the improvement of product quality.

4. Conclusions

This paper proposes a DMAIC model for water-efficient product quality improvement based on the DECIA of six sigma management, providing a theoretical guideline for manufacturers to improve product quality. Based on the DECIA model, water-saving appliance manufacturers, through the construction of evaluation systems and the formation of relevant policy support, etc., can promote the upgrading of production processes or technologies, forming a virtuous cyclical circulation, and ultimately achieve the goal of improving product quality and increasing market penetration. In addition, the means of pasting the water efficiency label will influence the consumer's purchase decision, promote a positive relationship between marketing and enterprise innovation, and achieve a cycle of quality improvement.

The DECIA model suggests improving the quality of water-efficient products in the following aspects. First, strengthen enterprises' awareness of quality enhancement. Quality is the soul of a product, so higher quality will increase the sales volume while improving the user experience, which further translates into economic benefits for enterprises. In light of this, enterprises should focus on the quality of their product through strategy, ideology, and business practices. Second, enhance R&D. Technology is the driving force behind product upgrades and iterations. Only through constant R&D and innovation can manufacturers be self-motivated to improve the quality of their products. Third, improve the checking methods and equipment. The checking accuracy and precision should be improved to push the quality improvement of water-efficient products. Fourth, improve the product evaluation system. The evaluation system for water efficiency, green, and leader should be continuously improved, and relevant standards should be revised from time to time to remain effective in the long term. Fifth, build a top-level design for quality improvement. Top-level policies and norms should be formulated to give full play to the efforts of the government and the market, promote the wide application of water-efficient products, enhance the market penetration of high-quality water-efficient products, and stimulate the production and R&D enthusiasm of manufacturers.

Author Contributions: Conceptualization, Y.B. and J.L.; methodology, R.Z.; formal analysis, Y.B.; writing—original draft preparation, Y.B.; writing—review and editing, X.B. All authors have read and agreed to the published version of the manuscript.

Funding: This research received no external funding.

Institutional Review Board Statement: Not applicable.

Informed Consent Statement: Not applicable.

Data Availability Statement: The data used to support the findings of this study are available from the corresponding author upon request.

Conflicts of Interest: The authors declare no conflict of interest.

References

1. Kaushik, P.; Khanduja, D. Developing a Six Sigma Methodology to Increase the Passing Rate of Student in Engineering Education. *J. Eng.* **2006**, *20*, 23–30.
2. Amrina, U.; Firmansyah, H. Analysis of defect and quality improvement for o ring product through applying DMAIC methodology. *Penelit. Dan Apl. Sist. Dan Tek. Ind.* **2019**, *13*, 136–148. [CrossRef]
3. Trimarjoko, A.; Hardi Purba, H.; Nindiani, A. Consistency of DMAIC phases implementation on Six Sigma method in manufacturingand service industry: A literature review. *Manag. Prod. Eng. Rev.* **2020**, *11*, 34–45.

4. Nagi, K.; Charmonman, S. Applying Six Sigma techniques to improving the quality of eLearning courseware components-A case study. In Proceedings of the 2010 IEEE International Conference on Management of Innovation & Technology, Singapore, 2–5 June 2010.

5. Chakraborty, A.; Tan, K.C. Case study analysis of Six Sigma implementation in service organisations. *Bus. Process Manag. J.* **2012**, *18*, 992–1019. [CrossRef]

6. Lynch, D.P.; Bertolino, S.; Cloutier, E. How to scope DMAIC projects. *Qual. Prog.* **2003**, *36*, 37–41.

7. Rana, P.; Kaushik, P. Initiatives of six-sigma in an automotive ancillary unit: A case study. *Manag. Sci. Lett.* **2018**, *8*, 569–580. [CrossRef]

8. Jaffal, M.S.; Korkmaz, I.; Özceylan, E. Critical success factors for Six Sigma implementation in Gaziantep carpet companies. *Ind. Eng. Lett.* **2017**, *7*, 83–92.

9. ALmasarweh, M.S.Y.; Rawashdeh, A.M. The effect of using Six Sigma methodologies on the quality of health service: A field study at Prince Hashem hospital/city of Aqaba. *J. Soc. Sci. (COESRJ-JSS)* **2016**, *5*, 396–407. [CrossRef]

10. Smętkowska, M.; Mrugalska, B. Using Six Sigma DMAIC to improve the quality of the production process: A case study. *Procedia-Soc. Behav. Sci.* **2018**, *238*, 590–596. [CrossRef]

11. Godina, R.; Silva, B.G.R.; Espadinha-Cruz, P. A DMAIC integrated fuzzy FMEA model: A case study in the Automotive Industry. *Appl. Sci.* **2021**, *11*, 3726. [CrossRef]

12. Abhilash, C.; Thakkar, J.J. Application of Six Sigma DMAIC methodology to reduce the defects in a telecommunication cabinet door manufacturing process: A case study. *Int. J. Qual. Reliab. Manag.* **2019**, *36*, 1540–1555.

13. Rahman, M.A.; Talapatra, S. Defects reduction in casting process by applying Six Sigma principles and DMAIC problem solving methodology (a case study). In Proceedings of the International Conference on Mechanical, Industrial and Materials Engineering 2015 (ICMIME2015), Rajshahi, Bangladesh, 11–13 December 2015.

14. More, C.S.; Nerkarm, A.A.; Khonde, R.V.; Tilekar, S. Study of application of six sigma of DMAIC in dent minimization in gear manufacturing. *Glob. Res. Dev. J. Eng.* **2017**, *2*, 82–87.

15. Qutait, M. The role of six sigma approach by applying (DMAIC) model and forming work team in improving banks performance: A survey study in syrian banks in damascus. *J. Soc. Sci. (COESRJ-JSS)* **2018**, *14*, 169–180. [CrossRef]

16. Aditya, G.M.; Irawan, H. Application of the six sigma method as a tool for management to improve quality of services (case Study On Bank Bni Syariah Bandung). *eProc. Manag.* **2020**, *7*, 1046–1056.

17. Antony, J.; Scheumann, T.; M., V.S.; Cudney, E.; Rodgers, B.; Grigg, N.P. Using Six Sigma DMAIC for Lean project management in education: A case study in a German kindergarten. *Total Qual. Manag. Bus. Excell.* **2021**, *33*, 1489–1509. [CrossRef]

18. Qureshi, M.I.; Janjua, S.Y.; Zaman, K.; Lodhi, M.S.; Tariq, Y.B. Internationalization of higher education institutions: Implementation of DMAIC cycle. *Scientometrics* **2014**, *98*, 2295–2310. [CrossRef]

19. Al Kuwaiti, A.; Subbarayalu, A.V. Reducing hospital-acquired infection rate using the Six Sigma DMAIC approach. *Saudi J. Med. Med. Sci.* **2017**, *5*, 260. [CrossRef]

20. Vijay, S.A. Reducing and optimizing the cycle time of patients discharge process in a hospital using six sigma dmaic approach. *Int. J. Qual. Res.* **2014**, *8*, 169–182.

21. Franchetti, M.; Yanik, M. Continuous improvement and value stream analysis through the lean DMAIC Six Sigma approach: A manufacturing case study from Ohio, USA. *Int. J. Six Sigma Compet. Advant.* **2011**, *6*, 278–300. [CrossRef]

22. He, Z.; Hu, H.; Liu, H.L.; Wang, W. A review on the theoretical research and application of lean six sigma. *Ind. Eng. J.* **2021**, *24*, 1–8, 54.

23. Schulz, M.; Short, M.D.; Peters, G.M. A streamlined sustainability assessment tool for improved decision making in the urban water industry. *Integr. Environ. Assess. Manag.* **2012**, *8*, 183–193. [CrossRef] [PubMed]

24. Todt, D.; Bisschops, I.; Chatzopoulos, P.; van Eekert, M.H.A. Practical performance and user experience of novel DUAL-Flush vacuum toilets. *Water* **2021**, *13*, 2228. [CrossRef]

25. Iyo, T.; Asakura, K.; Nakano, M.; Yamada, M.; Omae, K. BBidet toilet seats with warm-water tanks: Residual chlorine, microbial community, and structural analyses. *J. Water Health* **2016**, *14*, 68–80. [CrossRef]

26. Bao, K.; Shen, Q. Research and development of intelligent controller for high-grade sanitary ware//International Conference on Graphic and Image Processing (ICGIP 2012). *SPIE* **2013**, *8768*, 146–150.

27. Yuan, Q.; Mohajerani, A.; Kurmus, H.; Smith, J.V. Possible recycling options of waste materials in manufacturing ceramic tiles. *GEOMATE J.* **2021**, *20*, 73–80. [CrossRef]

28. Hoecker, J.; Bracciano, D. Passive conservation: Codifying the use of water-efficient technologies. *J.-Am. Water Work. Assoc.* **2012**, *104*, 93–99. [CrossRef]

Article

Impact of Subsidy Policy on Remanufacturing Industry's Donation Strategy

Xintong Chen [1], **Zonghuo Li** [2,*] **and Junjin Wang** [3]

1   School of Management, Nanjing University of Posts and Telecommunications, Nanjing 210003, China
2   School of Politics and Public Administration, Soochow University, Suzhou 510632, China
3   School of Maritime Economics and Management, Dalian Maritime University, Dalian 116026, China
*   Correspondence: zh.li@suda.edu.cn

Abstract: Motivated by the donation subsidy policy, this paper studies a supply chain consisting of a manufacturer and a remanufacturer. The manufacturer sells new and remanufactured products and can also donate two products. The remanufacturer can only sell and donate remanufactured products. Using the Stackelberg game model, we investigate the optimal production and donation strategies of two competing firms and discuss how the subsidy policy affects these strategies. Our main results include the following: First, the donation strategies of the two firms are not only affected by the subsidies but could also be influenced by the competitor's donation decision, especially when the subsidy is high. Second, the subsidized products for sale in the market will decline as the subsidy increases. Therefore, a high subsidy always causes insufficient market supply. Third, the first-mover advantage may not make the manufacturer avoid a dilemma; however, when the remanufacturer becomes the leader in the market, the first-mover advantage will help the remanufacturer prevent any competitor donation threats. Lastly, the scenario where the manufacturer donates nothing and the remanufacturer donates seems to be a Pareto improvement for two firms, but this scenario is not stable, and the last equilibrium is that both firms decide to donate remanufactured products.

Keywords: subsidy policy; remanufacturing industry; donation strategy; prisoner's dilemma

Citation: Chen, X.; Li, Z.; Wang, J. Impact of Subsidy Policy on Remanufacturing Industry's Donation Strategy. *Processes* **2023**, *11*, 118. https://doi.org/10.3390/pr11010118

Academic Editors: Conghu Liu, Xiaoqian Song, Zhi Liu and Fangfang Wei

Received: 21 November 2022
Revised: 27 December 2022
Accepted: 28 December 2022
Published: 1 January 2023

1. Introduction

Market competition is becoming fierce, and while companies are trying to make more money, they are also paying more and more attention to improving environmental benefits and social concerns. As an example, the Chinese smartphone manufacturer Meizu created a special website called "Aihuishou" to collect end-of-life smartphones (such as iPhone, Meizu, and Samsung) from consumers [1]. By collecting used phones, there will be less pollution in our environment. In 2020, the Meizu company donated some "Meizu X8" smartphones to students living in poverty in the Zhuhai High-tech Zone, so that these students could gain knowledge through online courses in today's epidemic age. This donation mainly solved the problem for students in poverty who could not access online material without phones. Through this donation, the Meizu company not only fulfilled its corporate social responsibility but also aroused social attention, which enhanced its brand awareness and reputation. Apart from the Meizu company, many other manufacturers are doing similar things, such as Apple, Samsung, Home Depot, and so on.

Remanufacturing is regarded as an environmentally friendly and energy-saving way to cope with end-of-life products by many companies [2,3]. Moreover, through remanufacturing, manufacturers (such as IBM, and Xerox) could produce remanufactured products for resale, though remanufactured products with low prices may compete with the new ones [4]. This huge profit cake not only makes manufacturers engage in remanufacturing but also attracts competition from many other enterprises (remanufacturers). Remanufacturing has brought huge benefits to manufacturers and remanufacturers, but at the same time, these companies have to face an inevitable problem [5]. The problem comes from consumers who always value less remanufactured products and have a low willingness to pay. To solve this, some companies try to donate remanufactured products to make them easier for consumers to accept. Meanwhile, a donation strategy could grasp the attention of consumers and recover the value of remanufactured products [6,7]. For example, in 2013, the "Beijing Tibet Trip" was held by the National Development and Reform Commission (NDRC, China). Plenty of remanufactured products were donated by remanufacturers to the Tibet Autonomous Region [8]. Therefore, in addition to sales, donations seem to be a way for enterprises to make full use of their remanufactured products. In this paper, we consider this point and start a further discussion on donating new or remanufactured products.

Since donations can help solve part of the market's demand problems, ease the relationship between supply and demand, and realize the secondary allocation of resources, the government usually offers subsidies or incentives to the companies that make donations [9]. For instance, the Congress of the United States set a series of tax reduction policies to encourage companies to donate [10]. However, the manufacturer, who sells new and remanufactured products at the same time, may encounter a selection issue, namely, which product is better for donation. Donating new products may have higher consumer acceptance and be good for promoting product brands. While the cost of remanufactured products is low, donating remanufactured products can save money and avoid resource waste. From the remanufacturer's perspective, selling remanufactured products is his main business, but as a competitor, the donations could be another way to market a competition strategy that would attract consumers' attention, especially when the manufacturer donates first. Thus, the second objective of this paper is to study the impact of donation subsidy on the remanufacturing industries and outline the optimal coping strategy for the remanufacturer.

The first-mover advantage and the second-mover advantage have different effects on decision-making. In a competitive market, if the leader of the market changes, the optimal strategies for the two parties always differ. The subsidy policy may lead manufacturers or remanufacturers to adopt different strategies. So, under the subsidy policy, whether one party's donation strategy will promote the other party to donate or will prevent the other party from donating is the third objective of this paper.

To solve the objectives above, we try to answer the following research questions: (1) What are the effects of government subsidies on the donation supply chain (price, production, and environmental impact)? (2) When the manufacturer chooses to donate, should the remanufacturer also choose to donate as a competitive strategy? (3) Under government subsidies, will the manufacturer and the remanufacturer have a prisoner's dilemma (lose-lose) or a Pareto improvement (win-win) in their donation strategies? (4) In the case of government subsidies, do the manufacturer and the remanufacturer have the first-mover advantage or the second-mover advantage?

To make the research questions above clear, we consider a manufacturer and a remanufacturer who are competing with each other in a market and where the government provides subsidies for donations. According to the real cases in practice and the previous literature, we conclude that the manufacturer may have three donation strategies: no-donation, donating new products, and donating remanufactured products. However, the remanufacturer only has two donation strategies: no-donation and donating remanufactured products. Therefore, there are six donation scenarios in total. At first, we characterize each scenario's optimal equilibrium and clarify the necessary conditions under each donation strategy. Next, we analyze the impact of the donation subsidy on the willingness of the two firms to donate and whether production and prices change when the subsidy increases. Further, we investigate the dilemma and the Pareto situations that the manufacturer and the remanufacturer may meet. Lastly, we discuss the situation where the remanufacturer moves first, and compare the findings with the results in the previous section.

This paper studies the donation strategy of the remanufacturing industry under a subsidy policy. As such, the conclusions bear strategic practical implications, such as suggestions to the remanufacturer on whether it is necessary to donate remanufactured products to cope with the manufacturer's donation strategy. Second, this paper provides policy insights on how to set donation subsidies to encourage the remanufacturing industry to donate. Third, this paper is the first to study the first-mover advantage of donations and the second-mover advantage of donations for remanufacturing industries.

The remainder of the paper proceeds as follows: Section 2 introduces the related literature and points out the research gaps. Section 3 presents the model and notions used in this paper. Section 4 outlines the optimal donation strategy for the manufacturer and remanufacturer and points out the dilemma where the two parties may encounter a lose-lose. Section 5 includes an extensive discussion concerning the scenario where the remanufacturer moves first and shows the first-mover's advantage. The last section concludes the paper and proposes managerial insights.

2. Literature Review

2.1. Remanufacturing

In practice, after collecting or recycling end-of-life products, enterprises always choose to make full use of them, for example through remanufacturing [11]. As the U.S. remanufacturing market value has reached USD 43 billion [12], there has been a lot of literature that focuses on remanufacturing [13–25]. For example, Chai et al. (2018) studied a monopoly market where a manufacturer sold new and remanufactured products at the same time. They found that remanufacturing could be environmentally friendly and profitable, and meanwhile, carbon cap and trade could be valuable for remanufacturing. However, green consumers also had great effects on remanufacturing, and excessive green consumers might not benefit the manufacturer, especially when the carbon trading price is high. Huang and Wang (2019) investigated a two-tier closed-loop supply chain in which the manufacturer granted the third-party remanufacturer a brand license. By discussing three remanufacturing scenarios: no-remanufacturing, partial-remanufacturing, and full-remanufacturing, they analyzed the impact of strategic consumer behavior and the remanufacturing cost of the third-party remanufacturer on the remanufacturing market. The results showed that, if the consumer became more strategic, the demand for new products would decline and the demand for remanufactured products would increase. Moreover, the manufacturer and a third-party remanufacturer could experience a win-win when the cost of the third-party remanufacturer is quite low. At this time, the manufacturer preferred the third-party remanufacturer to remanufacture rather than himself. Ma et al. (2020) discussed two options for consumers: buying new products or trading in for remanufactured products when there is only one manufacturer selling products in the market. Considering consumers' double reference effects, they also put the remanufacturing subsidy and consumer rebate ratio into the model. The results showed that the remanufacturing subsidy could improve the profits of the manufacturer. The consumer's rebate ratio only affected and increased the price of remanufactured products, but did not change the manufacturer's profit. McKie et al. (2018) studied when there are multiple conditions and generations of products available, how consumers would evaluate remanufactured products, and the risks that remanufactured products brought to the new products. By analyzing the real data from the iPad on eBay, they found that the production, conditions, and attributes of the product all had great influences on shaping consumers' purchasing decisions. The remanufactured products posed the same threat to new products as do used ones. The results provided insights on how to achieve more profitable remanufacturing for remanufacturing industries.

The remanufacturing literature suggests that more and more enterprises are engaging in remanufacturing to achieve a huge profit. Based on the previous studies, our paper focuses on the remanufacturing industries: the manufacturer sells new and remanufactured products together, and the remanufacturer only produces remanufactured products. To reflect the real market situation, except the product differentiation, this paper also considers the competition relationship between the manufacturer and the remanufacturer.

2.2. Donation

A donation usually seems to be a way for enterprises to achieve corporate social responsibility [26]. However, its social and practical significance is far-reaching. Many scholars have made great contributions to this topic [7,10,27–37]. For example, Arya and Mittendorf (2015) studied a firm's donation inventory strategy, and they found that incentives from the government could help the firm achieve societal objectives. While the firm donated inventory by forgoing potential profit, they would be sensitive to the wholesale price, which might make the retail price higher. The results showed that incentives from the government not only promoted the firm's donation but also had notable effects on supply chain members in the market. Modak et al. (2019) explored two corporate social responsibility tools consisting of donations and recycled investment. Considering the donation amount, recycling investment, pricing, and order quantity, they proposed a model for profit maximization under a carbon tax policy. The results showed that the consumer's social work donation elasticity had a positive effect on recycling and that it could help reduce the carbon tax. Heydari and Mosanna (2018) investigated a cause-related market where the manufacturer contributed to the market by donating a sum per sold product. They developed two models to analyze the two-echelon supply chain coordination problem. One was the donation size as a portion of the retail price, and the other was that the retail price was a decision parameter. In their discussion, they concluded that developed models could help the manufacturer make the correct decisions on the donation size. Song et al. (2020) set profit donations as a corporate social responsibility investment and analyzed three donation models: the centralized model, the manufacturer's donation model, and the retailer's donation model. They found that subsidies from the government could not only improve market demand but could also increase the recycling rate. If the manufacturer chose to make a corporate social responsibility investment, it would benefit many parties, such as the government, the environment, the economy, and society.

Our paper also studies the donation strategies that the manufacturer and the remanufacturer may adopt in operating. In particular, we focus on the changing of donation strategies under the subsidy policy.

2.3. Subsidy

A subsidy is an incentive method that the government always uses to encourage enterprises to engage in producing, donating, selling, or other actions. When the government needs the market to reach a certain state or needs companies to complete certain tasks, subsidy policies are usually useful. The literature contains many subsidy models to help the government regulate the behavior of the market [6,8,38–53]. For instance, Cao et al. (2020) studied the optimal production and pricing decisions of two firms in a dual-channel supply chain under a remanufacturing subsidy policy and a carbon tax policy. Through discussing which policy was better for firms, they found that the output of remanufactured products would increase with the subsidy rate of the remanufacturing subsidy and that the output of new products would decrease with the subsidy rate of the carbon tax. The remanufacturing subsidy policy was more effective in terms of its carbon emissions. Gu et al. (2021) used the battery industry as their research objective and assumed that the government would provide subsidies for secondary users. The results showed that, if the remanufacturing rate was low or the battery's remaining power was high, the government would begin to provide subsidies. Their findings helped firms manage closed-loop supply with secondary battery use and made the supply chain more efficient. Their study also contributed to the government improving the social economy and sustainability. Qiao and Su (2021) investigated how the subsidy affects remanufacturing activity and market competition. By building a game model, they discussed the impact of the government subsidy on production, profit, consumer surplus, and environmental benefits. They concluded that the profits of manufacturers and remanufacturers might decrease with the subsidy, so providing subsidies to improve environmental benefit and social surplus was not always a good choice. The government should make the competitive market and environmental impact of the per unit product clear before increasing the subsidy. Wang et al. (2018) have made a great contribution to the field of subsidy research, as they introduced donation subsidies for remanufactured products, which made the recovery of product value more meaningful. They studied the conditions under which a subsidy policy could improve environmental and economic performance. They found that the donation subsidy for remanufactured products could expand the demand for new and remanufactured products, and the subsidy could create a win-win situation, especially when the cost of the new product was low.

The literature on subsidies appears to focus on remanufactured and donated products, as the government always encourages firms to do this. Further, most of the prior work did not pay attention to donation strategies that the manufacturer may have in a competitive market when his competitor donates. Moreover, whether there could be a win-win donation strategy or a dilemma is still unclear. Thus, our paper tries to apply a Stackelberg game model to discuss two firms' donation strategies under the donation subsidies.

2.4. Research Gap

Table 1 contains the work to date on this problem. From the literature, donation subsidies could benefit firms, consumers, and society [8]. However, most of the literature focused on remanufacturing subsidies and treated donations as a form of corporate social responsibility. Only a few papers studied the donation subsidy. Extending the work of Wang et al. (2018), we consider the donation strategy as a competition method that could be affected by donation subsidies. Six donation scenarios are discussed in this paper to address the remanufacturing industries' optimal donation strategies, involving the Pareto improvement situation, where two firms could achieve a win-win; the single-win situation, where only one firm could achieve high profit; and the lose-lose situation, where both parties may lose advantages. While Qiao and Su (2021) examined the impact of subsidies on remanufacturing activity and market competition, they did not include the donation strategy, which can also serve as an effective means of competition. Wang et al. (2019) analyzed a manufacturer–retailer supply chain in which both the manufacturer and retailer may donate. Based on this, we extend this thinking and mainly explore the competition between a manufacturer and a remanufacturer. However, they marginalize the fact that the donation strategies of two firms may have different scenarios, and they did not distinguish between the first-mover advantage and the second-mover advantage. Thus, recognizing the difference that may be brought by different first-movers, we investigate two situations where the manufacturer acts as a first-mover and a second-mover, to show where the first-mover advantages and the second-mover advantages are, which could provide some useful managerial insights to firms.

Table 1. Overview of recent literature.

Source	Donation	Subsidy	Remanufacturing		Game Theory		Supply Chain	
			Manufacturer	Remanufacturer	Nash	Stackelberg	Closed	Not Closed
[18]			√		√		√	
[14]		√	√		√		√	
[26]	√		√			√	√	
[29]	√		√			√		√
[32]		√	√			√	√	

Table 1. *Cont.*

Source	Donation	Subsidy	Remanufacturing		Game Theory		Supply Chain	
			Manufacturer	Remanufacturer	Nash	Stackelberg	Closed	Not Closed
[42]		√	√			√	√	
[50]		√	√	√	√		√	
[39]		√	√	√		√	√	
[27]	√	√	√			√	√	
[8]	√	√	√		√		√	
This study	√	√	√	√		√	√	

3. Model Overview

Consider a manufacturer selling new and remanufactured products at the same time and a remanufacturer who only sells remanufactured products. There is a government that provides donation subsidies to encourage companies to donate, as highlighted in Figure 1.

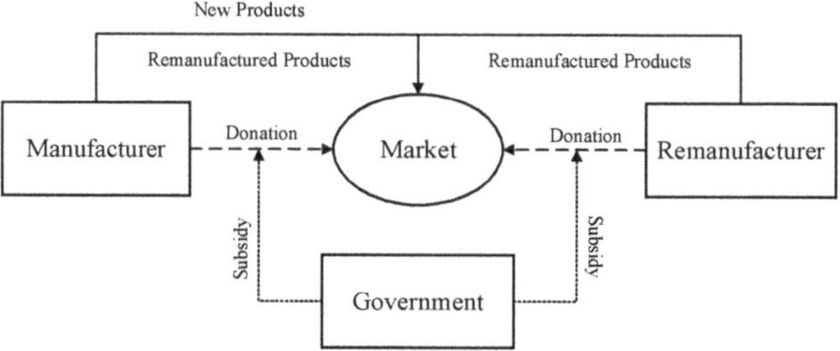

Figure 1. Model structure.

3.1. Notation

We first introduce our demand functions and two chain members' response and discuss the assumptions used. We frame the optimization problems of the manufacturer and the remanufacturer using the following notation: M and R denoting the manufacturer and the remanufacturer, respectively, subscripts n and r to denote new and remanufactured products, respectively, and subscript d to denote donation behavior. Table 2 contains the parameters, decision variables, and objectives used.

We assume that the product demand is for one cycle, that the product price and output are known, and that the donation subsidies (subsidy for donated new product and subsidy for donated remanufactured product) are unchanged. These assumptions are widely used in the recent remanufacturing and subsidy literature [27,32]. The manufacturer decides on how many new (q_n) and remanufactured (q_{Mr}) products to make, and how many new (q_{Mdn}) or remanufactured (q_{Mdr}) products to donate. The remanufacturer needs to decide how many remanufactured products (q_{Rr}) to produce, and how many products (q_{Rdr}) to donate. The government will provide one or two kinds of subsidies, one (s_n) for donated new products and one (s_r) for donated remanufactured products.

According to previous research, consumers could choose to buy new or remanufactured products or not at all to maximize their utility [54,55]. Consumers have a perceived value v of a new product at price p_n. We normalize this perception to $[0, 1]$ obtain the consumer's utility for a new product as $U_n = v - p_n$. Further, we assume that consumers set different perceived values for new and remanufactured products. When a consumer buys a remanufactured product, the consumer will value that product less, at price p_r [56–58]. Suppose a consumer is willing to pay αv, $\alpha \in (0, 1)$, for a remanufactured product, this yields a utility of $U_r = \alpha v - p_r$. Thus, based on the two utility functions, we can derive two products' price functions: $p_n = 1 - q_n - \alpha(q_{Mr} + q_{Rr})$, $p_r = \alpha(1 - q_n - q_{Mr} - q_{Rr})$, which are generally used in the literature [59,60].

Table 2. Notation used.

Parameter	
c_n	the unit cost of a new product for the manufacturer
c_{Mr}	the unit cost of a remanufactured product for the manufacturer
c_{Rr}	the unit cost of a remanufactured product for the remanufacturer
α	consumer's valuation of a remanufactured product as a fraction of the value of a new product
s_n	donation subsidy for unit new product
s_r	donation subsidy for unit remanufactured product
θ_n	reputational benefits brought by unit new product donation
θ_r	reputational benefits brought by unit remanufactured product donation
Decision Variable	
q_n	amount of new products
q_{Mr}	amount of collected products
q_{Rr}	amount of remanufactured products
q_{Mdn}	amount of donated new products
q_{Mdr}	amount of donated remanufactured products from the manufacturer
q_{Rdr}	amount of donated remanufactured products from the remanufacturer
p_n	the unit price of a new product
p_r	the unit price of a remanufactured product
Objective	
Π_M	manufacturer's profit
Π_R	remanufacturer's profit

3.2. Problem

We now state the problems that the manufacturer and the remanufacturer face separately. For the manufacturer, we denote the unit manufacturing cost by c_n, the unit remanufacturing cost c_{Mr}. We assume that the total manufacturing and remanufacturing costs are non-linear in the quantity produced, which is widely used in the previous literature [61–64]. While we separately consider the manufacturer's problem under different donation strategies, we present here only the problem formulation under donation strategies. Formulations for no donation, donating new products, and donating remanufactured products follow by setting (1) $q_{dn} = 0$, $q_{Mdr} = 0$; (2) $q_{dn} \neq 0$, $q_{Mdr} = 0$; (3) $q_{dn} = 0$, $q_{Mdr} \neq 0$ accordingly.

$$\Pi_M = p_n q_n + p_r q_{Mr} - c_n(q_n + q_{Mdn})^2 - c_{Mr}(q_{Mr} + q_{Mdr})^2 + s_n q_{Mdn} + \theta_n q_{Mdn} + s_r q_{Mdr} + \theta_r q_{Mdr} \tag{1}$$

The manufacturer maximizes profit by selling new and remanufactured products, receiving donation subsidies from the government, and earning a reputation through donations, rather than through manufacturing and remanufacturing costs.

Next, we define the remanufacturer's problem. Let c_{Rr} denote the unit cost of remanufacturing. While the remanufacturer has two donation options, the formulation for no donation follows by setting q_{Rdr} to zero.

$$\Pi_R = p_r q_{Rr} - c_{Rr}(q_{Rr} + q_{Rdr})^2 + s_r q_{Rdr} + \theta_r q_{Rdr} \tag{2}$$

The remanufacturer maximizes profit by selling remanufactured products, receiving donation subsidies from the government, and earning a reputation through donations, rather than through the remanufacturing cost. Finally, because the remanufacturer should collect end-of-life products first, the remanufacturer can be seen as a second-mover in the market, as shown in Figure 2.

To reveal different donation scenarios, we use Table 3 to show the manufacturer's and the remanufacturer's donation strategies. Let <0,0> denote the scenario where there is no donation, <n,0> only the manufacturer denoting new products, <r,0> only the manufacturer denoting remanufactured products, <0,r> only the remanufacturer denoting remanufactured products, <n,r> the manufacturer denoting new products and the remanufacturer denoting remanufactured products, <r,r> both the manufacturer the remanufacturer denoting remanufactured products.

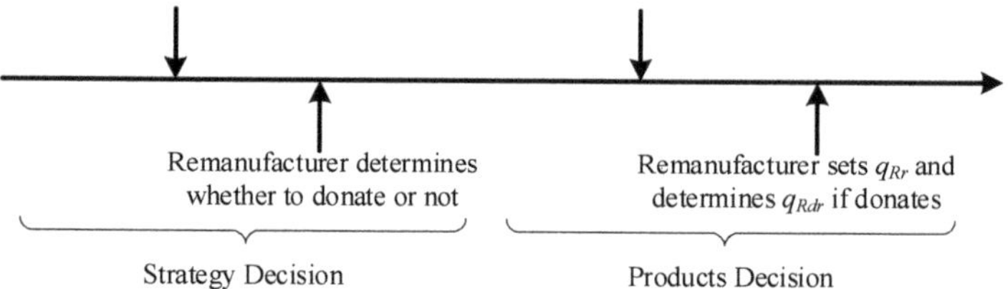

Figure 2. The game sequences.

Table 3. Donation strategies.

		Manufacturer		
		0	**n**	**r**
Remanufacturer		$q_{Mdn}=0$ $q_{Mdr}=0$	$q_{Mdn}\neq 0$ $q_{Mdr}=0$	$q_{Mdn}=0$ $q_{Mdr}\neq 0$
0	$q_{Rdr}=0$	<0,0>	<n,0>	<r,0>
r	$q_{Rdr}\neq 0$	<0,r>	<n,r>	<r,r>

We derive the Nash equilibrium of the Stackelberg game between the manufacturer and the remanufacturer by solving the first-order conditions.

4. Analysis

In this section, we first characterize the manufacturer's and the remanufacturer's optimal decisions in equilibrium under different donation scenarios. We then discuss the impact of a subsidy on products, prices, profits, and donation strategies for the two parties.

4.1. Characterization of Equilibrium

We summarize each donation scenario's optimal decisions in Table 4 (the related proofs can be found in Appendix A).

Table 4. Optimal decisions in each scenario.

	<0,0>	<n,0>	<r,0>
q_n	$\dfrac{\alpha(1-\alpha)T_3+T_1}{4c_{Mr}(1+c_n)(c_{Rr}+\alpha)-T_2}$	$-\dfrac{\alpha T_3(-1+\Delta_n+\alpha)+T_4}{2\alpha(1-\alpha)T_3+2T_1}$	$\dfrac{1+\Delta_r-\alpha}{2(1+c_n-\alpha)}$
q_{Mr}	$\dfrac{c_n\alpha T_3}{4c_{Mr}(1+c_n)(c_{Rr}+\alpha)-T_2}$	$\dfrac{\alpha T_3\Delta_n}{2\alpha(1-\alpha)T_3+2T_1}$	$\dfrac{2c_{Rr}(\Delta_r+c_n\Delta_r-c_n\alpha)-T_5}{2\alpha T_3(\alpha-1-c_n)}$
q_{Rr}	$\dfrac{1}{4}\left(T_8+\dfrac{2c_n c_{Mr}\alpha}{T_9 T_3+T_{10}}\right)$	$\dfrac{\alpha(\alpha(\alpha-1)T_3-T_7)}{4(c_{Rr}+\alpha)((\alpha-1)\alpha T_3-T_1)}$	$\dfrac{2c_{Rr}T_6+\alpha(2\Delta_r+\alpha)}{4T_3 c_{Rr}+4T_3\alpha}$
q_{Mdn}	-	$\dfrac{\Delta_n}{2c_n}+\dfrac{c_n(\alpha T_3 T_{12}+T_4)}{2c_n(\alpha(1-\alpha)T_3+T_1)}$	-
q_{Mdr}	-	-	$\dfrac{\Delta_r}{2c_{Mr}}-\dfrac{\eta_2+T_{11}}{2T_9 T_3}$

	<0,r>	<n,r>	<r,r>
q_n	$\dfrac{T_{16}+c_{Mr}\Delta_r}{2(T_{21}+T_9)}$	$\dfrac{\alpha T_{12}+c_{Mr}T_{14}}{2(c_{Mr}(\alpha-2)+(-1+\alpha)\alpha)}$	$\dfrac{1+\Delta_r-\alpha}{2+2c_n-2\alpha}$
q_{Mr}	$\dfrac{\Delta_r(1+c_n-\alpha)+\alpha c_n}{2(T_{21}+T_9)}$	$\dfrac{\Delta_r-\Delta_r\alpha+\Delta_n\alpha}{2(c_{Mr}(2-\alpha)+(1-\alpha)\alpha)}$	$-\dfrac{\Delta_r+c_n\Delta_r-c_n\alpha}{2T_9}$
q_{Rr}	$\dfrac{T_9(3\Delta_r-\alpha)+c_{Mr}\eta_1}{-4\alpha(T_{21}+T_9)}$	$\dfrac{(1-\alpha)\alpha(\alpha-3\Delta_r)+T_{15}}{4\alpha(c_{Mr}(2-\alpha)+(1-\alpha)\alpha)}$	$\dfrac{\alpha-\Delta_r}{4\alpha}$
q_{Mdn}	-	$\dfrac{\Delta_n}{2c_n}+\dfrac{c_n(\alpha T_{12}+c_{Mr}T_{14})}{2c_n T_{16}}$	-
q_{Mdr}	-	-	$\dfrac{T_9\Delta_r+T_{13}}{2c_{Mr}\alpha(1+c_n-\alpha)}$
q_{Rdr}	$\dfrac{T_9 T_{20}+T_{19}}{4\alpha c_{Rr}(T_{21}+T_9)}$	$\dfrac{\eta_2+c_{Mr}T_{18}}{4c_{Rr}\alpha T_{16}}$	$\dfrac{2\alpha\Delta_r+c_{Rr}(\Delta_r-\alpha)}{4c_{Rr}\alpha}$

Where $T_1 = c_{Mr}(2c_{Rr} + (2-\alpha)\alpha)$, $T_2 = 2c_{Mr}\alpha^2 - 2\alpha T_3(1 + c_n - \alpha)$, $T_3 = 2c_{Rr} + \alpha$, $T_4 = c_{Mr}(2c_{Rr}(-1 + \Delta_n) + \alpha(-2 + 2\Delta_n + \alpha))$, $T_5 = \alpha(\Delta_r(\alpha - 2 - 2c_n) + \alpha c_n)$, $T_6 = \Delta_r + \alpha$, $T_7 = c_{Mr}(2c_{Rr}(1 + \Delta_n) + \alpha(2 + 2\Delta_n - \alpha))$, $T_8 = \alpha/(c_{Rr} + \alpha)$, $T_9 = \alpha(1 + c_n - \alpha)$, $T_{10} = c_{Mr}(2c_{Rr}(1 + c_n) + \alpha(2 + 2c_n - \alpha))$, $T_{11} = \alpha(\Delta_r(\alpha - 2 - 2c_n) + \alpha c_n)$, $T_{12} = -1 + \Delta_n + \alpha$, $T_{13} = c_{Mr}(\Delta_r + c_n\Delta_r - c_n\alpha)$, $T_{14} = -2 + 2\Delta_n - \Delta_r + \alpha$, $T_{15} = c_{Mr}(\alpha\Delta_r - 4\Delta_r + \alpha(2 + 2\Delta_n - \alpha))$, $T_{16} = c_{Mr}(2 - \alpha) + (1 - \alpha)\alpha$, $T_{17} = 2(2 - \alpha)\alpha\Delta_r$, $T_{18} = T_{17} + c_{Rr}(4\Delta_r + \alpha T_{14})$, $T_{19} = c_{Mr}(2\alpha\Delta_r(2 + 2c_n - \alpha) + \eta_1 c_{Rr})$, $T_{20} = 2\alpha\Delta_r + c_{Rr}(3\Delta_r - \alpha)$, $T_{21} = c_{Mr}(2 + 2c_n - \alpha)$.

4.2. Impact of Subsidies on Donation Willingness

Comparing the equilibrium of the different donation scenarios, we can derive some insights regarding how the donation subsidies affect the manufacturer's and the remanufacturer's donation willingness, which are summarized in Proposition 1–3 (the related proofs can be found in Appendix B).

Proposition 1. *If the remanufacturer adopts no donation strategy.*

1. Under the <n,0> scenario, only when the subsidy is satisfied: $s_n > s_n^{<n,0>}$, the manufacturer will choose to donate new products;

2. Under the <r,0> scenario, only when the subsidy is satisfied: $s_r > s_r^{<r,0>}$, the manufacturer will choose to donate remanufactured products.

Proposition 1 shows the situation where the remanufacturer will not donate and the necessary conditions that encourage the manufacturer to donate. Proposition 1 (1) and (2) indicate that the donation subsidies could affect the manufacturer's donation willingness. Low subsidies will not make the manufacturer adopt a donation strategy, but high subsidies will. However, from Proposition 1 (2), we also notice that a high subsidy for donated remanufactured products could influence the manufacturer's donation decision. However, at this time, it is reasonable for the remanufacturer to change his donation strategy, especially when the high subsidy could bring more benefits. Next, we will discuss when the remanufacturer's willingness changes with an increasing subsidy.

Proposition 2. *The necessary condition for the remanufacturer to adopt a donation strategy is that the subsidy should meet:* $s_r > min\left\{ s_r^{<0,r>}, s_r^{<n,r>}, s_{Rr}^{<r,r>} \right\}$.

Proposition 2 reveals that there are three situations where the remanufacturer chooses to donate remanufactured products. Depending on the donation subsidy, the remanufacturer's willingness to donate may differ, while the manufacturer may have different donation strategies. From Proposition 2, we conclude the necessary condition for the remanufacturer to donate. As long as the subsidy is higher than $min\left\{ s_r^{<0,r>}, s_r^{<n,r>}, s_{Rr}^{<r,r>} \right\}$, the remanufacturer can choose a donation strategy. If the subsidy is higher than $max\left\{ s_r^{<0,r>}, s_r^{<n,r>}, s_{Rr}^{<r,r>} \right\}$, we find that the remanufacturer will donate products, regardless of whether the manufacturer donates or not. However, when the subsidy is higher than $min\left\{ s_r^{<0,r>}, s_r^{<n,r>}, s_{Rr}^{<r,r>} \right\}$, but lower than $max\left\{ s_r^{<0,r>}, s_r^{<n,r>}, s_{Rr}^{<r,r>} \right\}$, the donation decision for the remanufacturer may be complicated. In this situation, the manufacturer and the remanufacturer will compete in the donation field, and their donation strategies will interact with each other. We will explore this deeply in Sections 4.3 and 4.4. Because of their competitive donating relationship, we must make it clear when the manufacturer will donate new products and when he will donate remanufactured ones to cope with the remanufacturer's donation strategies.

Proposition 3. *If the remanufacturer adopts a donation strategy.*

1. The necessary condition for the manufacturer to donate new products ($<n,r>$ scenario) is that the subsidy should meet: $s_n > s_n^{<n,r>}$.

2. The necessary condition for the manufacturer to donate remanufactured products ($<r,r>$ scenario) is that the subsidy should meet: $s_r > s_{Mr}^{<r,r>}$.

Proposition 3 answers the question of what the manufacturer will do to compete with the remanufacturer under the different subsidy situations. Compared with the results from Proposition 1, we find that the manufacturer's willingness to donate changes (the thresholds of subsidy for the manufacturer to donate are different) when the remanufacturer participates in the donation. Further, we notice that $s_n^{<n,r>} > s_n^{<n,0>}$ and $s_{Mr}^{<r,r>} > s_{Mr}^{<r,0>}$, which means that, when the remanufacturer chooses to donate, only higher subsidies for the manufacturer will encourage him to donate.

4.3. Impact of Subsidies on Products and Prices

In Section 4.2, we have discussed the willingness of the manufacturer and the remanufacturer to donate, and we also notice that different donation strategies will bring different results in production decisions and pricing decisions, which deeply influence the market supply. Therefore, in this section, we will continue to investigate how the subsidies affect products and prices (the related proofs can be found in Appendix B).

Proposition 4. *The donation decisions of the manufacturer and the remanufacturer are affected by donation subsidies.*

1. The sales share of subsidized products (such as when the manufacturer donates new products and also sells new products) in the market will decrease with the increase in the subsidy;
2. However, other products' output will increase;
3. When two subsidies exist, the donated products will be negatively affected by the other subsidy's increase.

Proposition 4 reveals the relationships between subsidies and products. Except for the common sense that high subsidy leads to high donated products, the back effect of the subsidy is that the subsidized products for sale in the market will decline as the subsidy increases. From this, the management insight for policymakers is that products with a large demand in the market should not be given high donation subsidies to avoid insufficient market supply. However, from Proposition 4 (2), we also notice that products that compete with donated products will become highly productive (such as in the $<n,0>$ scenario, new products for sale in the market decline, but the remanufactured products become more). The management insight here indicates that if policymakers prefer to encourage competing products to be put on the market, a high subsidy for original products may be a good choice. Finally, Proposition 4 (3) shows the donation competition situation where the manufacturer donates new products and the remanufacturer donates remanufactured products. A typical subsidy dilemma has happened, and it becomes complicated for policymakers to decide how to set two subsidies because one subsidy may hurt another product's output. For this reason, we know that higher subsidies are not always better and that they are usually accompanied by negative effects. Policymakers should be aware of the duality of subsidies and set them properly.

Corollary 1. *When the government adopts subsidies strategies for new products (or remanufactured products), the total amount of new products (or remanufactured products), including for sale and donation, will increase, which means that donation subsidies have a positive stimulating effect on this product.*

Serving as a complementary conclusion of Proposition 4, Corollary 1 shows the impact of the subsidies policy on the total subsidized products. Taking scenario $<n,0>$ as an example, we can see that despite the new products for sale will decrease with the subsidy, the increase in donated new products could offset the decline in the former. This increases the total amount of new products by the manufacturer.

Due to the shortage of global chip supply, some automakers have been unable to guarantee the supply of cars on time and in quantity (such as the Tesla Model 3 and the BYD Song Plus DM-i). Sundar Kamak (Head of Manufacturing Solutions at Ivalua) suggested taking used chips back and remanufacturing key components. The management insight of Corollary 1 here is that the policymakers could provide subsidies for donated remanufactured products to achieve high production. For example, providing a subsidy on donated used chips may lead to chip remanufacturing, which could ease the woes of chip shortages in the market. The fundamental reason is that donation subsidies stimulate donation behavior. To enjoy such subsidies, firms are willing to donate products in their hands or collect used products in advance. In this way, the market demand for the product increases, which further stimulates the production of the manufacturer and the remanufacturer. This also indicates that the government tries to use the invisible power of donation subsidies to stimulate end-demand and drive upstream supply.

Proposition 5. *In scenario <0,r>, if and only if* $s_r \in \left(s_r^{<0,r>}, s_r^* \right)$, *both new and remanufactured products have price disadvantages, however, when* s_r *is increasing, the price advantages of the two products will gradually appear. In other donation scenarios, the new products and the remanufactured products always have price advantages.*

Proposition 5 shows the relationship between donation subsidies and product prices. In most cases, the products' prices are higher than those in scenario <0,0> and prices will increase with the subsidies. This is because high subsidies make the donated products more expensive (as mentioned in Proposition 4). However, donation is a non-profit strategy, so to offset this, the manufacturer or the remanufacturer needs to set high sale prices. However, if only the remanufacturer chooses to donate, the prices of two products could be lower than those in scenario <0,0>. The reason for this is that the remanufacturer (a follower in the market) tries to use the donation strategy as a competing method to gain market share. In the beginning, the subsidy for the remanufactured products improves the new products' output, and at the same time, the remanufacturer could gain more consumer traction with a low price. However, with the increase in the subsidy, the manufacturer and the remanufacturer will raise their prices to achieve more profits (as explained above).

Proposition 5 has compared prices in the <0,0> scenario with prices in other scenarios. Next, we will discuss the prices in each scenario and see what the difference in prices is and what the impact of the subsidies is on them. Figure 3 depicts the impacts of the subsidies on the prices of new and remanufactured products in scenario <n,0> and scenario <n,r>. Based on Figure 3, first, without considering that the manufacturer will change his donation strategy, the prices of two products in the market are always raised as the remanufacturer participates in donation, even if the donation strategy will make the remanufacturer's products for sale decline. It means that the remanufacturer could expand the price (remanufactured products' price) advantage through a donation strategy. This may help explain why the remanufacturer will follow the manufacturer's donation strategy. Second, as stated in Proposition 5, increasing subsidies can drive price advantages to appear. Figure 3 implies that both subsidies (the subsidy for new products and the subsidy for remanufactured products) have incentives for price advantages. It also shows that the subsidy for donated remanufactured products increases the price advantages of remanufactured products even more. Third, Figure 3 also tells us that adopting two subsidies' policies is more conducive to stimulating the prices of products. Consumers need to pay more in the market while receiving donations. One reason is that, after subsidizing the donated products, consumers have a high surplus value when they purchase the products, and meanwhile, consumers will achieve a high reserve price for the products, which leads to a high price for sale in the market.

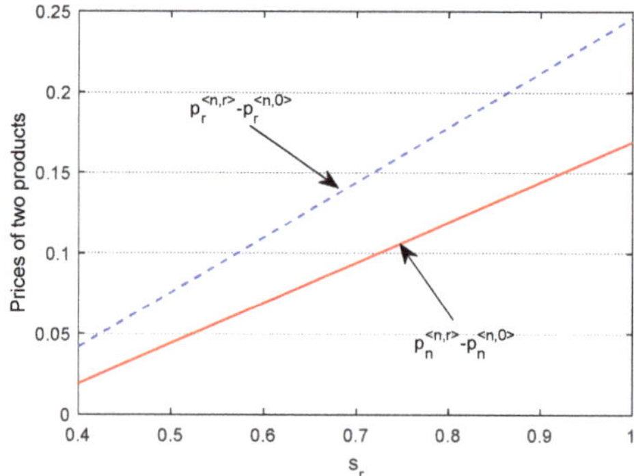

Figure 3. Price comparison between $<n,0>$ and $<n,r>$.

However, things are changed when $<n,r>$ scenario compares with $<0,r>$ and $<r,r>$ scenarios. Figure 4a reveals that, compared with the $<0,r>$ scenario, new products and remanufactured products have price advantages when the $<n,r>$ scenario is the two parties' donation strategy. However, with the increase in the subsidy for donated remanufactured products, the price advantages of the two products will decline until the advantages disappear. Figure 4b shows that compared with the $<r,r>$ scenario, new and remanufactured products have price disadvantages when the $<n,r>$ scenario is the two parties' donation strategy. With the increase in subsidy for donated remanufactured products, the price disadvantages of the two products will continue to expand. This means that consumers can pay less for new and remanufactured products while receiving donations.

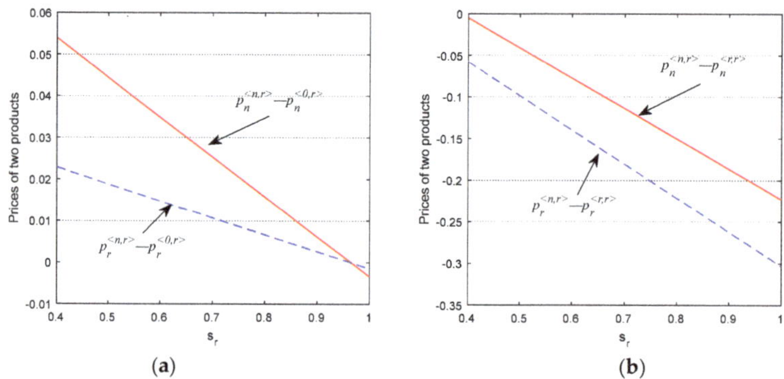

Figure 4. Price comparison under two subsidies policy. (**a**) $<n,r>$ and $<0,r>$. (**b**) $<n,r>$ and $<r,r>$.

This seems to be a paradox that conflicts with what Figure 3 shows. However, we notice that the obvious difference between Figures 3 and 4 is that the members of the new donation strategy are different. Figure 4a indicates that, when the manufacturer changes its donation strategy from no donation to donating new products, slight advantages can be found in both products. Figure 4b shows that, when the manufacturer changes its donation strategy from donating remanufactured products to donating new products, price disadvantages will appear. However, what the two figures have in common is that increasing the subsidy for remanufactured products will make the donation of new products not an ideal choice, especially when the remanufacturer adopts a donation strategy as a competing method in the market. This may explain why the manufacturer always loses his first-mover advantage when the subsidy for remanufactured products is very high and the remanufacturer is willing to donate.

Moreover, Figure 5a,b gives out price comparisons among the *<r,r>*, *<r,0>*, and *<0,r>* scenarios. Even though the manufacturer's donating remanufactured products strategy could make products on the market have price advantages, we notice that an increase in the subsidy for remanufactured products will result in the price of new products having the same price advantage as the price of the remanufactured products, significantly when the remanufacturer changes his strategy from no donation to donating, as shown in Figure 5a. Figure 5b indicates that, compared with the *<0,r>* scenario, both new and remanufactured products have price advantages, and the price advantage of remanufactured products is higher than that of new products in the *<r,r>* scenario. Different from what is shown in Figure 4, in Figure 5, the manufacturer adopts a donating remanufactured products strategy and seizes the first-mover advantage of donating remanufactured products. As a result, the remanufacturer's donation strategy no longer has a second-mover advantage. This implies that the decision-making sequence of the donation strategy determines the changes in products' prices in the market and also affects the decisions and advantages of the manufacturer and the remanufacturer. This will be further discussed in Section 5.

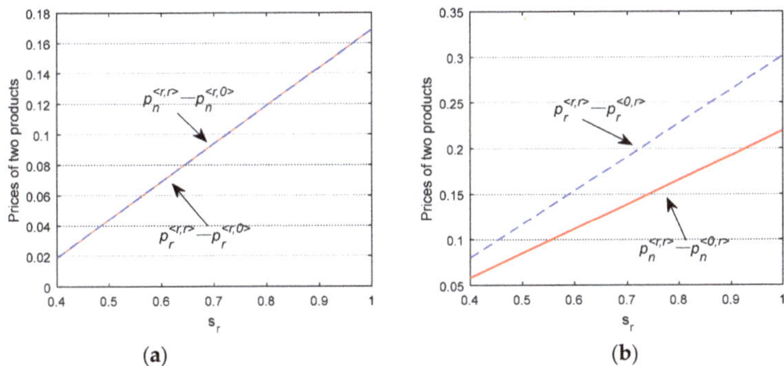

Figure 5. Price comparison under single subsidy policy. (**a**) *<r,r>* and *<r,0>*. (**b**) *<r,r>* and *<0,r>*.

4.4. Dilemma and Pareto Improvement

Different subsidies could lead to different product output and prices and could also change the profits of the manufacturer and the remanufacturer. To make the two parties' preferences on subsidy policy clear, this section will investigate whether the manufacturer and the remanufacturer may have a dilemma or Pareto improvement conditions under donation subsidies.

As stated in Section 4.1, the manufacturer is the leader in the market, and he has the first-mover advantage. Once the manufacturer decides to donate, the remanufacturer's decision will make the manufacturer's profit change. Proposition 6 below reveals the situation where the manufacturer may have a dilemma in the *<n,0>* scenario.

Proposition 6. *For the manufacturer in the <n,0> scenario, the manufacturer has a prisoner dilemma when there are two donation subsidies.*

Proposition 6 tells us that a low subsidy for the remanufactured products can incentivize the remanufacturer to engage in donating, which results in the manufacturer failing to achieve more profit. Especially when the manufacturer prefers to donate the new products and the subsidy for the new products is very high, the remanufacturer decides to donate. Therefore, the remanufacturer would choose to donate the remanufactured products to maximize his profit when $s_r^{<n,r>} < s_r^{**} < s_r < s_r^{<r,0>}$, as shown in Figure 6 where $\Pi_R^{<n,0>} < \Pi_R^{<n,r>}$. This will indirectly cause the donation strategy of both parties in the market to change from *<n,0>* to *<n,r>*, which also indicates that the *<n,0>* scenario has become a prisoner dilemma for the manufacturer.

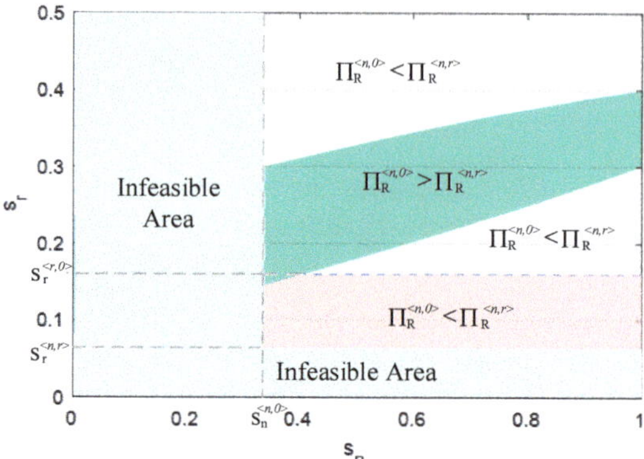

Figure 6. The manufacturer's prisoner dilemma in *<n,0>* scenario.

At this time, because of the changes in the remanufacturer's donation strategy, whether the manufacturer will gain or lose benefits is related to the government subsidies. We can see that, as the first-mover in the market, the manufacturer takes the lead in choosing the strategy of donating new products, that is, *<n,0>* is supposed to be a single-win situation for the manufacturer. However, with the remanufacturer's second-move choice, the donation equilibrium scenario turns out to be *<n,r>*. This also means that the manufacturer's benefits are affected by the remanufacturer's donation strategy, while the government's subsidy policy has a significant impact on the manufacturer's profits. Based on Figure 6, Figure 7 reveals a situation where the manufacturer is usually at a disadvantage in the market, especially when $\Pi_R^{\langle n,0\rangle} < \Pi_R^{\langle n,r\rangle}$. For example, the area "lose-win in *<n,r>*" is shown in Figure 7, $\Pi_M^{\langle n,0\rangle} > \Pi_M^{\langle n,r\rangle}$. This shows that the benefits of the manufacturer in *<n,0>* are due to the change in the remanufacturer's donation strategy, which leads to the decline of the manufacturer's profits in the *<n,r>* scenario. What's more, Figure 7 also verifies the result in Proposition 6 and tells us when $s_n > s_n^{\langle n,0\rangle}$ and $s_r^{\langle n,r\rangle} < s_r^{**} < s_r < s_r^{\langle r,0\rangle}$, the manufacturer's first-mover advantage does not work. Instead, he will be subject to the remanufacturer's second-mover advantage.

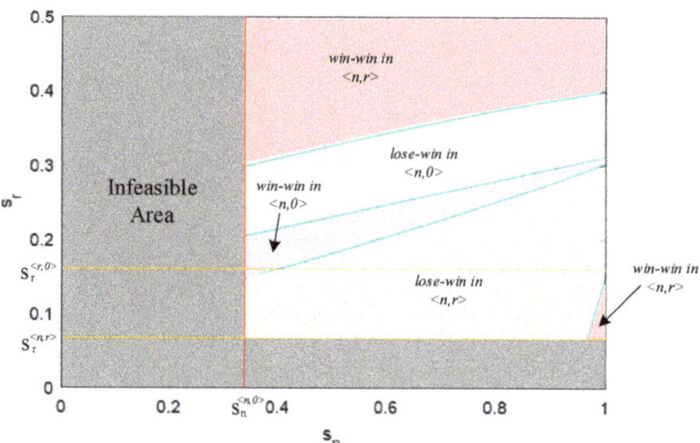

Figure 7. "Lose/win" in *<n,0>* and *<n,r>* scenarios.

However, we also find that when the government sets the two donation subsidies very high, it can often achieve a win-win situation for the manufacturer and the remanufacturer, as shown in "win-win in *<n,0>*" and "win-win in *<n,r>*" in Figure 7. From Figure 8 below, we notice that when the manufacturer chooses to donate new products, it does not mean that the higher the donation subsidy for the remanufactured products, the more incentive the remanufacturer will have to donate. As the area "lose-win in *<n,0>*" is shown in Figure 7, the remanufacturer could achieve a single-win situation, as long as he keeps his no-donation strategy. When the manufacturer realizes that the remanufacturer could benefit more without donating, this will weaken the manufacturer's incentives to donate. The management implication here is that when the government provides two subsidies for different donated products, it should scientifically and reasonably control the relationship between the subsidy for new products and the subsidy for remanufactured products to avoid the situation backfiring. In addition, with two subsidies existing in the market, both the manufacturer and the remanufacturer could not receive more benefits from donating, which indicates that the government should lower the subsidy for donated remanufactured products while keeping the subsidy for donated new products. In this way, a "win-win in $\langle n,0\rangle$" equilibrium will appear instead of a "lose-win in $\langle n,0\rangle$" situation. Or the government can increase the subsidy for donated new products while keeping the subsidy for donated remanufactured products unchanged, which also achieves the same effect.

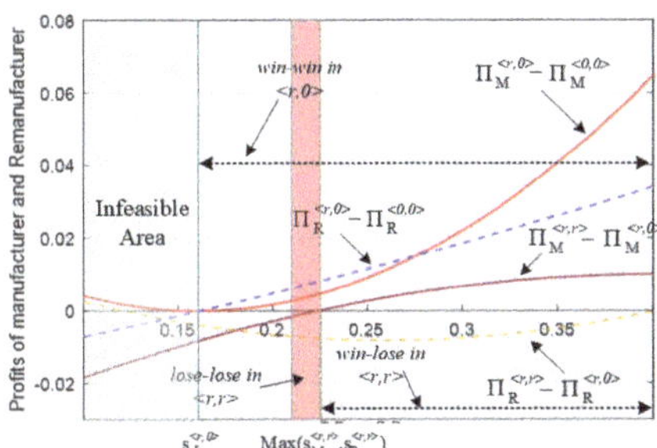

Figure 8. Comparison of profits between $\langle r,0\rangle$ and $\langle r,r\rangle$ scenarios.

The comparison of the manufacturer's profits between $\langle n,0\rangle$ and $\langle n,r\rangle$ scenarios has already proved that the remanufacturer's donation strategy may bring the manufacturer into a dilemma. Next, we will analyze the profits of the manufacturer and the remanufacturer between the $\langle r,0\rangle$ and $\langle r,r\rangle$ scenarios.

Figure 8 shows that when the manufacturer adopts a strategy of donating remanufactured products, it is the wisest choice for the remanufacturer to choose the no-donation strategy. Because in the $\langle r,0\rangle$ scenario, both of them could meet a win-win equilibrium. With an increase in the government subsidy for donated remanufactured products, the benefits to the manufacturer and the remanufacturer will gradually increase. If the remanufacturer also chooses to donate remanufactured products, that is, in the $\langle r,r\rangle$ scenario, when the subsidy for donated remanufactured products is low, there is a lose-lose ending, which makes the benefits of the manufacturer and the remanufacturer lower than those in the $\langle r,0\rangle$ scenario. When the subsidy for donated remanufactured products is set slightly higher, the manufacturer will gradually benefit from the $\langle r,r\rangle$ scenario. However, the remanufacturer will still be losing, which leads to a win-lose situation. Knowing that a win-win situation $\langle r,r\rangle$ only exists if the subsidy for donated remanufactured products is very high, our suggestions for the remanufacturer are as follows: (1) Suppose the subsidy for donated remanufactured products has met the prerequisites for donation. In that case, the remanufacturer should not immediately adopt the donation strategy and keep the no-donation strategy, as this may result in a lose-lose ending with the manufacturer or a situation where the remanufacturer loses but the manufacturer wins. (2) When the subsidy improves a lot, the remanufacturer can try to donate remanufactured products to achieve a win-win situation with the manufacturer. That is, conservative remanufacturers should adopt a non-donation strategy to benefit from the manufacturers' donation strategy; risky remanufacturers can choose to adopt the donation strategy to achieve more benefits when the subsidy for donated remanufactured products is high.

Proposition 7. *For the remanufacturer, when the government only provides the subsidy for donated remanufactured products, since the remanufacturer has a second-mover disadvantage, the <0,r> scenario will not be the optimal ending and this scenario is not stable. With the subsidy increasing, <r,r> will become the Nash equilibrium.*

Proposition 7 claims that the remanufacturer could achieve more benefits from donating remanufactured products because he is a follower in the market with a second-move disadvantage. However, meanwhile, the manufacturer will inevitably choose to donate to increase its profits. Therefore, the <0,r> scenario is a Pareto improvement for both the manufacturer and the remanufacturer, but it is also a feasible strategy for the remanufacturer. However, the <0,r> scenario is not stable. The reason for this is that the profits in the <r,r> scenario will be higher than those in the <0,r> scenario, and this can be predicted by the manufacturer, which will make the manufacturer decide to donate remanufactured products at the beginning to improve his benefits. As expected, this strategy will hurt the remanufacturer's benefits. However, there is nothing that the remanufacturer can do about this, and he can only choose to donate remanufactured products to prevent himself from losing more because his benefits in the <r,r> scenario are always higher than those in the <r,0> scenario.

Figure 9 indicates that when the government's subsidy for donated remanufactured products is low, it is unwise for the remanufacturer to donate because this will lead to a lose-lose ending for both the manufacturer and the remanufacturer in the <0,r> scenario. When the subsidy for donated remanufactured products is slightly increased, the manufacturer will benefit from the donation behavior of the remanufacturer. At this time, the remanufacturer still cannot achieve additional benefits from donating, and his donation strategy benefits the competitor (the manufacturer), which results in a win-lose ending. Only when the subsidy is high enough can the manufacturer and the remanufacturer achieve a win-win in the <0,r> scenario. From this, we know that the remanufacturer's willingness to donate would be affected by the donation subsidy policy. If the government wants to encourage the remanufacturer to donate, it needs to fully consider how to set up a subsidy to increase the remanufacturer's willingness to donate.

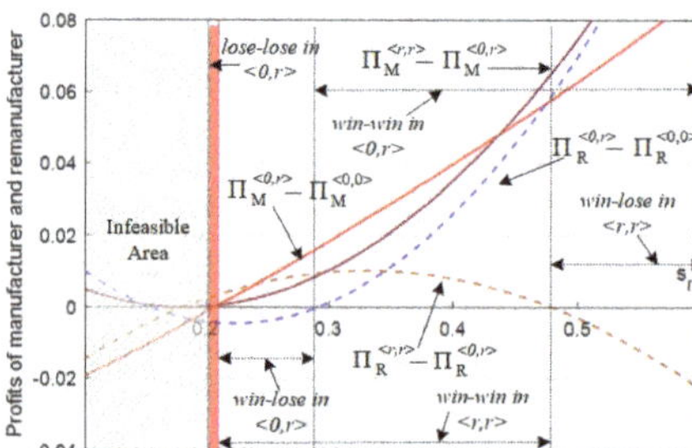

Figure 9. Dilemma for the remanufacturer in <0,r> and <r,r> scenarios.

We also find that when the remanufacturer adopts a donation strategy, the manufacturer will inevitably choose to donate remanufactured products because $\Pi_M^{<r,r>} > \Pi_M^{<0,r>}$. Figure 9 shows that the manufacturer and the remanufacturer may reach a win-win situation in the <r,r> scenario. However, when the subsidy for donated remanufactured products increases, we unexpectedly see that the manufacturer could still receive benefits but that the remanufacturer's benefits decline, and the final result is $\Pi_R^{<r,r>} < \Pi_R^{<0,r>}$. This shows that a high subsidy for donated remanufactured products will stimulate the manufacturer to participate in donating remanufactured products, but it will weaken the market competitiveness of the remanufacturer. Therefore, when the manufacturer and the remanufacturer adopt donation strategies at the same time, the government should pay attention to the subsidy to avoid a market imbalance. In the area of the "win-lose in

$<r,r>$" scenario, the remanufacturer would realize $\Pi_R^{<r,r>} < \Pi_R^{<0,r>}$, and also know that $\Pi_R^{<r,r>} > \Pi_R^{<r,0>}$ (as mentioned in Proposition 5). This means that the high subsidy for donated remanufactured products has effectively stimulated the manufacturer to donate. Under the first-mover advantage of the manufacturer, it is wise for the remanufacturer to donate too, otherwise, the remanufacturer will lose more benefits. Two management insights here are: (1) For the government, scientific and reasonable subsidy setting is of great significance. The too low subsidy will cause both the manufacturer and the remanufacturer to face a lose-lose ending and inhibit their enthusiasm for donation unexpectedly; however, the too-high subsidy will benefit the manufacturer but will weaken the remanufacturer's market position. (2) For the remanufacturer, $<0,r>$ is an optimal scenario lacking stability, and $<r,r>$ could be his Nash equilibrium. Thus, if the subsidy for donated remanufactured products is high and there is a threat from the manufacturer's donation strategy, the remanufacturer should donate products to protect himself from losing too much.

5. Extended Discussion and Managerial Implications

In this section, we discuss an extended situation where the remanufacturer moves first to decide whether to donate or not, and then the manufacturer chooses his donation strategy. To investigate and compare the impacts of the decision sequence on donation strategies, the strategy decision sequence and the product decision sequence are different from those in Figure 2. For the extended analysis, Proposition 8 highlights a new conclusion when the remanufacturer moves first. Compared with the results shown in Proposition 7, we can see a conclusion in favor of the remanufacturer. All the related proofs are given in Appendix C.

Proposition 8. *If the remanufacturer is the first-mover in the market, adopting a donation strategy in advance can improve his benefits.*

From Proposition 8, we can see that when the remanufacturer is the leader in the market, he will choose the donation strategy to maximize his profits because he could receive more in the $<0,r>$ scenario. Compared with Proposition 7, it can be seen that unlike the situation where the manufacturer is the leader in the market, the remanufacturer could avoid a lose-lose ending in the $<0,r>$ scenario, as in the $<0,r>$ scenario, the remanufacturer can still reach a single-win (lose-win) ending in the worst case. This is because, the first-mover advantage of the remanufacturer changes the former strategy decision-making sequence, thereby indirectly changing the former follower position of the remanufacturer, making him have the priority of donation. The remanufacturer could control the products' output and donation to reach his own optimal goals.

Figure 10a,b depict the profits of the manufacturer in the $<0,r>$, $<n,r>$, and $<r,r>$ scenarios. They illustrate that in the case of the remanufacturer as the market leader, there is no lose-lose ending in the $<0,r>$ scenario and a win-win situation may exist. Besides this, we notice that $\Pi_M^{<0,r>} > \Pi_M^{<r,r>}$ when the subsidy for donated remanufactured products is low and $\Pi_M^{<0,r>} > \Pi_M^{<n,r>}$ when the subsidy for donated new products is also low. This implies that $<0,r>$ is an optimal scenario for the manufacturer (adopting a no-donation strategy), no matter whether the government implements one or two subsidies for the donated products. This also shows that when the subsidy is low, the $<0,r>$ scenario is a stable win-win situation, which achieves the Pareto improvement. That is the biggest difference from the conclusion of Proposition 7. Furthermore, only under the condition of a win-win in $<0,r>$ scenario, it can be said that the remanufacturer's first-mover donation strategy can prevent the manufacturer's donation behavior. In other cases, the remanufacturer cannot prevent the manufacturer's donation.

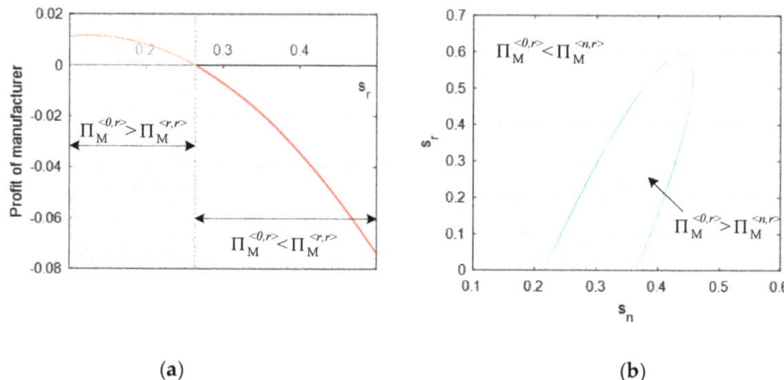

(a)   (b)

Figure 10. The profits of the manufacturer in the <0,r>, <n,r>, and <r,r> scenarios. (a) Profits in <0,r> and <r,r> scenarios. (b) Profits in <0,r> and <n,r> scenarios.

The analysis above has proved that when the manufacturer decides to donate nothing, the remanufacturer can receive greater benefits from donating. Next, we will discuss the situation where the manufacturer tries to donate new or remanufactured products and how will remanufacturer's the donation strategy change.

Proposition 9. *If the remanufacturer is the leader in the market, the donating remanufactured strategy of the manufacturer could not affect the remanufacturer's donation strategy.*

Proposition 9 indicates that when the remanufacturer has the first-mover advantage, the manufacturer's copycat strategy (donating remanufactured products) will not change (or threaten) the remanufacturer's donation behavior. First, this is because the government's subsidy policy has an incentive effect on the remanufacturer, which has prompted the remanufacturer to donate. Second, the remanufacturer has the first-mover advantage, which means he can predict the donation behavior of the manufacturer and make the optimal production and donation decisions for himself. Lastly, the amount of donated products by the manufacturer in the <r,r> scenario is lower than that in the <r,0> scenario, while the total output of remanufactured products in the market declines, which indirectly makes the prices of two products increase and expand the margin profit of the unit production of the remanufacturer. The management implication here is that if the remanufacturer is the leader in the market, the manufacturer cannot prevent the remanufacturer from adopting the donation strategy by donating remanufactured products. When comparing the <n,r> scenario and the <n,0> scenario, we find a similar conclusion, namely, that it is difficult for the manufacturer to prevent the remanufacturer's donation decision by donating new products, as shown in Figure 11.

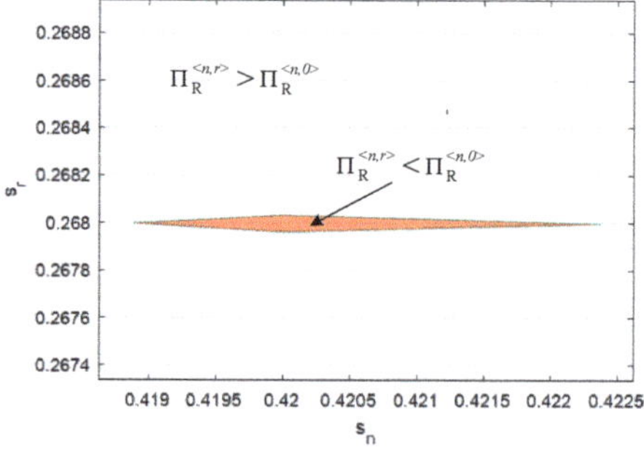

Figure 11. The profits of the remanufacturer in <n,0> and <n,r> scenarios.

In addition to the proof of Proposition 9, we can see that the donation strategy of the remanufacturer cannot bring more market share or enhance his market advantage, but it can increase his benefits. This shows that with the government subsidies, the donation is economical and can be an important method to enhance his brand awareness.

From the previous discussion, it can be seen that the profits of the remanufacturer in the *<r,r>* scenario are higher than those in the *<r,0>* scenario, and the profits in the *<n,r>* scenario are higher than those in the *<n,0>* scenario. Therefore, there are three main optimal scenarios for the remanufacturer: *<0,r>*, *<n,r>*, and *<r,r>*. Figure 12 characterizes the optimal scenario distribution of the remanufacturer when he is the leader in the market. In Figure 12, there is an interesting area, where the optimal scenario is *<r,0>* for the remanufacturer when the subsidies for new and remanufactured products are both high. This could be a counter-intuition phenomenon. In theory, if the government's donation subsidies are very high, the remanufacturer should be highly motivated to donate, but at this time, the remanufacturer chooses to give up the donation strategy to receive more benefits. The main reason for this is that, as the remanufacturer acts in a leading role in the market, he has the first-mover advantage and can predict the manufacturer's donation strategy when the subsidy is high. Therefore, after the manufacturer adopts the donation strategy under the stimulus of donation subsidies, the manufacturer's market investment will decrease, which will increase the prices of the products on the market and make the remanufacturer's benefits higher. This implies that the more the manufacturer donates, the more the remanufacturer will receive. This free-riding can also be predicted by the remanufacturer, which results in no donation from the remanufacturer. The management implication here is that the government's subsidies for donated products are not as high as possible. Too high donation subsidies will only lead some enterprises in the market to take "free-rider" behavior.

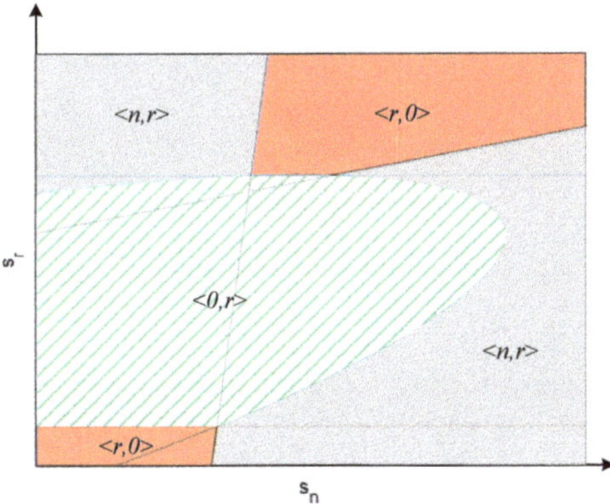

Figure 12. The optimal scenario distribution of the remanufacturer.

6. Conclusions and Remarks

In this paper, we have investigated how the subsidy policy affects the remanufacturing industry's donation strategy, which has not been discussed in the early literature. Specifically, we modelled a supply chain consisting of a manufacturer (M) and a remanufacturer (R). The manufacturer sells new and remanufactured products, but the remanufacturer only sells remanufactured products. In addition to the market competition for selling products, the two parties also compete in the donation field when the government provides subsidies for donated products. Our models for different scenarios are based on the Stackelberg game, which requires that strategic decisions and product decisions be made sequentially by the two parties.

To answer the research questions in Section 1, we summarize our main results and managerial insights as follows:

1. The sales share of subsidized products (such as when the manufacturer donates new products and also sells them) in the market will decrease with an increase in the subsidy. However, other products' output will increase. When two subsidies exist, the donated products will be negatively affected by the other subsidy's increase. Moreover, when the government adopts subsidy strategies for new products (or remanufactured products), the total amount of new products (or remanufactured products), including for sale and for donation, will increase, which means that donation subsidies have a positive stimulating effect on this product. For the prices, both new and remanufactured products may have price disadvantages; however, when the subsidy for donated remanufactured products is increasing, the price advantages of the two products will gradually appear. In other donation scenarios, the new products and the remanufactured products always have price advantages.

2. When the subsidies are high, the manufacturer and the remanufacturer will adopt a donation strategy in most cases. However, depending on the donation subsidy, the remanufacturer's willingness to donate may differ, while the manufacturer may have different donation strategies. The manufacturer and the remanufacturer will compete in the donation field. and their donation strategies will interact with each other. The manufacturer's willingness to donate also changes (the thresholds of subsidy for the manufacturer to donate are different) while the remanufacturer participates in the donation.

3. For the manufacturer in the $\langle n,0 \rangle$ scenario, the manufacturer has a prisoner dilemma when there are two donation subsidies. The low subsidy for donated remanufactured products can incentivize the remanufacturer to engage in donating, which results in the manufacturer failing to achieve more profit. Especially when the manufacturer prefers to donate the new product and when the subsidy for the new product is very high, the remanufacturer decides to donate. This will indirectly cause the donation strategy of both parties in the market to change from $\langle n,0 \rangle$ to $\langle n,r \rangle$, which also indicates that the $\langle n,0 \rangle$ scenario has become a prisoner dilemma for the manufacturer. However, for the remanufacturer, when the government only provides the subsidy for donated remanufactured products, since the remanufacturer has a second-mover disadvantage, the $\langle 0,r \rangle$ scenario will not be the optimal ending, and this scenario is not stable. With an increasing subsidy, $\langle r,r \rangle$ will become the Nash equilibrium finally. Thus, if the subsidy for donated remanufactured products is high and there is a threat to the manufacturer's donation strategy, the remanufacturer should donate products to protect himself from losing too much.

4. If the remanufacturer is the first-mover in the market, adopting a donation strategy in advance can improve his benefits. Especially when the manufacturer decides to donate nothing, the remanufacturer can achieve many more benefits from donating. However, the remanufactured strategy of the manufacturer could not affect the remanufacturer's donation strategy. Finally, the more the manufacturer donates, the more the remanufacturer will receive. This free-riding can also be predicted by the remanufacturer, which results in no donation from the remanufacturer.

There are some potential limitations to this study. First, we assume the manufacturer and the remanufacturer are in the same market. Thus, it is worthwhile to discuss whether the remanufacturer could encroach on the market where only the manufacturer exists by donating. Second, we examine donations as a competitive strategy for two parties. Future work can analyze donations as a cooperation strategy for the manufacturer to fulfill his extended producer responsibility with his partner remanufacturer. For example, the remanufacturer helps the manufacturer collect end-of-life products on the market and send them back to the manufacturer. The manufacturer then donates some remanufactured products to the remanufacturer for sale.

Author Contributions: Methodology, X.C.; software, J.W.; validation, X.C., Z.L. and J.W.; formal analysis, X.C.; investigation, Z.L.; resources, X.C.; data curation, J.W.; writing—original draft preparation, X.C.; writing—review and editing, X.C.; visualization, J.W.; supervision, Z.L.; project administration, J.W. All authors have read and agreed to the published version of the manuscript.

Funding: This research was funded by National Natural Science Foundation of China, grant number 72202103 and the Philosophy and Social Science Foundation of Jiangsu Higher Education Institutions of China, grant number 2020SJA0342.

Data Availability Statement: Data is unavailable due to privacy or ethical restrictions.

Conflicts of Interest: The authors declare no conflict of interest.

Appendix A

Scenario <0,0>

In this benchmark scenario, both the manufacturer and the remanufacturer choose no donation strategies. First, the manufacturer sets the output of new q_n and remanufactured products q_{Mr}, then the remanufacturer determines the output of remanufactured products q_{Rr}. According to Stackelberg game solving way, we can obtain Proposition A1.

Proposition A1. *Under scenario <0,0>, the equilibrium solutions are as follows:*

$$q_n^{<0,0>} = \frac{\alpha(1-\alpha)(\alpha+2c_{Rr})+c_{Mr}(2c_{Rr}+(2-\alpha)\alpha)}{4c_{Mr}(1+c_n)(c_{Rr}+\alpha)-2c_{Mr}\alpha^2+2\alpha(1+c_n-\alpha)(\alpha+2c_{Rr})};$$

$$q_{Mr}^{<0,0>} = \frac{c_n\alpha(2c_{Rr}+\alpha)}{4c_{Mr}(1+c_n)(c_{Rr}+\alpha)-2c_{Mr}\alpha^2+2\alpha(1+c_n-\alpha)(\alpha+2c_{Rr})};$$

$$q_{Rr}^{<0,0>} = \frac{1}{4}\left(\frac{\alpha}{c_{Rr}+\alpha}+\frac{2c_nc_{Mr}\alpha}{\alpha(1+c_n-\alpha)(\alpha+2c_{Rr})+c_{Mr}(2c_{Rr}(1+c_n)+\alpha(2+2c_n-\alpha))}\right).$$

Since no one would donate products to society, there will be no subsidy offering to enterprises and product outputs are mainly affected by companies' manufacturing costs and remanufacturing costs. Thus, as shown in Proposition A1, higher costs always decrease the profits of two firms.

Proof of Proposition A1. In <0,0> scenario, the manufacturer and remanufacturer will not choose to donate, which means $q_{Mdn} = 0$, $q_{Mdr} = 0$, $q_{Rdr} = 0$. To verify the existence of extreme value of (1), the Hessian matrix of (1) is obtain as follows:

$$H\left(\Pi_M^{<0,0>}\right) = \begin{bmatrix} \frac{\partial^2\Pi_M^{<0,0>}}{\partial q_n^2} & \frac{\partial^2\Pi_M^{<0,0>}}{\partial q_n\partial q_{Mr}} \\ \frac{\partial^2\Pi_M^{<0,0>}}{\partial q_{Mr}\partial q_n} & \frac{\partial^2\Pi_M^{<0,0>}}{\partial q_{Mr}^2} \end{bmatrix} = \begin{bmatrix} -2-2c_n & -2\alpha \\ -2\alpha & -2\alpha-2c_{Mr} \end{bmatrix}$$

From the Hessian matrix above, it is easy to find that $H_1\left(\Pi_M^{<0,0>}\right) = -2-2c_n < 0$, $H_2\left(\Pi_M^{<0,0>}\right) = 4\left((1+c_n)c_{Mr}+(1+c_n-\alpha)\alpha\right) > 0$. Therefore, the Hessian matrix is negative, indicating that formula (1) is a concave function about q_n and q_{Mr}, and there is a maximum value. Besides, from formula (2), we can know $\partial^2\Pi_R^{<0,0>}/\partial(q_{Rr})^2 = -2c_{Rr}-2\alpha < 0$, which indicates that formula (2) is a concave function about q_{Rr}, and there is a maximum value. Using the reverse order solution method, we first solve the formula (2) and let $\partial\Pi_R^{<0,0>}/\partial q_{Rr} = 0$. We can have $q_{Rr} = (1-q_n-q_{Mr})\alpha/(2(\alpha+c_{Rr}))$. And then we bring the value of q_{Rr} into formula (1), and make $\partial\Pi_M^{<0,0>}/\partial q_n = 0$ and $\partial\Pi_M^{<0,0>}/\partial q_{Mr} = 0$. We can get

$$q_n^{<0,0>} = \frac{\alpha(1-\alpha)(\alpha+2c_{Rr})+c_{Mr}(2c_{Rr}+(2-\alpha)\alpha)}{4c_{Mr}(1+c_n)(c_{Rr}+\alpha)-2c_{Mr}\alpha^2+2\alpha(1+c_n-\alpha)(\alpha+2c_{Rr})} \tag{A1}$$

$$q_{Mr}^{<0,0>} = \frac{c_n\alpha(2c_{Rr}+\alpha)}{4c_{Mr}(1+c_n)(c_{Rr}+\alpha)-2c_{Mr}\alpha^2+2\alpha(1+c_n-\alpha)(\alpha+2c_{Rr})} \tag{A2}$$

At last, we take (A1) and (A2) into q_{Rr} and we can have

$$q_{Rr}^{<0,0>} = \frac{1}{4}\left(\frac{\alpha}{c_{Rr}+\alpha} + \frac{2c_n c_{Mr}\alpha}{\alpha(1+c_n-\alpha)(\alpha+2c_{Rr})+c_{Mr}(2c_{Rr}(1+c_n)+\alpha(2+2c_n-\alpha))}\right) \tag{A3}$$

$\square$

Scenario *<n,0>*

In this scenario, the manufacturer decides to donate new products to society, while the government would provide a new products donation subsidy. Hence, the remanufacturer needs to set the output of remanufactured products q_{Rr} after the manufacturer determines the output of new q_n and remanufactured products q_{Mr}, and the number of donated ones q_{dn}. By solving the problems, we can obtain Proposition A2.

Proposition A2. *Under scenario <n,0>, the equilibrium solutions are as follows:*

$$q_n^{<n,0>} = -\frac{\alpha(2c_{Rr}+\alpha)(-1+\Delta_n+\alpha)+c_{Mr}(2c_{Rr}(-1+\Delta_n)+\alpha(-2+2\Delta_n+\alpha))}{2(\alpha(1-\alpha)(2c_{Rr}+\alpha)+c_{Mr}(2c_{Rr}+(2-\alpha)\alpha))};$$

$$q_{Mr}^{<n,0>} = \frac{\alpha(2c_{Rr}+\alpha)\Delta_n}{2\alpha(1-\alpha)(2c_{Rr}+\alpha)+c_{Mr}(4c_{Rr}+2\alpha(2-\alpha))};$$

$$q_{Mdn}^{<n,0>} = \frac{\Delta_n}{2c_n} + \frac{c_n(\alpha(2c_{Rr}+\alpha)(-1+\Delta_n+\alpha)+c_{Mr}(2c_{Rr}(-1+\Delta_n)+\alpha(-2+2\Delta_n+\alpha)))}{2c_n(\alpha(1-\alpha)(2c_{Rr}+\alpha)+c_{Mr}(2c_{Rr}+(2-\alpha)\alpha))};$$

$$q_{Rr}^{<n,0>} = \frac{\alpha(\alpha(\alpha-1)(2c_{Rr}+\alpha)-c_{Mr}(2c_{Rr}(1+\Delta_n)+\alpha(2+2\Delta_n-\alpha)))}{4(c_{Rr}+\alpha)\left((\alpha-1)\alpha(2c_{Rr}+\alpha)-c_{Mr}\left(2c_r^R+(2-\alpha)\alpha\right)\right)}.$$

where $\Delta_n = s_n + \theta_n$.

Proposition A2 shows that, once the manufacturer chooses to donate new products, the donation subsidy not only affects the manufacturer's donation decision on how much to donate but also has an impact on the output of remanufactured products from two firms.

Proof of Proposition A2. In *<n,0>* scenario, only the manufacturer donates the new products, which means $q_{Mdn} \neq 0$, $q_{Mdr} = 0$, $q_{Rdr} = 0$. To verify the existence of extreme value of (1), the Hessian matrix of (1) is obtain as follows:

$$H\left(\Pi_M^{<n,0>}\right) = \begin{bmatrix} \dfrac{\partial^2\Pi_M^{<n,0>}}{\partial q_n^2} & \dfrac{\partial^2\Pi_M^{<n,0>}}{\partial q_n\partial q_{Mr}} & \dfrac{\partial^2\Pi_M^{<n,0>}}{\partial q_n\partial q_{Mdn}} \\ \dfrac{\partial^2\Pi_M^{<n,0>}}{\partial q_n\partial q_{Mr}} & \dfrac{\partial^2\Pi_M^{<n,0>}}{\partial q_{Mr}^2} & \dfrac{\partial^2\Pi_M^{<n,0>}}{\partial q_{Mdn}\partial q_{Mr}} \\ \dfrac{\partial^2\Pi_M^{<n,0>}}{\partial q_n\partial q_{Mdn}} & \dfrac{\partial^2\Pi_M^{<n,0>}}{\partial q_{Mdn}\partial q_{Mr}} & \dfrac{\partial^2\Pi_M^{<n,0>}}{\partial q_{Mdn}^2} \end{bmatrix} = \begin{bmatrix} -2-2c_n & -2\alpha & -2c_n \\ -2\alpha & -2\alpha-2c_{Mr} & 0 \\ -2c_n & 0 & -2c_n \end{bmatrix}$$

From the Hessian matrix above, it is easy to find that $H_1\left(\Pi_M^{<n,0>}\right) = -2-2c_n < 0$, $H_2\left(\Pi_M^{<n,0>}\right) = 4\left((1+c_n)c_{Mr}+(1+c_n-\alpha)\alpha\right) > 0$, $H_3\left(\Pi_M^{<n,0>}\right) = -8c_{Mr}c_n - 8c_n\alpha(1-\alpha) < 0$. Thus, the Hessian matrix is negative, indicating that formula (1) is a concave function and there is a maximum value. Besides, from formula (2), we can know $\partial^2\Pi_R^{<n,0>}/\partial(q_{Rr})^2 = -2c_{Rr}-2\alpha < 0$. This shows that formula (2) is a concave function, and there is a maximum value. Using the reverse order solution method to solve the production decision of the remanufacturer, we let $\partial\Pi_R^{<n,0>}/\partial q_{Rr} = 0$ and we can have $q_{Rr} = \alpha(1-q_n-q_{Mr})/(2(\alpha+c_{Rr}))$. Then we take q_{Rr} into formula (1) and set $\partial\Pi_M^{<n,0>}/\partial q_n = 0$, $\partial\Pi_M^{<n,0>}/\partial q_{Mr} = 0$, and $\partial\Pi_M^{<n,0>}/\partial q_{Mdn} = 0$. And we can get

$$q_n^{<n,0>} = -\frac{\alpha(2c_{Rr}+\alpha)(-1+\Delta_n+\alpha)+c_{Mr}(2c_{Rr}(-1+\Delta_n)+\alpha(-2+2\Delta_n+\alpha))}{2(\alpha(1-\alpha)(2c_{Rr}+\alpha)+c_{Mr}(2c_{Rr}+(2-\alpha)\alpha))} \tag{A4}$$

$$q_{Mr}^{<n,0>} = \frac{\alpha(2c_{Rr}+\alpha)\Delta_n}{2\alpha(1-\alpha)(2c_{Rr}+\alpha)+c_{Mr}(4c_{Rr}+2\alpha(2-\alpha))} \tag{A5}$$

$$q_{Mdn}^{<n,0>} = \frac{\Delta_n}{2c_n} + \frac{c_n(\alpha(2c_{Rr}+\alpha)(-1+\Delta_n+\alpha)+c_{Mr}(2c_{Rr}(-1+\Delta_n)+\alpha(-2+2\Delta_n+\alpha)))}{2c_n(\alpha(1-\alpha)(2c_{Rr}+\alpha)+c_{Mr}(2c_{Rr}+(2-\alpha)\alpha))} \tag{A6}$$

At last, we take (A4), (A5) and (A6) into q_{Rr} and we can have

$$q_{Rr}^{<n,0>} = \frac{\alpha(\alpha(\alpha-1)(2c_{Rr}+\alpha)-c_{Mr}(2c_{Rr}(1+\Delta_n)+\alpha(2+2\Delta_n-\alpha)))}{4(c_{Rr}+\alpha)((\alpha-1)\alpha(2c_{Rr}+\alpha)-c_{Mr}(2c_r^R+(2-\alpha)\alpha))} \tag{A7}$$

$\square$

Scenario <r,0>

In this scenario, the manufacturer decides to donate remanufactured products to society, while the government would provide remanufactured products donation subsidy. Hence, the remanufacturer needs to set the output of remanufactured products q_{Rr} after the manufacturer determines the output of new q_n and remanufactured products q_{Mr}, and the number of donated ones q_{Mdr}. By solving the problems, we can obtain Proposition A3.

Proposition A3. *Under scenario <r,0>, the equilibrium solutions are as follows:*

$$q_n^{<r,0>} = \frac{1+\Delta_r-\alpha}{2(1+c_n-\alpha)};$$

$$q_{Mr}^{<r,0>} = \frac{2c_{Rr}(\Delta_r+c_n\Delta_r-c_n\alpha)-\alpha(\Delta_r(\alpha-2-2c_n)+\alpha c_n)}{2\alpha(\alpha-1-c_n)(2c_{Rr}+\alpha)};$$

$$q_{Mdr}^{<r,0>} = \frac{1}{2}\left(\frac{\Delta_r}{c_{Mr}} + \frac{\eta_2+\alpha(\Delta_r(\alpha-2-2c_n)+\alpha c_n)}{\alpha(\alpha-1-c_n)(2c_{Rr}+\alpha)}\right);$$

$$q_{Rr}^{<r,0>} = \frac{2c_{Rr}(\Delta_r+\alpha)+\alpha(2\Delta_r+\alpha)}{4(c_{Rr}+\alpha)(2c_{Rr}+\alpha)},$$

where $\Delta_r = s_r + \theta_r$.

Proposition A3 extends the conclusion of Proposition A2, revealing the situation where the manufacturer tries to donate remanufactured products and the government provides the remanufactured products donation subsidy. Similar to Proposition A2, the donation subsidy for remanufactured products also shows its impact on all products existing in the market.

Proof of Proposition A3. In <r,0> scenario, only the manufacturer donates the remanufactured products, which means $q_{Mdn} = 0$, $q_{Mdr} \neq 0$, $q_{Rdr} = 0$. To verify the existence of extreme value of (1), the Hessian matrix of (1) is obtain as follows:

$$H\left(\Pi_M^{<r,0>}\right) = \begin{bmatrix} \frac{\partial^2\Pi_M^{<r,0>}}{\partial q_n^2} & \frac{\partial^2\Pi_M^{<r,0>}}{\partial q_n\partial q_{Mr}} & \frac{\partial^2\Pi_M^{<r,0>}}{\partial q_n\partial q_{Mdr}} \\ \frac{\partial^2\Pi_M^{<r,0>}}{\partial q_n\partial q_{Mr}} & \frac{\partial^2\Pi_M^{<r,0>}}{\partial q_{Mr}^2} & \frac{\partial^2\Pi_M^{<r,0>}}{\partial q_{Mdr}\partial q_{Mr}} \\ \frac{\partial^2\Pi_M^{<r,0>}}{\partial q_n\partial q_{Mdr}} & \frac{\partial^2\Pi_M^{<r,0>}}{\partial q_{Mdr}\partial q_{Mr}} & \frac{\partial^2\Pi_M^{<r,0>}}{\partial q_{Mdr}^2} \end{bmatrix} = \begin{bmatrix} -2-2c_n & -2\alpha & 0 \\ -2\alpha & -2\alpha-2c_{Mr} & -2c_{Mr} \\ 0 & -2c_{Mr} & -2c_{Mr} \end{bmatrix}$$

From the Hessian matrix above, it is easy to find that $H_1\left(\Pi_M^{<r,0>}\right) = -2-2c_n < 0$, $H_2\left(\Pi_M^{<r,0>}\right) = 4\left((1+c_n)c_{Mr}+(1+c_n-\alpha)\alpha\right) > 0$, $H_3\left(\Pi_M^{<r,0>}\right) = -8c_{Mr}c_n\alpha - 8c_{Mr}\alpha(1-\alpha) < 0$. Thus, the Hessian matrix is negative, indicating that formula (1) is a concave function and there is a maximum value. Besides, from formula (2), we can know $\partial^2\Pi_R^{<r,0>}/\partial(q_{Rr})^2 = -2c_{Rr} - 2\alpha < 0$. This shows that formula (2) is a concave function, and there is a maximum value. Using the reverse order solution method to solve the production decision of the remanufacturer, we let $\partial\Pi_R^{<r,0>}/\partial q_{Rr} = 0$ and we can have $q_{Rr} = (1 - q_n - q_{Mr})\alpha/(2(\alpha + c_{Rr}))$. Then we take q_{Rr} into formula (1) and set $\partial\Pi_M^{<r,0>}/\partial q_n = 0$, $\partial\Pi_M^{<r,0>}/\partial q_{Mr} = 0$, and $\partial\Pi_M^{<r,0>}/\partial q_{Mdr} = 0$. And we can get

$$q_n^{<r,0>} = \frac{1+s_r-\alpha+\theta_r}{2(1+c_n-\alpha)} \tag{A8}$$

$$q_{Mr}^{<r,0>} = \frac{2c_{Rr}(s_r+c_ns_r-c_n\alpha+\theta_r+c_n\theta_r)-\alpha(s_r(\alpha-2-2c_n)+c_n(\alpha-2\theta_r)+\theta_r(\alpha-2))}{2\alpha(\alpha-1-c_n)(2c_{Rr}+\alpha)} \tag{A9}$$

$$q_{Mdr}^{<r,0>} = \frac{1}{2}\left(\frac{s_r + \theta_r}{c_{Mr}} + \frac{\eta_1 + \alpha(s_r(\alpha - 2 - 2c_n) + c_n(\alpha - 2\theta_r) + (\alpha - 2)\theta_r)}{\alpha(\alpha - 1 - c_n)(2c_{Rr} + \alpha)}\right) \tag{A10}$$

where $\eta_1 = -2c_{Rr}(s_r + c_n(s_r - \alpha + \theta_r) + \theta_r)$. At last, we take (A8), (A9) and (A10) into q_{Rr} and we can have

$$q_{Rr}^{<r,0>} = \frac{2c_{Rr}(s_r + \alpha + \theta_r) + \alpha(2s_r + \alpha + 2\theta_r)}{4(c_{Rr} + \alpha)(2c_{Rr} + \alpha)} \tag{A11}$$

□

Scenario <0,r>

In this scenario, the manufacturer decides not to donate any products, though the government would provide remanufactured products donation subsidy. However, the remanufacturer prefers to donate to society and gain his reputation. Thus, the remanufacturer needs to set the output of remanufactured products q_{Rr} and his donating quantity q_{Rdr} after the manufacturer determines the output of new q_n and remanufactured products q_{Mr}. By solving the problems, we can obtain Proposition A4.

Proposition A4. *Under scenario <0,r>, the equilibrium solutions are as follows:*

$$q_n^{<0,r>} = \frac{\alpha(1-\alpha) + c_{Mr}(2 + \Delta_r - \alpha)}{2(c_{Mr}(2 + 2c_n - \alpha) + \alpha(1 + c_n - \alpha))};$$

$$q_{Mr}^{<0,r>} = \frac{\Delta_r(1 + c_n - \alpha) + \alpha c_n}{2(c_{Mr}(2 + 2c_n - \alpha) + \alpha(1 + c_n - \alpha))};$$

$$q_{Rr}^{<0,r>} = \frac{\alpha(1 + c_n - \alpha)(3\Delta_r - \alpha) + c_{Mr}\eta_1}{4\alpha(c_{Mr}(-2 - 2c_n + \alpha) + \alpha(-1 - c_n + \alpha))};$$

$$q_{Rdr}^{<0,r>} = \frac{\alpha(1 + c_n - \alpha)(2\alpha\Delta_r + c_{Rr}(3\Delta_r - \alpha)) + c_{Mr}(2\alpha\Delta_r(2 + 2c_n - \alpha) + \eta_1 c_{Rr})}{4\alpha c_{Rr}(c_{Mr}(2 + 2c_n - \alpha) + \alpha(1 + c_n - \alpha))},$$

where $\eta_1 = \Delta_r(4 + 4c_n - \alpha) + \alpha^2 - \alpha(2 + 4c_n)$.

Proposition A4 complements Proposition A3 from the perspective of only the manufacturer donating. As the same as Proposition A3 states, the donation subsidy for remanufactured products will affect two firms' product decisions.

Proof of Proposition A4. In <0,r> scenario, only the remanufacturer donates the remanufactured products, which means $q_{Mdn} = 0$, $q_{Mdr} = 0$, $q_{Rdr} \neq 0$. To verify the existence of extreme value of (1), the Hessian matrix of (1) is obtain as follows:

$$H\left(\Pi_M^{<0,r>}\right) = \begin{bmatrix} \frac{\partial^2 \Pi_M^{<0,r>}}{\partial q_n^2} & \frac{\partial^2 \Pi_M^{<0,r>}}{\partial q_n \partial q_{Mr}} \\ \frac{\partial^2 \Pi_M^{<0,r>}}{\partial q_{Mr} \partial q_n} & \frac{\partial^2 \Pi_M^{<0,r>}}{\partial q_{Mr}^2} \end{bmatrix} = \begin{bmatrix} -2 - 2c_n & -2\alpha \\ -2\alpha & -2\alpha - 2c_{Mr} \end{bmatrix}$$

From the Hessian matrix above, it is easy to find that $H_1\left(\Pi_M^{<0,r>}\right) = -2 - 2c_n < 0$, $H_2\left(\Pi_M^{<0,r>}\right) = 4\left((1 + c_n)c_{Mr} + (1 + c_n - \alpha)\alpha\right) > 0$. Thus, the Hessian matrix is negative, indicating that formula (1) is a concave function and there is a maximum value. And the Hessian matrix of formula (2) is as follows:

$$H\left(\Pi_R^{<0,r>}\right) = \begin{bmatrix} \frac{\partial^2 \Pi_R^{<0,r>}}{\partial q_{Rr}^2} & \frac{\partial^2 \Pi_R^{<0,r>}}{\partial q_{Rr} \partial q_{Rdr}} \\ \frac{\partial^2 \Pi_R^{<0,r>}}{\partial q_{Rdr} \partial q_{Rr}} & \frac{\partial^2 \Pi_R^{<0,r>}}{\partial q_{Rdr}^2} \end{bmatrix} = \begin{bmatrix} -2\alpha - 2c_{Rr} & -2c_{Rr} \\ -2c_{Rr} & -2c_{Rr} \end{bmatrix}$$

We can see that $H_1\left(\Pi_R^{<0,r>}\right) = -2\alpha - 2c_{Rr} < 0$ and $H_2\left(\Pi_R^{<0,r>}\right) = 4\alpha c_{Rr} > 0$. So, the Hessian matrix is negative, indicating that formula (2) is a concave function and there is a maximum value. Using the reverse order solution method to solve the production decision of the remanufacturer, we first let $\partial \Pi_R^{<0,r>}/\partial q_{Rr} = 0$ and $\partial \Pi_R^{<0,r>}/\partial q_{Rdr} = 0$. Then we can have:

$$q_{Rr} = -\frac{s_r + \alpha(q_n + q_{Mr} - 1) + \theta_r}{2\alpha} \tag{A12}$$

$$q_{Rdr} = \frac{\alpha(s_r + \theta_r) + c_{Rr}(sr + \theta_r + \alpha(q_n + q_{Mr} - 1))}{2\alpha c_{Rr}} \tag{A13}$$

We take (A12) and (A13) into formula (1), and set $\partial \Pi_M^{<0,r>}/\partial q_n = 0$ and $\partial \Pi_M^{<0,r>}/\partial q_{Mr} = 0$. We can have

$$q_n^{<0,r>} = \frac{\alpha(1 - \alpha) + c_{Mr}(2 + s_r - \alpha + \theta_r)}{2(c_{Mr}(2 + 2c_n - \alpha) + \alpha(1 + c_n - \alpha))} \tag{A14}$$

$$q_{Mr}^{<0,r>} = \frac{s_r(1 + c_n - \alpha) + \theta_r(1 - \alpha) + c_n(\theta_r + \alpha)}{2(c_{Mr}(2 + 2c_n - \alpha) + \alpha(1 + c_n - \alpha))} \tag{A15}$$

We bring (A14) and (A15) back to (A12) and (A13). We can get

$$q_{Rr}^{<0,r>} = \frac{\alpha(1 + c_n - \alpha)(3\Delta_r - \alpha) + c_{Mr}\left(s_r(4 + 4c_n - \alpha) + \alpha^2 + 4\theta_r(1 + c_n) - \alpha(2 + 4c_n + \theta_r)\right)}{4\alpha(c_{Mr}(-2 - 2c_n + \alpha) + \alpha(-1 - c_n + \alpha))} \tag{A16}$$

$$q_{Rdr}^{<0,r>} = \frac{\alpha(1 + c_n - \alpha)(2\alpha\Delta_r + c_{Rr}(3\Delta_r - \alpha)) + c_{Mr}(2\alpha\Delta_r(2 + 2c_n - \alpha) + \eta_2 c_{Rr})}{4\alpha c_{Rr}(c_{Mr}(2 + 2c_n - \alpha) + \alpha(1 + c_n - \alpha))} \tag{A17}$$

where $\eta_2 = s_r(4 + 4c_n - \alpha) + \alpha^2 + 4(1 + c_n)\theta_r - \alpha(2 + 4c_n + \theta_r)$. $\square$

Scenario <n,r>

In this scenario, both the manufacturer and the remanufacturer decide to donate products, while the manufacturer chooses to donate new products and the remanufacturer donates remanufactured products. And the government would provide two kinds of donation subsidies for firms at the same time. Thus, the remanufacturer needs to set the output of remanufactured products q_{Rr} and his donating quantity q_{Rdr} after the manufacturer determines the output of new q_n and remanufactured products q_{Mr}, and the donating quantity of new products q_{dn}. By solving the problems, we can obtain Proposition A5.

Proposition A5. *Under scenario <n,r>, the equilibrium solutions are as follows:*

$$q_n^{<n,r>} = \frac{\alpha(-1 + \Delta_n + \alpha) + c_{Mr}(-2 + 2\Delta_n - \Delta_r + \alpha)}{2(c_{Mr}(-2 + \alpha) + (-1 + \alpha)\alpha)};$$
$$q_{Mr}^{<n,r>} = \frac{\Delta_r - \Delta_r\alpha + \Delta_n\alpha}{2(c_{Mr}(2 - \alpha) + (1 - \alpha)\alpha)};$$
$$q_{Mdn}^{<n,r>} = \frac{\Delta_n}{2c_n} + \frac{c_n(\alpha(-1 + \Delta_n + \alpha) + c_{Mr}(-2 + 2\Delta_n - \Delta_r + \alpha))}{2c_n(c_{Mr}(2 - \alpha) + (1 - \alpha)\alpha)};$$
$$q_{Rr}^{<n,r>} = \frac{(1 - \alpha)\alpha(\alpha - 3\Delta_r) + c_{Mr}(\alpha\Delta_r - 4\Delta_r + \alpha(2 + 2\Delta_n - \alpha))}{4\alpha(c_{Mr}(2 - \alpha) + (1 - \alpha)\alpha)};$$
$$q_{Rdr}^{<n,r>} = \frac{\eta_2 + c_{Mr}(2(2 - \alpha)\alpha\Delta_r + c_{Rr}(4\Delta_r - \alpha(2 + \Delta_r + 2\Delta_n - \alpha)))}{4c_{Rr}\alpha(c_{Mr}(2 - \alpha) + (1 - \alpha)\alpha)},$$

where $\eta_2 = (1 - \alpha)\alpha(2\alpha\Delta_r + c_{Rr}(3\Delta_r - \alpha))$.

From Proposition A5, we notice that the optimal product decisions are quite different from Proposition A3 and Proposition A4, while the government offers two donation subsidies at the same time. The output of each product would be affected by two donation subsidies together.

Proof of Proposition A5. In <n,r> scenario, the manufacturer donates the new products and the remanufacturer donates the remanufactured products, which means $q_{Mdn} \neq 0$, $q_{Mdr} = 0$, $q_{Rdr} \neq 0$. To verify the existence of extreme value of (1), the Hessian matrix of (1) is obtain as follows:

$$H(\Pi_M^{<n,r>}) = \begin{bmatrix} \frac{\partial^2 \Pi_M^{<n,r>}}{\partial q_n^2} & \frac{\partial^2 \Pi_M^{<n,r>}}{\partial q_n \partial q_{Mr}} & \frac{\partial^2 \Pi_M^{<n,r>}}{\partial q_n \partial q_{Mdn}} \\ \frac{\partial^2 \Pi_M^{<n,r>}}{\partial q_n \partial q_{Mr}} & \frac{\partial^2 \Pi_M^{<n,r>}}{\partial q_{Mr}^2} & \frac{\partial^2 \Pi_M^{<n,r>}}{\partial q_{Mdn} \partial q_{Mr}} \\ \frac{\partial^2 \Pi_M^{<n,r>}}{\partial q_n \partial q_{Mdn}} & \frac{\partial^2 \Pi_M^{<n,r>}}{\partial q_{Mdn} \partial q_{Mr}} & \frac{\partial^2 \Pi_M^{<n,r>}}{\partial q_{Mdn}^2} \end{bmatrix} = \begin{bmatrix} -2 - 2c_n & -2\alpha & -2c_n \\ -2\alpha & -2\alpha - 2c_{Mr} & 0 \\ -2c_n & 0 & -2c_n \end{bmatrix}$$

From the Hessian matrix above, it is easy to find that $H_1\left(\Pi_M^{<n,r>}\right) = -2 - 2c_n < 0$, $H_2\left(\Pi_M^{<n,r>}\right) = 4\left((1 + c_n)\,c_{Mr} + (1 + c_n - \alpha)\,\alpha\right) > 0$, $H_3\left(\Pi_M^{<n,r>}\right) = -8c_{Mr}c_n - 8c_n\alpha(1 - \alpha)$ < 0. Thus, the Hessian matrix is negative, indicating that formula (1) is a concave function and there is a maximum value. And the Hessian matrix of formula (2) is as follows:

$$H\left(\Pi_R^{<n,r>}\right) = \begin{bmatrix} \dfrac{\partial^2 \Pi_R^{<n,r>}}{\partial q_{Rr}^2} & \dfrac{\partial^2 \Pi_R^{<n,r>}}{\partial q_{Rr}\partial q_{Rdr}} \\ \dfrac{\partial^2 \Pi_R^{<n,r>}}{\partial q_{Rdr}\partial q_{Rr}} & \dfrac{\partial^2 \Pi_R^{<n,r>}}{\partial q_{Rdr}^2} \end{bmatrix} = \begin{bmatrix} -2\alpha - 2c_{Rr} & -2c_{Rr} \\ -2c_{Rr} & -2c_{Rr} \end{bmatrix}$$

We can see that $H_1\left(\Pi_R^{<n,r>}\right) = -2\alpha - 2c_{Rr} < 0$ and $H_2\left(\Pi_R^{<n,r>}\right) = 4\alpha c_{Rr} > 0$. So, the Hessian matrix is negative, indicating that formula (2) is a concave function and there is a maximum value. Using the reverse order solution method to solve the production decision of the remanufacturer, we first let $\partial\Pi_R^{<n,r>}/\partial q_{Rr} = 0$ and $\partial\Pi_R^{<n,r>}/\partial q_{Rdr} = 0$. Then we can have:

$$q_{Rr} = -\frac{s_r + \alpha(q_n + q_{Mr} - 1) + \theta_r}{2\alpha} \tag{A18}$$

$$q_{Rdr} = \frac{\alpha(s_r + \theta_r) + c_{Rr}(s_r + \alpha(q_n + q_{Mr} - 1) + \theta_r)}{2\alpha c_{Rr}} \tag{A19}$$

We take (A18) and (A19) into formula (1), and set $\partial\Pi_M^{<n,r>}/\partial q_n = 0$, $\partial\Pi_M^{<n,r>}/\partial q_{Mdn} = 0$ and $\partial\Pi_M^{<n,r>}/\partial q_{Mr} = 0$. We can have

$$q_n^{<n,r>} = \frac{\alpha(-1 + \Delta_n + \alpha) + c_{Mr}(-2 + 2\Delta_n - \Delta_r + \alpha)}{2(c_{Mr}(-2 + \alpha) + (-1 + \alpha)\alpha)} \tag{A20}$$

$$q_{Mr}^{<n,r>} = \frac{\Delta_r - s_r\alpha + \alpha(\Delta_n - \theta_r)}{2(c_{Mr}(2 - \alpha) + (1 - \alpha)\alpha)} \tag{A21}$$

$$q_{Mdn}^{<n,r>} = \frac{\Delta_n}{2c_n} + \frac{c_n(\alpha(-1 + \Delta_n + \alpha) + c_{Mr}(-2 + 2\Delta_n - \Delta_r + \alpha))}{2c_n(c_{Mr}(2 - \alpha) + (1 - \alpha)\alpha)} \tag{A22}$$

We bring (A20), (A21) and (A22) back to (A18) and (A19). We can get

$$q_{Rr}^{<n,r>} = \frac{(1 - \alpha)\alpha(\alpha - 3\Delta_r) + c_{Mr}(\alpha\Delta_r - 4\Delta_r + \alpha(2 + 2\Delta_n - \alpha))}{4\alpha(c_{Mr}(2 - \alpha) + (1 - \alpha)\alpha)} \tag{A23}$$

$$q_{Rdr}^{<n,r>} = \frac{\eta_3 + c_{Mr}(2(2 - \alpha)\alpha\Delta_r + c_{Rr}(4\Delta_r - \alpha(2 + \Delta_r + 2\Delta_n - \alpha)))}{4c_{Rr}\alpha(c_{Mr}(2 - \alpha) + (1 - \alpha)\alpha)} \tag{A24}$$

where $\eta_3 = (1 - \alpha)\alpha(2\alpha\Delta_r + c_{Rr}(3\Delta_r - \alpha))$. $\square$

Scenario <r,r>

In this scenario, both the manufacturer and the remanufacturer decide to donate remanufactured products, and the government would provide donation subsidies for firms. Thus, the remanufacturer needs to set the output of remanufactured products q_{Rr} and his donating quantity q_{Rdr} after the manufacturer determines the output of new q_n and remanufactured products q_{Mr}, and the donating quantity of remanufactured products q_{Mdr}. By solving the problems, we can obtain Proposition A6.

Proposition A6. *Under scenario <r,r>, the equilibrium solutions are as follows:*

$$q_n^{<r,r>} = \frac{1 + \Delta_r - \alpha}{2 + 2c_n - 2\alpha};$$
$$q_{Mr}^{<r,r>} = \frac{\Delta_r + c_n\Delta_r - c_n\alpha}{2\alpha(-1 - c_n + \alpha)};$$
$$q_{Mdr}^{<r,r>} = \frac{(1 + c_n - \alpha)\alpha\Delta_r + c_{Mr}(\Delta_r + c_n\Delta_r - c_n\alpha)}{2c_{Mr}\alpha(1 + c_n - \alpha)};$$
$$q_{Rr}^{<r,r>} = \frac{\alpha - \Delta_r}{4\alpha};$$
$$q_{Rdr}^{<r,r>} = \frac{2\alpha\Delta_r + c_{Rr}(\Delta_r - \alpha)}{4c_{Rr}\alpha}.$$

Proposition A6 reveals the situation where the manufacturer and the remanufacturer may have competition in the donation field, especially when they donate the same product. As the subsidy increases, the number of donations will increase, but the output of other products will not always increase.

Proof of Proposition A6. In $<r,r>$ scenario, the manufacturer and the remanufacturer both donate the remanufactured products, which means $q_{Mdn} = 0$, $q_{Mdr} \neq 0$, $q_{Rdr} \neq 0$. To verify the existence of extreme value of (1), the Hessian matrix of (1) is obtain as follows:

$$H\left(\Pi_M^{<r,r>}\right) = \begin{bmatrix} \frac{\partial^2 \Pi_M^{<r,r>}}{\partial q_n^2} & \frac{\partial^2 \Pi_M^{<r,r>}}{\partial q_n \partial q_{Mr}} & \frac{\partial^2 \Pi_M^{<r,r>}}{\partial q_n \partial q_{Mdr}} \\ \frac{\partial^2 \Pi_M^{<r,r>}}{\partial q_n \partial q_{Mr}} & \frac{\partial^2 \Pi_M^{<r,r>}}{\partial q_{Mr}^2} & \frac{\partial^2 \Pi_M^{<r,r>}}{\partial q_{Mdr} \partial q_{Mr}} \\ \frac{\partial^2 \Pi_M^{<r,r>}}{\partial q_n \partial q_{Mdr}} & \frac{\partial^2 \Pi_M^{<r,r>}}{\partial q_{Mdr} \partial q_{Mr}} & \frac{\partial^2 \Pi_M^{<r,r>}}{\partial q_{Mdr}^2} \end{bmatrix} = \begin{bmatrix} -2 - 2c_n & -2\alpha & 0 \\ -2\alpha & -2\alpha - 2c_{Mr} & -2c_{Mr} \\ 0 & -2c_{Mr} & -2c_{Mr} \end{bmatrix}$$

From the Hessian matrix above, it is easy to find that $H_1\left(\Pi_M^{<r,r>}\right) = -2 - 2c_n < 0$, $H_2\left(\Pi_M^{<r,r>}\right) = 4\left((1+c_n)\,c_{Mr}+(1+c_n-\alpha)\,\alpha\right) > 0$, $H_3\left(\Pi_M^{<r,r>}\right) = -8c_{Mr}c_n\alpha - 8c_{Mr}\alpha(1-\alpha) < 0$. Thus, the Hessian matrix is negative, indicating that formula (1) is a concave function and there is a maximum value. And the Hessian matrix of formula (2) is as follows:

$$H\left(\Pi_R^{<r,r>}\right) = \begin{bmatrix} \frac{\partial^2 \Pi_R^{<r,r>}}{\partial q_{Rr}^2} & \frac{\partial^2 \Pi_R^{<r,r>}}{\partial q_{Rr} \partial q_{Rdr}} \\ \frac{\partial^2 \Pi_R^{<r,r>}}{\partial q_{Rdr} \partial q_{Rr}} & \frac{\partial^2 \Pi_R^{<r,r>}}{\partial q_{Rdr}^2} \end{bmatrix} = \begin{bmatrix} -2\alpha - 2c_{Rr} & -2c_{Rr} \\ -2c_{Rr} & -2c_{Rr} \end{bmatrix}$$

We can see that $H_1\left(\Pi_R^{<r,r>}\right) = -2\alpha - 2c_{Rr} < 0$ and $H_2\left(\Pi_R^{<r,r>}\right) = 4\alpha c_{Rr} > 0$. So, the Hessian matrix is negative, indicating that formula (2) is a concave function and there is a maximum value. Using the reverse order solution method to solve the production decision of the remanufacturer, we first let $\partial\Pi_R^{<r,r>}/\partial q_{Rr} = 0$ and $\partial\Pi_R^{<r,r>}/\partial q_{Rdr} = 0$. Then we can have:

$$q_{Rr} = -\frac{s_r + \alpha(-1 + q_n + q_{Mr}) + \theta_r}{2\alpha} \tag{A25}$$

$$q_{Rdr} = \frac{\alpha(s_r + \theta_r) + c_{Rr}(s_r + \alpha(-1 + q_n + q_{Mr}) + \theta_r)}{2c_{Rr}\alpha} \tag{A26}$$

We take (A25) and (A26) into formula (1), and set $\partial\Pi_M^{<r,r>}/\partial q_n = 0$, $\partial\Pi_M^{<r,r>}/\partial q_{Mr} = 0$ and $\partial\Pi_M^{<r,r>}/\partial q_{Mdr} = 0$. We can have

$$q_n^{<r,r>} = \frac{1 + s_r - \alpha + \theta_r}{2 + 2c_n - 2\alpha} \tag{A27}$$

$$q_{Mr}^{<r,r>} = \frac{s_r + c_n s_r - c_n \alpha + \theta_r + c_n \theta_r}{2\alpha(-1 - c_n + \alpha)} \tag{A28}$$

$$q_{Mdr}^{<r,r>} = \frac{(1 + c_n - \alpha)\alpha(s_r + \theta_r) + c_{Mr}(s_r + c_n s_r - c_n \alpha + \theta_r + c_n \theta_r)}{2c_{Mr}\alpha(1 + c_n - \alpha)} \tag{A29}$$

We bring (A27), (A28) and (A29) back to (A25) and (A26). We can get

$$q_{Rr}^{<r,r>} = \frac{\alpha - s_r - \theta_r}{4\alpha} \tag{A30}$$

$$q_{Rdr}^{<r,r>} = \frac{2\alpha(s_r + \theta_r) + c_{Rr}(s_r - \alpha + \theta_r)}{4c_{Rr}\alpha} \tag{A31}$$

□

Appendix B

Proof of Proposition 1. According to the results in Proposition A3 and A5, as long as the manufacturer prefers donation strategy, his optimal donation decisions should satisfy: $q_{Mdn}^{<n,0>} > 0$ and $q_{Mdr}^{<r,0>} > 0$. So we can have:

$$s_n > s_n^{<n,0>} \tag{A32}$$

$$s_r > s_r^{<r,0>} \tag{A33}$$

where $s_n^{<n,0>} = \frac{((\alpha-1)\alpha(2c_{Rr}+\alpha)-c_{Mr}A_1)\theta_n - c_n(\alpha(2c_{Rr}+\alpha)(\alpha+\theta_n-1)-c_{Mr}(A_1-2(c_{Rr}+\alpha)\theta_n))}{(1+c_n-\alpha)\alpha(2c_{Rr}+\alpha)+c_{Mr}(2(1+c_n)c_{Rr}+(2+2c_n-\alpha)\alpha)}$,

$A_1 = 2c_{Rr} + (2-\alpha)\alpha$ and $s_r^{<r,0>} = \frac{c_n c_{Mr}\alpha(2c_{Rr}+\alpha)}{\alpha(1+c_n-\alpha)(2c_{Rr}+\alpha)+c_{Mr}(2c_{Rr}(1+c_n)+\alpha(2+2c_n-\alpha))} - \theta_r.$ $\square$

Proof of Proposition 2. According to the conclusions in Proposition A2, A4, and A6, the optimal product decisions of the remanufacturer should meet that $q_{Rdr}^{<0,r>} > 0$, $q_{Rdr}^{<n,r>} > 0$, and $q_{Rdr}^{<r,r>} > 0$, if he chooses to donate. Then we can have

$$s_r > s_r^{<0,r>} \tag{A34}$$

$$s_r > s_r^{<n,r>} \tag{A35}$$

$$s_r > s_{Rr}^{<r,r>} \tag{A36}$$

where $s_r^{<n,r>} = \frac{\alpha c_{Rr}(\alpha(\alpha-1)+c_{Mr}(\alpha-2-2s_n-2\theta_n))}{\alpha(\alpha-1)(3c_{Rr}+2\alpha)+c_{Mr}(c_{Rr}(\alpha-4)+2\alpha(\alpha-2))} - \theta_r$, $s_{Rr}^{<r,r>} = \frac{\alpha c_{Rr}}{c_{Rr}+2\alpha} - \theta_r$, $s_r^{<0,r>} = \frac{\alpha c_{Rr}(c_{Mr}(\alpha-2-4c_n)+\alpha(\alpha-1-c_n))}{\alpha(\alpha-1-c_n)(3c_{Rr}+2\alpha)+c_{Mr}(c_{Rr}(\alpha-4-4c_n)+2\alpha(\alpha-2-2c_n))} - \theta_r.$ $\square$

Proof of Proposition 3. According to the conclusions in Proposition A4 and A6, the optimal product decisions of the manufacturer should meet that $q_{Mdn}^{<n,r>} > 0$, and $q_{Mdr}^{<r,r>} > 0$, so we can have

$$s_n > s_n^{<n,r>} \tag{A37}$$

$$s_r > s_{Mr}^{<r,r>} \tag{A38}$$

where $s_n^{<n,r>} = \frac{\theta_n(c_{Mr}(\alpha-2)+\alpha(\alpha-1))+c_n\left(\alpha-\alpha(\alpha+\theta_n)+c_{Mr}\left(2+s_r^{<n,r>}-\alpha-2\theta_n+\theta_r\right)\right)}{c_{Mr}(2+2c_n-\alpha)+\alpha(1-\alpha+c_n)}.$ and $s_{Mr}^{<r,r>} = \frac{\alpha c_n c_{Mr}}{c_{Mr}+c_n c_{Mr}+\alpha+\alpha c_n-\alpha^2} - \theta_r.$ $\square$

Proof of Proposition 4. According to the conclusions in Proposition A2, A3, A4, A5, and A6, we analyze how each product changes with the subsidies in each scenario.

Scenario *<n,0>*

$$\frac{\partial q_n^{<n,0>}}{\partial s_n} = -\frac{2c_{Mr}(c_{Rr}+\alpha)+\alpha(2c_{Rr}+\alpha)}{2(1-\alpha)\alpha(2c_{Rr}+\alpha)+c_r^M(4c_{Rr}+2(2-\alpha)\alpha)} < 0 \tag{A39}$$

$$\frac{\partial q_{Mr}^{<n,0>}}{\partial s_n} = \frac{\alpha(2c_{Rr}+\alpha)}{2(1-\alpha)\alpha(2c_{Rr}+\alpha)+c_{Mr}(4c_{Rr}+2(2-\alpha)\alpha)} > 0 \tag{A40}$$

$$\frac{\partial q_{Mdn}^{<n,0>}}{\partial s_n} = \frac{(1+c_n-\alpha)\alpha(2c_{Rr}+\alpha)+c_{Mr}(2(1+c_n)c_{Rr}+(2+2c_n-\alpha)\alpha)}{2c_n((1-\alpha)\alpha(2c_{Rr}+\alpha)+c_{Mr}(2c_{Rr}+(2-\alpha)\alpha))} > 0 \tag{A41}$$

$$\frac{\partial q_{Rr}^{<n,0>}}{\partial s_n} = \frac{c_{Mr}\alpha}{2(1-\alpha)\alpha(2c_{Rr}+\alpha)+c_{Mr}(4c_{Rr}+2(2-\alpha)\alpha)} > 0 \tag{A42}$$

In scenario *<n,0>*, only new products will decline with the subsidy increasing. This indicates that the manufacturer's new products donation strategy may conflict with new product's marketing strategy.

Scenario <r,0>

$$\frac{\partial q_n^{<r,0>}}{\partial s_r} = \frac{1}{2 + 2c_n - 2\alpha} > 0 \tag{A43}$$

$$\frac{\partial q_{Mr}^{<r,0>}}{\partial s_r} = -\frac{1}{2}\left(\frac{1}{\alpha} + \frac{1}{1 - \alpha + c_n} + \frac{1}{2c_{Rr} + \alpha}\right) < 0 \tag{A44}$$

$$\frac{\partial q_{Mdr}^{<r,0>}}{\partial s_r} = \frac{1}{2}\left(\frac{1}{c_{Mr}} + \frac{1}{1 + c_n - \alpha} + \frac{1}{\alpha} + \frac{1}{2c_{Rr} + \alpha}\right) > 0 \tag{A45}$$

$$\frac{\partial q_{Rr}^{<r,0>}}{\partial s_r} = \frac{1}{4c_{Rr} + 2\alpha} > 0 \tag{A46}$$

In scenario <r,0>, (A44) shows the same situation in (A39), where the remanufactured products that the manufacturer produced will decrease with the subsidy increasing, while other products will increase.

Scenario <0,r>

$$\frac{\partial q_n^{<0,r>}}{\partial s_r} = \frac{c_{Mr}}{2(c_{Mr}(2 + 2c_n - \alpha) + \alpha(1 + c_n - \alpha))} > 0 \tag{A47}$$

$$\frac{\partial q_{Mr}^{<0,r>}}{\partial s_r} = \frac{1 + c_n - \alpha}{2(c_{Mr}(2 + 2c_n - \alpha) + \alpha(1 + c_n - \alpha))} > 0 \tag{A48}$$

$$\frac{\partial q_{Rdr}^{<0,r>}}{\partial s_r} = \frac{1}{4}\left(\frac{2}{c_{Rr}} + \frac{2}{\alpha} + \frac{1 + c_n + c_{Mr} - \alpha}{\alpha(1 + c_n - \alpha) + c_{Mr}(2 + 2c_n - \alpha)}\right) > 0 \tag{A49}$$

$$\frac{\partial q_{Rr}^{<0,r>}}{\partial s_r} = -\frac{1}{4}\left(\frac{2}{\alpha} + \frac{1 + c_n + c_{Mr} - \alpha}{\alpha(1 + c_n - \alpha) + c_{Mr}(2 + 2c_n - \alpha)}\right) < 0 \tag{A50}$$

Scenario <0,r> reveals that not only the manufacturer but also the remanufacturer will meet the donation problem that the subsidy will make his products (both sell and donate) in the market decline.

Scenario <n,r>

$$\frac{\partial q_n^{<n,r>}}{\partial s_n} = -\frac{2c_{Mr} + \alpha}{2(c_{Mr}(2 - \alpha) + (1 - \alpha)\alpha)} < 0 \tag{A51}$$

$$\frac{\partial q_n^{<n,r>}}{\partial s_r} = \frac{c_{Mr}}{2(c_{Mr}(2 - \alpha) + (1 - \alpha)\alpha)} > 0 \tag{A52}$$

$$\frac{\partial q_{Mr}^{<n,r>}}{\partial s_n} = \frac{\alpha}{2(c_{Mr}(2 - \alpha) + (1 - \alpha)\alpha)} > 0 \tag{A53}$$

$$\frac{\partial q_{Mr}^{<n,r>}}{\partial s_r} = \frac{1 - \alpha}{2(c_{Mr}(2 - \alpha) + (1 - \alpha)\alpha)} > 0 \tag{A54}$$

$$\frac{\partial q_{Mdn}^{<n,r>}}{\partial s_n} = \frac{1}{2}\left(\frac{1}{c_n} + \frac{2c_{Mr} + \alpha}{c_{Mr}(2 - \alpha) + \alpha(1 - \alpha)}\right) > 0 \tag{A55}$$

$$\frac{\partial q_{Mdn}^{<n,r>}}{\partial s_r} = -\frac{c_{Mr}}{2(c_{Mr}(2 - \alpha) + (1 - \alpha)\alpha)} < 0 \tag{A56}$$

$$\frac{\partial q_{Rr}^{<n,r>}}{\partial s_n} = \frac{c_{Mr}}{2(c_{Mr}(2 - \alpha) + (1 - \alpha)\alpha)} > 0 \tag{A57}$$

$$\frac{\partial q_{Rr}^{<n,r>}}{\partial s_r} = -\frac{c_{Mr}(4-\alpha)+3(1-\alpha)\alpha}{4\alpha(c_{Mr}(2-\alpha)+(1-\alpha)\alpha)} < 0 \tag{A58}$$

$$\frac{\partial q_{Rdr}^{<n,r>}}{\partial s_n} = -\frac{c_{Mr}}{2(c_{Mr}(2-\alpha)+(1-\alpha)\alpha)} < 0 \tag{A59}$$

$$\frac{\partial q_{Rdr}^{<n,r>}}{\partial s_r} = \frac{1}{4}\left(\frac{2}{c_{Rr}}+\frac{2}{\alpha}+\frac{1+c_{Mr}-\alpha}{c_{Mr}(2-\alpha)+(1-\alpha)\alpha}\right) > 0 \tag{A60}$$

Scenario $<n,r>$ shows a complicated situation where two subsidies exist. From (A51) and (A58), we find that two formulations here confirm the previous results. And we also notice that donated new products will decline with the increase of remanufactured products donation subsidy in (A56) and the similar situation for donated remanufactured products in (A59). This indicates that, when two subsidies exist, the donated products will be negatively affected by the other subsidy's increase.

Scenario $<r,r>$

$$\frac{\partial q_n^{<r,r>}}{\partial s_r} = \frac{1}{2+2c_n-2\alpha} > 0 \tag{A61}$$

$$\frac{\partial q_{Mr}^{<r,r>}}{\partial s_r} = -\frac{1+c_n}{2\alpha(1+c_n-\alpha)} < 0 \tag{A62}$$

$$\frac{\partial q_{Mdr}^{<r,r>}}{\partial s_r} = \frac{1}{2}\left(\frac{1}{c_{Mr}}+\frac{1}{1+c_n-\alpha}+\frac{1}{\alpha}\right) > 0 \tag{A63}$$

$$\frac{\partial q_{Rr}^{<r,r>}}{\partial s_r} = -\frac{1}{4\alpha} < 0 \tag{A64}$$

$$\frac{\partial q_{Rdr}^{<r,r>}}{\partial s_r} = \frac{1}{4}\left(\frac{2}{c_{Rr}}+\frac{1}{\alpha}\right) > 0 \tag{A65}$$

Scenario $<r,r>$ shows the situation where the manufacturer and the remanufacturer donate the same products. (A62) and (A64) explain that, even though two parties may compete in the donation field, the subsidy still has a negative effect on remanufactured products in the market. $\square$

Proof of Corollary 1. According to the conclusions in Proposition A2, A3, A4, A5, and A6, we analyze how each product's total amount changes with the subsidies in each scenario.

Scenario $<n,0>$

$$\frac{\partial\left(q_n^{<n,0>}+q_{Mdn}^{<n,0>}\right)}{\partial s_n} = \frac{1}{2c_n} > 0 \tag{A66}$$

Scenario $<r,0>$

$$\frac{\partial\left(q_{Mr}^{<r,0>}+q_{Mdr}^{<r,0>}+q_{Rr}^{<r,0>}\right)}{\partial s_r} = \frac{c_{Mr}+\alpha+2c_{Rr}}{2\alpha c_{Mr}+4c_{Rr}c_{Mr}} > 0 \tag{A67}$$

Scenario $<0,r>$

$$\frac{\partial\left(q_{Mr}^{<0,r>}+q_{Rr}^{<0,r>}+q_{Rdr}^{<0,r>}\right)}{\partial s_r} = \frac{1}{2}\left(\frac{1}{c_{Rr}}+\frac{1+c_n-\alpha}{c_{Mr}(2+2c_n-\alpha)+\alpha(1+c_n-\alpha)}\right) > 0 \tag{A68}$$

Scenario $<n,r>$

$$\frac{\partial\left(q_n^{<n,r>} + q_{Mdn}^{<n,r>}\right)}{\partial s_n} = \frac{1}{2c_n} > 0 \tag{A69}$$

$$\frac{\partial\left(q_{Mr}^{<n,r>} + q_{Rdr}^{<n,r>} + q_{Rr}^{<n,r>}\right)}{\partial s_r} = \frac{1}{2}\left(\frac{1}{c_{Rr}} + \frac{1-\alpha}{c_{Mr}(2-\alpha) + (1-\alpha)\alpha}\right) > 0 \tag{A70}$$

Scenario <r,r>

$$\frac{\partial\left(q_{Mr}^{<r,r>} + q_{Mdr}^{<r,r>} + q_{Rr}^{<r,r>} + q_{Rdr}^{<r,r>}\right)}{\partial s_r} = \frac{c_{Mr} + c_{Rr}}{2c_{Rr}c_{Mr}} > 0 \tag{A71}$$

.$\square$

Proof of Proposition 5. According to the conclusions in Proposition A2, A3, A4, A5, A6, and 1, we can see that, in scenario <n,0>, the new product subsidy should satisfy $s_n > s_n^{<n,0>}$. Besides, we can have $p_n^{<n,0>} > p_n^{<0,0>}$ and $p_r^{<n,0>} > p_r^{<0,0>}$, and at the same time we find that $\partial\left(p_n^{<n,0>} - p_n^{<0,0>}\right)/\partial s_n > 0$ and $\partial\left(p_r^{<n,0>} - p_r^{<0,0>}\right)/\partial s_n > 0$. This indicates that, in scenario <n,0>, The prices of new products and remanufactured products in the market are higher than products in scenario <0,0>, so both products have price advantages in the market, and these advantages are more and more obvious with the increase of subsidy. Similar situations can be found in other scenarios (such as <r,0>, <n,r>, and <r,r>) and these proofs will be omitted.

However, in scenario <0,r>, the remanufactured product subsidy should satisfy $s_r > s_r^{<0,r>}$. We find that there exists a s_r^*. When $s_r \in \left(s_r^{<0,r>}, s_r^*\right)$, $p_n^{<0,r>} < p_n^{<0,0>}$ and $p_r^{<0,r>} < p_r^{<0,0>}$, which mean that two products have price disadvantages in the market; and when $s_r \in (s_r^*, 1)$, $p_n^{<0,r>} > p_n^{<0,0>}$ and $p_r^{<0,r>} > p_r^{<0,0>}$, which indicate that two products now have price advantages in the market.

Where $s_r^* = \frac{c_{Mr}(4+4c_n-3\alpha)+\alpha(1+c_n-\alpha)}{4(c_{Mr}(2+2c_n-\alpha)+\alpha(1+c_n-\alpha))}$. $\square$

Proof of Proposition 6. According to the results in Proposition 1, the necessary condition for the manufacturer to choose to donate the new products is that the subsidy for donated new products should meet $s_n > s_n^{<n,0>}$ and $s_r < s_r^{<r,0>}$. As a follower in the market, the remanufacturer has second-mover advantages. He could make his decisions after knowing that the manufacturer will adopt donating new products strategy. And then the remanufacturer makes a comparison between <n,0> and <n,r> decisions and he can find that there exists a s_r^{**}, when $s_r^{<n,r>} < s_r^{**} < s_r < s_r^{<r,0>}$, $\Pi_R^{<n,0>} < \Pi_R^{<n,r>}$ which means that he can get more profits in <n,r> scenario.

Where $s_r^{**} = \dfrac{-\sqrt{c_{Rr}(c_{Mr}(\alpha-2)+\alpha(\alpha-1))^2\left(A\left(B_1+B_2+c_{Mr}^2 B_3\right)\right)-4\alpha^3 B_4^2}-B_5+B_6 B_7}{B_1+B_2+c_{Mr}^2 B_3}$,

$A_1 = c_{Mr}(2c_{Rr}(1+s_n+\theta_n)+\alpha(2+2s_n-\alpha+2\theta_n))$,

$A_2 = \alpha(\alpha-1)(2c_{Rr}+\alpha)+c_{Mr}(\alpha(\alpha-2)-2c_{Rr})$,

$A = \dfrac{\alpha^3(\alpha(\alpha-1)(2c_{Rr}+\alpha)-A_1)}{A_2^2(c_{Rr}+\alpha)}$, $B_1 = \alpha^2(1-\alpha)^2(9c_{Rr}+4\alpha)$,

$B_2 = 2c_{Mr}\alpha(\alpha-1)(3c_{Rr}(\alpha-4)+4\alpha(\alpha-2))$, $B_3 = c_{Mr}^2\left(c_{Rr}(4-\alpha)^2+4\alpha(2-\alpha)^2\right)$,

$B_4 = \alpha(\alpha-1)+c_{Mr}(-2-2s_n+\alpha-2\theta_n)$, $B_5 = 4\alpha\theta_r(c_{Mr}(\alpha-2)+\alpha(\alpha-1))^2$,

$B_6 = c_{Rr}(c_{Mr}(\alpha-4)+3\alpha(\alpha-1))$,

$B_7 = c_{Mr}\alpha(\alpha-2-2s_n-2\theta_n)+\alpha(\alpha-1)(\alpha-3\theta_r)-c_{Mr}\theta_r(\alpha-4)$. $\square$

Proof of Proposition 7. According to the results in Proposition A2, A3, A4, A5, A6, and 1, when the manufacturer chooses no-donation and the remanufacturer decides to denote, that is <0,r> scenario, we can get $s_r \geq max\left\{s_r^{<0,r>}, s_{Rr}^{<0,r>}\right\}$, through solving $\Pi_R^{<0,r>} -$

$\Pi_R^{<0,0>} > 0$. At the same time, under this subsidy condition, we find that $\Pi_M^{<0,r>} > \Pi_M^{<0,0>}$, $\Pi_M^{<r,r>} > \Pi_M^{<0,r>}$, and $\Pi_R^{<r,r>} > \Pi_R^{<0,r>}$. This shows that high subsidy for donated remanufactured products stimulate the enthusiasm of the manufacturer to donate. Therefore, the manufacturer has reasons to adopt the strategy of donating remanufactured products, which makes the *<0,r>* scenario change to the *<r,r>* scenario.

And then, we discuss the profit of the remanufacturer in *<r,r>* scenario and let $\partial(\Pi_R^{<r,r>} - \Pi_R^{<0,r>})/\partial s_r > 0$. After calculating, we know that if $s_r \leq max\{s_{Rr1}^{<r,r>}, s_{Rr2}^{<r,r>}\}$, $\partial(\Pi_R^{<r,r>} - \Pi_R^{<0,r>})/\partial s_r > 0$; and if $s_r \geq max\{s_{Rr1}^{<r,r>}, s_{Rr2}^{<r,r>}\}$, $\partial(\Pi_R^{<r,r>} - \Pi_R^{<0,r>})/\partial s_r < 0$. This shows that with the increase of subsidy for donated remanufactured products, the profit of the remanufacturer will gradually be lower than that under the *<0,r>* scenario, that is, the market share of the remanufacturer will decrease slowly and his market position will be lost. This also implies that the subsidy for donated remanufactured products will widen the benefits gap between the manufacturer and the remanufacturer. However, the fundamental reason for this is that the remanufacturer is a follower in the market and has a second-move disadvantage. And, according to the conclusion of Proposition A5, it can be found that in the *<r,r>* scenario, if the remanufacturer gives up his donation strategy, this will make the market equilibrium scenario become *<r,0>*. Moreover, it will only lead the remanufacturer lose more benefits, because of $\Pi_R^{<0,r>} > \Pi_R^{<r,r>} > \Pi_R^{<r,0>}$. It can be seen that when the remanufacturer faces the manufacturer's donation strategy, he can only follow the donation to ensure the least damage to its own benefits, that is, although donating remanufactured products is not the optimal decision of the remanufacturer, it is indeed a Nash equilibrium for two parties.

Where $C_1 = (c_{Rr} + \alpha)(C_5(2c_{Rr} + \alpha) + c_{Mr}(2c_{Rr}(1 + c_n) + \alpha(2 + 2c_n - \alpha)))^2$,
$C_2 = \alpha(c_{Mr}(2 + 2c_n - \alpha) + C_5)^2$,
$C_3 = \alpha^2(C_5(2c_{Rr} + \alpha) + c_{Mr}(c_{Rr}(2 + 4c_n) + (2 + 4c_n - \alpha)\alpha))^2$,
$C_4 = c_{Mr}(4 + 4c_n - \alpha)$, $C_5 = \alpha(1 + c_n - \alpha)$,
$C_6 = c_{Rr}(C_4 + 3C_5)(C_4 - 2c_{Mr}\alpha + C_5(\alpha - 3\theta_r) - C_4\theta_r)$,
$C_7 = \alpha^2 + 4\theta_r(1 + c_n) - \alpha(2 + 4c_n + \theta_r)$,

$$s_{Rr}^{<0,r>} = \frac{\sqrt{C_1 C_2 c_{Rr}\left(-4C_1\alpha^2(C_4 - 2c_{Mr} + C_5)^2 + C_3\left(4C_2 + c_{Rr}(C_4 + 3C_5)^2\right)\right)} + C_1(-4C_2\theta_r + C_6)}{C_1\left(4C_2 + c_{Rr}(C_4 + 3C_5)^2\right)},$$

$$s_{Rr1}^{<r,r>} = \frac{-\left(\alpha(-3C_5 - c_{Mr}C_4)(C_7 c_{Mr} - C_5(\alpha - 3\theta_r)) + C_2(\theta_r - \alpha) + \sqrt{C_2\alpha(C_7 c_{Mr} + 2C_5\alpha - C_4(\theta_r - \alpha))^2}\right)}{C_2 + \alpha(3C_5 + C_4)^2},$$

$$s_{Rr2}^{<r,r>} = \frac{-\alpha(-3C_5 - c_{Mr}C_4)(C_7 c_{Mr} - C_5(\alpha - 3\theta_r)) + C_2(\theta_r - \alpha) + \sqrt{C_2\alpha(C_7 c_{Mr} + 2C_5\alpha - C_4(\theta_r - \alpha))^2}}{C_2 + \alpha(3C_5 + C_4)^2}. \quad \square$$

Appendix C

Proof of Proposition 8. According to the formulations (1) and (2) in Section 3.2, when the remanufacturer is the first-mover, we first solve the manufacturer's optimal decisions and then solve the remanufacturer's optimal decisions, by using the reverse order solution method. The proof process is similar to the proof of Proposition A1.

First, we solve the optimal solutions in *<0,0>* scenario.

$$q_n^{<0,0>*} = \frac{c_{Mr} + \alpha - \alpha^2 - \dfrac{c_{Mr}\alpha^2(c_{Mr} + 2c_n c_{Mr} + C_5)}{4c_{Mr}(1 + c_n)(c_{Rr} + \alpha) - 2c_{Mr}\alpha^2 + 2C_5(2c_{Rr} + \alpha)}}{2(c_{Mr}(1 + c_n) + C_5)} \tag{A72}$$

$$q_{Mr}^{<0,0>*} = \frac{\alpha\left(4c_n c_{Mr} c_{Rr}(1 + c_n) + \alpha(1 + c_n)D_1 + \alpha^2\left(c_n^2 - 1 + c_{Mr} - 4c_n c_{Rr}\right) + 2\alpha^3 - \alpha^4\right)}{4((1 + c_n)c_{Mr} + C_5)(C_5(2c_{Rr} + \alpha) + c_{Mr}(2c_{Rr}(1 + c_n) + \alpha(2 + 2c_n - \alpha)))} \tag{A73}$$

$$q_{Rr}^{<0,0>*} = \frac{\alpha(c_{Mr} + 2c_n c_{Mr} + C_5)}{4c_{Mr}(1 + c_n)(c_{Rr} + \alpha) - 2c_{Mr}\alpha^2 + 2C_5(2c_{Rr} + \alpha)} \tag{A74}$$

According to (A72), (A73), and (A74), we can get the remanufacturer's maximum profit in <0,0> scenario is

$$\Pi_R^{<0,0>*} = \frac{\alpha^2(c_{Mr} + 2c_n c_{Mr} + C_5)}{8((c_{Mr}(1+c_n) + C_5)(C_5(2c_{Rr} + \alpha) + c_{Mr}(2c_{Rr}(1+c_n) + \alpha(2 + 2c_n - \alpha))))} \tag{A75}$$

Then we solve the optimal solutions in <0,r> scenario.

$$q_n^{<0,r>*} = \frac{c_{Mr} + \alpha - \alpha^2 - \frac{c_{Mr}(C_5(\alpha - 2s_r - 2\theta_r) + c_{Mr}(2s_r(1+c_n) - \alpha(1+2c_n) + 2\theta_r(1+c_n)))}{2(c_{Mr}(2+2c_n - \alpha) + C_5)}}{2(c_{Mr}(1+c_n) + C_5)} \tag{A76}$$

$$q_{Mr}^{<0,r>*} = \frac{C_5(D_2 + \alpha(\alpha + c_n - 1)) + c_{Mr}(D_2(1+c_n) + \alpha(\alpha + c_n - 1 + 2c_n^2))}{4(c_{Mr}(1+c_n) + C_5)(c_{Mr}(2 + c_n - \alpha) + C_5)} \tag{A77}$$

$$q_{Rr}^{<0,r>*} = \frac{C_5(2(s_r + \theta_r) - \alpha) + c_{Mr}(2(s_r + \theta_r)(1+c_n) - \alpha(1 + 2c_n))}{2\alpha(c_{Mr}(\alpha - 2 - 2c_n) - C_5)} \tag{A78}$$

$$q_{dRr}^{<0,r>*} = \frac{C_5(\alpha(s_r + \theta_r) + c_{Rr}(2(s_r + \theta_r) - \alpha)) - c_{Mr}(\alpha(s_r + \theta_r)(\alpha - 2c_n - 2) + c_{Rr}D_3)}{2c_{Rr}\alpha(c_{Mr}(2c_n + 2 - \alpha) + C_5)} \tag{A79}$$

According to (A76), (A77), (A78), and (A79), we can get the remanufacturer's maximum profit in <0,r> scenario is

$$\Pi_R^{<0,r>*} = \frac{-\left(C_5^2\left(c_{Rr}(\alpha - 2(s_r + \theta_r))^2 + 2\alpha(s_r + \theta_r)^2\right) + c_{Mr}^2 D_4 + 2c_{Mr}C_5(D_5 + D_6)\right)}{8c_{Rr}\alpha(c_{Mr}(1+c_n) - C_5)(c_{Mr}(2 + 2c_n - \alpha) + C_5)} \tag{A80}$$

Through comparing, we find that

$$\Pi_R^{<0,R>*} - \Pi_R^{<0,0>*} = \frac{(-C_5(\alpha(s_r + \theta_r) + D_7) + c_{Mr}(\alpha(s_r + \theta_r)(\alpha - 2 - 2c_n) + c_{Rr}D_3))^2}{4c_{Rr}\alpha(c_{Mr}(2 + 2c_n - \alpha) + C_5)(C_5(2c_{Rr} + \alpha) + D_8)} \geq 0 \tag{A81}$$

where $C_5 > 0$ and $D_8 > 0$.

Therefore, if the remanufacturer is the firs-mover in the market, adopting donation strategy is a wise choice.

Where $D_1 = (2c_n - 1)c_{Mr} + 4c_n c_{Rr}$, $D_2 = 2(s_r + \theta_r)(1 + c_n - \alpha)$,
$D_3 = \alpha + 2c_n\alpha - 2(s_r + \theta_r)(1 + c_n)$,
$D_4 = 2(1 + c_n)(2 + 2c_n - \alpha)(s_r + \theta_r)^2 + c_{Rr}D_3^2$, $D_5 = \alpha(3 + 3c_n - \alpha)(s_r + \theta_r)^2$,
$D_6 = c_{Rr}(2(s_r + \theta_r) - \alpha)(2(s_r + \theta_r)(1 + c_n) - \alpha(1 + 2c_n))$, $D_7 = c_{Rr}(2(s_r + \theta_r) - \alpha)$,
$D_8 = c_{Mr}(2c_{Rr}(1 + c_n) + (2 + 2c_n - \alpha)\alpha)$. $\square$

Proof of Proposition 9. Based on the decision sequence mentioned in proof of Proposition 8, we first solve the optimal solutions in the <r,r> scenario.

$$q_n^{<r,r>*} = \frac{1 + s_r - \alpha + \theta_r}{2(1 + c_n - \alpha)} \tag{A82}$$

$$q_{Mr}^{<r,r>*} = \frac{\alpha + s_r(1 + c_n + \alpha) - (c_n + \alpha)(\alpha - \theta_r) + \theta_r}{4\alpha(\alpha - 1 - c_n)} \tag{A83}$$

$$q_{Mdr}^{<r,r>*} = \frac{1}{2}\left(\frac{s_r + \theta_r}{c_{Mr}} + \left(\frac{\alpha + s_r(1 + c_n - \alpha) - (c_n + \alpha)(\alpha - \theta_r) + \theta_r}{2C_5}\right)\right) \tag{A84}$$

$$q_{Rr}^{<r,r>*} = \frac{\alpha - s_r - \theta_r}{2\alpha} \tag{A85}$$

$$q_{Rdr}^{<r,r>*} = \frac{1}{2}\left(\frac{(c_{Rr}+\alpha)(s_r+\theta_r)}{c_{Rr}\alpha} - 1\right)$$ (A86)

According to the solutions above, we can get the remanufacturer's profit as below:

$$\Pi_R^{<r,r>*} = \frac{2\alpha(s_r+\theta_r)^2 + c_{Rr}(s_r+\theta_r-\alpha)^2}{8c_{Rr}\alpha}$$ (A87)

Next, we calculate the optimal solutions in the <r,0> scenario.

$$q_n^{<r,0>*} = \frac{1+s_r-\alpha+\theta_r}{2(1+c_n-\alpha)}$$ (A88)

$$q_{Mr}^{<r,0>*} = \frac{4c_{Rr}((s_r+\theta_r)(1+c_n)-\alpha c_n)-\alpha((s_r+\theta_r)(\alpha-3-3c_n)+\alpha(\alpha-1+c_n))}{-4C_5(2c_{Rr}+\alpha)}$$ (A89)

$$q_{Mdr}^{<r,0>*} = \frac{1}{2}\left(\frac{s_r+\theta_r}{c_{Mr}} - \frac{1}{\alpha} + \frac{(1+c_n)(1+s_r+\theta_r-\alpha)}{C_5} + \frac{s_r+\theta_r+\alpha}{4c_{Rr}+2\alpha}\right)$$ (A90)

$$q_{Rr}^{<r,0>*} = \frac{\alpha+s_r+\theta_r}{4c_{Rr}+2\alpha}$$ (A91)

According to the solutions above, we can get the remanufacturer's profit as below:

$$\Pi_R^{<r,0>*} = \frac{(s_r+\theta_r+\alpha)^2}{8(2c_{Rr}+\alpha)}$$ (A92)

Comparing the remanufacturer's profits in two scenarios, we can have

$$\Pi_R^{<r,r>*} - \Pi_R^{<r,0>*} = \frac{(c_{Rr}(s_r+\theta_r-\alpha)+\alpha(s_r+\theta_r))^2}{4c_{Rr}\alpha(2c_{Rr}+\alpha)} \geq 0$$ (A93)

So, we can know that when the remanufacturer is the leader in the market, the manufacturer's donation strategy will not affect the remanufacturer's decision. $\square$

References

1. Xue, D.; Teunter, R.H.; Zhu, S.X.; Zhou, W. Entering the High-End Market by Collecting and Remanufacturing a Competitor's High-End Cores. *Omega* **2021**, *99*, 102168. [CrossRef]
2. Giuntini, R.; Gaudette, K. Remanufacturing: The next Great Opportunity for Boosting US Productivity. *Bus. Horiz.* **2003**, *46*, 41–48. [CrossRef]
3. Ostojic, P. Pumps and Circular Economy. *World Pumps* **2016**, *2016*, 30–33. [CrossRef]
4. Ayres, R.; Ferrer, G.; Van Leynseele, T. Eco-Efficiency, Asset Recovery and Remanufacturing. *Eur. Manag. J.* **1997**, *15*, 557–574. [CrossRef]
5. Habibi, M.K.K.; Battaïa, O.; Cung, V.D.; Dolgui, A. Collection-Disassembly Problem in Reverse Supply Chain. *Int. J. Prod. Econ.* **2017**, *183*, 334–344. [CrossRef]
6. Zhu, X.; Wang, Z.; Wang, Y.; Li, B. Incentive Policy Options for Product Remanufacturing: Subsidizing Donations or Resales? *Int. J. Environ. Res. Public Health* **2017**, *14*, 1496. [CrossRef]
7. Li, B.; Wang, Z.; Wang, Y.; Tang, J.; Zhu, X.; Liu, Z. The Effect of Introducing Upgraded Remanufacturing Strategy on OEM's Decision. *Sustainability* **2018**, *10*, 828. [CrossRef]
8. Wang, Z.; Li, B.; Zhu, X.; Xin, B.; Wang, Y. The Impact of Donation Subsidy of Remanufactured Products on Manufacturer's Pricing-Production Decisions and Performances. *J. Clean. Prod.* **2018**, *202*, 892–903. [CrossRef]
9. Chen, S.; Yi, Y. The Manufacturer Decision Analysis for Corporate Social Responsibility under Government Subsidy. *Math. Probl. Eng.* **2021**, *2021*, 6617625. [CrossRef]
10. Arya, A.; Mittendorf, B. Supply Chain Consequences of Subsidies for Corporate Social Responsibility. *Prod. Oper. Manag.* **2015**, *24*, 1346–1357. [CrossRef]
11. Xintong, C.; Bangyi, L.; Zonghuo, L.; Goh, M.; Shanting, W. Take-Back Regulation Policy on Closed Loop Supply Chains: Single or Double Targets? *J. Clean. Prod.* **2021**, *283*, 124576. [CrossRef]

12. USITC. *Remanufactured Goods: An Overview of the U.S. and Global Industries, Markets, and Trade*; USITC: Washington, DC, USA, 2012.
13. Huang, Y.; Wang, Z. Pricing and Production Decisions in a Closed-Loop Supply Chain Considering Strategic Consumers and Technology Licensing. *Int. J. Prod. Res.* **2019**, *57*, 2847–2866. [CrossRef]
14. Ma, P.; Gong, Y.; Mirchandani, P. Trade-in for Remanufactured Products: Pricing with Double Reference Effects. *Int. J. Prod. Econ.* **2020**, *230*, 107800. [CrossRef]
15. Zheng, M.; Shi, X.; Xia, T.; Qin, W.; Pan, E. Production and Pricing Decisions for New and Remanufactured Products with Customer Prejudice and Accurate Response. *Comput. Ind. Eng.* **2021**, *157*, 107308. [CrossRef]
16. McKie, E.C.; Ferguson, M.E.; Galbreth, M.R.; Venkataraman, S. How Do Consumers Choose between Multiple Product Generations and Conditions? An Empirical Study of IPad Sales on EBay. *Prod. Oper. Manag.* **2018**, *27*, 1574–1594. [CrossRef]
17. Abbey, J.D.; Guide, V.D.R. A Typology of Remanufacturing in Closed-Loop Supply Chains. *Int. J. Prod. Res.* **2018**, *56*, 374–384. [CrossRef]
18. Chai, Q.; Xiao, Z.; Lai, K.; Zhou, G. Can Carbon Cap and Trade Mechanism Be Beneficial for Remanufacturing? *Int. J. Prod. Econ.* **2018**, *203*, 311–321. [CrossRef]
19. Subramanian, R.; Subramanyam, R. Key Factors in the Market for Remanufactured Products. *Manuf. Serv. Oper. Manag.* **2012**, *14*, 315–326. [CrossRef]
20. Liu, Z.; Li, K.W.; Tang, J.; Gong, B.; Huang, J. Optimal Operations of a Closed-Loop Supply Chain under a Dual Regulation. *Int. J. Prod. Econ.* **2021**, *233*, 107991. [CrossRef]
21. Liu, Z.; Zheng, X.X.; Li, D.F.; Liao, C.N.; Sheu, J.B. A Novel Cooperative Game-Based Method to Coordinate a Sustainable Supply Chain under Psychological Uncertainty in Fairness Concerns. *Transp. Res. E Logist. Transp. Rev.* **2021**, *147*, 102237. [CrossRef]
22. Li, X.; Li, Y.; Saghafian, S. A Hybrid Manufacturing/Remanufacturing System with Random Remanufacturing Yield and Market-Driven Product Acquisition. *IEEE Trans. Eng. Manag.* **2013**, *60*, 424–437. [CrossRef]
23. Raz, G.; Ovchinnikov, A.; Blass, V. Economic, Environmental, and Social Impact of Remanufacturing in a Competitive Setting. *IEEE Trans. Eng. Manag.* **2017**, *64*, 476–490. [CrossRef]
24. Xia, X.; Zhang, C. The Impact of Authorized Remanufacturing on Sustainable Remanufacturing. *Processes* **2019**, *7*, 663. [CrossRef]
25. Tan, H.; Cao, G.; He, Y.; Lu, Y. Channel Structure Choice for Remanufacturing under Green Consumerism. *Processes* **2021**, *9*, 1985. [CrossRef]
26. Modak, N.M.; Kelle, P. Using Social Work Donation as a Tool of Corporate Social Responsibility in a Closed-Loop Supply Chain Considering Carbon Emissions Tax and Demand Uncertainty. *J. Oper. Res. Soc.* **2021**, *72*, 61–77. [CrossRef]
27. Wang, Y.; Wang, Z.; Li, B.; Liu, Z.; Zhu, X.; Wang, Q. Closed-Loop Supply Chain Models with Product Recovery and Donation. *J. Clean. Prod.* **2019**, *227*, 861–876. [CrossRef]
28. Haeri, A.; Hosseini-Motlagh, S.M.; Ghatreh Samani, M.R.; Rezaei, M. A Mixed Resilient-Efficient Approach toward Blood Supply Chain Network Design. *Int. Trans. Oper. Res.* **2020**, *27*, 1962–2001. [CrossRef]
29. Heydari, J.; Mosanna, Z. Coordination of a Sustainable Supply Chain Contributing in a Cause-Related Marketing Campaign. *J. Clean. Prod.* **2018**, *200*, 524–532. [CrossRef]
30. Modak, N.M.; Kazemi, N.; Cárdenas-Barrón, L.E. Investigating Structure of a Two-Echelon Closed-Loop Supply Chain Using Social Work Donation as a Corporate Social Responsibility Practice. *Int. J. Prod. Econ.* **2019**, *207*, 19–33. [CrossRef]
31. Patil, A.; Shardeo, V.; Dwivedi, A.; Madaan, J.; Varma, N. Barriers to Sustainability in Humanitarian Medical Supply Chains. *Sustain. Prod. Consum.* **2021**, *27*, 1794–1807. [CrossRef]
32. Song, L.; Yan, Y.; Yao, F. Closed-Loop Supply Chain Models Considering Government Subsidy and Corporate Social Responsibility Investment. *Sustainability* **2020**, *12*, 2045. [CrossRef]
33. Tat, R.; Heydari, J.; Rabbani, M. Corporate Social Responsibility in the Pharmaceutical Supply Chain: An Optimized Medicine Donation Scheme. *Comput. Ind. Eng.* **2021**, *152*, 107022. [CrossRef]
34. Gilani Larimi, N.; Yaghoubi, S. A Robust Mathematical Model for Platelet Supply Chain Considering Social Announcements and Blood Extraction Technologies. *Comput. Ind. Eng.* **2019**, *137*, 106014. [CrossRef]
35. Mallidis, I.; Sariannidis, N.; Vlachos, D.; Yakavenka, V.; Aifadopoulou, G.; Zopounidis, K. Optimal Inventory Control Policies for Avoiding Food Waste. *Oper. Res.* **2022**, *22*, 685–701. [CrossRef]
36. Zou, J.; Tang, Y.; Qing, P.; Li, H.; Razzaq, A. Donation or Discount: Effect of Promotion Mode on Green Consumption Behavior. *Int. J. Environ. Res. Public Health* **2021**, *18*, 1912. [CrossRef]
37. Ross, A.; Khajehnezhad, M.; Otieno, W.; Aydas, O. Integrated Location-Inventory Modelling under Forward and Reverse Product Flows in the Used Merchandise Retail Sector: A Multi-Echelon Formulation. *Eur. J. Oper. Res.* **2017**, *259*, 664–676. [CrossRef]
38. Wang, Y.; Chang, X.; Chen, Z.; Zhong, Y.; Fan, T. Impact of Subsidy Policies on Recycling and Remanufacturing Using System Dynamics Methodology: A Case of Auto Parts in China. *J. Clean. Prod.* **2014**, *74*, 161–171. [CrossRef]
39. Qiao, H.; Su, Q. Impact of Government Subsidy on the Remanufacturing Industry. *Waste Manag.* **2021**, *120*, 433–447. [CrossRef]
40. Mitra, S.; Webster, S. Competition in Remanufacturing and the Effects of Government Subsidies. *Int. J. Prod. Econ.* **2008**, *111*, 287–298. [CrossRef]
41. Zhao, S.; Zhu, Q.; Cui, L. A Decision-Making Model for Remanufacturers: Considering Both Consumers' Environmental Preference and the Government Subsidy Policy. *Resour. Conserv. Recycl.* **2018**, *128*, 176–186. [CrossRef]

42. Cao, K.; He, P.; Liu, Z. Production and Pricing Decisions in a Dual-Channel Supply Chain under Remanufacturing Subsidy Policy and Carbon Tax Policy. *J. Oper. Res. Soc.* **2020**, *71*, 1199–1215. [CrossRef]
43. Miao, Z.; Mao, H.; Fu, K.; Wang, Y. Remanufacturing with Trade-Ins under Carbon Regulations. *Comput. Oper. Res.* **2018**, *89*, 253–268. [CrossRef]
44. He, P.; He, Y.; Xu, H. Channel Structure and Pricing in a Dual-Channel Closed-Loop Supply Chain with Government Subsidy. *Int. J. Prod. Econ.* **2019**, *213*, 108–123. [CrossRef]
45. Hong, Z.; Zhang, Y.; Yu, Y.; Chu, C. Dynamic Pricing for Remanufacturing within Socially Environmental Incentives. *Int. J. Prod. Res.* **2020**, *58*, 3976–3997. [CrossRef]
46. Han, X.; Shen, Y.; Bian, Y. Optimal Recovery Strategy of Manufacturers: Remanufacturing Products or Recycling Materials? *Ann. Oper. Res.* **2020**, *290*, 463–489. [CrossRef]
47. Sun, J.; Xiao, Z. Channel Selection for Automotive Parts Remanufacturer under Government Replacement-Subsidy. *Eur. J. Ind. Eng.* **2018**, *12*, 808–831. [CrossRef]
48. Zhang, Y.; Hong, Z.; Chen, Z.; Glock, C.H. Tax or Subsidy? Design and Selection of Regulatory Policies for Remanufacturing. *Eur. J. Oper. Res.* **2020**, *287*, 885–900. [CrossRef]
49. Zhang, F.; Zhang, R. Trade-in Remanufacturing, Customer Purchasing Behavior, and Government Policy. *Manuf. Serv. Oper. Manag.* **2018**, *20*, 601–616. [CrossRef]
50. Gu, X.; Zhou, L.; Huang, H.; Shi, X.; Ieromonachou, P. Electric Vehicle Battery Secondary Use under Government Subsidy: A Closed-Loop Supply Chain Perspective. *Int. J. Prod. Econ.* **2021**, *234*, 108035. [CrossRef]
51. Xie, X.; Huo, J.; Qi, G.; Zhu, K.X. Green Process Innovation and Financial Performance in Emerging Economies: Moderating Effects of Absorptive Capacity and Green Subsidies. *IEEE Trans. Eng. Manag.* **2016**, *63*, 101–112. [CrossRef]
52. Jiang, T.; He, Z.; Xiang, G.; Hu, T. The Impact of Government Subsidies on Private Pension Enterprises: A Decision-Making Model Based on Their CSR Levels. *IEEE Access* **2021**, *9*, 167190–167203. [CrossRef]
53. Raz, G.; Ovchinnikov, A. Coordinating Pricing and Supply of Public Interest Goods Using Government Rebates and Subsidies. *IEEE Trans. Eng. Manag.* **2015**, *62*, 65–79. [CrossRef]
54. Esenduran, G.; Kemahlıoğlu-Ziya, E.; Swaminathan, J.M. Take-Back Legislation: Consequences for Remanufacturing and Environment. *Decis. Sci.* **2016**, *47*, 219–256. [CrossRef]
55. Atasu, A.; Özdemir, Ö.; Van Wassenhove, L.N. Stakeholder Perspectives on E-Waste Take-Back Legislation. *Prod. Oper. Manag.* **2013**, *22*, 382–396. [CrossRef]
56. Ferrer, G.; Swaminathan, J.M. Managing New and Remanufactured Products. *Manag. Sci.* **2006**, *52*, 15–26. [CrossRef]
57. Ferguson, M.E.; Toktay, L.B. The Effect of Competition on Recovery Strategies. *Prod. Oper. Manag.* **2006**, *15*, 351–368. [CrossRef]
58. Galbreth, M.R.; Boyac, T.; Verter, V. Product Reuse in Innovative Industries. *Prod. Oper. Manag.* **2013**, *22*, 1011–1033. [CrossRef]
59. Esenduran, G.; Kemahlıoğlu-Ziya, E.; Swaminathan, J.M. Impact of Take-Back Regulation on the Remanufacturing Industry. *Prod. Oper. Manag.* **2017**, *26*, 924–944. [CrossRef]
60. Atasu, A.; Sarvary, M.; Wassenhove, L.N.V. Remanufacturing as a Marketing Strategy. *Manag. Sci.* **2008**, *54*, 1731–1746. [CrossRef]
61. Esenduran, G.; Kemahlıoğlu-Ziya, E.; Swaminathan, J.M. *Product Take-Back Legislation and Its Impact on Recycling and Remanufacturing Industries*; Springer: New York, NY, USA, 2012; Volume 174.
62. Kovach, J.J.; Atasu, A.; Banerjee, S. Salesforce Incentives and Remanufacturing. *Prod. Oper. Manag.* **2018**, *27*, 516–530. [CrossRef]
63. Zheng, X.; Li, D.-F.; Liu, Z.; Jia, F.; Lev, B. Willingness-to-cede behaviour in sustainable supply chain coordination. *Int. J. Prod. Econ.* **2021**, *240*, 108207. [CrossRef]
64. Zhang, X.; Li, Q.-w.; Liu, Z.; Chang, C.-T. Optimal pricing and remanufacturing mode in a closed-loop supply chain of WEEE under government fund policy. *Comput. Ind. Eng.* **2021**, *151*, 106951. [CrossRef]

 *processes*

Article

Analysis Method and Case Study of the Lightweight Design of Automotive Parts and Its Influence on Carbon Emissions

Qiang Li [1,2], Yu Zhang [3,*], Cuixia Zhang [1,*], Xiang Wang [1] and Jianqing Chen [4]

1 School of Mechanical and Electronic Engineering, Suzhou University, Suzhou 234000, China
2 School of Mechatronic Engineering, China University of Mining and Technology, Xuzhou 221116, China
3 School of Mechanical Engineering, Anyang Institute of Technology, Anyang 455000, China
4 QiuZhen School, Huzhou University, Huzhou 313000, China
* Correspondence: z13464238735@sina.com (Y.Z.); cuixiazhang@126.com (C.Z.); Tel.: +86-15105576286 (C.Z.)

Abstract: The automobile industry, as a representative in pursuing the goals of "emission peak" and "carbon neutrality", has made low carbon a new industrial practice. With regard to low carbon, the lightweight design proves to be an effective approach to reducing carbon emissions from automobiles. Given the state of research, in which the existing lightweight design schemes of automobiles seldom consider the impact of the lightweight quality on carbon emissions during the whole life cycle of the automobiles, this paper proposes a more comprehensive lightweight design method for automobiles in regard to carbon emissions. First, the finite element method was adopted to analyze the stress, strain and safety factors of the automobile parts based on their stress, so as to identify the positions where the lightweight design was applicable. Subsequently, a lightweight scheme was designed accordingly. Next, the finite element method was re-applied to the parts whose weights had been reduced. In this way, the feasibility of the lightweight scheme was verified. In addition, a method of calculating the carbon emissions produced by changes in the mass, manufacturing processes, application and recycling of automobile parts after the application of the lightweight design was also presented. The method can be used for evaluating the low carbon benefits of the lightweight design scheme. To prove the feasibility of the method, the ZS061750-152101 wheel hub designed and manufactured by Anhui Axle Co., Ltd. was taken as an example for the case analysis. The lightweight design changes three structures of the wheel hub, reducing its weight by 1.4 kg in total. For a single wheel hub, the carbon emissions are reduced by 51.22 kg altogether. That is to say, if the lightweight scheme were to be applied to all the wheels produced by Anhui Axle Co., Ltd. (about 500,000 per year), the carbon emissions from the wheel production, application and recycling could be cut by 2.56×10^7 kg, marking a favorable emission reduction effect. The proposed method can not only provide insight into the lightweight design of automobiles and other equipment against the background of low carbon but also provide a channel for calculating the carbon emission changes in the whole process after the application of the lightweight design.

Keywords: lightweight; carbon emissions; low carbon benefits; optimization design; wheel hub

Citation: Li, Q.; Zhang, Y.; Zhang, C.; Wang, X.; Chen, J. Analysis Method and Case Study of the Lightweight Design of Automotive Parts and Its Influence on Carbon Emissions. *Processes* **2022**, *10*, 2560. https://doi.org/10.3390/pr10122560

Academic Editor: Duhwan Mun

Received: 30 October 2022
Accepted: 24 November 2022
Published: 1 December 2022

1. Introduction

In recent years, automobile production and sales have continued to rise at high rates, resulting in the fast growth of automobile ownership [1]. On the one hand, this phenomenon brings enormous development opportunities to the automobile industry and related industries. On the other hand, it incurs various challenges [2,3], such as environmental problems caused by greenhouse gas emissions and resource depletion induced by energy consumption [4]. The energy saving and emission reduction of automobiles have become a focus of the automobile industry [5]. The question of how to incorporate a low-carbon design of automobiles so as to develop them into internationally certified low-carbon products, keep pace with the trends of the consumer market and seize the green automobile consumption market has become the long-term development strategy of automobile manufacturing enterprises [6]. The "Made in China 2025" initiative stated specific demands for the lightweight quality of automobiles, including the realization of the lightweight property of automobile parts and optimization of the structure of automobiles [7]. Therefore, applying a low-carbon design to automobiles is essential for minimizing the negative impact of automobile manufacturing on the environment and improving the environmental friendliness of automobile products. Specifically, such an aim can be realized by supporting the low-carbon design and manufacturing of automobile products, quantitatively assessing the carbon emissions of their typical parts in the early stage of the design and optimizing their carbon emission performance through structure design and favorable processes.

In terms of the lightweight design, scholars generally rely on structure design, advanced manufacturing processes and new materials to reduce the weight of automobile parts. To name a few, Baskin et al. [8] applied the variable density topology optimization technology to the body-in-white development process. The application accomplished the lightweight property of the automobile while ensuring the enhancement of its various performance aspects. Shimoda et al. [9] adopted the parameter-free shape optimization method to redesign the shape of the weight-reducing hole of the main bearing beam of the automobile from the perspective of strength constraint, which successfully reduced the weight of the structure. Gauchia et al. [10] employed the genetic algorithm to optimize the key dimensions of the bus body, effectively reducing its weight and strengthening its torsional stiffness. Xiao et al. [11] suggested a multi-objective optimization method based on programming work for steel wheels, in which the design variable is the thickness of the spoke. Xiao et al. [12] built a static and dynamic multi-objective topology optimization mathematical model for spokes by utilizing the compromise programming method. With the aid of this model, Xiao et al. performed topology optimization on the spokes when they were running under bending fatigue conditions, which realized the lightweight quality of the hubs. Zhang et al. [13] performed single-objective topology optimization on aluminum alloy wheel spokes using the variable density topology optimization method. Pang et al. [14] established a response surface model between dimensions and structural properties of the spokes and hubs for the purpose of parameter optimization to achieve the lightweight property. Hu et al. [15] optimized the parameters of a certain type of aluminum alloy wheel hub by regarding the thicknesses of the wheel and the rim as the design variables and the minimum mass of the wheel as the goal. Chai et al. [16] designed a lightweight wheel hub formed of a composite whose thermoplasticity was enhanced using long glass fiber and studied its fatigue life by means of simulation and experiments. Yi et al. [17] pointed out that compared with the aluminum alloy hubs, the green effect produced by magnesium alloy hubs is mainly due to the reduction in fuel consumption caused by weight loss. With respect to the carbon emissions of automobiles, Zhang et al. [18] studied the design process of automobile engines. From the results of the analysis of the integrated system, their study reported that the raw material procurement stage of the engine leads to the largest amount of carbon emissions, with the cylinder block being the chief contributor. Based on the established evaluation model, they also proposed optimizations for the structure and materials and verified the feasibility of the integrated system. Hao et al. [19] established a bottom-up model to consider the energy and environmental impacts of reducing greenhouse gas emissions from a passenger vehicle fleet. Wu et al. [20] investigated the identification of barriers, analysis and solutions for hydrogen fuel cell vehicles designed for deployment in China with the goal of carbon neutrality. Du et al. [21] conducted a study that focused on policy making and its impact on the automotive industry in the context of carbon neutrality. Du et al. [22] investigated the costs and potential of reducing CO_2 emissions in the Chinese transportation sector by means of energy system analysis. Jhaveri et al. [23] developed a parametric life cycle model to assess the life cycle performance of TWDCI compared to conventional cast iron and cast aluminum in terms of energy (MJ) and GHGs (as carbon dioxide equivalents: kg CO_2e).

In summary, the lightweight design of automobile parts is often accompanied by the optimization of their shapes and sizes. However, the optimization scheme can only be obtained through an iterative search based on a rigorous mathematical model, which proves to be a cumbersome process. As a consequence, there is an urgent need to establish a popular, simple and practical lightweight optimization method for automobile parts. Meanwhile, in terms of automobile carbon emissions, scholars have mostly focused on calculating the total carbon emissions of the automobile industry, attached importance to technologies concerning carbon emission reduction and investigated the impacts of carbon policies on the automobile industry. Unfortunately, limited attention has been devoted to automobile carbon emissions in the initial design and optimization of automobile parts.

After the weight of automobile parts is decreased, some manufacturing processes will inevitably be added, which will certainly result in extra carbon emissions. This begs the questions: how can we calculate the reduced carbon emissions resulting from the decreased mass and the newly increased carbon emissions arising from the additional processes, and what are the comprehensive carbon emission benefits of the two?

To answer the above questions, this paper proposed a lightweight design method for the carbon emissions of automobile systems. The specific research scheme is described as follows: First, the finite element method was adopted to analyze the stress, strain and safety factors of the automobile parts based on their stress, so as to identify the positions where the lightweight design was applicable. Subsequently, a lightweight scheme was designed accordingly. Next, the finite element method was re-applied to the parts whose weights had been reduced for the purpose of analyzing their stress, strain and safety factors. In this way, the feasibility of the lightweight scheme was verified. In addition, a method for calculating the carbon emissions of automobile parts with lightweight scores was also presented, through which the variation in the carbon emissions produced by the mass reduction, manufacturing processes, application and recycling of the parts could be measured. Based on these data, the low-carbon benefits of the lightweight scheme were comprehensively evaluated. Moreover, the ZS061750-152101 wheel hub designed and manufactured by Anhui Axle Co., Ltd. was taken as an example for the case analysis. The lightweight design changes three structures of the wheel hub, reducing its weight by 1.4 kg in total. The carbon emissions resulting from the saved material, changed manufacturing processes and application and recycling of the parts are reduced by 51.22 kg altogether. That is to say, if the lightweight scheme were to be applied to all the wheels produced by Anhui Axle Co., Ltd. (about 500,000 per year), the carbon emissions from the wheel production, application and recycling could be cut by 2.56×10^7 kg, marking a favorable emission reduction effect.

This study provides both theoretical and practical contributions. In theory, on the basis of its structural analysis, the modal, fatigue strength and life of the hub were simulated using the finite element method, and the lightweight design of the hub was verified. This method provides a useful reference for the lightweight design of automotive parts. In engineering practice, this method can clearly inform the engineering designers about how to efficiently and quickly lighten the mechanical parts of the automobile and to calculate the reduced carbon emissions resulting from the decreased mass and the newly increased carbon emissions arising from the additional processes. It can improve the structure, save materials, reduce the production costs, shorten the design cycle and meet the low carbon demands of customers for manufacturing enterprises.

The remaining content of this paper is arranged in the following manner: A lightweight design method for the carbon emissions of automobile systems is provided, and the low carbon benefits of the lightweight scheme are comprehensively evaluated in Section 2. The ZS061750-152101 wheel hub designed and manufactured by Anhui Axle Co., Ltd. is taken as an example for the case analysis, and a comparative analysis of the mechanical properties of the wheel hub before and after the lightweight design and an analysis of the emission reduction benefits after the lightweight design are presented in Section 3. The main innovations are as follows: This study proposed a lightweight design method for the carbon emissions of automobile parts considering the reduction in carbon emissions arising from the mass reduction, manufacturing process and recycling. The carbon emission benefits of each link caused by lightweight design were considered and calculated.

2. Method

2.1. Flowchart of the Method

We built a lightweight design method for the carbon emissions produced in the automobile life cycle. Specifically, the finite element method was adopted to analyze the stress, strain and safety factors of the automobile parts based on their actual stress and obtain the corresponding contours. Afterwards, the positions with the lowest stress and strain and the largest safety factors were identified on the basis of these contours, which are crucial for the determination of the positions that can achieve the lightweight property. Then, considering the actual working conditions of the parts, a lightweight solution was proposed. After the implementation of the scheme on the parts, the finite element method was re-applied to the lightweight parts to explore their changes in stress, strain and the safety factor, as well as the mechanical properties, hence verifying the feasibility of the lightweight scheme. Finally, the variations in the carbon emissions caused by the mass reduction, manufacturing processes, application and recycling of the parts were calculated to evaluate the carbon emissions generated by the lightweight design and assess the low carbon benefits. The flowchart of the proposed method is presented in Figure 1.

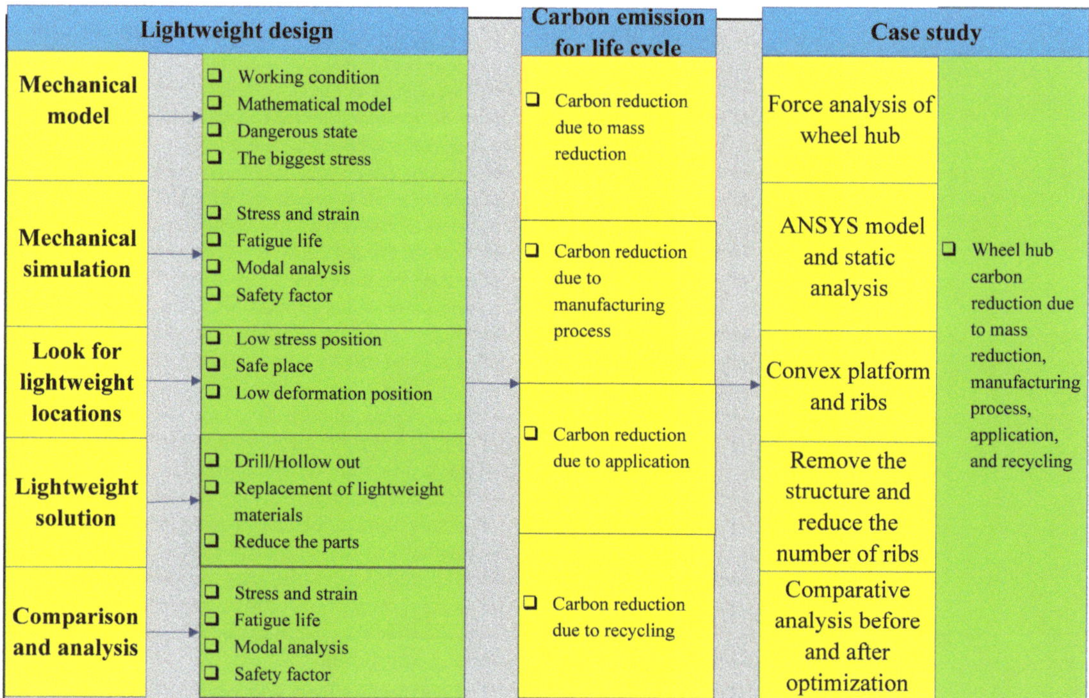

Figure 1. Flowchart of the proposed lightweight design method for carbon emissions.

2.2. Finite Element Analysis and Lightweight Scheme

The primary research object of this study is the automobile industry. Considering that the stress states of automobile parts vary significantly due to their abundant types, we conducted finite element analysis in accordance with the actual stress state. The procedure of the analysis is described as follows:

(1) Build the 3D model. According to the actual parameters of the part, the 3D modeling software is used to establish the 3D model of the part.

(2) Import the 3D model into the finite element software (ANSYS). Convert the format of the 3D model (generally, prt, sldprt, etc.) to the format that can be recognized by the finite element analysis software (generally stl, igs, stp, etc.).

(3) Apply the materials. Select materials in the material library of the finite element analysis software or add material attributes as needed to define the materials.

(4) Perform meshing: The higher the node number and the cell number are, the more accurate the calculation results will be, but more computation time will be required. The grid size needs to be selected or divided according to the actual application requirements.

(5) Add the constraint. According to the actual constraint status of the components, simulate their constraint status in the software and add appropriate constraints according to the actual situation

(6) Add the loads. Simplify the actual stress state through force analysis and add forces, pressures and other loads that can be simulated in the software.

(7) Simulation calculation. Set the simulation step size, simulation time, etc., as required. After setting the above process, use computers for the solution and post-processing. The cloud diagram of the stress value, shape variables and safety factor of the brake hub in this working environment can been obtained through a simulation calculation.

(8) Identify lightweight positions. Analyze the simulation results to acquire contours of the stress, strain and safety factor of the parts. Afterwards, identify lightweight positions, i.e., positions with the lowest stress and strain and the largest safety factor, according to the above contours.

(9) Design the corresponding lightweight scheme. Design the corresponding lightweight scheme based on the structures and actual working conditions of the parts. In the lightweight design scheme, some materials are removed. For example, holes, chamfers and grooves are added, while the number of ribs and the thickness of the plates are reduced.

The process of finite element analysis and the lightweight scheme are shown in Figure 2.

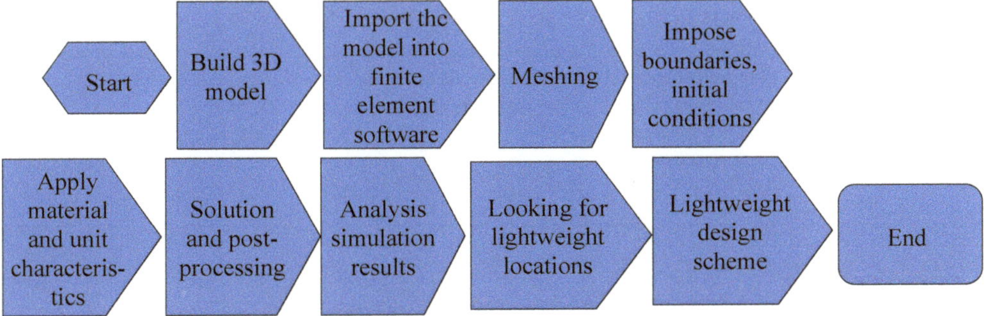

Figure 2. Finite element analysis process and lightweight scheme.

2.3. Comparison and Analysis

A 3D model of the parts developed according to the lightweight design scheme is built. On this basis, the weights of the parts before and after the application of the lightweight design can be compared to reveal whether the mass of the parts is reduced. Then, the above finite element analysis is performed on the parts developed according to the lightweight design scheme to obtain contours of the stress, strain and safety factor of the parts, and the procedure is the same as that stated in Section 2.2. Next, these contours are comparatively analyzed with those acquired prior to the lightweight design to determine whether the stress, strain and safety factor remain within the reasonable ranges.

2.4. Carbon Emissions Caused by the Lightweight Design

The carbon emissions of the parts developed according to the lightweight design are calculated to evaluate the carbon reduction benefits of the lightweight scheme. In particular, the carbon reductions resulting from the mass reduction, manufacturing processes, use and recycling of the parts are calculated.

(1) Carbon emissions from the mass reduction after the lightweight design

After the lightweight design is applied, the mass of the parts is reduced, which means that a smaller amount of raw materials (such as steel and aluminum) are needed. As a result, the carbon emissions produced in the mining and processing of these less-used materials decrease. The decreased carbon emissions can be calculated by Equation (1) [24]. Carbon impact factors are used in this calculation. The carbon emission factor generally refers to the statistical average of the amount of carbon dioxide produced (or emitted) by a unit of production under normal conditions [25]. It can also be converted according to standard coal. The emission factor method is a common method used to calculate carbon emissions based on the empirical emission factors and yields of products:

$$C_{\Delta m} = \Delta m \times f_m \tag{1}$$

where $C_{\Delta m}$ is the carbon emissions from the lightweight-induced mass reduction, kg; Δm is the reduced mass, kg; and f_m is the carbon emission factor of the material flow, kg/kg.

(2) Carbon emissions from changes in the manufacturing processes after the lightweight design

The calculation of the carbon emissions produced in the production and processing stages should take all the inputs and outputs involved into account. The former include the inputs of the materials (raw materials, auxiliary materials, etc.) and energy (electricity, coal, natural gas, etc.), while the latter cover the outputs of the energy (electricity and heat) and emissions (gas, waste and liquid waste). Therefore, the carbon emissions of the entire production and processing stages can be calculated from the perspective of the "three flows" (energy flow, material flow and environmental emission flow). The "three-flow" model includes both the direct carbon emissions from material production and processing and the indirect carbon emissions from energy consumption and waste disposal.

The production of the parts generally involves complex processes, each of which emits carbon. Therefore, the material flow consumption M, the energy flow consumption E and the environmental emission flow consumption W in each process i ($i = 1, 2, \ldots, i_0$)) should all be comprehensively considered. The consumptions of the three flows include various types of materials j ($j = 1, 2, \ldots, j_0$), energy k ($k = 1, 2, \ldots, k_0$) and pollutants l ($l = 1, 2, \ldots, l_0$), respectively:

$$C_m = \sum_{i=1}^{i_n} \sum_{j=1}^{j_n} M_i \times f_m^N \tag{2}$$

$$C_e = \sum_{i=1}^{i_n} \sum_{k=1}^{k_n} E_i \times f_e^N \tag{3}$$

$$C_w = \sum_{i=1}^{i_n} \sum_{l=1}^{l_n} W_i \times f_w^N \tag{4}$$

where i, j, k and l are all natural numbers.

The analysis of the material flow, energy flow and environmental emission flow in the production and processing stages reveals that the lightweight design of the parts is accompanied by changes in the processes. Once the processes are changed, the amount of carbon emissions also changes. The carbon emissions resulting from changes in the production and processing stages after the lightweight design can be expressed as:

$$C_{\Delta p} = \Delta C_m + \Delta C_e + \Delta C_w$$
$$= -\sum_{i=1}^{i_n}\sum_{j=1}^{j_n} \Delta M_i \times f_m^j - \sum_{i=1}^{i_n}\sum_{k=1}^{k_n} \Delta E_i \times f_e^k - \sum_{i=1}^{i_n}\sum_{l=1}^{l_n} \Delta W_i \times f_w^l \tag{5}$$

where $C_{\Delta p}$ is the variation in carbon emissions due to changes in the production and processing stages after the lightweight design, kg; C_m is the carbon emissions of the material flow, kg; C_e is the carbon emissions of the energy flow, kg; C_w is the carbon emissions of the environmental emission flow, kg; f_m^N is the carbon emission factor of the material flow, kg/kg; f_e^N is the carbon emission factor of the energy flow, kg/kg; and f_w^N is the carbon emission factor of the environmental emission flow, kg/kg. It should be noted that the negative sign in Equation (5) denotes the negative carbon emissions produced by the additional processes after the lightweight design.

(3) Carbon emissions from the use of the parts after the lightweight design

The application of the lightweight design to the parts can directly influence the carbon emissions of a car. The relationship between the carbon emissions of a car and the reduction in the vehicle mass can be expressed as:

$$C_l = l\eta_w \Delta m \tag{6}$$

where C_l is the carbon emissions from the vehicle exhaust caused by the application of the lightweight design on the brake hub, kg; η_w is the emission coefficient per kilometer, km^{-1}; and l is the vehicle kilometers traveled, km.

(4) Carbon emissions from the recycling of the parts after the lightweight design

Recycling, here, refers to the process of reusing waste automobile parts for blank production through remanufacturing, recycling and other means, which can effectively reduce the carbon emissions in the production and processing stages. During recycling, some materials of the parts are treated and recycled in the production stage [26,27], and others are returned to the processing stage for material reuse. Hence, the carbon emissions from the recycling of the parts after the lightweight design can be expressed as:

$$\begin{cases} C_r = -\eta_1 \Delta m_{production} f_m - \eta_2 \Delta m_{process} f_m \\ \Delta m_{production} + \Delta m_{process} = \eta_m \Delta m \\ \eta_1 + \eta_2 = 1 \end{cases} \tag{7}$$

where C_r is the carbon emissions in the recycling stage of the parts after the lightweight design; $\Delta m_{production}$ is the mass of the parts reused in the production process; $\Delta m_{process\,sin\,g}$ is the mass of the parts reused in the processing stage; η_1 is the utilization rate of the hub material used for reproduction; and η_2 is the utilization rate of the hub material used for reprocessing.

Therefore, the comprehensive carbon emission benefits yielded by the lightweight design of the parts consist of the carbon emissions from the four parts mentioned above, which can be expressed as:

$$C_{LD} = C_{\Delta m} + C_{\Delta p} + C_l + C_r$$
$$= \Delta m \times f_m - \sum_{i=1}^{i_n}\sum_{j=1}^{j_n} \Delta M_i \times f_m^j - \sum_{i=1}^{i_n}\sum_{k=1}^{k_n} \Delta E_i \times f_e^k - \sum_{i=1}^{i_n}\sum_{l=1}^{l_n} \Delta W_i \times f_w^l \tag{8}$$
$$+ l\eta_w \Delta m - \eta_1 \Delta m_{production} f_m - \eta_2 \Delta m_{process} f_m$$

where C_{LD} is the comprehensive carbon emission benefits of the parts after the lightweight design.

3. Case Study

3.1. Background

Wheel hubs are one of the important components of a car [28]. As of the first half of 2022, the number of motor vehicles in China has reached 406 million, and each automobile requires several wheels. In 2022, the domestic demand for passenger vehicles in China has reached 21.1 million, an annual increase of 4.4%, and the predicted value at the beginning of the year scored an annual increase of 3%. The large number of annual sales of automobiles have induced a great demand for wheel hubs [29,30]. Experience indicates that for every 1 kg mass reduction of the rotating parts, such as automobile wheel hubs, the mass of the other non-rotating parts will be cut by 1.2–1.5 kg per year. Clearly, the mass reduction of wheel hubs is essential for the application of the lightweight design to automobiles. To date, the lightweight design of wheel hubs has become one of the key technologies for wheel hubs and even automobile manufacturers aiming to improve their core competitiveness.

Therefore, the proposed lightweight brake hub can reduce the carbon emissions produced by brake hub manufacturers and operating vehicles.

3.2. Finite Element Analysis of the Wheel Hub

1. Building a 3D model and performing the meshing

The lightweight design takes the ZS061750-152101 wheel hub designed and manufactured by Anhui Axle Co., Ltd. as the prototype. The wheel hub has a load of 6T and is extensively used. Its 3D model, established by 3D modeling software, is shown in Figure 3.

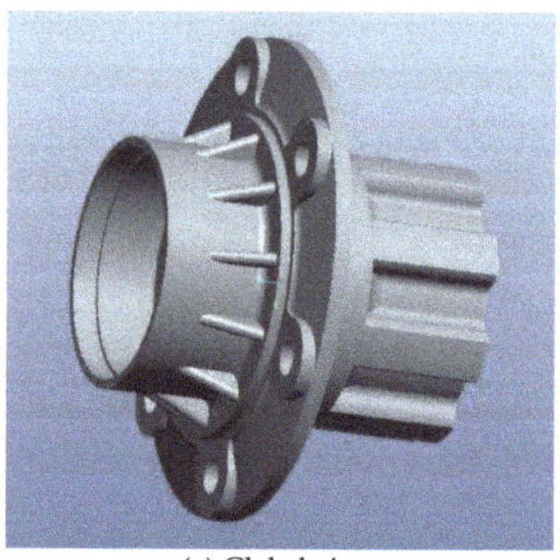 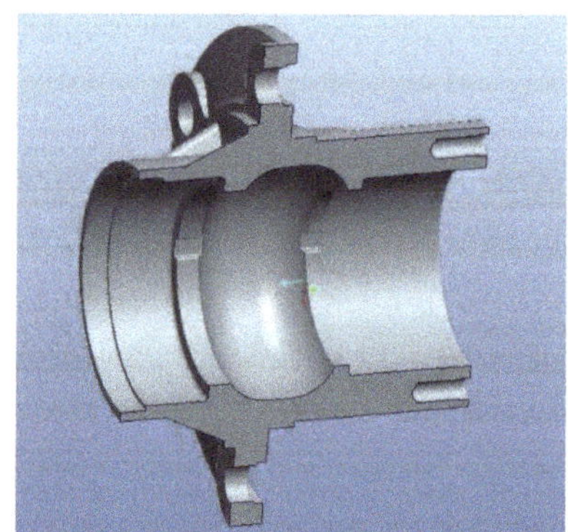

(**a**) Global view   (**b**) Half sectional view

Figure 3. 3D model of the ZS061750-152101 wheel hub.

To reduce the number of unnecessary calculations, the 3D model of the wheel hub is simplified by removing the rounded corners, chamfers, surfaces and irregular entities without violating the actual requirements of ANSYS Workbench for finite element analysis. The simplified 3D model imported into ANSYS is presented in Figure 4.

After the model is imported into the ANSYS software, the meshing is conducted. Concretely, through the Sizing function in the Mesh program, fine meshing is performed on the part where the stress is concentrated, i.e., on the installation rabbet, while sparse meshing is performed on the other parts that are slightly stressed or not stressed. The results of the meshing are exhibited in Figure 5. A total of 145,211 3D tetrahedral solid units and 244,181 nodes are divided.

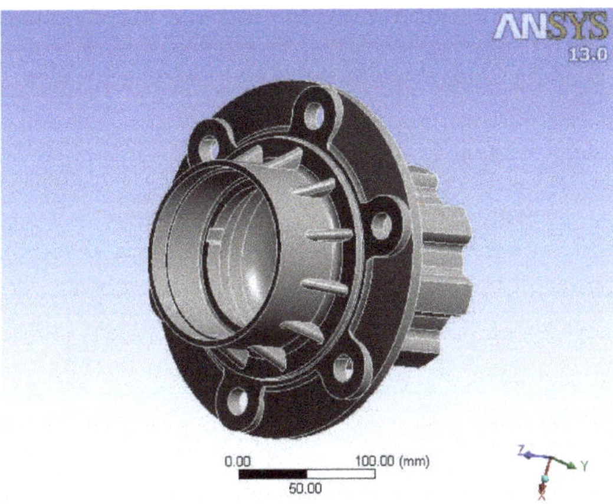

Figure 4. Simplified 3D model of the wheel hub.

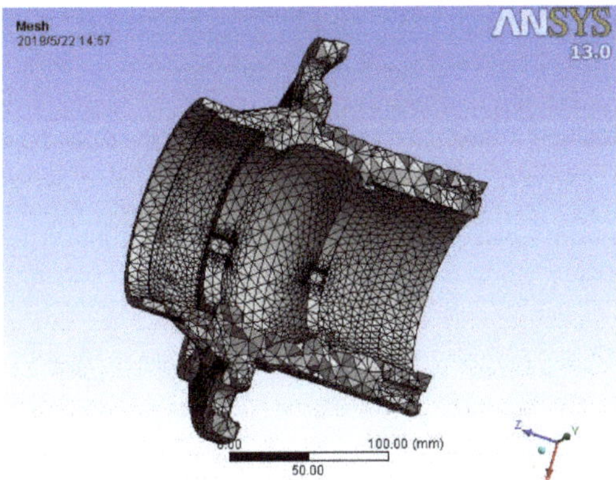

(**a**) Internal mesh units

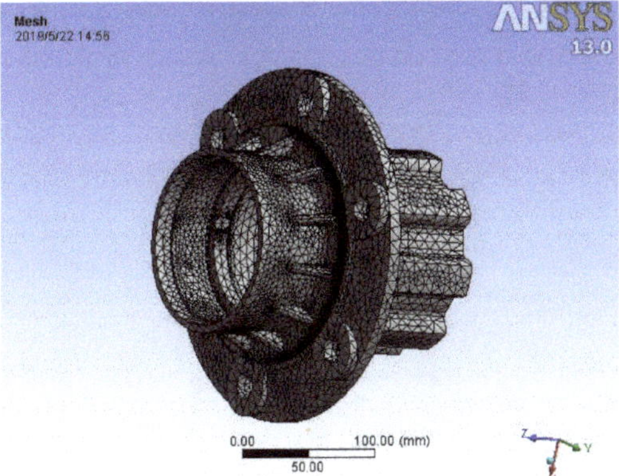

(**b**) Whole mesh

Figure 5. Meshing model of the wheel hub.

2. Adding the material properties

The wheel hub is formed of ductile iron QT500-10, whose properties are listed in Table 1. In this step, these material parameters are added, correspondingly, to the ANSYS software in turn.

Table 1. Mechanical parameters of ductile iron QT500-10.

Symbol	Property	Parameter
E	Elastic modulus (MPa)	1.7×10^5
v	Poisson's ratio	0.29
S_{Yt}	Yield strength (Pa)	3.6×10^8
S_{Ut}	Tensile strength (Pa)	5×10^8
ρ	Density (kg/m^3)	7100

3. Applying the loads

The research object in this paper is the 6T wheel hub. Its maximum load is 6 tons under normal working conditions, and the pressure on a single hub is 30,000N and is distributed along the lower semicircular axis of the wheel hub. Considering that the impact coefficient is 1.67, the rated load of the unilateral hub is 50,000 N. During loading, the pressure load increases linearly to 50,000 N within 1 s and maintains this value for 1–2 s. The tool of ANSYS that we used was a static structural tool. The installation rabbet, i.e., the flange, is the position where the stress is concentrated.

4. Applying the constraints

During working conditions, the hub is subject to 3 degrees of freedom in the translation direction and 2 degrees of freedom in the rotation direction. However, the degree of freedom with the central axis of the hub as the rotation direction is not constrained, and the constraint positions are located at the bearing installation points. Thus, the constraints are set as fully fixed supports (Figure 6).

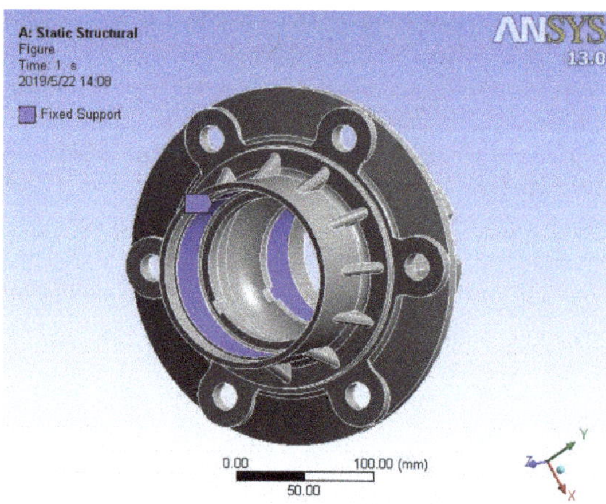

Figure 6. Positions of the constraints applied to the wheel hub.

5. Finite element analysis and results

After running the simulation program, the calculation results are obtained. The contours of the stress, deformation (i.e., displacement) and safety factor of the wheel hub are displayed in Figure 7.

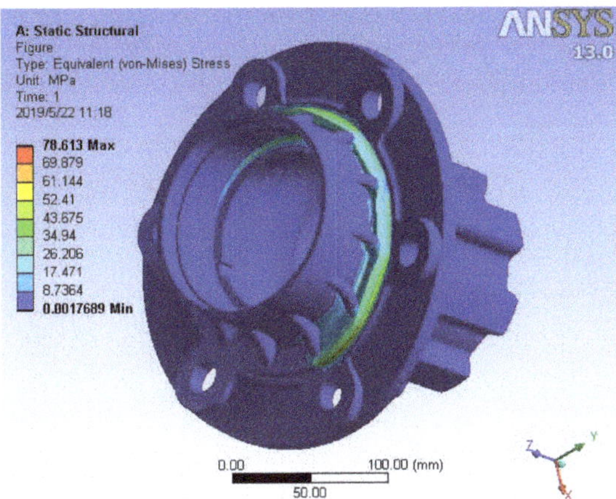

(**a**) Contour of the stress

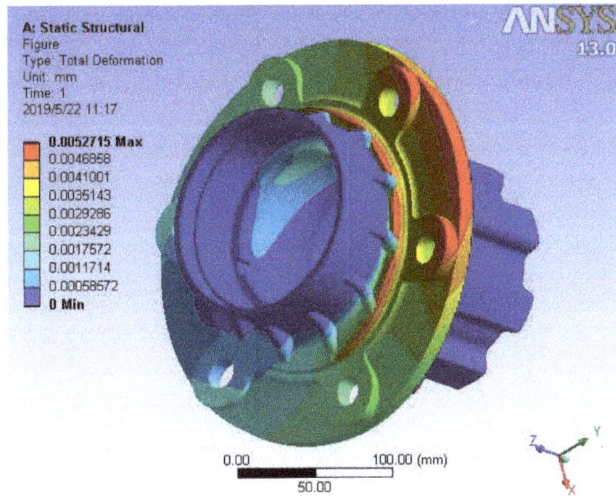

(**b**) Contour of the displacement

Figure 7. Finite element analysis results.

As can be observed from Figure 7a, the stress is relatively concentrated in the hub installation rabbet, which is in line with the predicted stress and strain positions in practice. In addition, the maximum stress value is 78.613 MPa, which is much lower than its yield limit, i.e., 360 MPa. According to Figure 7b, the maximum deformation is 0.0052715 mm, which is far lower than the specified value, 1 mm. The safety factor is 7.9503, which is greater than the required range, 2–2.5.

6. Fatigue Life Analysis

The S–N curve can be used to describe the relationship between the fatigue strength and service life of the parts.

Its mathematical expression is:

$$NS^m = C \tag{9}$$

where N is the life at the time of failure, as well as the number of cycles, and C is a constant.

The wheel hub of an automobile is subjected to unstable variable stress in the period of work, and the theoretical basis for calculating its fatigue strength is the cumulative fatigue damage hypothesis (namely, Miner's law). The mathematical expression is:

$$\sum_{i=1}^{z} \frac{n_i}{N_i} = 1 \tag{10}$$

where n_i is the number of stress cycles at the i-th grade, and N_i is the fatigue life at the i-th grade.

Based on this hypothesis, the engineers and researchers performed numerous experiments and arrived at Equation (11):

$$\sum_{i=1}^{z} \frac{n_i}{N_i} = 0.7 \tag{11}$$

In the Statics module, the number of life cycles of the wheel hub and the fatigue factor for the safe use of the structure are added to the Fatigue Toolbar. Since the fatigue of the wheel hub is defined as high-cycle fatigue, the number of its life cycles is set to be over 10^6. The specific formula of the fatigue notch factor K_f is shown in Equation (12). The value of K_f is calculated to be 0.35, and the value of the fatigue strength factor is 1:

$$K_f = \frac{\sigma_{max}}{\sigma_{-1}} \tag{12}$$

where K_f is the fatigue notch factor; σ_{max} is the maximum local elastic stress; and σ_{-1} is the fatigue limit value.

The service life of the wheel hub is calculated using the Fatigue module in the ANSYS Workbench software, and the fatigue properties are added to the material for the analysis. The load spectrum used for the fatigue life analysis of the wheel hub is consistent with the load added in the mechanical analysis. During loading, the pressure load increases linearly to 50,000 N within 1 s and maintains this value for 1–2 s.

For the wheel hub cast with ductile iron, its number of life cycles N should exceed 10^6. According to this standard, the fatigue notch factor is calculated to be 0.35. Then, the life, safety factor and fatigue sensitivity in the solution are solved to obtain the result of the life analysis (Figure 8).

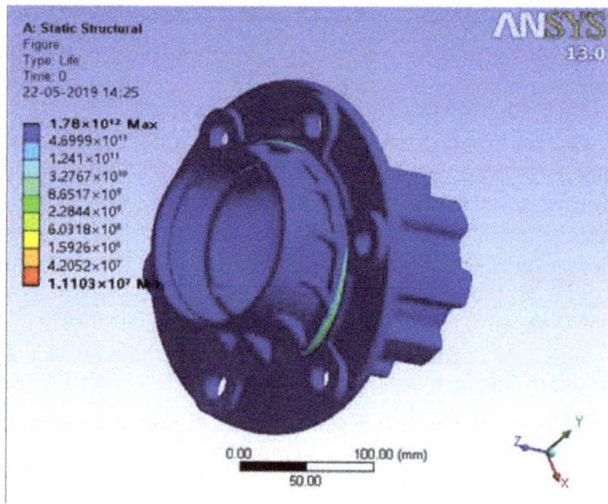

Figure 8. Analysis on the number of life cycles.

From the perspective of the service life, the number of cycles for the safe use of the wheel hub is 1.1103×10^7 (Figure 8), which is much larger than the set number, 1×10^6. This indicates that the service life of the wheel hub meets the requirements.

The above analysis proves that the design of the wheel hub is reasonable and leaves sufficient room for lightweight improvement. For this purpose, further lightweight design strategies can be applied to the wheel hub to render it more economical and efficient.

3.3. Lightweight Design

(1) Determination of the lightweight positions

According to the above analysis, the maximum displacement of the wheel hub is 0.0052715 mm, far smaller than the specified value of 1 mm. The stress is concentrated in the installation rabbet (i.e., the flange), with a maximum value of 78.613 MPa, far smaller than the yield strength (360 MPa) of the wheel hub material, QT500-10. The minimum safety factor is 7.9503, much larger than the required minimum safety factor of 2.0–2.5. The minimum number of life cycles is 1.1103×10^7, much greater than the set number of life cycles of 10^6. The fatigue factor for safe use is 1.2327, which is greater than 1. All these data suggest that the wheel hub has considerable potential for lightweight improvement, which can be realized at such safe positions as the ribs, large cylindrical structure and bosses.

(2) Lightweight design scheme

In the lightweight design, the strength, stiffness and fatigue life of the wheel hub are selected as the constraints. After the lightweight design is applied, the maximum deformation of the wheel hub should be less than 1 mm; the maximum stress should be lower than the yield limit of the raw material (360 MPa); the number of life cycles should exceed the predicted value; and the fatigue factor for safe use should be greater than 1. Only when all these requirements are met can the safety and reliability of a lightweight wheel hub during its use be guaranteed.

From the finite element analysis results, the ribs, large cylindrical structure and bosses are safe positions that can be selected for lightweight design. The mass can be reduced by adding holes, reducing the auxiliary structure, adding grooves or thinning. The designed lightweight scheme is described as follows: the structure at the axle neck is removed (in Figure 9, position 1, adding grooves); the width of the reinforcing rib is reduced from 33 mm to 30 mm (in Figure 9, position 2, thinning), and its number is decreased from 12 to 10 (reducing auxiliary structure); and a groove with a depth of 2 mm is dug into the installation rabbet (in Figure 9, position 3, adding grooves). The details of the lightweight design scheme modifications are presented in Figure 9.

As exhibited in Figure 9, the lightweight design also leads to some changes in the process: at Position 1, lathe cutting to a depth of 60 mm and a width of 5 mm is required; at Position 2, the time required to create the sand mold in the casting process is shortened; and at Position 3, the milling cutter must be used to dig one more groove that is 2 mm deep and 50 mm wide.

3.4. Carbon Emission Results

(1) Comparative analysis of the mechanical properties of the wheel hub before and after the lightweight design

With the aid of the ANSYS Workbench software, the wheel hub after the lightweight design is analyzed in accordance with the procedure described in Section 3.2. The meshing results of the wheel hub after the lightweight design are displayed in Figure 10. A total of 226,716 nodes and 136,773 solid units are divided.

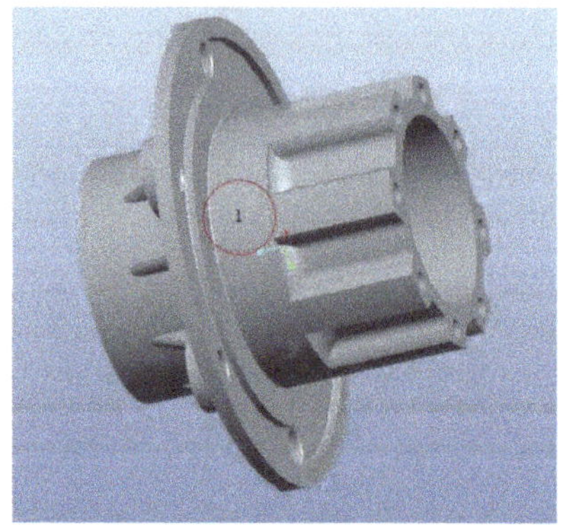

(**a**) After the lightweight design (Position 1)

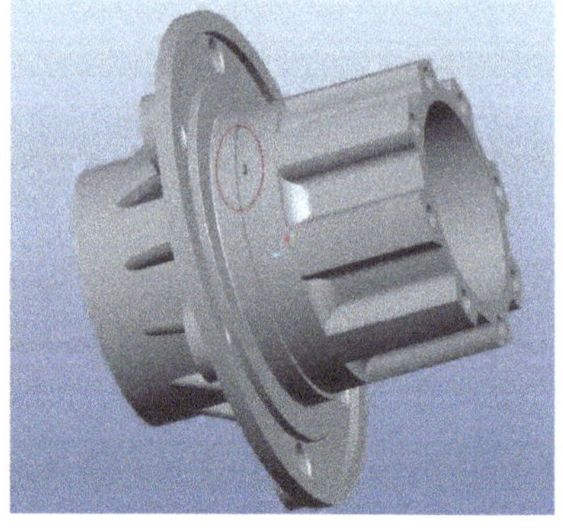

(**b**) Before the lightweight design (Position 1)

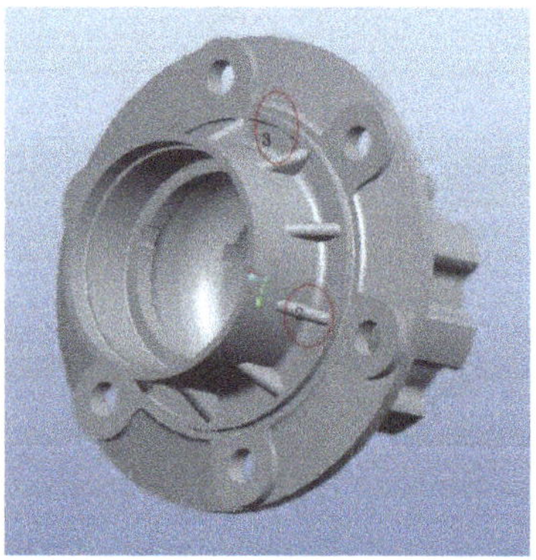

(**c**) After the lightweight design (Positions 2 and 3)

(**d**) Before the lightweight design (Positions 2 and 3)

Figure 9. Lightweight design scheme.

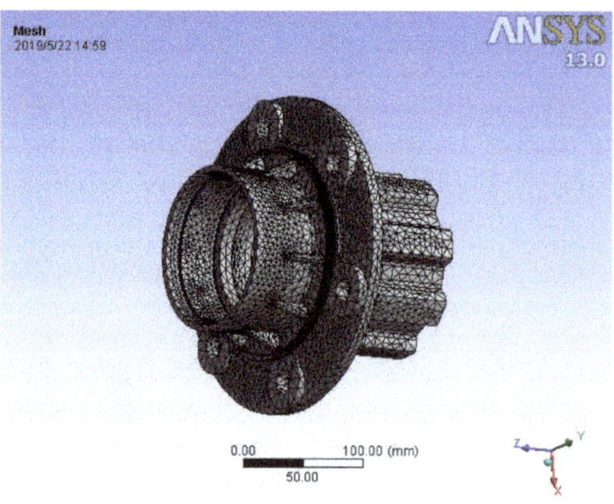

Figure 10. Meshing of the wheel hub after the lightweight design.

The contours of the stress and displacement of the wheel hub based on the simulation analysis and calculations are depicted in Figure 11.

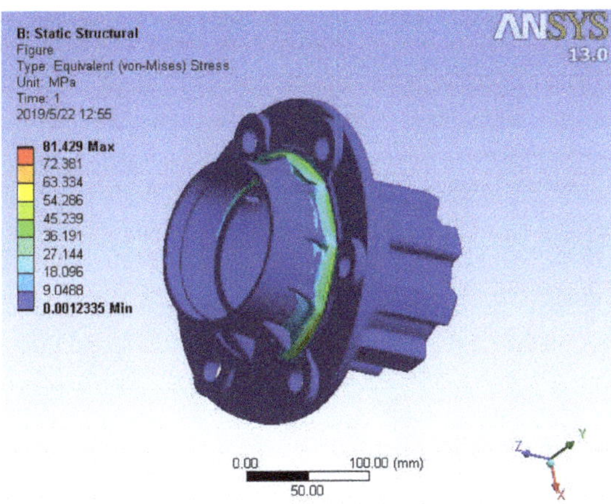

(**a**) Contour of the stress after the lightweight design

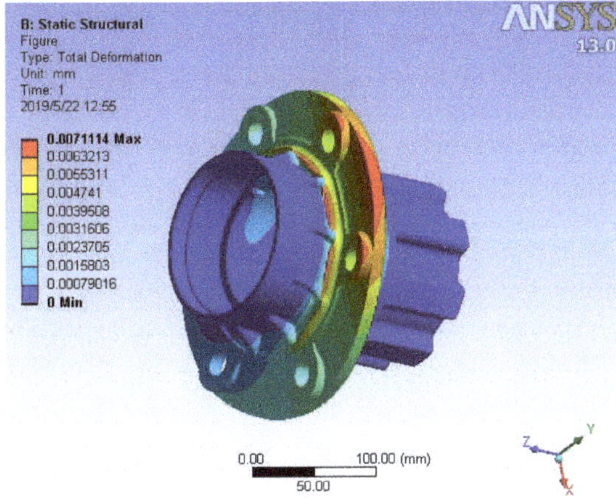

(**b**) Contour of the displacement after the lightweight design

Figure 11. Simulation analysis results of the wheel hub after the lightweight design.

The number of life cycles of the wheel hub after the lightweight design yielded from the life fatigue analysis is displayed in Figure 12.

As illustrated in Figure 11, after the lightweight design, the maximum stress value of the wheel hub is increased to 81.429 MPa, but it remains lower than the yield limit value (360 MPa) of its raw material. The deformation at the installation rabbet increases slightly to 0.0071114 mm (compared with Figures 5 and 6), which is still less than 1 mm, suggesting that the deformation remains within the required range. The safety factor is 7.6754, exceeding the required minimum value of 2–2.5, that is, it is up to standard. As indicated in Figure 12, the minimum number of life cycles for safe use is 8.2088×10^6, which exceeds the set number of 1×10^6; thus, it satisfies the requirements. In conclusion, the wheel hub succeeds in realizing the lightweight quality while guaranteeing its required working performance.

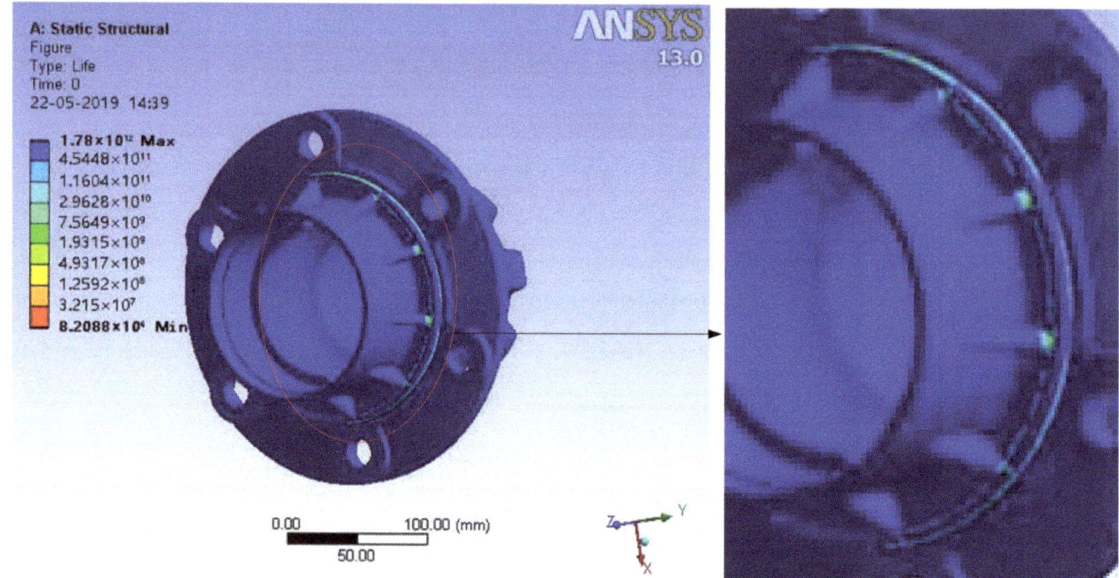

Figure 12. Number of life cycles of the wheel hub after the lightweight design (including a partial enlarged view).

The simulation results of the mechanical properties of the wheel hub before and after the lightweight design are compared in Table 2.

Table 2. Comparison of the mechanical properties of the wheel hub before and after the lightweight design.

No.	Parameter	Before the Lightweight Design	After the Lightweight Design	Value Change	Proportion of Change (%)
1	Stress (MPa)	78.613	81.429	2.816	+3.58
2	Displacement (mm)	0.0052715	0.0071114	0.0018399	+34.9
3	Safety factor	7.9503	7.6754	0.2749	−3.46
4	Fatigue life	1.1103×10^7	8.2088×10^6	2.823×10^6	−26
5	Gross mass (kg)	11.895	10.495	1.4	−11.77

(2) Analysis of the emission reduction benefits after the lightweight design

The lightweight design inevitably increases processes such as the turning and milling. With respect to the turning process, the processing equipment is a cylindrical lathe, CA6140. The parameters commonly used by this company include the following: in rough turning, the spindle speed of 150 r/min, the feed rate of 0.4 mm/r and the cutting depth of 3 mm; in semi-rough turning, the spindle speed of 200 r/min, the feed rate of 0.3 mm/r and the cutting depth of 1.5 mm; and in finish turning, the spindle speed of 450 r/min, the feed rate of 0.2 mm/r and the cutting depth of 0.5 mm. When the lathe is not loaded, its average power is 5350 W, and the non-load duration is 10 s. At the time of processing, its average power is 6450 W, and the processing lasts for 180 s in total. As for the milling process, the equipment used is a SINMERIK808D CNC milling machine, and the rod-shaped milling cutter is used for the plane milling. The specific parameters of the milling machine include: the feed rate of 1 mm/min, the rotational speed of 1500 r/min, the processing time of 120 s, the idle duration of 10 s, the idle power of 150 W and the average power during processing of 1750 W.

The carbon emissions added in the turning and milling processes originate from the consumed electric energy, cutting tool and cutting fluid. In the light of Equations (2)–(8), these emissions can be calculated according to Equation (13):

$$\begin{cases} \Delta C_m = \Delta m_1 f_m^1 + \Delta m_2 f_m^2 + \Delta m_3 f_m^3 + \Delta m_4 f_m^4 \\ \Delta C_e = E_1 f_e^1 + E_2 f_e^2 \end{cases} \tag{13}$$

where Δm_t is the mass of the cutting tool used in the added processes after the lightweight design, kg; f_t is the carbon emission factor of the cutting tool, kg/kg; Δm_l is the mass of the cutting fluid used in the added processes after the lightweight design, kg; and f_l is the carbon emission factor of the cutting fluid, kg/L.

The casting process before and after the lightweight design barely changes, except for the increased amount of silica sand. According to Equations (2)–(8), the carbon emission caused by the increased amount of silica sand can be determined using Equation (14):

$$C_{si} = \frac{\Delta m}{\rho} \rho_{si} f_m^{si} \tag{14}$$

where ρ is the density of the hub material; ρ_{si} is the density of the silica sand, kg/m^3; and f_{si} is the carbon emission factor of the silica sand, kg/kg.

Survey results of Anhui Axle Co., Ltd. demonstrate that about 70% of the parts are recycled in the forging and casting stage after they are scrapped, and approximately 30% are recycled in the processing stage. With reference to the literature [31,32], the meanings and values of the symbols in Equations (1)–(14) can be determined (Table 3).

Table 3. Meanings and values of the symbols [33–35].

No.	Meaning	Symbol	Value	Unit
1	Mass variation arising from the lightweight design	Δm	1.4	kg
2	Carbon emission factor of the hub material	f_m	2.69	kg/kg
3	Kilometers traveled	l	7×10^5	Km
4	Emission coefficient per kilometer	η_w	5×10^{-5}	Km^{-1}
5	Utilization rate of the hub material for reproduction	η_1	0.7	/
6	Utilization rate of the hub material for reprocessing	η_2	0.3	/
7	Utilization rate of the parts	η_m	0.6	/
8	Mass of the parts reused in the production process	$\Delta m_{production}$	0.588	kg
9	Mass of the parts reused in the processing stage	$\Delta m_{process}$	0.252	kg
10	Mass of the cutting tool used in the added turning process	Δm_1	2.7×10^{-4}	kg
11	Mass of the cutting fluid used in the added turning process	Δm_2	3.8×10^{-3}	L
12	Mass of the cutting tool used in the added milling process	Δm_3	2.2×10^{-4}	kg
13	Mass of the cutting fluid used in the added milling process	Δm_4	4.2×10^{-3}	kg
14	Energy consumed in the added turning process	E_1	0.34	Kwh
15	Energy consumed in the added milling process	E_2	0.059	Kwh
16	Grid emission factor of turning	f_e^1	0.499	kg/Kwh
17	Grid emission factor of milling	f_e^2	0.499	kg/Kwh
18	Carbon emission factor of the cutting tool in turning	f_m^1	29.6	kg/kg
19	Carbon emission factor of the cutting fluid in turning	f_m^2	2.85	kg/L
20	Carbon emission factor of the cutting tool in milling	f_m^3	29.9	kg/kg
21	Carbon emission factor of the cutting fluid in milling	f_m^4	2.95	kg/L
22	Carbon emission factor of the silica sand	f_m^{si}	7×10^{-3}	kg/kg
23	Density of the hub material	ρ	7.85	kg/m^3
24	Density of the silica sand	ρ_{si}	2.65	kg/m^3

By substituting the values in Table 3 into Equations (1)–(14), the comprehensive carbon emission benefit of the wheel hub after the application of the lightweight design can be determined, as shown in Equation (15). Meanwhile, the carbon emission benefits of each part caused by the lightweight design can be obtained, as shown in Table 4.

designed accordingly. Next, the finite element method was re-applied to the parts whose weights had been reduced for the purpose of analyzing their stress, strain and safety factor. In this way, the feasibility of the lightweight scheme was verified.

(2) A method for calculating the carbon emissions caused by the lightweight design based on the whole life cycle of the hub was proposed. The method can be adopted to measure the carbon emissions arising from changes in the mass, process and whole life cycle after the application of the lightweight design and then to evaluate the low carbon benefits.

(3) The feasibility of the lightweight design scheme was verified by taking the ZS061750-152101 wheel hub designed and manufactured by Anhui Axle Co., Ltd. as an example for the case analysis. The lightweight design changes three structures of the wheel hub, reducing its weight by 1.4 kg in total. For a single wheel hub, the carbon emissions resulting from the saved material, changed manufacturing processes and application and recycling of the parts are reduced by 51.22 kg in total. That is to say, if the lightweight scheme were to be applied to all the wheels produced by Anhui Axle Co., Ltd. (about 500,000 per year), the carbon emissions from the wheel production, application and recycling processes could be cut by 2.56×10^7 kg, marking a favorable emission reduction effect.

The proposed method serves as a reference for the lightweight design of the spare parts of automobiles and other equipment, and it also provides a new method for calculating the carbon emission changes resulting from the design, use and recycling of the parts after the application of the lightweight design. However, there are still some limitations.

Firstly, in the case analysis, although each component of the parts still satisfies the operating requirements, the lightweight property is achieved at the expense of some of the mechanical properties of the parts.

Secondly, for a total wheel hub carbon emission reduction of 51.22 kg, the greatest loss in the mechanical performance (displacement) reached 34.9%, and the minimum performance loss (safety factor) has reached 3.46%. The relationship between the emission reduction benefits and loss in the mechanical properties has a positive correlation. As the global response to climate change continues to grow, carbon peak and carbon neutrality have become a global focus. China named carbon peak and carbon neutrality as two of the eight major tasks of the central economic work undertaken in 2021 [1]. Against such a background, decision makers will adopt methods that can reduce carbon emissions while the mechanical properties meet the requirements [36]. The quantitative relationship between the loss in the mechanical properties and lightweight quality has not been clarified, and the emission reduction benefits have not been comparatively analyzed with respect to the loss in the mechanical properties. In the follow-up research, efforts will be made to investigate the relationship between the emission reduction benefits and the loss in the mechanical properties in the hope of providing greater convenience for decision makers.

Author Contributions: The author contributions are as follows: writing—original draft, review, and editing, Q.L.; data acquisition, Y.Z., X.W. and J.C.; investigation, Q.L.; methodology, C.Z.; validation, Q.L.; writing—review and editing, Q.L. All authors have read and agreed to the published version of the manuscript.

Funding: This work was supported by the Social Science Innovation and Development in Anhui Province (2021CX069); the Scientific Research Project of Anhui Universities (KJ2021A1106); the Henan Province Scientific and Technological Research (222102220020); and the Scientific Research Platform Open Project of Suzhou University (2020ykf12); the Teaching Research Project of Anhui Province (2021jyxm1502).

Institutional Review Board Statement: Not applicable.

Informed Consent Statement: Not applicable.

Data Availability Statement: Not applicable.

Conflicts of Interest: The authors declare no conflict of interest.

References

1. Wu, N.; Zhao, S.C.; Zhang, Q. A study on the determinants of private car ownership in China: Findings from the panel data. *Transp. Res. Part A Policy Pract.* **2016**, *85*, 186–195. [CrossRef]
2. Liu, Y.; Liu, Y.; Chen, J. The Impact of the Chinese Automotive Industry: Scenarios Based on the National Environmental Goals. *J. Clean. Prod.* **2015**, *96*, 102–109. [CrossRef]
3. Yang, Z.S.; Jia, P.; Liu, W.D.; Yin, H.C. Car ownership and urban development in Chinese cities: A panel data analysis. *J. Transp. Geogr.* **2017**, *58*, 127–134. [CrossRef]
4. Zhao, X.D.; Mu, X.H.; Du, F.P.; Guo, H.L.; Chen, J.H. Research and development of automotive structure lightweight. *Modern Manuf. Eng.* **2017**, *10*, 157–162. [CrossRef]
5. Xu, P.F.; Duan, S.Y.; Wang, F. Reverse Modeling and Topological Optimization for Lightweight Design of Automobile Wheel Hubs with Hollow Ribs. *Int. J. Comput. Methods* **2020**, *17*, 1–12. [CrossRef]
6. Wan, X.F.; Liu, X.D.; Shan, Y.C.; Jiang, E.; Yuan, H.W. Numerical and Experimental Investigation on the Effect of Tire on the 13 degrees Impact Test of Automotive Wheel. *Adv. Eng. Softw.* **2019**, *133*, 20–27. [CrossRef]
7. Central People's Government of China. Notice of the Three Departments on the Issuance of "Implementation Plan of Made in China 2025-Energy Equipment". Available online: http://www.gov.cn/xinwen/2016-06/21/content_5084099.htm (accessed on 21 June 2022).
8. Baskin, D.M.; Reed, D.B.; Seel, T.N.; Hunt, M.N.; Oenkal, M.; Takacs, Z.; Vollmer, A.B. A Case Study in Structural Optimization of an Automotive Body-In-White Design. *SAE World Congr. Exhib.* **2008**, *1*, 1–12.
9. Shimoda, M.; Tsuji, J. Shape Optimization for Weight Reduction of Automotive Shell Structures Subject to a Strength Constraint. *SAE Tech. Pap.* **2007**, *1*, 186–191.
10. Gauchia, A.; Diaz, V.; Boada, M.J.L.; Boada, B.L. Torsional Stiffness and Weight Optimization of a Real Bus Structure. *Int. J. Automot. Technol.* **2010**, *11*, 41–47. [CrossRef]
11. Xiao, D.H.; Zhang, H.; Liu, X.D.; He, T.; Shan, Y.C. Novel steel wheel design based on multi-objective topology optimization. *J. Mech. Sci. Technol.* **2014**, *28*, 1007–1016. [CrossRef]
12. Jang, G.W.; Kim, Y.Y. Topology optimization with displacement-based nonconforming finite elements for incompressible materials. *J. Mech. Sci. Technol.* **2009**, *23*, 442–451. [CrossRef]
13. Zhang, Z.J.; Jia, H.L.; Sun, J.Y.; Wang, M.M. Application of Topological optimization on Aluminum Alloy Automobile Wheel Designing. *Adv. Mater. Res.* **2012**, *562–564*, 705–708. [CrossRef]
14. Pang, W.; Wang, W.P.; Zhang, W.H.; Wang, X. Modeling and Optimization for Lightweight Design of Aluminum Alloy Wheel Hub. *Key Eng. Mater.* **2016**, *723*, 322–328. [CrossRef]
15. Hu, J.H.; Liu, X.X.; Sun, H.X.; Zhu, Z.H.; Li, B.H. Development and Application of Light-Weight Design of the Aluminum Alloy Wheel. *Appl. Mech. Mater.* **2013**, *310*, 253–257. [CrossRef]
16. Chai, W.H.; Liu, X.D.; Shan, Y.C.; Wan, X.F.; Jiang, E. Research on Simulation of the Bending Fatigue Test of Automotive Wheel Made of Long Glass Fiber Reinforced Thermoplastic Considering Anisotropic Property. *Adv. Eng. Softw.* **2018**, *116*, 1–8. [CrossRef]
17. Yi, Q.P.; Tang, C.P. Environmental impact assessment of magnesium alloy automobile hub based on life cycle assessment. *J. Cent. South Univ.* **2018**, *25*, 1870–1878. [CrossRef]
18. Zhang, L.; Jiang, R.; Jin, Z.; Huang, H.; Li, X.; Zhong, Y. CAD-Based Identification of Product Low-Carbon Design Optimization Potential: A Case Study of Low-Carbon Design for Automotive in China. *Int. J. Adv. Manuf. Technol.* **2019**, *100*, 751–769. [CrossRef]
19. Hao, H.; Liu, F.; Sun, X.; Liu, Z.; Zhao, F. Quantifying the Energy, Environmental, Economic, Resource Co-benefits and Risks of GHG Emissions Abatement: The Case of Passenger Vehicles in China. *Sustainability* **2019**, *11*, 1344. [CrossRef]
20. Wu, Y.; Liu, F.; He, J.; Wu, M.; Ke, Y. Obstacle Identification, Analysis and Solutions of Hydrogen Fuel Cell Vehicles for Application in China Under the Carbon Neutrality Target. *Energy Policy* **2021**, *159*, 112643. [CrossRef]
21. Du, W.; Huang, Q.; Grasso, S. Analysis of Global Policy and Impact on Automobile Industry Under Carbon Neutrality. In Proceedings of the 2021 6th International Conference on Materials Science, Energy Technology and Environmental Engineering (MSETEE 2021), Hangzhou, China, 13–15 August 2021; p. 308.
22. Du, H.; Li, Q.; Liu, X.; Peng, B.; Southworth, F. Costs and Potentials of Reducing CO_2 Emissions in China's Transport Sector: Findings From an Energy System Analysis. *Energy* **2021**, *234*, 121163. [CrossRef]
23. Jhaveri, K.; Lewis, G.M.; Sullivan, J.L.; Keoleian, G.A. Life cycle assessment of thin-wall ductile cast iron for automotive lightweighting applications. *Sustain. Mater. Technol.* **2018**, *15*, 1–8. [CrossRef]
24. Liu, Z.; Guan, D.B.; Wei, W.; Davis, S.J.; Ciais, P.; Bai, J.; Peng, S.S.; Zhang, Q.; Hubacek, K.; Marland, G.; et al. Reduced carbon emission estimates from fossil fuel combustion and cement production in China. *Nature* **2015**, *524*, 335–338. [CrossRef] [PubMed]
25. Shan, Y.L.; Liu, J.H.; Liu, Z.; Xu, X.W.H.; Shao, S.; Wang, P.; Guan, D.B. New provincial CO_2 emission inventories in China based on apparent energy consumption data and updated emission factors. *Appl. Energy* **2016**, *184*, 742–750. [CrossRef]
26. Liu, C.H.; Cai, W.; Jia, S.; Zhang, M.Y.; Guo, H.Y.; Hu, L.K.; Jiang, Z.G. Energy-based evaluation and improvement for sustainable manufacturing systems considering resource efficiency and environment performance. *Energy Convers. Manag.* **2018**, *177*, 176–189. [CrossRef]
27. Liu, C.H.; Zhu, Q.H.; Wei, F.F.; Rao, W.Z.; Liu, J.J.; Hu, J.; Cai, W. An integrated optimization control method for remanufacturing assembly system. *J. Clean. Prod.* **2020**, *248*, 119261. [CrossRef]

28. Zhang, C.X.; Dong, Q.D.; Zhang, L.X.; Li, Q.; Chen, J.Q.; Shen, X.H. Modeling analysis and lightweight design for an axle hub considering stress and fatigue life. *Sci. Prog.* **2020**, *103*, 1–14. [CrossRef]
29. Li, C.; Liu, P.; Li, Z.A. Long-Term Decarbonisation Modelling and Optimisation Approach for Transport Sector Planning Considering Modal Shift and Infrastructure Construction: A Case Study of China. *Processes* **2022**, *10*, 1371. [CrossRef]
30. Liu, C.H.; Cai, W.; Dinolov, O.; Zhang, C.X.; Rao, W.Z.; Jia, S.; Li, L.; Chan, F.T.S. Energy based sustainability evaluation of remanufacturing machining systems. *Energy* **2018**, *150*, 670–680. [CrossRef]
31. Wan, C.; Zhou, D.; Xue, B. LCA-Based Carbon Footprint Accounting of Mixed Rare Earth Oxides Production from Ionic Rare Earths. *Processes* **2022**, *10*, 1354. [CrossRef]
32. Li, Y.; Sharaai, A.H.; Ma, S.; Wafa, W.; He, Z.; Ghani, L.A. Quantification of Carbon Emission and Solid Waste from Pottery Production by Using Life-Cycle Assessment (LCA) Method in Yunnan, China. *Processes* **2022**, *10*, 926. [CrossRef]
33. Cao, H.; Li, H.; Cheng, H.; Luo, Y.; Yin, R.; Chen, Y. A carbon efficiency approach for life-cycle carbon emission characteristics of machine tools. *J. Clean. Prod.* **2012**, *37*, 19–28. [CrossRef]
34. He, Y.; Liu, B.; Zhang, X.D.; Gao, H.; Liu, X.H. A modeling method of task-oriented energy consumption for machining manufacturing system. *J. Clean. Prod.* **2012**, *23*, 167–174. [CrossRef]
35. China Energy and CO_2 Emission Research Group. *2050 China Energy and CO_2 Emission Report*; Science Press: Beijing, China, 2009. Available online: https://www.china5e.com/m/news/news-52098-1.html (accessed on 14 October 2009).
36. Zhang, C.X.; Li, Q.; Zhang, X.X.; Wang, X.Y.; Chen, J.Q.; Shen, X.H. Lightweight Design of Special Axle Hub Based on ANSYS Consider Different Working Conditions for Low Carbon. *Math. Probl. Eng.* **2022**, *2022*, 1–8. [CrossRef]
37. Li, Q.; Tong, M.; Jia, M.; Yang, J. Towards Low Carbon: A Lightweight Design of Automotive Brake Hub. *Sustainability* **2022**, *14*, 15122. [CrossRef]
38. China Business Information Network. Statistics of Automobile and New Energy Vehicle Ownership in China in the First Quarter of 2022. Available online: https://baijiahao.baidu.com/s?id=1729560645146103844&wfr=spider&for=pc (accessed on 14 June 2022).
39. Garwood, T.L.; Hughes, B.R.; Oates, M.R.; O'Connor, D.; Hughes, R. A Review of Energy Simulation Tools for the Manufacturing Sector. *Renew. Sustain. Energy Rev.* **2018**, *81*, 895–911. [CrossRef]
40. Tian, G.D.; Yuan, G.; Aleksandrov, A.; Zhang, T.Z.; Li, Z.W.; Fathollahi-Fard, A.M.; Ivanov, M. Recycling of spent Lithium-ion Batteries: A comprehensive review for identification of main challenges and future research trends. *Sustain. Energy Technol. Assess.* **2022**, *53*, 102447. [CrossRef]
41. Zhao, F.Q.; Liu, X.L.; Zhang, H.Y.; Liu, Z.W. Automobile Industry Under China's Carbon Peaking and Carbon Neutrality Goals: Challenges, Opportunities, and Coping Strategies. *J. Adv. Transp.* **2022**, *2022*, 1–13. [CrossRef]
42. Tian, G.; Zhang, C.; Fathollahi-Fard, A.M.; Li, Z.; Zhang, C.; Jiang, Z. An Enhanced Social Engineering Optimizer for Solving an Energy-Efficient Disassembly Line Balancing Problem Based on Bucket Brigades and Cloud Theory. *IEEE Trans. Ind. Inform.* **2022**, 1–11. [CrossRef]

Article

Sustainable Operations of Last Mile Logistics Based on Machine Learning Processes

Jerko Oršič [1], Borut Jereb [2] and Matevž Obrecht [2,*]

[1]   Epilog d.o.o., 1000 Ljubljana, Slovenia
[2]   Faculty of Logistics, University of Maribor, 3000 Celje, Slovenia
*   Correspondence: matevz.obrecht@um.si

Abstract: The last-mile logistics is regarded as one of the least efficient, most expensive, and polluting part of the entire supply chain and has a significant impact and consequences on sustainable delivery operations. The leading business model in e-commerce called Attended Home Delivery is the most expensive and demanding when a short delivery window is mutually agreed upon with the customer, decreasing possible optimizing flexibility. On the other hand, last-mile logistics is changing as decisions should be made in real time. This paper is focused on the proposed solution of sustainability opportunities in Attended Home Delivery, where we use a new approach to achieve more sustainable deliveries with machine learning forecasts based on real-time data, different dynamic route planning algorithms, tracking logistics events, fleet capacities and other relevant data. The developed model proposes to influence customers to choose a more sustainable delivery time window with important sustainability benefits based on machine learning to predict accurate time windows with real-time data influence. At the same time, better utilization of vehicles, less congestion, and fewer failures at home delivery are achieved. More sustainable routes are selected in the preplanning process due to predicted traffic or other circumstances. Increasing time slots from 2 to 4 h makes it possible to improve travel distance by about 5.5% and decrease cost by 11% if we assume that only 20% of customers agree to larger time slots.

Keywords: supply chain management; real-time; home delivery; business modeling; e-commerce; time window

Citation: Oršič, J.; Jereb, B.; Obrecht, M. Sustainable Operations of Last Mile Logistics Based on Machine Learning Processes. *Processes* **2022**, *10*, 2524. https://doi.org/10.3390/pr10122524

Academic Editors: Conghu Liu, Xiaoqian Song, Zhi Liu and Fangfang Wei

Received: 7 November 2022
Accepted: 24 November 2022
Published: 28 November 2022

1. Introduction

1.1. Challenges of Last Mile Logistics

Online shopping is growing fast, providing customers with fast delivery to their home door, avoiding traveling to shop and carrying items to the home. The online store has 24 h opening, and delivery can be the best time for the customer. On the other side, all benefits for the customer impose different challenges to achieve optimized and efficient logistics where the sustainable side effect must also be considered.

Several negative effects on environmental and social dimensions of sustainability are represented by greenhouse gas emissions (e.g., CO_2 and others), traffic congestion, accidents, noise, and direct consequences on human health and urban life quality [1].

World economic forum [2] predicts that urban last-mile delivery emissions will increase by 30% by 2030 globally in the top 100 cities. Emissions could reach 25 million tons of CO_2 annually by 2030, and traffic congestion is expected to rise by over 21%, meaning 11 min longer for each passenger's daily commute. Last-mile logistics (LML) is becoming one of the most expensive and complex segments of logistics [3]. It has a significant impact and consequences on sustainable operations [4], where real-time data can affect, for example, the energy efficiency of scheduled transport in the LML [5]. With increasing customer demands for punctuality, traffic problems, constant changes in delivery methods, and real-time decisions, the requirements of sustainable operation are becoming an even greater challenge [6].

As LML has raised essential issues in increasing the efficiency of delivery systems and reducing related externalities, different solutions were proposed in the scientific literature, from introducing new technologies, transport means, innovative techniques and organizational strategies to changing LML through economically, environmentally and socially sustainable logistics schemes [7].

Bosona [6] investigated the challenges and opportunities of improving sustainability in urban freight last-mile logistics environmental, economic, and societal sustainability dimensions. In e-commerce, third-party logistics providers (3PL) [8] are also very important as, in most cases, they should execute sustainable delivery.

Many issues mentioned in the environmental dimension, like innovative vehicles from electric vehicles to drones, collaborative and cooperative urban logistics where resources are shared between different LML actors, creative depot stations where goods can be stored when the customers are not at home until they can pick them up, are, in many cases, connected with the support of advanced decision systems using real-time data optimization techniques and prediction methods. Arnold et al. [9] used different parameters to determine operational and external costs by delivery in Antwerp with vans compared to e-bikes. They used parameters of vehicles like capacity limit, operating fee, driving speed and average time per delivery with labor costs for drivers and pollution external costs for emissions, noise, and congestion. Olson et al. [10] argued that a better understanding of the last mile is required to enhance sustainability. City logistics are mainly driven by growing urbanization, sharing economy, e-commerce development, consumers' desire to increase delivery speed, and increased attention to sustainability.

Significant challenges for researchers are the improvement of last-mile logistics with significant externalities reduction. From the economic dimension, lowering the delivery costs is very close to sustainable progress. Ranieri et al. [11] suggest that retailers should evaluate their performances using attributes in the context of their services, distances to travel from depot to customer, traveling speeds, city size and delivery area, population density, available infrastructure, and congestion levels. The customers with their orders define freight volume and weight, volume delivery location, and time slot [10,12].

1.2. Attended Home Delivery with Time Window and Time Slot

Attended Home Delivery (AHD) is now the leading business model in e-commerce as the most challenging part of last-mile logistics. A time window (TW) is an interval during which an activity can or must occur. Within the logistics sector, it means the period during which a shipment is delivered or picked up. A time slot (TS) is a time assigned on a schedule or agenda when something can happen or is planned to happen, especially when it is one of possible several times. The time window is usually composed of many time slots. Time windows vary depending on the activity being performed, while time slots are of fixed length. Usually, a time window consists of several time slots.

In (AHD) customers typically interact with an online booking system about delivery TWs possibilities for their shopping [13]. If a retailer can provide narrow time slots when customers agree to be at home is a great advantage as he can better adjust their schedule.

Most e-commerce AHD business approaches are taking the increasing customer expectations for faster and narrower TW delivery as a given. When trying to make last-mile delivery more sustainable, the retailers focus on improving sustainability in different aspects, e.g., using no-emissions vehicles, improving the routing, and similar with clear improving limitations. However, the most challenging part, the time window/time slots are not adequately addressed.

The first step in the fulfillment process for attended home deliveries in e-commerce is delivery window offer and confirmation. The customer closes the basket on the retailer's website and reveals the delivery location. Selection of possible TWs is offered to the customer, and when the customer chooses his most convenient TW is confirmed by the retailer immediately.

The retailer must provide a wide selection of relatively narrow delivery TWs to ensure customer satisfaction and minimize undeliverable orders. Customers arrive dynamically; the location of future customer requests and the delivery size are unknown, so offering a specific TW to a customer in a static way may not be feasible or accurate enough.

For each customer, a valid delivery schedule must be constructed depending on delivery location, locations, and capacity of existing orders, considering different constraints like size and weight of the shopping freight. Providing the input data for the offering feasible time slots process requires communication across several services, and many database queries require a significant amount of time. There is even less time to solve the actual optimization problem [14].

Creating delivery window sets for customers is challenging, as the offer to the requesting customer must be made immediately and almost in real-time, so using complex optimization algorithms is not possible [15].

Another issue is confirming customer choice in the order acceptance phase, as simultaneous customer interactions may unavoidably create invalid offers with possibly longer customer waiting times [16]. Web performance analytics show that even 100-millisecond delays can impact customer engagement and online revenue [17].

1.3. Machine Learning Forecast

Creating or computing a feasible offer in real-time, notwithstanding that reaching the "absolute optimum" is a problem that is otherwise not computable in polynomial time (NP-hard problem). An exact algorithm should find all permutations and combinations of routes and compare each to extract possible TW. Therefore, another possible way is to use a machine learning forecast to develop a "near optimal" solution. Mitchell [18] covers the field of machine learning, the study of algorithms that allow computer programs to improve through experience and automatically infer general laws from specific data. Later interpreted as [19], Machine learning is an essential component of the growing field of data science. Using statistical methods, algorithms are trained to make classifications or predictions and to uncover critical insights in data mining projects [20]. Different learning algorithms and methods could be used for predictive analysis. Supervised learning methods are suitable with the dedicated training of a model on historical data and then application to predict the target variable [21]. The goal in the modeling phase is to obtain information about the behavior to be expected from the historical data with different data mining approaches to obtain the meaning variables [22].

The design of efficient last-mile delivery prediction Machine learning methods has been proposed to solve combinatorial optimization problems [23]; continuous learning is needed for the model to stay fresh to reflect the changing input data patterns. Necessary mechanisms for automatically retraining the models with new data are critical to deploying machine learning models, primarily to address the LML problem [24].

Different models for machine learning, simulation, and optimization for sustainable transportation systems could be used [25]. Majumdar et al. [26] showed that ML using Long Short-Term Memory networks is a powerful method for predicting the evolution of vehicle speed and congestion over short timescales on real-world road networks. Improving such predictions allow better management of smart transport systems, improving travel times and reducing transport-generated pollution within the urban environment. Shang et al. [27] explain that machine learning methods are the most effective solution for dynamic and non-linear traffic problems with the ability to approximate any degree of complexity of traffic flow and gain knowledge from training data without a predefined model structure with good robustness. Florio [28] presents a machine-learning framework for a data-driven method for designing high-quality last-mile routes pre-classified by domain experts (e.g., drivers) concerning their quality.

The concept of retraining ML models in real time and using a feedback loop to refine them as it happens is used in different real-time predictions [29]. To keep the model up to date, new real-time data has to be incorporated, and the model has to be retrained [21].

1.4. Predicting Time Slots with ML

Dynamic TWs slotting lets the retailer control the offered time slots per arriving customer by reducing the delivery capacity in the selected time slot. Lang et al. [30] argued that solution approaches must quickly compare the immediate reward from accepting an order to the opportunity cost; for predicting the delivery success, [31] trained several machine learning models based on historical order delivery data supplied by retailer website applications. The order delivery data consists of order-level attributes like the product category, attempt time to delivery, delivery speed, expected delivery date, delay, order value, and payment type [32].

1.5. Sustainability in AHD

In the e-commerce AHD business models, customers expect limited delivery times to suit their schedules. To respond to that expectations, more and more individualized delivery options are offered regarding fast delivery in narrow TWs, hardly affecting the delivery's sustainability.

Some retail companies claim that almost 80% of customers indicated that they consider sustainability for their purchases [33]. Of consumers, 86% are willing to delay deliveries and sacrifice time and money for sustainability, and 50% said that their decision to make sustainable purchases is directly related to concern for the planet [34]. At the same time, many customers are becoming more environmentally conscious, as argued by [35]. Retailers use the advantage of increasing the environmental awareness of customers by asserting the benefit of online shopping for the environment [6].

Boyer et al. [12] as well as Olson et al. [10], researched the effects of greater customer density and delivery window length in a given region. The travel distance is more than twice longer in 1 h TW compared to no window time restriction, facilitating greater delivery efficiency. The total number of kilometers per day is calculated to analyze the impact of changing the width of the TWs. If time slots are increased from 2 to 4 h, the improvement is 55%, and from 2 to 6 h, the improvement is about 68%. The reason is that narrow time slot limits solution space for delivery planning [1]. Some research concludes that if customers agree to a longer delivery window, the CO_2 emissions generated by the delivery fulfillment decrease considerably, thus leading to a more environmentally sustainable delivery [6,33,35]. In that way, encouraging customers to accept longer delivery windows could reduce the delivery's environmental impact.

Ignat and Chankov [36] proved that sharing information about all sustainability pillars (economic, environmental, and social) leads customers to choose a more sustainable delivery. Once made customers aware of the effect of these issues, customers are willing to contribute by creating economic and convenient sacrifices in the name of saving the planet's sustainability.

Social media's influence is also significant, as 52% of consumers indicate that social media influences shopping frequency at sustainable retailers. Facebook was the most popular for ages 45 and older, and Instagram for ages 18–44. Nearly a third of the TikTok votes (28%) were from ages 18–29 [34]. With customer knowledge, he chose to do something for sustainable delivery instead of a suitable delivery time to his schedule. The sustainability impact could also be presented on social media and used for public relations and branding since the public perception of green transportation and sustainable business operations is gaining increasing attention.

1.6. Gap, Aim, and Objective

Concerning the literature and practice, the most challenging part of last mile logistics–TW is not adequately addressed. Existing studies also show lacking potential for integrating real-time data into the decision-making process. As customers in AHD can place their orders any time around the clock with any possible location for delivery with significantly different sizes and weights of freight is very challenging to calculate the feasible time slots while the customer is waiting for the offer since all decisions must be made in real-time. As last-mile delivery efficiency depends on multiple factors, such as consumer density, depo capacity and location, shipment size fragmentation, homogeneity, TWs, and congestions, the only solution is to make a good machine learning prediction for new incomers what they would wish to order and where and when the delivery should be executed. Primarily the core of last-mile logistics optimization is identified in cost optimization and not environmental impact optimization.

Sustainable AHD Framework is the name of a new model to improve sustainability in AHD for large retailer companies as well as for logistics service providers is being developed based on real-time data and machine learning. At the start, we try to optimize the AHD delivery process in the traditional way to improve routing with better algorithms, choosing a more appropriate fleet with better efficiency and better environmental footprint, and other known solutions. The main improvement is, however, on TW optimization based on more sustainable operations that also result in lower costs. Therefore, the main novelty is optimizing the environmental impact of ADH. Consequently, cost savings can be achieved, and sustainable operations attract new customers. The shift toward sustainability as the priority of business operations in selecting TW instead of focusing primarily on cost optimization is an integral part of the novel approach.

In our simulations of AHD deliveries carried out for developed model validation, it was found that offering TW or time slots to customers is a critical stage to finding feasible TW and simultaneously achieving better sustainability goals. To find workable delivery time slots, all delivery constraints should be applied to machine learning, which also contains the retailer's business policy, as all subsequent decisions depend on it. When we try to implement narrower time slots to create a feasible TW, such as 1 h, all indicators for more economically and sustainable deliveries deteriorate.

In this article, the impact of a more sustainable AHD Framework is researched based on a machine learning forecast where different sustainable delivery slots are predicted. To obtain accurate prediction results, historical data from track and trace systems at execution delivery time are used to train the appropriate model. A permanent retraining model is proposed with actual data about traffic loads, weather, fleet capabilities, and delivery capacities, and special region features are added, including actual execution tracking where every logistics event is monitored for every delivery. The rest of the paper is organized as follows. Section 2 describes Model conceptualization with an innovative approach to improving LML sustainability in AHD. In Section 3, we describe the validation of the model through a real-life experiment. A Discussion section and Conclusion follow this.

2. Model Conceptualization

2.1. Attended Home Delivery

For AHD delivery, the process is started in the depot where the vehicle is loaded, in most cases, for the predefined area where the delivery will be conducted. Area for delivery could be statically predefined like ZIP code or defined in the prediction process by the density of delivery locations. Figure 1 represents the Area, a Depot, and their mutual dependence.

The area where to deliver is defined by:

- Distance from Depot;
- Size of delivery Area;
- The shape of the Area;
- Stop density (delivery locations by region);
- Average distance between stops and;
- Max capacity to be served.

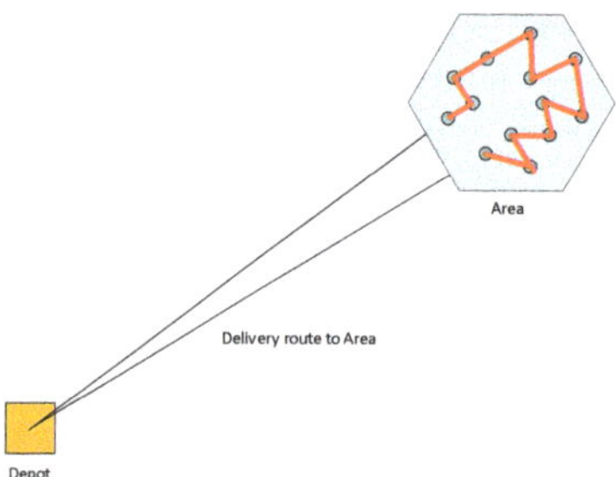

Figure 1. Delivery Route to Area.

When we look at the timeline represented in Figure 2 for the delivery route, we see that first is loading time, then driving to the delivery Area after driving to the first delivery location, and then spending servicing time for parking, finding the customer, and handover the delivery. Service time may last longer than driving from one delivery location to another.

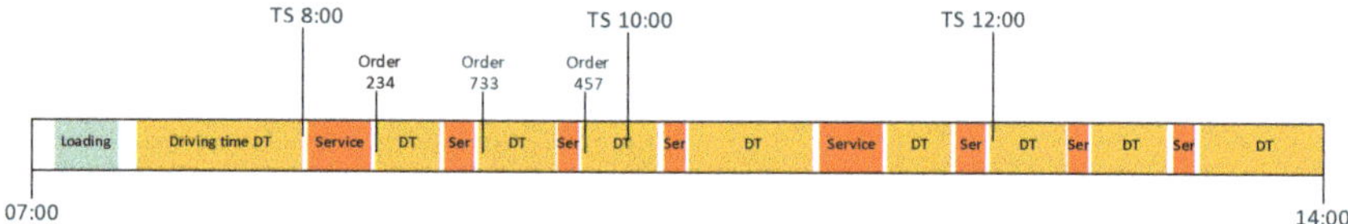

Figure 2. Timeline for a delivery route.

To successfully execute deliveries, exact preparation processes for all activities must be calculated. In the first phase of our model prediction based on AHD data, the transport capacity of the vehicle can be calculated for each area and TW. As a result of the prediction, feasible time slots for each arriving customer are calculated and offered to the customer. The customer chooses one of the time slots and then approves it in our model. Time slots are open until capacity for delivery is available. In the second step, optimized planning of routes and all delivery events is performed and finished before loading the deliveries on vehicles. The third phase is the delivery execution after the vehicle is loaded and the handover of deliveries is performed. It means that by these three steps, we allow and offer the calculation of vehicle transport capacity to the customer based on AHD.

2.2. Management of Offered Time Slots to the Customer with Sustainability Evaluation

In the developed model, we are explaining our approach to improving the costliest part of AHD, narrow TW, when the customer expects the delivery at home. The classical solution for available time slots is statical with a fixed capacity for deliveries based on available resources, regardless of real-life circumstances.

Dynamical time slot management in real-time sustainable influence is introduced to show the customer the possibility of choosing more sustainable and feasible time slots for delivery and reducing sustainable impacts with the customer's consent.

Our approach combines feasibility, opportunity cost, and the sustainability aspect of delivery. From the results of forecasts and real-time data on incoming requirements, it is known which TWs are better for delivery even with sustainability variables of the appropriate fleet, full vehicle load, delivery area specifics, traffic jams when delivery is prepared in the depot, and similar.

2.3. Proposed Prediction of Accurate Delivery TSs with Sustainability Preference

The requests for delivery in AHD are constantly coming into the system; decisions must be made in real-time so that retailers can offer personalized time slots for each incoming customer.

We used Machine Learning to enable capacity estimation to ensure fast response and scalability on simultaneous customer interactions based on historical data from previous deliveries to achieve our goal. The machine learning training process is continuously run to allow interference with real data from the next steps in the delivery process and other external data. In our framework, we use delivery prediction analysis methods of supervised learning for model creation.

To keep the model up to date, new real-time data must be incorporated, and the model must be retrained. Constant model retraining with real-time exceptions is used to adapt to the new reality and to refine the results.

In addition to obtaining fast results, fast algorithms for calculating distances are used, and historical driving speeds in the area are used to calculate the delivery schedule and add additional sustainability issues such as real-time traffic or weather conditions.

We are determining the input variables based on standard LML activities for the machine learning model to learn patterns and dependencies between these variables and the delivery time, as the target represents in Figure 3. For the exact delivery location, a professional geolocating platform is used, and external data from different sources are used for delivery prediction.

A set of variables based on standard LML activities is used:

- Locations of deliveries;
- Size and weight of cargo;
- The region is a representation of the different areas of delivery according to the number of stops and distances;
- Delivery time depends on region and cargo;
- Available fleet;
- Weather conditions and traffic affecting the duration of delivery; and
- Sustainable constraints.

The first step in our AHD Prediction is Data preparation, as shown in the first yellow box in Figure 3. The source of data to fill the above-mentioned variables are different resources from historical data, an e-commerce platform with customer orders, and additional delivery data to weather, traffic, and execution track and trace systems and must be prepared for use in the machine learning system. Data cleaning Includes basic operations such as removing noise or outliers and mapping missing or unknown values. In advance of data preparation, a strategy for dealing with missing data fields is prepared by considering information about the time sequence of the retrieved data.

After Data Preparation (next yellow box in Figure 3), Machine Learning Algorithms are used to create a prediction for an available time slot for each customer order. Machine learning algorithms are developed with machine learning techniques, and we know how accurately the predicted values are from the learning phase of algorithms where testing data with the known outcome are used.

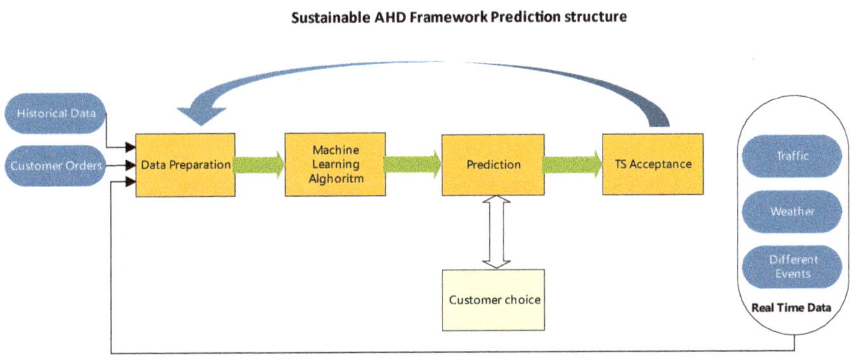

Figure 3. Sustainable AHD Framework Prediction structure.

The training of the ML algorithms is performed based on the processed data from the past "Historical Data," as shown on the left side in Figure 3.

As input to AHD Prediction, customer orders with additional logistic data, such as the size and weight of packages, are used. Data sources are from input variables, tracking system of the logistics service provider, external weather systems that provide weather data for the time of delivery, external traffic systems providing traffic loads and mapping services to take advantage of real-life data for minimizing route duration, reduce the reliance on expert heuristics, and exploit similarities between instance characteristics.

Our ML algorithms estimate travel time from the depot to each delivery location for each area. The time slot is calculated with the maximum capacity of available vehicles with their available cargo size and weight. Variations of our algorithms could be run in a parallel mode, so the best ones are selected. Machine learning model patterns and dependencies between variables are learned to obtain delivery TWs as the target.

Prediction of available time slots with sustainable evaluation because of Machine learning Algorithms is presented to the customer shown in Figure 3 as "Prediction." Customers choose one of the given time slots delivered as a green box named "Customer Choice." It is possible to select a set of feasible time slots with the influence of retailers' business rules with some benefits for the customer to encourage the Customer to choose more sustainable TS. In the following "TS acceptance" step, the time slot is approved in case the data could be changed in the customer's decision time for his TS and TS is no longer available. Roving process, the capacity for delivery in the area with that slot influence the data preparation with accepted TS, as shown with a blue arrow in Figure 3.

The validation of our Model is described in the next section.

3. Model Validation and Its Use in a Real Environment

3.1. Model Validation Process

Our Sustainable AHD Model uses available data for all three delivery phases: prediction, optimized planning, and execution. The prediction phase relates to the planning and execution delivery phase to increase or lower delivery capacity divided by areas and time slots. On the other hand, with more excellent reliability, we avoid late or not possible deliveries, which could be a problem for retailers. With the help of customers selecting broader and/or not-so-busy time slots, sustainable delivery results are possible. In our testing environment, we simulate customer cooperation based on declared sustainability sensibility and the fact that some customers choose larger, like 6 h time slots based on retailers' historical data.

In the Prediction phase, the first delivery areas are calculated depending on delivery similarity parameters such as average delivery distances between customers, average vehicle speed, and similarities. The delivery capacity forecast is made some time in advance for each area. It is not very easy to predict how many customers will show up in one place, so a static solution is to close the area for time slots where the delivery capacity is exceeded. If we use the back loop of customer accurate tome order data in prediction, we can change the time slot windows size or move some capacity from one area to another. With the help of the next arriving customers, we can ease the time slot limitation and provide a possible solution for the next customers' requests.

The optimized planning phase depends on time slot acceptance in the prediction phase. If capacity data are not correctly updated, all deliveries cannot be planned to fulfill all customer requests. In the planning phase, a complete vehicle should be loaded and, after all scheduled deliveries are executed, returned to the depot when the next loading begins. Planning is also essential to obtain real-time data from the next execution phase if some delay is expected and will also influence delivery capacity. Our model's real-time data loop is integrated and continuously used to make and modify optimized planning decisions. Using this concept, the sustainable benefit of fewer vehicles is needed, and better routing planning is achieved.

The last and most important customer is the execution phase, where the customer expects the promised delivery in an agreed time slot. Almost all retailers in e-commerce use third-party logistics providers with their fleet and personnel for the execution phase, where everything about deliveries, routing, and TW should be known in advance. The driver and possible delivery assistant use our model track and trace application. All delivery events are monitored and managed in real time with precise exception handling. All exceptions range by severity as closed road, heavy traffic, vehicle accident, bad weather, in the range up to customer not at home or unhappy with delivered goods. All those events are important for the prediction phase as some mean capacity reduction or closing next time slots.

3.2. Time Efficiency and Benefits of Model Applications

As an example, what happens when time slots are used? We consider all services and delivery times to be the same. In Figure 4, the vehicle has a capacity for 22 deliveries and is fully loaded in the depot. Then the vehicle is first driven to the delivery area, and the handover for each delivery begins. A possible capacity limitation is the time spent on each delivery in the area. So only 13 deliveries can be made in this time window. Other deliveries must be made in the next TW, or the vehicle must move to the next area. If there is a need for 22 deliveries in the area, another vehicle must be used, or 9 deliveries must reduce the available capacity for that slot.

When TS is narrowing, some deliveries are costly while the vehicle is not loaded enough, or we need many vehicles. If there is predicted congestion on the road, the delivery efficiency worsens again.

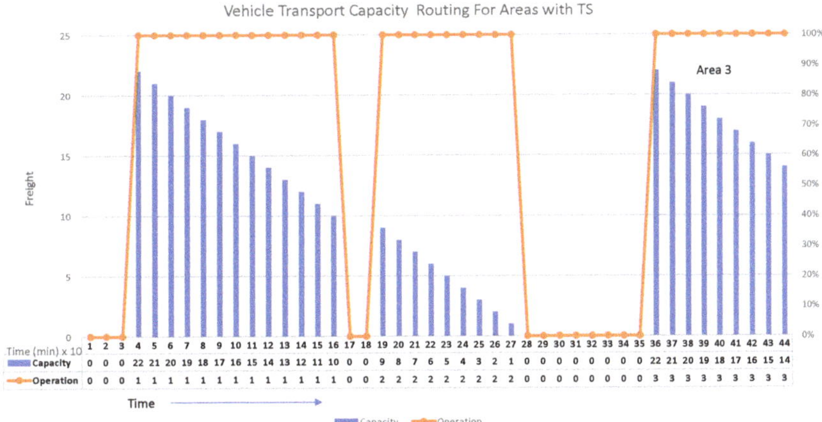

Figure 4. Vehicle transport capacity over the delivery time.

The shorter the delivery time slot, the higher the delivery price and the increased environmental externalities. For obvious reasons, the longer the time interval, the greater the flexibility in vehicle routing and the better the environmental sustainability.

It can be argued that more improved calculations and optimization of deliveries can help achieve better performance on sustainable issues; much more could be done if the customers consider the sustainability impact when making their decisions. As we make an offer in real-time, the problem needs to be addressed at the source, namely a potential change in customer behavior regarding sustainability issues that shows the customer option of choosing from a selection of more sustainable TWs.

3.3. Acceptance Phase of Chosen Time Slot

In prediction time sustainability related dimension of offered available TSs can be calculated. With a proper explanation to the customer, the retailer could agree that TS is wider, delivery is not executed in congestion times with dense traffic or when the congestions are in the customer delivery region, etc. The next suggestion could be that if the customer would not be at home at the agreed TS, the alternative delivery location is chosen, like a neighbor door, store, or other location close to the customer to fulfill booked TS.

When a new customer is closing his shopping basket, the new business model shows the customer feasible time slots when delivery is possible; after choosing the slot in the order acceptance procedure, we confirm the customer's choice. That step is needed, especially when several customers choose the time slots simultaneously, as shown in Figure 5.

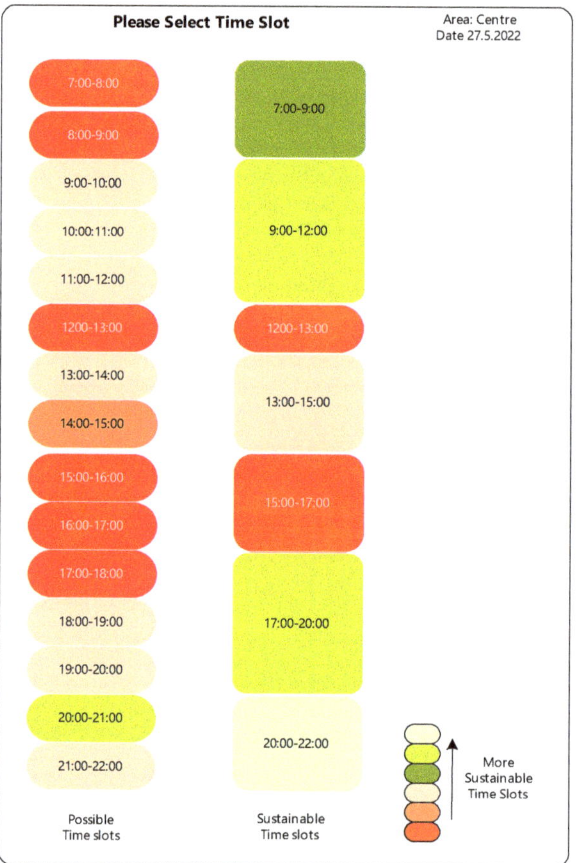

Figure 5. Time Slots Offer to Customer.

In the acceptance phase, we are using predictions to recalculate the feasibility of delivery for each customer delivery request to find semi-exact routes for capacity management for delivery resources. Quick filtering criteria will be run first, such as routes that match the desired delivery TW, routes that still have sufficient capacity in total and per time slot, temperature regime restrictions, etc. By inserting the order into the route, a preliminary dispatch time of the vehicle from the depot is calculated. In this phase, all other data from the next phase of optimized route planning and execution are considered, and the capacity can also be reduced for sustainable reasons.

3.4. Sustainable AHD Framework

Our work is focused on a decision support system as a Sustainable AHD framework for prediction, optimized planning, and execution that can be used to improve performance in all dimensions of sustainability in AHD as the most inefficient and demanding last mile case. The delivery process is performed in three steps:

1. First, the prediction of feasible time slots for delivery in real-time;
2. Second, the optimized planning in cut-off time before loading the vehicles with deliveries;
3. Third, the delivery execution.

In all three steps, additional data influences the delivery plan, so a backward loop is necessary to adapt the real-life conditions. Figure 6 shows real-time data dependencies in the Sustainable AHD Framework.

"Prediction" as a starting phase (shown on the left side of Figure 6) is essential for the subsequent phases where prediction output data is used in "Optimized Planning," and the result of planning is used for exact-driven "Execution" for loading the vehicle, routing instructions and handover the deliveries to customers. On the other side, some events in the "Execution" phase can be delayed or were not carried out. Those events could include traffic loads, weather, fleet capabilities, delivery capacities, special area features, and execution tracking using low-emissions vehicles, as shown in Figure 6 at the bottom. In the "Execution" phase, we can obtain the information that one bridge will be closed for the next 24 h, so that information should immediately influence "Prediction" and "Optimized planning," as shown with blue arrows in Figure 6. Similarly, when information in "Optimized Planning" about the planned vehicle is out of order and cannot be replaced by influence on "Prediction".

Predictions included sustainability related to dimension and obtaining real-time data from all available resources, allowing fast response to actual life events and the opportunity to find better and more sustainable delivery, including with the help of customers.

Optimized Planning is performed before the cut-off time before loading the deliveries on vehicles. Final routes are prepared considering no broken constraints, the actual traffic, weather, pickup and delivery TWs, capacities, temperature regime, and other sustainability-related indicators and data coming from execution time. Moreover, in this phase, the model uses some predictions over congestion, travel time in the area, weather conditions, and similar.

The delivery Execution enroute is monitored via a track and trace system handling all events affecting the transport and handover during the execution. Last-mile delivery tracking can be defined as a system allowing monitoring of the executing delivery process. At every step of delivery, all logistic events are recorded based on tracking standards. In AHD, all events are sequenced from loading, driving, finding the customer, unloading, and handover, typically executed repetitively. For track and trace, special applications are used by delivery personnel where the dimensions of the event are tracked. First, the business step behind the event is to define the context, information on the action performed on delivery items, location, and event time. All exceptions are also tracked, and the controlling system is informed, so real-time tracking enables communications between all stakeholders in LML, including the customer.

The empirical results from the real-world data confirmed the importance of the proper prediction and optimized planning traditionally adopted in vehicle routings, such as total travel time, TW performance, and routing consistency.

The proposed model novelty is to create a sustainable-driven AHD framework that will improve fast time slots offering, including sustainable data from various resources to obtain a better perspective on what delivery means for retailers and customers not only in terms of feasibility but also from a sustainability perspective. The impact on a more sustainable ADH framework is researched based on different machine learning forecasts. Different, more or less sustainable delivery slots are predicted since the quantity of data is too complex for optimization based on manual data processing. To obtain accurate prediction results, historical data from track and trace systems at execution delivery time are used to train the appropriate model. A permanent retraining model is proposed with actual data about traffic loads, weather, fleet capabilities, and delivery capacities; special area features are added, including execution tracking, where every logistics event is monitored for every delivery.

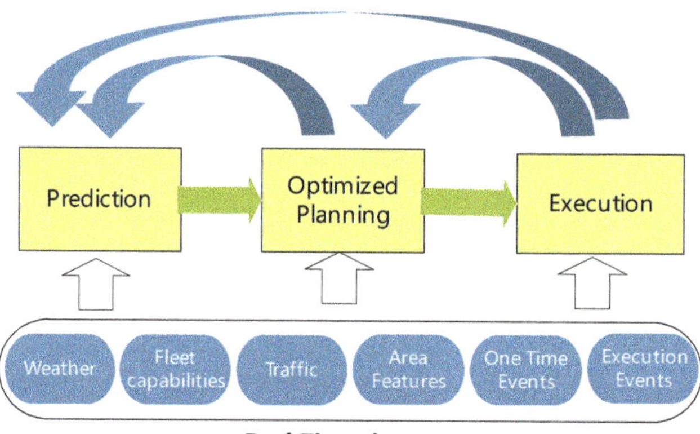

Figure 6. Real-Time Data Dependences in Sustainable AHD Framework.

The sustainable AHD Framework combines all mostly different solutions to obtain better results in all three pillars of sustainability. The full power of artificial intelligence is used to manage the whole AHD delivery process from prediction to optimized planning and execution phase to achieve the following objectives:

1. Predicting accurate, sustainable TWs alternatives for attended home delivery in real-time;
2. Improve the machine learning predictions with real-time data for offering TS, optimized planning and execution phase of AHD delivery;
3. To influence the customer to choose a more sustainable way for AHD and use all other sustainable improvements.

4. Discussion

AHD as the most expensive and demanding business model in e-commerce is also very questionable in terms of sustainability requirements. Customer wishes for the desired narrow delivery time result in less flexibility and consequently less scope for sustainable improvements.

The general recommendations for actions to improve sustainable operations in AHD emerged from the present study. Predicting more accurate alternatives for AHD based on machine learning enhanced with a real-time data feedback loop from different resources increases sustainability. With real-time data knowledge, it can be influenced customers to choose more sustainable TW for delivery. For example, 86% of the people in the mentioned research were willing to wait longer for delivery once they understood that a fast delivery or shorter time slot might have negative environmental impacts [34]. The conclusion is that participants are willing to wait longer, pay more, or choose a less convenient location to achieve a more sustainable delivery. With the help of customers, more sustainable decision and a valid delivery schedule are constructed with better utilization of vehicles avoiding congestion in the customer area and more exact delivery.

When the AHD delivery processes are connecting with real data in a reverse loop, many obstacles in delivery can be predicted and avoided. One of the most critical issues is synchronization with the depot, as all delivery events are very time connected. If the vehicle is not loaded at a predicted time in the time slot world, everything will be late, maybe also loading.

As time slot delivery is very demanding and has a vast sustainability impact, all retailers should be concerned about less harmful deliveries. The most important thing for the customer in the AHD world is that the retailer delivers the order in a mutually agreed time slot. If the feasibility is miscalculated or time slotting delivery is not working correctly, the customer will abandon the retailer. Many deliveries could not be redirected or later collected (e.g., freshly prepared food medicals, items that need additional service, precious items). Another challenge is that the customer is not at home when the handover should be done. Most such events occur in real time, so the supporting system must respond flexibly. With our Sustainable AHD Framework, where real-time influencing loops supported by ML are used, we can avoid most such scenarios. Still, the most critical issue is that the system can learn from a similar situation to react better next time.

On the other hand, in our system, real-time data from different sources is always used. It is also possible to manually add important (missing) data like unique events (e.g., football match next evening, road works for three days, a massive sale like Black Friday).

Prediction models assume that patterns learned from the training data will provide an appropriate basis for forecasts of future development. In the real world, many changes can happen on selected variables like a road in construction, so relationships between features learned in advance no longer apply. For retraining, data from the planning and execution process in the developed model for AHD can be used with all external resources.

It is not easy to calculate feasible delivery routes when the customer expects to receive an offer of delivery times if the proper tools are not used. Many real-time computing attempts with fixed algorithms have been made, but the solutions are not scalable. ML prediction with real-time data influence has, of course, some limitations. The most visible is if formal learning is not successful and inadequate reaction to different events is executed.

In addition, it is essential to consider all other circumstances relevant to delivery. Traffic takes place in an everyday world where various constraints affect delivery times, and it has been proven that ML can significantly improve outcomes [37]. There are also some limitations in identifying all possible situations, as many different events affect delivery times. Another issue is real-time data, which can affect forecasts and how they are presented.

There is also a chance that the customers do not want to contribute to more sustainable decisions about the time interval. However, according to customer trends, it can be speculated that the share of green customers will increase [38]. Consequently, the percentage of doses not interested in more sustainable operations will eventually be significantly lower or even irrelevant. Customers who make purchases online stop going to the stores. This also reduces the pollution that they would bring otherwise. Since e-commerce is increasing fast, distribution optimization should be the core of future process improvements.

It is evaluated that increasing time slots from 2 to 4 h makes possible improvement of travel distance by about 55% [1]. Suppose we assume that only 20 percent of customers agree to larger time slots. In that case, the costs decrease by 11% in real situations when other circumstances must be considered, such as traffic congestion or some other events that have a significant impact on delivery times where the prediction result shows that delivery capacity, in such a situation, is much lower and also increasing the vehicle fleet or adding more other resources is not a feasible solution.

As the costliest part of the whole supply chain with a meaningful, sustainable impact, AHD could be improved with awareness and action from all stakeholders, retailers, logistics providers, and customers. For retailers with sustainable business rules, 3PL is taking into account sustainable indicators [39]. The trend of development and the basis of business models requires considering sustainable indicators that place suppliers of logistics services at a certain level of quality, thereby affecting their desirability among all stakeholders in supply chains.

5. Conclusions

The proposed framework can improve predictions using real-time data from the planning and execution phase of delivery and other sources and offer personalized, accurate, sustainable TWs alternatives for AHD in real-time.

With proposed knowledge about offered time slots when the customer chooses the suitable delivery time for his schedule, the sustainability impact could also be presented to the customers to impact their decision on time slot selection based on the sustainability impact of delivery. As other researchers argued that sharing information about sustainability impacts leads to customers being more likely to choose a more sustainable delivery, we are sure that LML-related companies can impact their decision. It is expected that many customers will consider choosing a better one from the environmental sustainability perspective. When customers are aware of the ecological impact of LML and AHD challenge challenges and potential improvements, they will be more willing to contribute to the reduction of externalities. Therefore, the most significant reduction of environmental impacts is likely to be achieved in AHD. This is our contribution to building a more sustainable LML using artificial intelligence to encourage the cooperation of all stakeholders in achieving this goal.

This paper's framework contributes to understanding AHD's sustainable improvements by identifying the significant impacts of real-time decisions and explaining their interrelationships. The possibilities of continuous machine learning with the influence of real-time data contribute to developing a more coherent body of knowledge and provide information to customers who are becoming increasingly environmentally aware. The findings of this work suggest that further research is needed in various areas to improve all sustainable aspects of AHD logistics.

As AHD sustainability has many dimensions, other issues should also be inserted in the Sustainable AHD framework, many from social and Environmental pillars as more sustainable packaging of goods, joining orders in one delivery, combinations of different depots, delivery safety, and working conditions.

There are additional possibilities to improve the performance of the proposed Framework further. The complexity of AHD, which considers real-time data and decisions with economic, social, and environmental sustainability factors, is unlikely that our solution will cover all the challenges. First, there are new sources of data that we are not using yet as connected vehicles network that is used for autonomous driving CAV), where real-time data are interchanged with other vehicles, infrastructure, and data in the cloud. Moreover, the prediction algorithms could be improved to be more accurate. Social media could be better informed with real-time data about sustainable customer decisions about AHD, showing the environmental, economic, and social benefits. Due to very topical issues related to fuel economy, efficiency, and optimizing transportation, and especially LML, we expect that many researchers will contribute to the proposed framework in light of the new green deal and contribution to following 17 UN sustainable development goals. Required knowledge of future to be logistic experts will definitely contribute to the transformation of digitally supported and sustainability-oriented education in logistics that will be more related to the Green deal as well as consistent with the idea of EU taxonomy.

Author Contributions: Conceptualization, J.O., B.J. and M.O.; methodology, B.J. and J.O.; software, J.O.; validation, B.J.; formal analysis, J.O.; investigation, B.J.; resources, B.J. and M.O.; data curation, J.O.; writing—original draft preparation, J.O., B.J. and M.O.; essay—review and editing, B.J. and M.O.; visualization, J.O.; supervision, M.O. and B.J.; project administration, B.J. and M.O.; funding acquisition, M.O. All authors have read and agreed to the published version of the manuscript.

Funding: M.O. and B.J. received funding from the European Union-Next Generation EU & The Ministry of Education, Science and Sport. Research was carried out within the project entitled "Establishing an environment for green and digital logistics and supply chain education".

Institutional Review Board Statement: Not applicable.

Informed Consent Statement: Not applicable.

Data Availability Statement: Not applicable.

Acknowledgments: The authors would like to thank the funder for the research funding and for covering APC.

Conflicts of Interest: The authors declare no conflict of interest. The funders had no role in the design of the study; in the collection, analyses, or interpretation of data; in the writing of the manuscript; or in the decision to publish the results.

References

1. Manerba, D.; Mansini, R.; Zanotti, R. Attended Home Delivery: Reducing last-mile environmental impact by changing customer habits. *IFAC-PapersOnLine* **2018**, *51*, 55–60. [CrossRef]
2. World Economic Forum. The Future of the Last-Mile Ecosystem Transition Roadmaps for Public-and Private-Sector Players. Available online: www.weforum.org (accessed on 25 February 2022).
3. Mangano, G.; Zenezini, G. The Value Proposition of innovative Last-Mile delivery services from the perspective of local retailers. *IFAC-PapersOnLine* **2019**, *52*, 2590–2595. [CrossRef]
4. Rai, H.B.; Verlinde, S.; Macharis, C. The 'next day, free delivery' myth unravelled: Possibilities for sustainable last mile transport in an omnichannel environment. *Int. J. Retail Distrib. Manag.* **2019**, *47*, 39–54. [CrossRef]
5. Bányai, T. Real-time decision making in first mile and last mile logistics: How smart scheduling affects energy efficiency of hyperconnected supply chain solutions. *Energies* **2018**, *11*, 1833. [CrossRef]
6. Bosona, T. Urban freight last mile logistics—Challenges and opportunities to improve sustainability: A literature review. *Sustainability* **2020**, *12*, 8769. [CrossRef]
7. Marcucci, E.; Gatta, V.; Lozzi, G. City Logistics landscape in the era of on-demand economy Main challenges, trends and factors influencing city logistics. *Deliverable* **2020**, *1*, H2020.
8. Oršič, J.; Rosi, B.; Jereb, B. Sustanability Evaluation for the Distribution of Good. 2017. Available online: http://www.fpz.unizg. hr/zirp-lst/assets/files/ZIRP-2017-conference-proceedings.pdf (accessed on 10 August 2022).
9. Arnold, F.; Cardenas, I.; Sörensen, K.; Dewulf, W. Simulation of B2C e-commerce distribution in Antwerp using cargo bikes and delivery points. *Eur. Transp. Res. Rev.* **2018**, *10*, 2. [CrossRef]
10. Olsson, J.; Hellström, D.; Pålsson, H. Framework of last mile logistics research: A systematic review of the literature. *Sustainability* **2019**, *11*, 7131. [CrossRef]
11. Ranieri, L.; Digiesi, S.; Silvestri, B.; Roccotelli, M. A review of last mile logistics innovations in an externalities cost reduction vision. *Sustainability* **2018**, *10*, 782. [CrossRef]
12. Boyer, K.K.; Prud'homme, A.M.; Chung, W. The Last Mile Challenge: Evaluating the Effects of Customer Density and Delivery Window Patterns. *J. Bus. Logist.* **2009**, *30*, 185–201. [CrossRef]
13. Amorim, P.; DeHoratius, N.; Eng-Larsson, F.; Martins, S. Customer Preferences for Delivery Service Attributes in Attended Home Delivery. *SSRN Electron. J. (Work. Pap.)* **2020**. [CrossRef]
14. Truden, C.; Maier, K.; Jellen, A.; Hungerländer, P.; Kolleg, M.K.P. Computational Approaches for Grocery Home Delivery Services. *Algorithms* **2021**, *15*, 125. [CrossRef]
15. Ehmke, J.F.; Campbell, A.M. Customer acceptance mechanisms for home deliveries in metropolitan areas. *Eur. J. Oper. Res.* **2013**, *233*, 193–207. [CrossRef]
16. Visser, T.; Agatz, N.A.H.; Spliet, R. *Simultaneous Customer Interaction in Online Booking Systems for Attended Home Delivery*; Erim Report Series Reference Forthcoming; Erasmus Research Institute of Management: Rotterdam, The Netherlands, 2019. [CrossRef]
17. Akamai. Akamai Online Retail Performance Report: Milliseconds Are Critical. 2017. Available online: https://www. ir.akamai.com/news-releases/news-release-details/akamai-online-retail-performance-report-milliseconds-are (accessed on 31 March 2022).
18. Mitchell, T. *Machine Learning*; McGraw Hill: New York, NY, USA, 1997.
19. What Is Machine Learning? | Domino Data Science Dictionary. Available online: https://www.dominodatalab.com/data-science-dictionary/machine-learning (accessed on 24 July 2022).
20. What Is Machine Learning? | BM. Available online: https://www.ibm.com/cloud/learn/machine-learning (accessed on 8 August 2022).
21. Kelleher, J.D.; Namee, B.M.; D'Arcy, A. *Fundamentals of Machine Learning for Predictive Data Analytics: Algorithms, Worked Examples, and Case Studies*; The MIT Press: Cambridge, MA, USA, 2015.

22. Mariscal, G.; Marbán, Ó.; Fernández, C. A survey of data mining and knowledge discovery process models and methodologies. *Knowl. Eng. Rev.* **2010**, *25*, 137–166. [CrossRef]
23. Özarık, S.S.; da Costa, P.; Florio, A.M. *Machine Learning for Data-Driven Last-Mile Delivery Optimization*; Elsevier: Amsterdam, The Netherlands, 2022. [CrossRef]
24. Akkiraju, R. Are We There Yet? The First-Mile and Last-Mile Problem with Machine Learning Models. *IBM Cloud* **2021**. Available online: https://community.ibm.com/community/user/aiops/blogs/rama-akkiraju/2021/11/29/are-we-there-yet?CommunityKey=6e6a9ff2-b532-4fde-8011-92c922b61214 (accessed on 6 November 2022).
25. De la Torre, R.; Corlu, C.G.; Faulin, J.; Onggo, B.S.; Juan, A.A. Simulation, optimization, and machine learning in sustainable transportation systems: Models and applications. *Sustainability* **2021**, *13*, 1551. [CrossRef]
26. Majumdar, S.; Subhani, M.M.; Roullier, B.; Anjum, A.; Zhu, R. Congestion prediction for smart sustainable cities using IoT and machine learning approaches. *Sustain. Cities Soc.* **2021**, *64*, 102500. [CrossRef]
27. Shang, Q.; Lin, C.; Yang, Z.; Bing, Q.; Zhou, X. A Hybrid Short-Term Traffic Flow Prediction Model Based on Singular Spectrum Analysis and Kernel Extreme Learning Machine. *PLoS ONE* **2016**, *11*, e0161259. [CrossRef] [PubMed]
28. Florio, A.; Da Costa, P.; Özarik, S.S. *A Machine Learning Framework for Last-Mile Delivery Optimization*; Technical Proceeding of the 2021 Last Mile Routing Research Challenge; MIT Libraries: Cambridge, MA, USA, 2021.
29. Jones, S. Creating Scalable Machine Learning Systems for Analyzing Real-Time Data in Python—Part 1 I by Sahela Jones I Towards Data Science. Available online: https://towardsdatascience.com/creating-scalable-machine-learning-systems-for-analyzing-real-time-data-in-python-part-1-c303fbf79424 (accessed on 18 April 2022).
30. Lang, M.A.K.; Cleophas, C.; Ehmke, J.F. Multi-criteria decision making in dynamic slotting for attended home deliveries. *Omega* **2021**, *102*, 102305. [CrossRef]
31. Kandula, S.; Krishnamoorthy, S.; Roy, D. A prescriptive analytics framework for efficient E-commerce order delivery. *Decis. Support Syst.* **2021**, *147*, 113584. [CrossRef]
32. Rosendorff, A.; Hodes, A.; Fabian, B. Artificial intelligence for last-mile logistics—Procedures and architecture. *Online J. Appl. Knowl. Manag.* **2021**, *9*, 46–61. [CrossRef]
33. Sensormatic. Sustainability Survey Blog—Sensormatic. Sensormatic News Desk. 2022. Available online: https://www.sensormatic.com/resources/ar/2022/sustainability-survey-blog?utm_campaign=FY22-Sensormatic-Sustainability&utm_medium=BusinessWire&utm_source=FY22Q206_US-Sustainability-Consumer-Sentiment-Survey-Press-Release-Thought-Leadership-Article (accessed on 31 March 2022).
34. Blueyonder. 2022 Consumer Sustainability Survey: Consumers Pave the Way for Sustainable Retail and E-commerce. 2022. Available online: https://blueyonder.com/knowledge-center/collateral/consumers-pave-the-way-for-sustainable-retail-and-ecommerce (accessed on 10 August 2022).
35. Trapp, M.; Luttermann, S.; Rippel, D.; Kotzab, H.; Freitag, M. Modeling Individualized Sustainable Last Mile Logistics. *Dyn. Logist.* **2021**, 277–293. [CrossRef]
36. Ignat, B.; Chankov, S. Do e-commerce customers change their preferred last-mile delivery based on its sustainability impact? *Int. J. Logist. Manag.* **2020**, *31*, 521–548. [CrossRef]
37. Chu, H.; Zhang, W.; Bai, P.; Chen, Y. Data-driven optimization for last-mile delivery. *Complex Intell. Syst.* **2021**, *3*. [CrossRef]
38. Knez, M.; Jereb, B.; Gago, E.J.; Rosak-Szyrocka, J.; Obrecht, M. Features influencing policy recommendations for the promotion of zero-emission vehicles in Slovenia, Spain, and Poland. *Clean Technol. Environ. Policy* **2021**, *23*, 749–764. [CrossRef]
39. Oršič, J.; Rosi, B.; Jereb, B. Measuring Sustainable Performance among Logistic Service Providers in Supply Chains. *Technol. Gaz.* **2019**, *26*, 1478–1485. [CrossRef]

Article

Pricing and Return Strategy Selection of Online Retailers Considering Consumer Purchasing Behavior

Xinggang Shu * and **Zhenhua Hu**

Business School, Central South University, Changsha 410083, China
* Correspondence: sxg2021@126.com

Abstract: This article mainly considers the coexistence of physical sales channels and online sales channels. Online retailers with online sales channels consider whether to provide return policies and whether to provide consumers with return insurance. The research established four return strategy models that: do not provide returns; provide returns but do not provide return insurance; provide return insurance, but the cost is borne by online retailers; and provide return insurance, but the cost is borne by consumers. The authors then studied the online retailers' optimal return and shipping insurance selection strategies. The results show that when the proportion of residual return value after the value reduction of unit returned products was large, online retailers set higher sales prices and provided return policies, while offline retailers needed to reduce sales prices in order to attract more consumers. When the consumer unit product return compensation was relatively large, online retailers chose to provide consumers with free return insurance; otherwise, it was more beneficial for online retailers not to provide return insurance. Further research found that although the cost of online retailers increased when freight insurance was taken, it could better attract consumers, which was more beneficial to online retailers.

Keywords: online retailer; consumer purchase behavior; return strategy; return insurance

1. Introduction

The rapid development of the Internet has changed the traditional sales model based on physical retailing, and online shopping has become the norm. According to The National Bureau of Statistics of China, online retail sales in China reached CNY 13.1 trillion in 2021 and accounted for 24.5% of total retail sales of consumer goods [1]. However, compared to brick-and-mortar retailing, e-tailing has information-asymmetries problems owing to spatial distance, lack of "touch and experience," and a "Money-Back Guarantee (MBG)" consumer protection clause, resulting in a higher volume of returns. MBGs are popular because consumers can return unsatisfactory products and receive a full refund [2,3]. For instance, Taobao.com and JD.com e-merchants have adopted a "seven-day no questions asked" full refund return strategy. Securities Daily reported that during the most famous shopping festival in China, "Double 11" in 2021, the return rate of Taobao.com exceeded 20%, that of JD.com was approximately 10%, and that of e-commerce live streaming reached 60% [4]. Such lenient return terms have led to a shift away from quality-only returns, prompting online retailers to wonder whether they should offer returns on all products. Currently, online retailers only offer conditional returns on products such as skincare and fresh produce, which means that unless the product is defective, the online retailer usually requires the consumer to cover the cost of the return. Therefore, the main reason for online returns is the failure to meet consumer expectations rather than product defects, and the resulting return reverse logistics costs (return shipping) have become a significant barrier for consumers when purchasing online [5]. Thus, online retailers must consider return strategies carefully.

Return shipping insurance ("Return Insurance") was introduced in response to the cost of returning goods. Huatai Insurance Group first offered return insurance to consumers

Citation: Shu, X.; Hu, Z. Pricing and Return Strategy Selection of Online Retailers Considering Consumer Purchasing Behavior. *Processes* **2022**, *10*, 2490. https://doi.org/10.3390/pr10122490

Academic Editors: Conghu Liu, Xiaoqian Song, Zhi Liu, Fangfang Wei and Tsai-Chi Kuo

Received: 4 November 2022
Accepted: 22 November 2022
Published: 23 November 2022

and merchants on Taobao.com in 2010 [6]. Then, in 2013, PICC and China Life launched similar return insurance on B2C platforms such as JD.com and Dangdang [7]. One of the issues is whether the online retailer or the consumer should purchase the return insurance, which is a contentious point. Suppose that online retailers provide return insurance. This can increase consumers' willingness to purchase the product, while the low return cost caused by the return insurance compensation may lead to more returns [8,9] and thus increase the return cost for the retailers. As a result, online retailers offer return insurance purchased by consumers, which may prompt consumers to choose products more carefully and provide a return guarantee for returns that are not due to quality. However, this may also dissuade some consumers. Especially when competing with offline retailers, this may prompt online retailers to make return insurance decisions by weighing return costs more thoroughly against consumer demand.

In a competitive scenario with offline retailers, the ability of return insurance to improve the profitability of online retailers and increase consumers' willingness to pay has become the focus of current researchers. Moreover, adopting pricing decisions and return strategies is also vital for online retailers. This paper examines online retailers' pricing decisions, return strategy choices, and return insurance strategies. Specifically, it considers the following questions:

1. When should an online retailer choose to offer a return service, and what return strategy can increase the online retailer's market share?
2. Should online retailers provide free return insurance to consumers? If not, should consumers purchase their return insurance?
3. How do consumer returns and the cost of return insurance affect the performance of online retailers?

In order to solve these questions, this research constructed a game model of duopoly channel competition between an offline retailer and an online retailer. We explored the pricing and return decisions under the competition of different retail channels through a comparative analysis of online and offline sales channels. Furthermore, we studied the choice and impact of multiple return strategies based on a dual-channel perspective to provide a decision basis for manufacturers' channel development. In addition, from the standpoint of online retailers, we also compared four return strategies: not providing a return channel, providing a return channel but not providing return insurance, providing return insurance but at the cost of the online retailer, and providing return insurance but at the cost of the consumer. The aim was to provide a reference for online retailers to develop optimal pricing and return strategies. The study made a vital contribution to the existing literature. First, the author identified the pricing strategies by analyzing the competition between online and offline retailers. Second, the author provided the conditions for the online retailer's return policy. Finally, the author provided online retailers with a decision basis for pricing and return insurance in the presence of consumer unit product return compensation.

The paper is structured as follows: (Section 1) Introduction and presentation of the topic. (Section 2) Literature Review, including the definitions of main terms and related literature about (Section 2.1) Return Motivation and Return Policies and (Section 2.2) Return Insurance. (Section 3) Problem Description and Assumptions, concluding the description of the conceptual research model and hypotheses. (Section 4) Model Construction and Analysis, investigating the optimal pricing decisions under four scenarios: *NN, MN, MR,* and *MC* strategy. (Section 5) Strategy Selection, which compared the online retailer's optimal return and shipping insurance selection strategies. (Section 6) Numerical Analysis, reflecting the effect of the residual value ratio of returned products on pricing, demand, and profitability. (Section 7) Conclusions. All proofs are included in the Appendix A.

2. Literature Review

This study was closely relevant to two main streams of literature: (1) return motivation and return policies and (2) return insurance.

2.1. Return Motivation and Return Policies

Consumer returns are a common phenomenon in the retail industry. The main reason for returning products online is the virtuality of the online shopping environment [10,11], where consumers are uncertain about the price, demand, and quality of the product [12–14]. However, a loose return policy can reduce consumers' risk perception and stimulate a purchase's emotional response, thereby increasing consumers' willingness to buy and, ultimately, their willingness to pay [15–19]. For example, Griffs et al. [16] used empirical methods to verify that customer returns could significantly improve repeat purchase behavior.

With limited information available to consumers in online consumption, the decision made by consumers under high uncertainty improves the return rate of online sales [20]. How to help retailers prevent (or reduce) such return risks has become an issue of concern [21]. In terms of the application of a return fee, Hess et al. [22] believed that the practice of charging a return fee can effectively control the return rate; Shulman et al. [23] believed that the return rate could be reduced to some extent by setting the corresponding return cost and return period; further, Akcay et al. [2] reduced the return rate by controlling the selling price while considering the use of a return fee. Additionally, some studies start from the aspect of improving the disclosure of online product information. For example, considering the way to help customers better see the details of products through technology investment [24], using the offline store assistance [25,26], or arranging a physical exhibition hall for the online sales of products [27,28] to control the return rate of online sales. In addition, Lee [29] and Walsh et al. [30] believe that improving the supporting services of online sales, improving the quality of online sales, and constantly enhancing the reputation of online sales stores can greatly reduce the return rate.

Some studies have focused on the impact of the return policy on operations strategies. Xia et al. [26] demonstrated that product returns and retailer-assisted investment strongly impacted a manufacturer's decision on whether to increase online channels. Cai et al. [31] found that the choice of refund policy had a decisive impact on the competitiveness of new-entrant retailers. Batarfi et al. [32] and Guo et al. [33] found that the sales price of products in physical and online channels was only related to whether they provided MBG services in that channel and had nothing to do with whether another channel provided MBGs. Letizia et al. [34] investigated the impact of product returns on the multichannel sales strategies of manufacturers. Zhang et al. [35] found that a buyback policy helped the manufacturer obtain returns-rate information for free when the salvage value was the same for the manufacturer and the retailer. Ertekin and Agrawal [36] assessed the impact of a return period policy change on a multichannel retailer's performance.

Some studies have focused on the problem of return policy choice. Davis et al. [37] developed a model to determine the conditions under which MBGs are most effective in improving profits and social welfare. Guo [38] and Chen and Grewal [39] studied the effect of duopoly competition on retailers' return strategy choices. McWilliams [40] explored the problem of return strategy selection of competitive retailers in a competitive environment between high-quality and low-quality retailers. Nasiry and Popescu [41] studied how sellers choose an appropriate presale strategy based on consumer regret behavior and analyzed the defect-free return strategy in the case of limited capacity. Chen and Chen [42] analyzed two retailers' personalized pricing strategies and return policies with different customer satisfaction rates through a duopoly model. Li and Liu [43] investigated both return policies and found that a manufacturer's return policy could induce the retailer to adopt a return policy. Mondal and Giri [44] investigated two types of return policies in a green e-commerce supply chain: refund and replacement policies. Wang and He [45] established a dual-channel supply chain composed of one manufacturer and one retailer under mass customization and examined the manufacturer's channel strategy and return policy decisions.

The above research discussed the main reasons for returning products, the impact of the return policy on operations strategies, and the problem of return policy choice. However, researchers have not yet considered the impact of return insurance. Therefore, this research compared four types of returns: no return channel; a return channel but no return insurance; return insurance, but the cost is borne by the online retailer; and return insurance, but the cost is borne by the consumer. The strategies can provide a reference for online retailers.

2.2. Return Insurance

Some scholars have paid attention to the research on return insurance, but most of them have focused on traditional offline insurance, such as the early research of [46,47] and the recent research by Geng et al. [48]. Geng et al. [48] considered internet return insurance to deal with return risks and implemented a generous and economically viable return policy for online products. Marotta et al. [49] summarized the basic understanding of Internet insurance, discussed the uniqueness of return insurance, and analyzed the market reaction to Internet insurance through various empirical methods. Lin [50] showed that whether return insurance benefits retailers depended on the unit premium and return shipping costs.

Some studies have focused on the impact of return insurance on operations strategies. Chen et al. [51] proved that offering a return insurance strategy did not necessarily expand demand, and e-sellers with different qualities would increase selling prices when offering return insurance. Lin et al. [52] uncovered that a retailer who purchased RFI for consumers did not necessarily charge a higher price. A few scholars have focused on the question of when to introduce return insurance. Fan and Chen [53] studied the question of when e-tailers should offer free return insurance. Ren et al. [54] showed that when the net salvage value of the product is greater than or equal to zero, the online retailer should provide a refund guarantee return policy. Chen et al. [55] investigated an e-seller's strategy of offering return-freight insurance in the reselling and agency selling formats and proved that offering return-freight insurance may narrow the consumer market. Yang and Ji [56] assessed the impact of cross-selling on managing consumer returns under these three innovative mechanisms in omnichannel operations, that is, buy online and return to the store, return insurance, and a virtual try-on experience.

3. Problem Description and Assumptions

3.1. Problem Description

The paper considered a selling system comprising an offline retailer (represented by subscript "r"), an online retailer (represented by subscript "e"), and customers. In this system, there are two types of retailers from a duopoly competition, where the ordering cost per unit of product is c, and the same products are sold to consumers at prices p_r and p_e because of the different operation modes and costs. It is common for online and offline retailers to adopt different pricing strategies [57]. Zhang et al. [58] and Zhang et al. [59] provided a competitive model between the retailers. The paper assumed that the offline retailer and the online retailer are located at the two ends of the Hotelling line segment (i.e., 0 and 1) and that consumers are uniformly distributed on the line segment, with their locations denoted by $x \in [0,1]$. x is the transportation distance for the consumer to purchase from the offline retailer, and $(1 - x)$ is the transportation distance for the consumer to purchase from the online retailer. The online retailer transportation distance is the invisible distance which includes searching for products on the website and the time spent on transportation.

As consumers cannot experience the products while shopping online, the author assumed that when a product does not match a consumer's expectations, the consumer returns the product to the online retailer and receives a full refund. The online retailer can make secondary sales of those products with a discount price. Because those products have been used or damaged during the return process, for each product returned, this research assumed that the impairment value proportion is $(1 - k)$ and that the secondary sales price of each returned product is kp_e. The paper also assumed that the residual value of the retailers' unsold surplus product at the end of the selling season is 0.

In practice, if the product itself does not have defects, consumers generally need to pay the return-freight fee (s) when returning [11]. To improve the customers' experience and the adverse effect of reducing the return-freight fee, the online retailer could choose to spend t to afford the return insurance for customers. In addition, if the online retailer does not provide this, customers can purchase the insurance themselves. Regardless of who pays the return insurance, customers can obtain the return-freight compensation $r(r \leq s)$ when they return the product. The online retailer decides whether to provide the return strategy and whether to pay for the return-freight fee. Customers determine whether they buy this product and whether they pay for the return insurance. Based on this, the paper focused on following these four scenarios:

1. Scenario *NN*, where article 25 of the Consumer Protection Law stipulates the types of goods that cannot be returned or exchanged without reason for 7 days. In this scenario, the online retailer can refuse the return application if the customers are unsatisfied with the goods received;
2. Scenario *MN*, where the online retailer provides the return strategy, that is, allows customers to return the goods with which they are unsatisfied, whatever the product and without reason, for 7 days. In this scenario, the online retailer offers a return policy, but the consumer must bear the shipping costs if customers need to return the product;
3. Scenario *MR*, where the online retailer offers a return policy and provides free return insurance to consumers. In this scenario, the online retailer offers free return insurance to consumers, alleviating a certain degree of uncertainty about the consumer's purchase, which may encourage the consumer to purchase;
4. Scenario *MC*, where the online retailer offers a return policy but does not provide free return insurance to consumers. In this scenario, the consumer purchases return insurance, which significantly reduces the return shipping costs faced by the consumer when they return an unsatisfactory good.

3.2. Benchmark

The following assumptions were made here to analyze the subsequent studies in this section.

Assumption 1. *Whether customers buy the product depends on their purchase intention and utility. To illustrate the demand rate of the products purchased by consumers, the paper assumes that the potential market demand is 1 [54].*

Assumption 2. *Customers only buy products from the online retailer or offline retailer, and they do not buy multiple goods.*

Assumption 3. *When customers choose the online retailer, they cannot experience products, and the goods they receive may not match prior expectations. The paper assumes that the probability of the products matching the consumers' expectation is $\theta(0 < \theta < 1)$ [60,61]. When customers purchase from the offline retailer, this can reduce their uncertainty about the experience. Usually, the probability of consumers' expectation match is low. To highlight the differences, this research assumes that when consumers choose the offline retailer, $\theta = 1$.*

Assumption 4. *The paper assumes that consumers are completely rational people, they do not keep useless products, and there is no speculation about getting a return. If the online retailer provides return services, then when the goods do not match the customers' experience, they return the product to the online retailer and obtain a full refund [53,54].*

Assumption 5. *While differences in order quantities between online and offline retailers may affect ordering costs, to highlight the impact of the different effects of online and offline return insurance, the paper assumes that the ordering costs (c) of retailers are the same.*

The symbols of the relevant parameters involved in this paper are shown in Table 1.

Table 1. Symbol definition.

Symbol	Definition
v	Product valuation for consumers matched to products purchased by consumers at online retailers
p_r	Offline retailer sales price
p_e	Online retailer sales price
θ	Matching rate of online purchase products and consumer needs
s	The cost of returning a consumer unit of a product
r	Consumer unit product return compensation ($r \le s$)
t	The cost of purchasing return insurance for unit products ($t \le r$)
k	Percentage of residual value after impairment of the importance of returned products per unit
c	Unit product ordering cost
h	Consumer travel cost per unit distance
x	The transportation distance
U_r, U_e	Net utility of products purchased by consumers at offline retailers and online retailers
D_r, D_e	The demand for offline retailers and online retailers
π_r, π_e	Profits of offline retailers and online retailers

4. Model Construction and Analysis

4.1. NN Strategy

At this time, if the consumer is satisfied with the product purchased online, that is, it meets the consumer's expectations, they obtain utility $(v - p_e)$. However, if the customer is not satisfied with the product, they obtain utility $(0 - p_e)$. The preference cost of a consumer in position x to purchase products from online retailers is $h(1 - x)$. Therefore, in case *NN*, the utility of a consumer to purchase products from online retailers is

$$U_e^{NN} = \theta(v - p_e) + (1 - \theta)(0 - p_e) - (1 - x)h \tag{1}$$

In addition, a consumer can purchase products from offline retailers, and the preference cost of a consumer in position x to purchase products from offline retailers is hx. Therefore, the utility of a consumer to purchase products from offline retailers is

$$U_r^{NN} = v - p_r - hx \tag{2}$$

By comparing the consumer's utility U_e^{NN} and U_r^{NN}, the demand of online retailers and offline D_e^{NN} retailers can be obtained, which are D_e^{NN} and D_r^{NN}, respectively, as shown in Lemma 1.

Lemma 1. $D_r^{NN} = \frac{h + p_e - p_r + v(1 - \theta)}{2h}$; $D_e^{NN} = \frac{h - p_e + p_r - v(1 - \theta)}{2h}$.

By comparing the consumer's utility U_e^{NN} and U_r^{NN}, and the market demands D_r^{NN} and D_e^{NN}, the paper can obtain the profit functions of offline retailers and online retailers as follows:

$$\pi_r^{NN} = (p_r - c)D_r^{NN} \tag{3}$$

$$\pi_e^{NN} = (p_e - c)D_e^{NN} \tag{4}$$

The optimal sales price can be calculated by combining the profit functions of offline retailers and online retailers. As online retailers and offline retailers are in perfect competition, it is assumed that they both decide the sales price (p_e and p_r) at the same time. The result is shown in Proposition 1.

Proposition 1. $p_r^{NN} = c + h + \frac{(1 - \theta)v}{3}$, $p_e^{NN} = c + h - \frac{(1 - \theta)v}{3}$; $D_r^{NN} = \frac{3h + v(1 - \theta)}{6h}$, $D_e^{NN} = \frac{3h - (1 - \theta)v}{6h}$; $\pi_r^{NN} = \frac{(3h + v(1 - \theta))^2}{18h}$, $\pi_e^{NN} = \frac{(3h - v(1 - \theta))^2}{18h}$.

Proposition 1 gives the optimal selling price, demand function, and profit of offline retailers and online retailers under the *NN* strategy. Next, the paper analyzes the influence of parameters θ and h on the optimal sales price, demand function, and optimal profit of offline retailers and online retailers.

Corollary 1.

(1) $\frac{\partial p_r^{NN}}{\partial \theta} < 0,\ \frac{\partial p_e^{NN}}{\partial \theta} > 0;\ \frac{\partial D_r^{NN}}{\partial \theta} < 0,\ \frac{\partial D_e^{NN}}{\partial \theta} > 0;$

(2) $\frac{\partial \pi_r^{NN}}{\partial \theta} < 0;$ *when* $0 < \theta < \frac{v-3h}{v},\ \frac{\partial \pi_e^{NN}}{\partial \theta} < 0;$ *when* $\max\{0, \frac{v-3h}{v}\} < \theta < 1,\ \frac{\partial \pi_e^{NN}}{\partial \theta} > 0;$

(3) $\frac{\partial p_r^{NN}}{\partial h} > 0,\ \frac{\partial p_e^{NN}}{\partial h} > 0,\ \frac{\partial D_r^{NN}}{\partial h} < 0,\ \frac{\partial D_e^{NN}}{\partial h} > 0;$

(4) *when* $0 < h < \frac{v}{3}(1-\theta),\ \frac{\partial \pi_r^{NN}}{\partial h} < 0,\ \frac{\partial \pi_e^{NN}}{\partial h} < 0;$ *when* $\frac{v}{3}(1-\theta) < h,\ \frac{\partial \pi_r^{NN}}{\partial h} > 0,\ \frac{\partial \pi_e^{NN}}{\partial h} > 0.$

Corollary 1 shows that under the *NN* strategy, the greater the probability that the consumers purchase products online to match their expectation, the lower the risk of the consumers' online purchase, which causes customers' online purchase intention to increase. At this time, the increase in D_e^{NN} makes online retailers set higher sales prices, while D_r^{NN} decreases, and they should reduce sales prices to attract consumers to buy from offline retailers. Finally, as the matching rate increases, the profit of offline retailers gradually decreases. π_e^{NN} is related to the matching rate. When the matching rate is low, the U_e^{NN} is small and h is higher, so the consumer tends to choose offline retailers rather than online retailers. When the matching rate between online purchase products and consumer expectation is greater, the consumer's value of products is greater and the h is smaller, so more consumers choose to buy products from online retailers. When h is relatively small, consumers can choose to buy products from online retailers or offline retailers at the same time. Therefore, the two parties adopt the strategy of reducing the selling price to compete for the market, which reduces profits. However, the higher h is, the higher cost of purchasing products is from offline retailers. More consumers choose to purchase products from online retailers, so online retailers may set higher sales prices, and offline retailers correspondingly increase their sales prices. π_e^{NN} increases with the increase in h under the combined effect of increased sales price and demand.

4.2. MN Strategy

At this time, if the consumer is satisfied with the products, the consumer can obtain utility $v - p_e$. On the contrary, if the consumers are not satisfied with the products, the utility is $0 - s$, where s is the return freight cost of the consumer. Therefore, in the case of MN, the utility of a consumer to buy products from online and offline retailers is

$$U_e^{MN} = \theta(v - p_e) + (1 - \theta)(0 - s) - h(1 - x) \tag{5}$$

$$U_r^{MN} = v - p_r - hx \tag{6}$$

Lemma 2. $D_r^{MN} = \frac{h - p_r + \theta p_e + s(1-\theta) + v(1-\theta)}{2h};\ D_e^{MN} = \frac{h + p_r - \theta p_e - s(1-\theta) - v(1-\theta)}{2h}.$

Thus, the profit functions of offline retailers and online retailers are

$$\pi_r^{MN} = (p_r - c)D_r^{MN} \tag{7}$$

$$\pi_e^{MN} = (\theta p_e - c + (1 - \theta)kp_e)D_e^{MN} \tag{8}$$

By comparing the U_e^{MN} and U_r^{MN}, and D_r^{MN} and D_e^{MN}, the paper can obtain Proposition 2 by combining the profit functions of offline retailers and online retailers.

Proposition 2. $p_r^{MN} = \frac{c(2k(1-\theta)+3\theta)}{3k(1-\theta)+3\theta} + \frac{(k(1-\theta)+\theta)(3h+(1-\theta)(s+v))}{3k(1-\theta)+3\theta},$

$p_e^{MN} = \frac{c(k(1-\theta)+3\theta)}{3\theta(k(1-\theta)+\theta)} + \frac{(3h-(1-\theta)(s+v))}{3\theta};\ D_r^{MN} = \frac{(k(1-\theta)+\theta)(3h+(1-\theta)(s+v))}{6h(k(1-\theta)+\theta)} \frac{ck(1-\theta)}{6h(k(1-\theta)+\theta)},$

$$D_e^{MN} = \frac{ck(1-\theta)}{6h(k(1-\theta)+\theta)} + \frac{3h-(1-\theta)(s+v)}{6h}; \quad \pi_r^{MN} = \frac{((k(1-\theta)+\theta)(3h+(1-\theta)(s+v))-ck(1-\theta))^2}{18h(k(1-\theta)+\theta)^2},$$

$$\pi_e^{MN} = \frac{ck(1-\theta)+(k(1-\theta)+\theta)(3h-(1-\theta)(s+v))}{3\theta}.$$

Proposition 2 gives the optimal selling price, demand function, and profit of offline retailers and online retailers under the *MN* strategy. Next, the paper analyzes the effects of parameters θ and h on the optimal sales price, demand function, and optimal profit of offline retailers and online retailers.

Corollary 2.

(1) When $0 < s < \frac{ck}{(k(1-\theta)+\theta)^2} - v$, $\frac{\partial p_r^{MN}}{\partial\theta} > 0$; when $s > \max\{0, \frac{ck}{(k(1-\theta)+\theta)^2} - v\}$, $\frac{\partial p_r^{MN}}{\partial\theta} < 0$; When $0 < s < \frac{k^2c+c\theta(1-k)((3-k)\theta+2k)}{(k(1-\theta)+\theta)^2} + 3h - v$, $\frac{\partial p_e^{MN}}{\partial\theta} < 0$; when $s > \max\{0, \frac{k^2c+c\theta(1-k)((3-k)\theta+2k)}{(k(1-\theta)+\theta)^2} + 3h - v\}$, $\frac{\partial p_e^{MN}}{\partial\theta} > 0$.

(2) When $0 < s < \frac{ck}{(k(1-\theta)+\theta)^2} - v$, $\frac{\partial D_r^{MN}}{\partial\theta} > 0$, $\frac{\partial D_e^{MN}}{\partial\theta} < 0$; when $s > \max\{0, \frac{ck}{(k(1-\theta)+\theta)^2} - v\}$, $\frac{\partial D_r^{MN}}{\partial\theta} < 0$, $\frac{\partial D_e^{MN}}{\partial\theta} > 0$.

(3) When $\frac{(k(1-\theta)+\theta)^2 v}{k} < c \leq \frac{(k(1-\theta)+o)(3h+v(1-\theta))}{k(1-\theta)}$ and $0 < s < \frac{ck}{(k(1-\theta)+\theta)^2} - v$ or $c > \frac{(k(1-\theta)+o)(3h+v(1-\theta))}{k(1-\theta)}$ and $\frac{ck}{(k(1-\theta)+\theta)^2} - \frac{3h+v(1-\theta)}{1-\theta} < s < \frac{ck}{(k(1-\theta)+\theta)^2} - v$, $\frac{\partial \pi_r^{MN}}{\partial\theta} > 0$; otherwise, $\frac{\partial \pi_r^{MN}}{\partial\theta} < 0$.

(4) When $0 < s < \frac{(c+3h)k}{k(1-\theta^2)+\theta^2} - v$, $\frac{\partial \pi_e^{MN}}{\partial\theta} < 0$; when $s > \max\{0, \frac{(c+3h)k}{k(1-\theta^2)+\theta^2} - v\}$, $\frac{\partial \pi_e^{MN}}{\partial\theta} > 0$.

(5) $\frac{\partial p_r^{MN}}{\partial h} > 0$; $\frac{\partial p_e^{MN}}{\partial h} > 0$.

(6) When $0 < s < \frac{ck}{k(1-\theta)+\theta} - v$, $\frac{\partial D_r^{MN}}{\partial h} > 0$, $\frac{\partial D_e^{MN}}{\partial h} < 0$; when $s > \max\{0, \frac{ck}{k(1-\theta)+\theta} - v\}$, $\frac{\partial D_r^{MN}}{\partial h} < 0$, $\frac{\partial D_e^{MN}}{\partial h} > 0$.

(7) When $0 < h < \max\{\frac{(s+v)(1-\theta)}{3} - \frac{ck(1-\theta)}{3(k(1-\theta)+\theta)}, \frac{ck(1-\theta)}{3(k(1-\theta)+\theta)} - \frac{(s+v)(1-\theta)}{3}\}$, $\frac{\partial \pi_r^{MN}}{\partial h} < 0$; when $h > \max\{\frac{(s+v)(1-\theta)}{3} - \frac{ck(1-\theta)}{3(k(1-\theta)+\theta)}, \frac{ck(1-\theta)}{3(k(1-\theta)+\theta)} - \frac{(s+v)(1-\theta)}{3}\}$, $\frac{\partial \pi_r^{MN}}{\partial h} > 0$; $\frac{\partial \pi_e^{MN}}{\partial h} > 0$.

Corollary 2 shows that under the *MN* strategy, p_r^{MN} and p_e^{MN} is related to s. If c is small, the online retailers choose to reduce p_e^{MN}, as the demand probability increases in order to attract consumers to buy products. However, as the matching rate between offline products and consumer demand increases, D_r^{MN} naturally declines, so π_e^{MN} declines. When c is large, with the increase in matching rate, online retailers set higher sales prices. This is because when s is large, if the demand probability of consumers to purchase products is still large, online retailers naturally choose to increase the price of products. As the matching rate between online products and consumer demand increases, the demand for online products increases, so π_e^{MN} rises accordingly. In addition, π_r^{MN} is also related to the ordering cost per unit product. When unit ordering cost is greater than a certain value and s is less than a certain value, the profit of offline retailers increases with the matching rate. In the opposite scenario, it continues to decrease. In addition, with the increase in consumers' unit travel cost, p_r^{MN} and p_e^{MN} increases, but D_r^{MN} decreases while D_e^{MN} increases. This is because the increase in h leads the consumer to buy products online. Therefore, even if p_e^{MN} increases, the sales quantity still increases.

Corollary 3.

(1) $\frac{\partial p_r^{MN}}{\partial k} < 0$, $\frac{\partial p_e^{MN}}{\partial k} < 0$; $\frac{\partial D_r^{MN}}{\partial k} < 0$, $\frac{\partial D_e^{MN}}{\partial k} > 0$.

(2)   When $0 < s < \frac{ck}{k(1-\theta)+\theta} - \frac{(1-\theta)v+3h}{1-\theta}$, $\frac{\partial \pi_r^{MN}}{\partial k} > 0$; when $s > \max\{0, \frac{ck}{k(1-\theta)+\theta} - \frac{(1-\theta)v+3h}{1-\theta}\}$, $\frac{\partial \pi_r^{MN}}{\partial k} < 0$.

(3)   When $0 < s < \frac{c+3h}{1-\theta} - v$, $\frac{\partial \pi_e^{MN}}{\partial k} > 0$; when $s > \max\{0, \frac{c+3h}{1-\theta} - v\}$, $\frac{\partial \pi_e^{MN}}{\partial k} < 0$.

Corollary 3 shows that under the *MN* strategy, the greater the proportion of the surplus saleable value of product returns, the lower the impact of the consumers' return behavior on retailers' secondary sales. For example, the value of products that are easy to deteriorate or have a shorter shelf life is greatly discounted after a return. Usually, online and offline retailers can only reduce their selling prices to deal with them. Because online retailers provide a return strategy, as the surplus saleable value increases, the consumers often choose to buy online rather than offline. The profit of retailers is related to the unit return cost. If s is small, retailers choose to reduce the product price with the increase in the surplus saleable value to attract consumers to buy products. At this time, the increase in sales volume is greater than the decrease in product price, and the profit of online retailers continues to rise. On the contrary, when the s is large, offline retailers set a lower sales price with the increase in the surplus saleable value. At this time, with the continuous expansion of the online retail market, π_r^{MN} continues to decline.

4.3. MR Strategy

Under the *MR* strategy, online retailers buy return insurance for consumers and pay the cost of purchasing return insurance for unit products t. In case of a return, the online retailer refunds the product to the consumer (p_e), and the insurance company reimburses the consumer for the return freight, which is called return freight compensation (r). If the consumer is satisfied with the product, they can derive utility $(v - p_e)$. On the contrary, if they are not satisfied, then the product is returned, so they can derive utility $(r - s)$. Therefore, in the case of MR, the utility function of consumers buying products from online retailers is

$$U_e^{MR} = \theta(v - p_e) + (1 - \theta)(r - s) - h(1 - x) \tag{9}$$

In addition, consumers can buy products from offline retailers. The travel cost of consumers in location x to buy products from offline retailers is hx. Therefore, the utility of consumers buying products from offline retailers is

$$U_r^{MR} = v - p_r - hx \tag{10}$$

According to Equations (9) and (10), the demand of offline and online retailers is calculated to obtain Lemma 3.

Lemma 3. $D_r^{MR} = \frac{h - p_r + \theta p_e + (s+v-r)(1-\theta)}{2h}$, $D_e^{MR} = \frac{h + p_r - \theta p_e - (s+v-r)(1-\theta)}{2h}$.

By comparing consumers' utility of purchasing products from offline and online retailers, this research can obtain the demand of offline and online retailers. Thus, the profit functions of offline and online retailers are

$$\pi_r^{MR} = (p_r - c)D_r^{MR} \tag{11}$$

$$\pi_e^{MR} = (\theta p_e - (c + t) + (1 - \theta)kp_e)D_e^{MR} \tag{12}$$

The following Proposition 3 can be obtained by combining the profit functions of offline and online retailers.

Proposition 3. $p_r^{MR} = \frac{2c+3h+(s+v-r)(1-\theta)}{3} + \frac{\theta(c+t)}{3(k(1-\theta)+\theta)}$, $p_e^{MR} = \frac{c+3h-(s+v-r)(1-\theta)}{3\theta} + \frac{2\theta(c+t)}{3\theta(k(1-\theta)+\theta)}$; $D_r^{MR} = \frac{\theta(c+t)}{6h(k(1-\theta)+\theta)} + \frac{(s+v-r)(1-\theta)+3h-c}{6h}$, $D_e^{MR} = \frac{3h-(s+v-r)(1-\theta)+c}{6h} -$

$$\frac{\theta(c+t)}{6h(k(1-\theta)+\theta)}; \qquad \pi_r^{MR} \quad = \quad \frac{(\theta(c+t)+(k(1-\theta)+\theta)(3h-c+(s+v-r)(1-\theta)))^2}{18h(k(1-\theta)+\theta)^2},$$

$$\pi_e^{MR} = \frac{(\theta(c+t)+(k(1-\theta)+\theta)((s+v-r)(1-\theta)-3h-c))^2}{18h(k(1-\theta)+\theta)\theta}.$$

Proposition 3 gives the optimal selling price, demand function, and optimal profit of offline and online retailers under strategy *MR*. In the following, the parameters θ and h as well as their impact on the optimal selling price, demand function, and optimal profit of offline and online retailers are analyzed.

Corollary 4.

(1) When $0 < s < \frac{k(c+t)}{(k(1-\theta)+\theta)^2} + (r-v)$, $\frac{\partial p_r^{MR}}{\partial \theta} > 0$; when $s > \max\{0, \frac{k(c+t)}{(k(1-\theta)+\theta)^2} + (r-v)\}$, $\frac{\partial p_r^{MR}}{\partial \theta} < 0.$

(2) When $0 < s < \min\{\frac{\theta(1-k)(2ck(1-\theta)+c(1+k)\theta+2\theta(c+t))+ck^2}{(k(1-\theta)+\theta)^2} + 3h + r - v, \infty\}^+$, $\frac{\partial p_e^{MR}}{\partial \theta} < 0$; when $s > \max\{0, \frac{\theta(1-k)(2ck(1-\theta)+c(1+k)\theta+2\theta(c+t))+ck^2}{(k(1-\theta)+\theta)^2} + 3h + r - v\}$, $\frac{\partial p_e^{MR}}{\partial \theta} > 0.$

(3) When $0 < s < \frac{k(c+t)}{(k(1-\theta)+\theta)^2} + (r-v)$, $\frac{\partial D_e^{MR}}{\partial \theta} > 0$, $\frac{\partial D_r^{MR}}{\partial \theta} < 0$; when $s > \max\{0, \frac{k(c+t)}{(k(1-\theta)+\theta)^2} + (r-v)\}$, $\frac{\partial D_e^{MR}}{\partial \theta} < 0$, $\frac{\partial D_r^{MR}}{\partial \theta} > 0.$

(4) $\frac{\partial \pi_r^{MR}}{\partial \theta} < 0$; $\frac{\partial \pi_e^{MR}}{\partial \theta} > 0.$

(5) $\frac{\partial p_r^{MR}}{\partial h} > 0$; $\frac{\partial p_e^{MR}}{\partial h} > 0;$

(6) When $0 < s < \frac{ck(1-\theta)-\theta t}{(k(1-\theta)+\theta)(1-\theta)} + r - v$, $\frac{\partial D_r^{MR}}{\partial h} > 0$, $\frac{\partial D_e^{MR}}{\partial h} < 0$; when $s > \max\{0, \frac{ck(1-\theta)-\theta t}{(k(1-\theta)+\theta)(1-\theta)} + r - v\}$, $\frac{\partial D_r^{MR}}{\partial h} < 0$, $\frac{\partial D_e^{MR}}{\partial h_r} > 0.$

(7) When $\frac{\theta(c+t)}{(k(1-\theta)+\theta)(1-\theta)} - \frac{c+3h-(1-\theta)v}{1-\theta} < r \leq \frac{\theta(c+t)}{(k(1-\theta)+\theta)(1-\theta)} - \frac{c-3h-(1-\theta)v}{1-\theta}$ and $0 < s < \frac{ck(1-\theta)-\theta t}{(k(1-\theta)+\theta)(1-\theta)} + \frac{3h+(1-\theta)(r-v)}{1-\theta}$ or $r > \frac{\theta(c+t)}{(k(1-\theta)+\theta)(1-\theta)} - \frac{c-3h-(1-\theta)v}{1-\theta}$ and $\frac{ck(1-\theta)-\theta t}{(k(1-\theta)+\theta)(1-\theta)} + \frac{(1-\theta)(r-v)-3h}{1-\theta} < s < \frac{ck(1-\theta)-\theta t}{(k(1-\theta)+\theta)(1-\theta)} + \frac{(1-\theta)(r-v)+3h}{1-\theta}$, $\frac{\partial \pi_r^{MR}}{\partial h} > 0$, $\frac{\partial \pi_e^{MR}}{\partial h} > 0$; otherwise $\frac{\partial \pi_r^{MR}}{\partial h} < 0$, $\frac{\partial \pi_e^{MR}}{\partial h} < 0.$

Corollary 4 shows that the optimal selling price, demand, and profit under the *MR* strategy depends on the probability that consumers match demand for products purchased online θ and the travel cost per unit distance of consumers h. In particular, when the unit product return cost of consumers s is high, with the increase in θ, the online retailers whose cost increases owing to the shipping insurance choose to raise the selling price p_e^{MR} of products, and the online demand D_e^{MR} decreases accordingly. Meanwhile, offline retailers choose to reduce product prices p_r^{MR} to ensure the demand of offline retailers D_r^{MR} in order to attract consumers who are sensitive to price increases. When s is low, with the increase in θ, the online retailers should set a lower p_e^{MR}, and the D_e^{MR} increases. Meanwhile, the offline retailers also increase p_r^{MR} in order to ensure the realization of profits. Corollary 4 also shows that as h rises, the selling prices of both online and offline retailers should be set higher. In this situation, when s is low, D_r^{MR} increases with the rise of the consumers' unit distance travel cost h, while D_e^{MR} decreases with the rise of h. When s is high, D_r^{MR} decreases with the rise of h, while D_e^{MR} increases with the rise of h. In addition, when the consumer unit product return compensation r is within the range stipulated in the above formula, and the s is small or the r is high and s is within the range limited in the formula, with the increase in h, both online and offline profits increase as h increases. Conversely, in other cases, both online and offline profits decrease as h increases.

In the following, the parameters and their impact on the optimal selling price, demand function, and optimal profit of offline and online retailers are analyzed.

Corollary 5.

(1) $\frac{\partial p_r^{MR}}{\partial t} > 0,\ \frac{\partial p_r^{MR}}{\partial r} < 0;\ \frac{\partial p_e^{MR}}{\partial t} > 0,\ \frac{\partial p_e^{MR}}{\partial r} > 0;$

(2) $\frac{\partial D_r^{MR}}{\partial t} > 0,\ \frac{\partial D_r^{MR}}{\partial r} < 0;\ \frac{\partial D_e^{MR}}{\partial t} < 0,\ \frac{\partial D_e^{MR}}{\partial r} > 0.$

(3) When $0 < s < \frac{ck(1-\theta)-3h(k(1-\theta)+\theta)-\theta t}{(k(1-\theta)+\theta)(1-\theta)} + r - v,\ \frac{\partial \pi_r^{MR}}{\partial t} < 0,\ \frac{\partial \pi_r^{MR}}{\partial r} > 0;$ when $s >$ $\max\{0, \frac{ck(1-\theta)-3h(k(1-\theta)+\theta)-\theta t}{(k(1-\theta)+\theta)(1-\theta)} + r - v\},\ \frac{\partial \pi_r^{MR}}{\partial t} > 0,\ \frac{\partial \pi_r^{MR}}{\partial r} < 0.$

(4) When $0 < s < \frac{ck(1-\theta)-3h(k(1-\theta)+\theta)-\theta t}{(k(1-\theta)+\theta)(1-\theta)} + \frac{(r-v)(1-\theta)+6h}{1-\theta},\ \frac{\partial \pi_e^{MR}}{\partial t} < 0,\ \frac{\partial \pi_e^{MR}}{\partial r} > 0;$ when $s > \max\{0, \frac{ck(1-\theta)-3h(k(1-\theta)+\theta)-\theta t}{(k(1-\theta)+\theta)(1-\theta)} + \frac{(r-v)(1-\theta)+6h}{1-\theta}\},\ \frac{\partial \pi_e^{MR}}{\partial t} > 0,\ \frac{\partial \pi_e^{MR}}{\partial r} > 0.$

Corollary 5 shows that the higher the cost of purchasing return insurance for unit products t under MR strategies, the higher the cost of online retailers. Therefore, the online retailers increase the p_e^{MR}, and D_e^{MR} decreases. Meanwhile, more consumers buy products from offline retailers, and the offline retailers improve p_r^{MR}. The impact of t on profit depends on s. When s is high, the profit of online and offline retailers increases with t. On the contrary, when s is low, the profit of online and offline retailers decreases as t increases. Corollary 5 also shows that the higher the r, the lower the extra cost of return, which leads to the stronger willingness of consumers to buy products online and increase D_e^{MR}. However, as t increases the cost of online retailers, the online retailers increase p_e^{MR}, and the simultaneous increase in demand and price increases the profits of online retailers. From the perspective of offline retailers, the increase in r makes online sales more attractive, and offline retailers adopt the strategy of cutting prices owing to the decline in demand. In addition, the impact of r on the profits of offline retailers also depends on s; that is, when s is high, the profits of offline retailers decrease with the increase in r. When s is low, and the profits of offline retailers increase with r. This is because when s is low, because the compensation amount r is less than the return freight s, consumers are less sensitive to the return freight, and consumers who prefer offline experience choose to buy offline, so the increase in consumer return compensation per unit product improves the profits of offline retailers.

Corollary 6.

(1) $\frac{\partial p_r^{MR}}{\partial k} > 0,\ \frac{\partial p_e^{MR}}{\partial k} < 0.$

(2) $\frac{\partial D_r^{MR}}{\partial k} < 0,\ \frac{\partial D_e^{MR}}{\partial k} > 0.$

(3) When $0 < s < \frac{ck(1-\theta)-\theta t}{(k(1-\theta)+\theta)(1-\theta)} + \frac{(r-v)(1-\theta)-3h}{1-\theta},\ \frac{\partial \pi_r^{MR}}{\partial k} > 0;$ when $s > \max\{0, \frac{ck(1-\theta)-\theta t}{(k(1-\theta)+\theta)(1-\theta)} + \frac{(r-v)(1-\theta)-3h}{1-\theta}\},\ \frac{\partial \pi_r^{MR}}{\partial k} < 0.$

Under the MR strategy, a higher proportion of available value is returned per unit of product, increasing sales opportunities for online retailers. Thus, as the unit to return products increases, the proportion of remaining available value decreases, and more consumers are attracted to buying products online. Thus, offline demand decreases, and offline retailers raise the p_r^{MR} because offline consumers are not sensitive to prices or are unfamiliar with online products. When the return cost is high, the increase in the proportion of the residual saleable value of a unit product return makes online retailers willing to spend more in return-insurance costs to attract more consumers to buy products online. As D_e^{MR} increases and D_r^{MR} decreases, the profits of offline retailers also decrease.

4.4. MC Strategy

In the MC strategy, the consumer needs to purchase return insurance and gain utility $(v - p_e - t)$ if the consumer is satisfied with the product purchased; conversely, if the consumer is not satisfied and needs to return the product, utility $(r - s - t)$ is gained.

Therefore, in the MC strategy, the consumer's utility for purchasing a product from an online retailer is

$$U_e^{MC} = \theta(v - p_e - t) + (1 - \theta)(r - s - t) - h(1 - x) \tag{13}$$

In addition, consumers can also purchase products from offline retailers, and when a consumer in position x purchases a product from offline retailers, the cost of travel is hx; therefore, the utility of a consumer purchasing a product from an offline retailer is

$$U_r^{MC} = v - p_r - hx \tag{14}$$

According to Equations (13) and (14), we calculate the demand of offline retailers and online retailers to obtain Lemma 4.

Lemma 4. $D_r^{MC} = \frac{h - p_r + \theta p_e + (s + v - r)(1 - \theta) + t}{2h}$, $D_e^{MC} = \frac{h + p_r - \theta p_e - (s + v - r)(1 - \theta) - t}{2h}$.

By comparing the consumers' utility of purchasing products from offline retailers and online retailers, the paper can derive the demand of offline retailers and online retailers, which yields the profit functions of offline retailers and online retailers, respectively:

$$\pi_r^{MR} = (p_r - c)D_r^{MC} \tag{15}$$

$$\pi_e^{MC} = (\theta p_e - c + (1 - \theta)k p_e)D_e^{MC} \tag{16}$$

By combining the offline retailer and online retailer profit functions (15) and (16), the paper can derive Proposition 4.

Proposition 4. $p_r^{MC} = \frac{(c+t)\theta + kt(1-\theta)}{3(k(1-\theta)+\theta)} + \frac{(s+v-r)(1-\theta)+3h+2c}{3}$, $p_e^{MC} = \frac{2c}{3(k(1-\theta)+\theta)} + \frac{c-t+3h-(s+v-r)(1-\theta)}{3\theta}$; $D_r^{MC} = \frac{(s+v-r)(1-\theta)+3h}{6h} + \frac{k(1-\theta)(t-c)+t\theta}{6h(k(1-\theta)+\theta)}$, $D_e^{MC} = \frac{3h-(s+v-r)(1-\theta)}{6h} - \frac{k(1-\theta)(t-c)+t\theta}{6h(k(1-\theta)+\theta)}$; $\pi_r^{MC} = \frac{((k(1-\theta)+\theta)(3h+(s+v-r)(1-\theta)+t)-ck(1-\theta))^2}{18h(k(1-\theta)+\theta)^2}$, $\pi_e^{MC} = \frac{((k(1-\theta)+\theta)((s+v-r)(1-\theta)+t-3h)-ck(1-\theta))^2}{18h(k(1-\theta)+\theta)\theta}$.

Proposition 4 shows the optimal price, demand, and profit for the offline retailer and the online retailer in the MC strategy. Below, the paper analyzes the effect of parameters θ and h on the optimal price, demand, and profit for offline and online retailers.

Corollary 7.

(1) *When* $0 < s < \frac{ck}{(k(1-\theta)+\theta)^2} + r - v$, $\frac{\partial p_r^{MC}}{\partial \theta} > 0$, $\frac{\partial D_r^{MC}}{\partial \theta} > 0$, $\frac{\partial D_e^{MC}}{\partial \theta} < 0$; *when* $s > \max\{0, \frac{ck}{(k(1-\theta)+\theta)^2} + r - v\}$, $\frac{\partial p_r^{MC}}{\partial \theta} < 0$, $\frac{\partial D_r^{MC}}{\partial \theta} < 0$, $\frac{\partial D_e^{MC}}{\partial \theta} > 0$.

(2) *When* $0 < s < \frac{ck^2 + c\theta(1-k)((3-k)\theta + 2k)}{(k(1-\theta)+\theta)^2} + 3h + r - t - v$, $\frac{\partial p_e^{MC}}{\partial \theta} < 0$; *when* $s > \max\{0, \frac{ck^2 + c\theta(1-k)((3-k)\theta + 2k)}{(k(1-\theta)+\theta)^2} + 3h + r - t - v\}$, $\frac{\partial p_e^{MC}}{\partial \theta} > 0$.

(3) *When* $0 < s < \frac{ck(1-\theta)-ot}{(k(1-\theta)+\theta)(1-\theta)} + \frac{(1-\theta)(r-v)-3h}{1-\theta}$, $\frac{\partial \pi_r^{MC}}{\partial \theta} > 0$; *when* $s > \max\{0, \frac{ck(1-\theta)-ot}{(k(1-\theta)+\theta)(1-\theta)} + \frac{(1-\theta)(r-v)-3h}{1-\theta}\}$, $\frac{\partial \pi_r^{MC}}{\partial \theta} < 0$.

(4) *When* $0 < s < \frac{ck(1-\theta)-ot}{(k(1-\theta)+\theta)(1-\theta)} + \frac{(1-\theta)(r-v)+3h}{1-\theta}$, $\frac{\partial \pi_e^{MC}}{\partial \theta} > 0$; *when* $s > \max\{0, \frac{ck(1-\theta)-ot}{(k(1-\theta)+\theta)(1-\theta)} + \frac{(1-\theta)(r-v)+3h}{1-\theta}\}$, $\frac{\partial \pi_e^{MC}}{\partial \theta} < 0$.

(5) $\frac{\partial p_r^{MC}}{\partial h} > 0$; $\frac{\partial p_e^{MC}}{\partial h} > 0$;

(6) *When* $0 < s < \frac{ck}{k(1-\theta)+\theta} + \frac{(1-\theta)(r-v)-t}{1-\theta}$, $\frac{\partial D_r^{MC}}{\partial h} > 0$, $\frac{\partial D_e^{MC}}{\partial h} < 0$; *when* $s > \max\{0, \frac{ck}{k(1-\theta)+\theta} + \frac{(1-\theta)(r-v)-t}{1-\theta}\}$, $\frac{\partial D_r^{MC}}{\partial h} < 0$, $\frac{\partial D_e^{MC}}{\partial h} > 0$.

(7) When $\frac{v(1-\theta)+t-3h}{1-\theta} - \frac{ck}{k(1-\theta)+\theta} < r \leq \frac{v(1-\theta)+t+3h}{1-\theta} - \frac{ck}{k(1-\theta)+\theta}$ and $0 < s < \frac{ck}{k(1-\theta)+\theta} + \frac{3h-t}{1-\theta} + r - v$ or $r > \frac{v(1-\theta)+t+3h}{1-\theta} - \frac{ck}{k(1-\theta)+\theta}$ and $\frac{ck}{k(1-\theta)+\theta} - \frac{3h+t}{1-\theta} + r - v < s < \frac{ck}{k(1-\theta)+\theta} + \frac{3h-t}{1-\theta} + r - v$, $\frac{\partial \pi_r^{MC}}{\partial h} > 0$, $\frac{\partial \pi_e^{MC}}{\partial h} > 0$; otherwise, $\frac{\partial \pi_r^{MC}}{\partial h} < 0$, $\frac{\partial \pi_e^{MC}}{\partial h} < 0$.

Corollary 7 suggests that under the *MC* strategy, the higher the return shipping cost, the higher the probability of consumers to purchase products matching online, and the stronger the consumers' willingness to purchase products from online retailers, which increases D_e^{MC} and decreases D_r^{MC}. On this basis, online retailers choose to increase p_e^{MC} to gain more profit, and offline retailers lower p_r^{MC} to attract more consumers to purchase offline. The effect of h on the sales price, demand, and profit of online and offline retailers also relies on consumer travel cost per unit distance and return compensation per unit product; the findings are similar to the MR strategy and are not repeated here.

Corollary 8.

(1) $\frac{\partial p_r^{MC}}{\partial t} > 0$, $\frac{\partial p_r^{MC}}{\partial r} < 0$; $\frac{\partial p_e^{MC}}{\partial t} < 0$, $\frac{\partial p_e^{MC}}{\partial r} > 0$.

(2) $\frac{\partial D_r^{MC}}{\partial t} > 0$, $\frac{\partial D_r^{MC}}{\partial r} < 0$; $\frac{\partial D_e^{MC}}{\partial t} < 0$, $\frac{\partial D_e^{MC}}{\partial r} > 0$.

(3) When $0 < s < \frac{ck}{k(1-\theta)+\theta} + \frac{(r-v)(1-\theta)-t-3h}{1-\theta}$, $\frac{\partial \pi_r^{MC}}{\partial t} < 0$, $\frac{\partial \pi_r^{MC}}{\partial r} > 0$; when $s > \max\{0, \frac{ck}{k(1-\theta)+\theta} + \frac{(r-v)(1-\theta)-t-3h}{1-\theta}\}$, $\frac{\partial \pi_r^{MC}}{\partial t} > 0$, $\frac{\partial \pi_r^{MC}}{\partial r} < 0$.

(4) When $0 < s < \frac{ck}{k(1-\theta)+\theta} + \frac{(r-v)(1-\theta)-t+3h}{1-\theta}$, $\frac{\partial \pi_e^{MC}}{\partial t} < 0$, $\frac{\partial \pi_e^{MC}}{\partial r} > 0$; when $s > \max\{0, \frac{ck}{k(1-\theta)+\theta} + \frac{(r-v)(1-\theta)-t+3h}{1-\theta}\}$, $\frac{\partial \pi_e^{MC}}{\partial t} > 0$, $\frac{\partial \pi_e^{MC}}{\partial r} < 0$.

Corollary 8 indicates that under the *MC* strategy, the higher the t, the more consumers are inclined to purchase products from offline channels, and D_r^{MC} also increases. Therefore, offline retailers choose a higher p_r^{MC}. Consumers' willingness to purchase offline forces online retailers to slow the decline in online demand by lowering sales price. The increase in r reduces the risk of consumers purchasing products from online retailers, and more consumers purchase products from online retailers, which induces D_e^{MC} to increase, so online retailers raise the selling price. In addition, as D_r^{MC} continues to decrease with the increase in r, offline retailers choose lower sales prices to attract more consumers to purchase products offline. Corollary 8 also shows that the impact of the t and r on the profits of both online and offline retailers depends on the return freight cost s. When s is high, the profits of both online and offline retailers increase with the increase in t and decrease with the increase in r.

Corollary 9.

(1) $\frac{\partial p_r^{MC}}{\partial k} < 0$, $\frac{\partial p_e^{MC}}{\partial k} < 0$.

(2) $\frac{\partial D_r^{MC}}{\partial k} < 0$, $\frac{\partial D_e^{MC}}{\partial k} > 0$.

(3) When $0 < s < \frac{ck}{k(1-\theta)+\theta} - \frac{3h+t}{1-\theta} + r - v$, $\frac{\partial \pi_r^{MC}}{\partial k} > 0$; when $s > \max\{0, \frac{ck}{k(1-\theta)+\theta} - \frac{3h+t}{1-\theta} + r - v\}$, $\frac{\partial \pi_r^{MC}}{\partial k} < 0$.

Under the *MC* strategy, the higher the proportion of the residual saleable value of unit product return k is, the greater the possibility for online retailers to resell. Therefore, the online sales price decreases with the increase in k. In this way, more consumers are attracted to purchase products online, which leads D_e^{MC} to increase, so D_r^{MC} decreases, leaving offline retailers reduce the sales price of offline products in order to ensure the offline demand. When the return cost s is high, consumers need to pay more costs and turn to offline purchases. The profits of offline retailers increase with the increase in the k. On the contrary, when s is low, there is a decrease in consumers' return cost, an increase in their willingness to consume online, an increase in their online demand, and a decrease in their offline demand. The profits of offline retailers decrease with the increase in k.

5. Strategy Selection

5.1. Return Strategy Selection

This section examines the selling prices, market shares, and profitability of online retailers under the NN and MN strategies and reveals managerial insights to investigate whether they ought to provide a return strategy.

Proposition 5.

(1) When $0 < s < \frac{ck}{k(1-\theta)+\theta}$, the paper has $p_r^{MN} < p_r^{NN}, D_r^{MN} < D_r^{NN}, D_e^{MN} > D_e^{NN}$; When $s > \frac{ck}{k(1-\theta)+\theta}$, the paper has $p_r^{MN} > p_r^{NN}, D_r^{MN} > D_r^{NN}, D_e^{MN} < D_e^{NN}$.

(2) When $0 < s < \frac{c(k+3\theta(1-k))}{k(1-\theta)+\theta} + 3h - (1-\theta)v$, $p_e^{MN} > p_e^{NN}$; when $s > \max\{0, \frac{c(k+3\theta(1-k))}{k(1-\theta)+\theta} + 3h - (1-\theta)v\}$, $p_e^{MN} < p_e^{NN}$.

(3) When $0 < s < \max\{0, \frac{ck}{k(1-\theta)+\theta} - \frac{6h+2v(1-\theta)}{1-\theta}\}$ or $s > \frac{ck}{k(1-\theta)+\theta}$, the paper has $\pi_r^{MN} > \pi_r^{NN}$; when $\max\{0, \frac{ck}{k(1-\theta)+\theta} - \frac{6h+2v(1-\theta)}{1-\theta}\} < s < \frac{ck}{k(1-\theta)+\theta}$, $\pi_r^{MN} < \pi_r^{NN}$.

(4) When $0 < s < \frac{3h-2(1-\theta)v+2c}{2(1-\theta)} - \frac{(1-\theta)\theta v^2 - 9h^2 k - 6hv\theta}{6h(k(1-\theta)+\theta)} - \frac{c\theta}{(k(1-\theta)+\theta)(1-\theta)}$, $\pi_e^{MN} > \pi_e^{NN}$; when $s > \max\{0, \frac{3h-2(1-\theta)v+2c}{2(1-\theta)} - \frac{(1-\theta)\theta v^2 - 9h^2 k - 6hv\theta}{6h(k(1-\theta)+\theta)} - \frac{c\theta}{(k(1-\theta)+\theta)(1-\theta)}\}$, $\pi_e^{MN} < \pi_e^{NN}$.

Three implications can be drawn from Proposition 5: First, the offline retailer should set a higher p_r^{NN} when s is relatively low and $D_r^{MN} < D_r^{NN}$, while D_e^{NN} is the opposite. The offline retailer should set a lower p_r^{NN} but only if s is sufficiently high. The demand for the online channel is higher under the NN strategy. Second, based on the extent of s, online retailers should set different p_e^{NN} and p_e^{MN}. The number of s is a third factor that affects the online retailer's strategy. When s is low, online sellers in particular should provide consumers with return strategies.

5.2. Return Insurance Strategy Selection

The selling prices, market shares, and profit margin of online retailers under the MN, MR, and MC strategies are now compared and analyzed with a view to providing further managerial implications on the issues of whether online retailers offer free return insurance, whether online retailers buy return insurance, or whether online retailers make the decision to purchase return insurance.

Proposition 6.

(1) $p_e^{MR} > p_e^{MN}$; when $0 < r < \frac{\theta t}{(k(1-\theta)+\theta)(1-\theta)}$, $p_r^{MR} > p_r^{MN}, D_r^{MR} > D_r^{MN}, D_e^{MR} < D_e^{MN}$; when $r > \frac{\theta t}{(k(1-\theta)+\theta)(1-\theta)}$, $p_r^{MR} < p_r^{MN}, D_r^{MR} < D_r^{MN}, D_e^{MR} > D_e^{MN}$.

(2) When $0 < c < \frac{(k(1-\theta)+\theta)(3h+(1-\theta)(s+v))}{k(1-\theta)}$ or $c > \frac{(k(1-\theta)+\theta)(3h+(1-\theta)(s+v))}{k(1-\theta)}$ and $t > \frac{2}{\theta}(ck(1-\theta) - (k(1-\theta)+\theta)(3h+(1-\theta)(s+v)))$, if $0 < r < \frac{\theta t}{(k(1-\theta)+\theta)(1-\theta)}$ or $r > \frac{ot-2ck(1-\theta)}{(k(1-\theta)+\theta)(1-\theta)} + \frac{6h+2(s+v)(1-\theta)}{1-\theta}$, $\pi_r^{MR} > \pi_r^{MN}$; and if, $\frac{\theta t}{(k(1-\theta)+\theta)(1-\theta)} < r < \frac{ot-2ck(1-\theta)}{(k(1-\theta)+\theta)(1-\theta)} + \frac{6h+2(s+v)(1-\theta)}{1-\theta}$, the paper has $\pi_r^{MR} < \pi_r^{MN}$. When $c > \frac{(k(1-\theta)+\theta)(3h+(1-\theta)(s+v))}{k(1-\theta)}$ and $0 < t < \frac{2}{\theta}(ck(1-\theta) - (k(1-\theta)+\theta)(3h+(1-\theta)(s+v)))$, if $0 < r < \frac{\theta t}{(k(1-\theta)+\theta)(1-\theta)}$, $\pi_r^{MR} < \pi_r^{MN}$; and if $r > \frac{\theta t}{(k(1-\theta)+\theta)(1-\theta)}$, $\pi_r^{MR} > \pi_r^{MN}$.

(3) When $0 < h < \bar{h}_{R-N}$, $\pi_e^{MR} > \pi_e^{MN}$; when $h > \bar{h}_{R-N}$, $\pi_e^{MR} < \pi_e^{MN}$, where $\bar{h}_{R-N}$ is the solution to the equation $\frac{1}{18\theta}\left(\frac{((s-r)(k(1-\theta)+\theta)(1-\theta)+\theta t+(1-\theta)(k+(1-k)\theta)v-(c+3h)k(1-\theta)-3h\theta)^2}{/(hk+h(1-k)\theta)+6(k(1-\theta)+\theta)((1-\theta)(s+v)-3h)-6ck(1-\theta)} \right) = 0.$

It is evident from Proposition 6 that p_e is continuously greater under the *MR* strategy than under the *MN* strategy. This is because of the platform retailing's free return strategy, which raises p_e. Additionally, Proposition 6 demonstrates that when r is relatively small, owing to the discrepancy between consumers buying products from online retailers and the smaller compensation received by consumers for returning products, there is less incentive for consumers to buy products from online retailers, and fewer consumers buy products from online retailers, so online retailers gain more profit under the *MN* strategy. At the same time, the offline retailer also has a greater market share under the *MR* strategy compared to the *MN* strategy, so at this point, the offline retailer should set a greater p_r, which ultimately results in the offline retailer gaining greater profits under the *MR* strategy.

According to Proposition 6, consumers may be more inclined to purchase items from online retailers when they receive a greater r. This can stimulate D_e under the *MR* strategy compared to the *MN* strategy and boost their profitability. In order to draw in more customers, offline retailers use the *MR* strategy to offer lower p_r, but this lowers their profitability. Online retailers are more profitable under the *MR* strategy than the *MN* strategy when consumers' travel costs per unit distance is lower, while the *MN* strategy is more profitable when x is greater. These data indicate that x is a significant factor affecting both the *MN* strategy and the *MR* strategy. In contrast to the results of Ren et al. [54], here, when the online retailer offers a return strategy, r is a key factor influencing whether the online retailer offers return insurance.

Proposition 7.

(1) *When $0 < r < \frac{t}{1-\theta}$, $p_r^{MC} > p_r^{MN}$, $p_e^{MC} < p_e^{MN}$, $D_r^{MC} > D_r^{MN}$, $D_e^{MC} < D_e^{MN}$; when $r > \frac{t}{1-\theta}$, $p_r^{MC} < p_r^{MN}$, $p_e^{MC} > p_e^{MN}$, $D_r^{MC} < D_r^{MN}$, $D_e^{MC} > D_e^{MN}$.*

(2) *When $0 < c < \frac{(k(1-\theta)+\theta)(3h+(1-\theta)(s+v))}{k(1-\theta)}$ or $c > \frac{(k(1-\theta)+\theta)(3h+(1-\theta)(s+v))}{k(1-\theta)}$ and $t > \frac{2ck(1-\theta)}{k(1-\theta)+\theta} - 2(1-\theta)(s+v) - 6h$, if $0 < r < \min\{\frac{t}{1-\theta}, \frac{6h+t+2(v+s)(1-\theta)}{1-\theta} - \frac{2ck}{k(1-\theta)+\theta}\}$ or $r > \max\{\frac{t}{1-\theta}, \frac{6h+t+2(v+s)(1-\theta)}{1-\theta} - \frac{2ck}{k(1-\theta)+\theta}\}$, the paper has $\pi_r^{MC} > \pi_r^{MN}$; if $\min\{\frac{t}{1-\theta}, \frac{6h+t+2(v+s)(1-\theta)}{1-\theta} - \frac{2ck}{k(1-\theta)+\theta}\} < r < \max\{\frac{t}{1-\theta}, \frac{6h+t+2(v+s)(1-\theta)}{1-\theta} - \frac{2ck}{k(1-\theta)+\theta}\}$, we have $\pi_r^{MC} < \pi_r^{MN}$.*

  When $c > \frac{(k(1-\theta)+\theta)(3h+(1-\theta)(s+v))}{k(1-\theta)}$ and $0 < t < \frac{2ck(1-\theta)}{k(1-\theta)+\theta} - 2(1-\theta)(s+v) - 6h$, if $0 < r < \frac{t}{1-\theta}$, $\pi_r^{MC} < \pi_r^{MN}$; if $r > \frac{t}{1-\theta}$, $\pi_r^{MC} > \pi_r^{MN}$.

(3) *When $0 < h < \bar{h}_{C-N}$, $\pi_e^{MC} > \pi_e^{MN}$; when $h > \bar{h}_{C-N}$, $\pi_e^{MC} < \pi_e^{MN}$, where $\bar{h}_{C-N}$ is the solution of $\frac{1}{180}\left(\begin{array}{c} (ck(1-\theta) + (k(1-\theta)+\theta)(3h+(1-\theta)r-s-t-v+\theta(s+v)))^2/(h(k(1-\theta)+\theta)) \\ -6ck(1-\theta) - 6(k(1-\theta)+\theta)(3h-(1-\theta)(s+v)) \end{array} \right) = 0$*

From Proposition 7, the offline retailer should establish a higher p_r when the offline demand is larger under the *MC* strategy, and the online retailer should set a lower p_e when the c is modest. Under the *MC* strategy compared to the *MN* strategy, the offline retailer should set a lower p_r, and the online retailer should set a higher p_e when the r is greater. The profit of offline retailers under the *MC* strategy is greater than that under the *MN* strategy in the case of a smaller or larger c and larger t, which indicates that c and r are favorable for offline retailers under the *MC* strategy, and vice versa under the *MN* strategy. However, the profit of the offline retailer under the *MN* strategy is better than that under the *MC* strategy if r is lower when c is higher but t is lower, and vice versa for the offline retailer under the *MC* strategy. The online retailer is more profitable under the *MC* strategy than the *MN* strategy when h is below a critical value, indicating that when the h is low, a critical value can be found for the online retailer, making the online retailer better under the strategy where the consumer buys their own return insurance. However, when h exceeds a critical value, the *MN* strategy is more profitable.

Proposition 8.

(1) $p_r^{MC} > p_r^{MR}$, $p_e^{MC} < p_e^{MR}$; $D_r^{MC} > D_r^{MR}$, $D_e^{MC} < D_e^{MR}$.

(2) When $0 < t < \dfrac{2ck(1-\theta)-2(k(1-\theta)+\theta)(3h-(1-\theta)(r-s-v))}{k(1-\theta)+2\theta}$, $\pi_r^{MC} < \pi_r^{MR}$; when $t > \max\{0, \dfrac{2ck(1-\theta)-2(k(1-\theta)+\theta)(3h-(1-\theta)(r-s-v))}{k(1-\theta)+2\theta}\}$, $\pi_r^{MC} > \pi_r^{MR}$.

(3) When $0 < t < \dfrac{2ck(1-\theta)+2(k(1-\theta)+\theta)(3h+(1-\theta)(r-s-v))}{k(1-\theta)+2\theta}$, $\pi_e^{MC} < \pi_e^{MR}$; when $t > \max\{0, \dfrac{2ck(1-\theta)+2(k(1-\theta)+\theta)(3h+(1-\theta)(r-s-v))}{k(1-\theta)+2\theta}\}$, $\pi_e^{MC} > \pi_e^{MR}$.

According to Proposition 8, the *MR* strategy has higher pricing and market demand for online retailers than the *MC* strategy. This is because under the *MR* strategy, the online retailer offers insurance services to customers. These services generate premiums while safeguarding the customer's return, which raises the demand for the market as a whole. The truth that online retailers offer insurance is an effective form of competition for offline retailers, forcing them to compete only by raising prices. However, at the moment, the market demand for offline retailers is still declining and is below the overall demand in the *MC* strategy. Because the online retailer covers t under the *MR* strategy, the online retailer covers less when t is modest and generates more profits; nevertheless, when t is high, the *MC* strategy is preferable because the customer covers t. With regard to offline retailers, while t is relatively low, the new customer attraction of the *MR* strategy in which online retailers offer insurance is lower. At this time, offline retailers see higher profits when utilizing the *MR* strategy of online retailers. However, as t starts to climb, the customer appeal of online retailers' *MR* strategies also increases, which causes a drop in their profitability.

6. Numerical Analysis

Owing to the impact of the transportation process and after-sales, the residual value of products is different. This section analyzes the effect of the residual value ratio of returned products on pricing, demand, and profitability through numerical experiments. Referring to the numerical analysis framework of Fan and Chen [53] and Ren et al. [54], the parameters were set as follows: $v = 1$, $c = 0.25$, $r = 0.2$, $t = 0.02$, $\theta = 0.5$, $s = 0.5$, and $h = 0.25$. The paper analyzed the impact of the proportion of residual value after the reduction of unit returned product value on the optimal decision.

Figure 1 analyzes the effect of the proportion of residual value on the most available demand by setting different values of r for the diminished value of the returned product per unit. Comparing the different strategies, the paper found that the demand for all strategies except under the *NN* strategy was influenced by the consumer unit product return compensation and the proportion of residual value per unit of returned product value after impairment; regardless of the value of consumer unit product return compensation, as the proportion of residual value per unit of returned product value after impairment increased, the online demand for the three strategies *MN*, *MR*, and *MC* increased, and the offline demand decreased.

When the consumer unit product return compensation was low, at 0.2, the initial demand for the *MR* strategy was the highest, close to 1; the initial demand for *MR* and *MC* was slightly less, at around 0.95; and the initial demand for *NN* was the lowest, slightly over 0.8. As the unit product return compensation rose, the initial demand for both the *MR* and *NN* strategies remained the same, and the initial demand for *MR* and *MC* fell to the same level as that for the *NN* initial demand.

By comparing online demand with offline demand, the paper found that consumers preferred to buy products offline. With the highest proportion of residual value still occupying more than four-fifths of the overall market demand, online demand had less than one-fifth of the share. As the proportion of residual value increased, online demand tended to decrease, and offline demand tended to increase; when consumer unit product returns were compensated differently, the *MR* and *MC* strategy demand changed more significantly with the proportion of residual value per unit of returned product value after impairment.

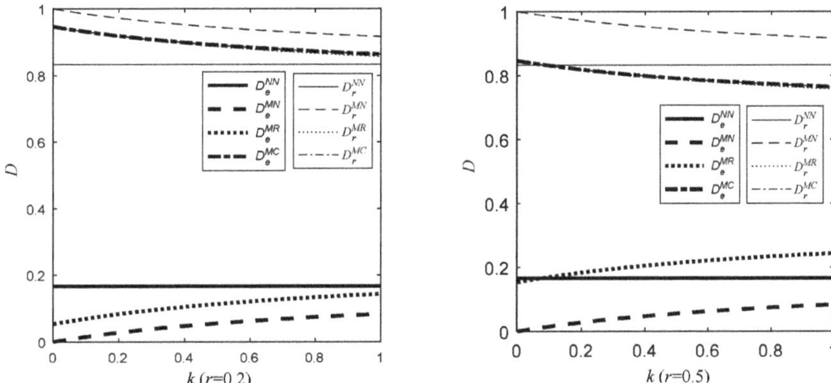

Figure 1. The impact of the proportion of residual value on optimal demand after the diminution of the value of the returned product per unit.

Figure 2 analyzes the impact of the proportion of residual value per unit of returned product value impairment and return compensation on the profit of offline retailers. The profit of offline retailers decreased as the loss of returned products became smaller; that is, when the residual value per unit of returned product value after impairment became larger, the advantage of offline retailers not having to return products decreased at this time, which led to a decrease in profit. Comparing the profits under different strategies, the paper found that the profits of offline retailers were the highest under the *MN* strategy, at which time the competitive advantage of online retailers was reduced, owing to the returns being borne by consumers, which was beneficial to offline retailers. At $r = 0.2$, profits under the *MC* strategy were higher than those under the *NN* strategy, and at $r = 0.5$, profits under the *NN* strategy were higher than those under the *MC* strategy.

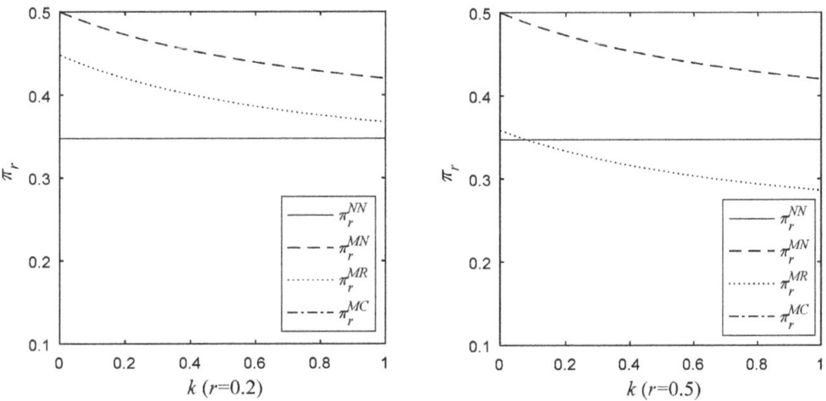

Figure 2. The impact of the percentage of residual value per unit of returned product value impairment on the profit of offline retailers.

Figure 3 analyzes the impact of the percentage of residual value per unit of returned product value impairment and return compensation on the profit of the online retailers. As the residual value per unit of returned product value increased, the profit of the online retailers increased, and it was more beneficial for the online retailers when the loss from the returned product was lower. Comparing the four strategies, the paper showed that the profit of online retailers was highest under the *NN* strategy when k was extremely low, indicating that the online retailers prefer not to offer returns when the loss from returns is extremely high (i.e., the *NN* strategy). At $r = 0.2$, the profits of online retailers under the *MN* strategy were higher than those under the *MR* and *MC* strategies, indicating that when the compensation given to consumers for returns is low, it is much less attractive to consumers, so choosing the *MN* strategy instead saves the loss of paying return insurance, and profits are higher. In contrast, the profit under the *MR* strategy was higher than that under the *MC* strategy because the return insurance was borne by the online retailers and could attract more consumers. At $r = 0.5$, when k was lower, the profit of online retailers under the *MR* and *MC* strategies was higher than under the *MN* strategy, indicating that despite the lower value of impaired unit returns, it is more attractive to offer returns to consumers at this time owing to the higher compensation to consumers under the *MR* and *MC* strategies. As k increased, profits under the *MN* strategy increased significantly above profits under the *MR*, *MC*, and *NN* strategies, indicating that online retailers prefer to return products rather than offer a return fee when the unit residual value after impairment of the returned product is higher.

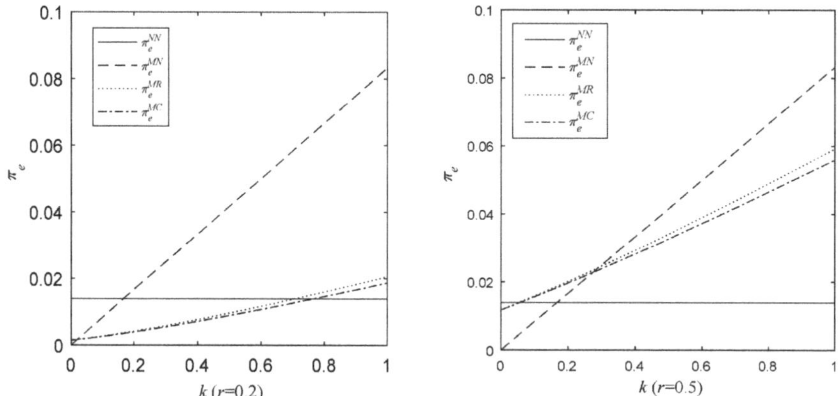

Figure 3. The impact on online retailers of the proportion of residual value after reduction of unit returned product value.

7. Conclusions

This paper constructed a game model of duopoly competition with channel competition between offline retailers and online retailers and investigated the optimal pricing decisions under four scenarios: online retailers do not offer returns; online retailers offer returns but do not provide return insurance; online retailers offer returns and provide return insurance; and online retailers offer returns, but consumers purchase return insurance. It also answered the three core questions of whether to offer returns, whether to provide return insurance, and who should bear the return insurance. Specifically, the main findings are reflected in the following three issues.

First is the problem of whether to provide return service. The statute of limitations for sales varies by product, such as cosmetics and food. Online retailers tend not to offer returns for products that cannot be resold after opening or whose value decreases significantly immediately after the sale when the return loss is too high. Then, online retailers may offer returns for products such as clothing that do not affect secondary sales after return, given the higher residual value per unit of returned product after impairment. The findings of this study are consistent with reality.

Second is the problem of whether to provide return insurance. The provision of return insurance allows consumers to purchase products without worries when there is a return policy in place and when the insurance company has the maximum amount to cover the shipping costs. Moreover, when the e-retailer provides free return insurance to consumers, it can increase the sales of products and contribute to the profits of the e-retailer. Therefore, online retailers should offer free return insurance to consumers when the compensation per unit of product returned is considerable.

Third is the issue of who should purchase the return insurance. Suppose online retailers are willing to provide return and return shipping insurance services. In that case, customers can return the product at no additional cost if the product does not match after purchase, reducing the uncertainty of consumers' purchase and thus attracting more consumers. The increase in sales improves online retailers' profits. This is consistent with the fact that in reality, most online merchants offer shipping insurance.

Our article provides an analytical framework for online retailers' pricing strategies and return strategies in competitive situations and reveals the conditions for online retailers to provide return strategies and return insurance strategies, which is similar to the research of Lin et al. [50], Fan and Chen [53], and Ren et al. [54]. However, we have expanded the above research. In particular, our paper considers such factors as consumer distance cost and return insurance compensation and expands the research situation to the competition situation of online and offline retailers. The research conclusion is more universal. At the same time, the research can provide management insights for online retailers. Under the competition with offline retailers, online retailers need to design return strategies for different products, and they do not provide return for products with large losses after return, which explains why Pinduoduo (www.pinduoduo.com) does not provide return for fresh products but only refund. In order to reduce consumers' concern about product matching, online retailers should provide return service to consumers, but whether it is free should be further considered. This explains why online stores of well-known brands, such as Uniqlo (www.uniqlo.cn), Decathlon (www.decathlon.com), and JD (www.jd.com), do not provide free return insurance to consumers but provide 7 days' return service without any reason. Under the return insurance service, the purchase of consumers can be improved. Many enterprises such as Nike will provide a return insurance service to share the return cost.

However, our research had some limitations, which can be further considered in depth in future research. First, the paper did not consider the impact of consumer return services. The impact of introducing return services on consumer purchase intentions and the interaction behavior between consumer purchase intentions and return services for different types of products in the study of return issues will be a topic for future research. Second, online retailers did not consider their own sales model when formulating their return strategy. Thus, issues such as the offline return model for online sales or the impact of different sales channels online on return policies are topics worthy of future research.

Author Contributions: X.S.: conceptualization, methodology, writing—review and editing; Z.H.: supervision, writing—review and editing. All authors have read and agreed to the published version of the manuscript.

Funding: This research received no external funding.

Institutional Review Board Statement: Not applicable.

Informed Consent Statement: Informed consent was obtained from all subjects involved in the study.

Data Availability Statement: Not applicable.

Acknowledgments: Many thanks to Conghu Liu of Suzhou University for inviting us and suggesting valuable revisions to our paper.

Conflicts of Interest: The authors declare no conflict of interest.

Appendix A

Proof of Lemma 1. When $x < x_r^{NN} = \frac{v-p_r}{h}$, there is $U_e^{NN} \geq 0$, and when $x > x_e^{NN} = \frac{h+p_e-\theta v}{h}$, there is $U_r^{NN} \geq 0$. When $x_{re}^{NN} = \frac{h+p_e-p_r+v(1-\theta)}{2h}$, there is $U_e^{NN} = U_r^{NN}$, i.e., the consumer's utility of purchasing a product from an offline retailer and an online retailer is the same, since the consumer will purchase the product at the offline retailer when and only when $U_r^{NN} \geq 0$ and $U_r^{NN} \geq U_e^{NN}$. Similarly, when $U_e^{NN} \geq 0$ and $U_e^{NN} \geq U_r^{NN}$, consumers will purchase the product at the online retailer. Without loss of generality, suppose $v \geq \frac{h+p_e+p_r}{1+\theta}$, getting $x_e^{NN} < x_{re}^{NN} < x_r^{NN}$; then, both the offline and online markets exist.

Since when $U_r^{NN} \geq 0$ and $U_r^{NN} \geq U_e^{NN}$, that is, consumers in the interval $\left[0, x_{re}^{NN}\right]$ will buy from an offline retailer and consumers in the interval $\left[x_{re}^{NN}, 1\right]$ will buy from an online retailer, it follows that the demands for online retailer and offline retailer are $D_r^{NN} = x_{re}^{NN} = \frac{h+p_e-p_r+v(1-\theta)}{2h}$ and $D_e^{NN} = 1 - x_{re}^{NN} = \frac{h-p_e+p_r-v(1-\theta)}{2h}$, respectively. The same can be proofed for Lemmas 2–4. $\square$

Proof of Proposition 1. Second-order derivatives of the profit functions of the offline and online retailers with respect to p_r and p_e, respectively, are $\frac{\partial^2 \pi_r^{NN}}{\partial (p_r^{NN})^2} = -\frac{1}{h} < 0$ and $\frac{\partial^2 \pi_e^{NN}}{\partial (p_e^{NN})^2} = -\frac{1}{h} < 0$. It is easy to prove that the Hessian matrix is $\begin{vmatrix} \partial^2 \pi_r^{NN}/\partial (p_r^{NN})^2 & \partial^2 \pi_r^{NN}/\partial p_r^{NN}\partial p_e^{NN} \\ \partial^2 \pi_e^{NN}/\partial p_e^{NN}\partial p_r^{NN} & \partial^2 \pi_e^{NN}/\partial (p_e^{NN})^2 \end{vmatrix} = \frac{3}{4h^2} > 0$. From the first-order derivative conditions $\frac{\partial \pi_r^{NN}}{\partial p_r^{NN}} = 0$ and $\frac{\partial \pi_e^{NN}}{\partial p_e^{NN}} = 0$, it can obtain $p_r^{NN} = \frac{3c+3h+(1-\theta)v}{3}$ and $p_e^{NN} = \frac{3c+3h-(1-\theta)v}{3}$. The optimal demand and profit functions are obtained by substituting the offline and online retailers' optimal selling prices into the demand and profit functions. Propositions 2–4 can be proven in the same way. End proof. $\square$

Proof of Corollary 1. Optimal selling price, demand, and optimal profit for offline and online retailers can be obtained by taking derivatives with respect to θ, respectively: $\frac{\partial p_r^{NN}}{\partial \theta} = -\frac{v}{3} < 0$, $\frac{\partial p_e^{NN}}{\partial \theta} = \frac{v}{3} > 0$, $\frac{\partial D_r^{NN}}{\partial \theta} = -\frac{v}{6h} < 0$, $\frac{\partial D_e^{NN}}{\partial \theta} = \frac{v}{6h} > 0$, $\frac{\partial \pi_r^{NN}}{\partial \theta} = -\frac{v(3h+v-\theta v)}{9h} < 0$, and $\frac{\partial \pi_e^{NN}}{\partial \theta} = \frac{v(3h-(1-\theta)v)}{9h}$.

The optimal selling price, demand and optimal profit for offline and online retailers can be obtained by taking derivatives with respect to h, respectively: $\frac{\partial p_r^{NN}}{\partial h} = 1 > 0$, $\frac{\partial p_e^{NN}}{\partial h} = 1 > 0$, $\frac{\partial D_r^{NN}}{\partial h} = -\frac{(1-\theta)v}{6h^2} < 0$, $\frac{\partial D_e^{NN}}{\partial h} = \frac{(1-\theta)v}{6h^2} 0$, $\frac{\partial \pi_r^{NN}}{\partial h} = \frac{1}{2} - \frac{(1-\theta)^2 v^2}{18h^2}$, and $\frac{\partial \pi_e^{NN}}{\partial h_r} = \frac{1}{2} - \frac{(1-\theta)^2 v^2}{18h^2}$. Corollaries 3, 5, 6, and 8 can be proven in the same way. End proof. $\square$

Proof of Corollary 2. Optimal selling price, demand, and optimal profit for offline and online retailers can be obtained by taking derivatives with respect to θ, respectively:
$$\frac{\partial p_r^{MN}}{\partial \theta} = -\frac{1}{3}\left(\frac{ck}{(k(1-\theta)+\theta)^2} - s - v\right), \quad \frac{\partial p_e^{MN}}{\partial \theta} = \frac{s+v-3h}{3\theta^2} - \frac{c(k^2(1-\theta)^2+2k(1-2\theta)\theta+3\theta^2)}{3\theta^2(k(1-\theta)+\theta)^2},$$
$$\frac{\partial D_r^{MN}}{\partial \theta} = \frac{ck-(k(1-\theta)+\theta)^2(s+v)}{6h(k(1-\theta)+\theta)^2}, \quad \frac{\partial D_e^{MN}}{\partial \theta} = \frac{(k(1-\theta)+\theta)^2(s+v)-ck}{6h(k(1-\theta)+\theta)^2} > 0, \quad \frac{\partial \pi_r^{MN}}{\partial \theta} =$$
$$\left(\frac{-ck+(k(1-\theta)+\theta)^2(s+v))(ck(1-\theta)-(k(1-\theta)+\theta)(3h+(1-\theta)(s+v)))}{9h(k(1-\theta)+\theta)^3}\right), \text{ and } \frac{\partial \pi_e^{MN}}{\partial \theta} = \frac{(k+(1-k)\theta^2)(s+v)-ck-3hk}{3\theta^2}.$$

The optimal selling price, demand, and optimal profit for the offline and online retailers can be obtained by taking derivatives with respect to h, respectively: $\frac{\partial p_r^{MN}}{\partial h} = 1 > 0$, $\frac{\partial p_e^{MN}}{\partial h} = \frac{1}{\theta} > 0$, $\frac{\partial D_r^{MN}}{\partial h} = \frac{(1-\theta)(ck-(k(1-\theta)+\theta)(s+v))}{6h^2(k(1-\theta)+\theta)}$, $\frac{\partial D_e^{MN}}{\partial h} = \frac{(1-\theta)(ck-(k(1-\theta)+\theta)(s+v))}{6h^2(k(1-\theta)+\theta)}$, $\frac{\partial \pi_r^{MN}}{\partial h} = \frac{(k(1-\theta)+\theta)(3h+(1-\theta)(s+v))-ck(1-\theta)(ck(1-\theta)+(k(1-\theta)+\theta)(3h-(1-\theta)(s+v)))}{18h^2(k(1-\theta)+\theta)^2}$, and $\frac{\partial \pi_e^{MN}}{\partial h} = \frac{k(1-\theta)+\theta}{\theta} > 0$. Corollaries 4 and 7 can be proven in the same way. End proof. $\square$

Proof of Proposition 5. According to Propositions 1 and 2, the difference in sales price between offline and online retailers in the MN and NN cases are $p_r^{MN} - p_r^{NN} = \frac{(1-\theta)(k(1-\theta)s-ck+\theta s)}{3(k(1-\theta)+\theta)}$ and $p_e^{MN} - p_e^{NN} = \frac{c(k(1-\theta)+3\theta)}{3\theta(k(1-\theta)+\theta)} + \frac{3h-(1-\theta)(s+v)}{3\theta} - \frac{3(c+h)-v(1-\theta)}{3}$, respectively, the difference in market share in the two cases are $D_r^{MN} - D_r^{NN} = -\frac{(1-\theta)(ck-s(k(1-\theta)+\theta))}{6h(k(1-\theta)+\theta)}$ and $D_e^{MN} - D_e^{NN} = \frac{(1-\theta)(ck-s(k(1-\theta)+\theta))}{6h(k(1-\theta)+\theta)}$, respectively, and the difference in optimal profits are $\pi_r^{MN} - \pi_r^{NN} = \frac{((k(1-\theta)+\theta)(3h+(1-\theta)(s+v))-ck(1-\theta))^2-(3h+v(1-\theta))^2(k(1-\theta)+\theta)^2}{18h(k(1-\theta)+\theta)^2}$ and $\pi_e^{MN} -$

$\pi_e^{NN} = \frac{ck(1-\theta)+(k(1-\theta)+\theta)(3h-(1-\theta)(s+v))}{3\theta} - \frac{(3h-(1-\theta)v)^2}{18h}$, respectively. Propositions 6–8 can be proven in the same way. $\square$

References

1. National Bureau of Statistical. *Annual Data of China in 2021*; National Bureau of Statistical: Beijing, China, 2021.
2. Akçay, Y.; Boyacı, T.; Zhang, D. Selling with money-back guarantees: The impact on prices, quantities, and retail proftability. *Prod. Oper. Manag.* **2013**, *22*, 777–791. [CrossRef]
3. Heydari, J.; Choi, T.M.; Radkhah, S. Pareto improving supply chain coordination under a money-back guarantee service program. *Serv. Sci.* **2017**, *9*, 91–105. [CrossRef]
4. Securities Daily. The Chaos behind the High Return Rate of "Double 11". 2021. Available online: https://baijiahao.baidu.com/s?id=1717297600588183567&wfr=spider&for=pc (accessed on 17 November 2022).
5. Li, Y.M.; Li, G.; Cheng, T.C.E. Return freight insurance: Implications for online platforms, third-party retailers and consumers. In Proceedings of the 8th International Conference on Logistics, Informatics and Service Sciences, Toronto, ON, Canada, 3–6 August 2018; IEEE: Piscataway, NJ, USA, 2018; pp. 1–6. [CrossRef]
6. Peking University Finance Law Research Venter. Birth, Advance and Development—Analysis of Taobao Return Freight Insurance. 2017. Available online: https://www.finlaw.pku.edu.cn/zxzx/zxwz/239619.htm (accessed on 17 November 2022).
7. The Return and Exchange Freight Insurance of JD, PICC Successfully Won the Bid to Take Advantage of the Opportunity. 2013. Available online: http://money.sohu.com/20131125/n390715093.shtml (accessed on 17 November 2022).
8. Gu, Z.J.; Tayi, G.K. Consumer mending and online retailer fit-uncertainty mitigating strategies. QME-Quant. *Mark. Econ.* **2015**, *13*, 251–282. [CrossRef]
9. Geng, S.D.; Li, W.L.; Chen, H. Complimentary return-freight insurance serves as quality signal or noise? In Proceedings of the 28th Australasian Conference on Information Systems, Sydney, Australia, 3–6 December 2017; pp. 1–10. Available online: https://aisel.aisnet.org/acis2017/47 (accessed on 17 November 2022).
10. Hess, J.D.; Mayhew, G.E. Modeling merchandise returns in direct marketing. *J. Direct Mark.* **1997**, *11*, 20–35. [CrossRef]
11. Ma, S.; Li, G.; Sethi, S.P.; Zhao, X. Advance selling in the presence of market power and risk-averse consumers. *Decision Sci.* **2019**, *50*, 142–169. [CrossRef]
12. Wood, S.L. Remote purchase environments: The influence of return policy leniency on two-stage decision processes. *J. Mark. Re.* **2001**, *38*, 157–169. [CrossRef]
13. Minnema, A.; Bijmolt, T.H.A.; Gensler, S.; Wiesel, T. To keep or not to keep: Effects of online customer reviews on product returns. *J. Retail.* **2016**, *92*, 253–267. [CrossRef]
14. Ambilkar, P.; Dohale, V.; Gunasekaran, A.; Bilolikar, V. Product returns management: A comprehensive review and future research agenda. *Int. J. Prod. Res.* **2022**, *60*, 3920–3944. [CrossRef]
15. Suwelack, T.; Hogreve, J.; Hoyer, W.D. Understanding money-back guarantees: Cognitive, affective, and behavioral outcomes. *J. Retail.* **2011**, *87*, 462–478. [CrossRef]
16. Griffis, S.E.; Rao, S.; Goldsby, T.J.; Niranjan, T.T. The customer consequences of returns in online retailing: An empirical analysis. *J. Oper. Manag.* **2012**, *30*, 282–294. [CrossRef]
17. Chen, J.; Chen, B.T. Competing with customer returns policies. *Int. J. Prod. Res.* **2016**, *54*, 2093–2107. [CrossRef]
18. Jeng, S.P. Increasing customer purchase intention through product return policies: The pivotal impacts of retailer brand familiarity and product categories. *J. Retail. Consum. Serv.* **2017**, *39*, 182–189. [CrossRef]
19. Oghazi, P.; Karlsson, S.; Hellström, D.; Hjor, K. Online purchase return policy leniency and purchase decision: Mediating role of consumer trust. *J. Retail. Consum. Serv.* **2018**, *41*, 190–200. [CrossRef]
20. Shulman, J.D.; Coughlan, A.T.; Savaskan, R.C. Optimal restocking fees and in-formation provision in an integrated demand-supply model of product returns. *Manuf. Serv. Oper. Manag.* **2009**, *11*, 577–594. [CrossRef]
21. Thieme, J. *Versandhandelsmanagement*; Gabler Verlag Springer Fachmedien Wiesbaden GmbH: Wiesbaden, Germany, 2003.
22. Hess, J.D.; Chu, W.; Gerstner, E. Controlling product returns in direct marketing. *Market. Lett.* **1996**, *7*, 307–317. [CrossRef]
23. Shulman, J.D.; Coughlan, A.T.; Savaskan, R.C. Managing consumer returns in a competitive environment. *Manage. Sci.* **2011**, *57*, 347–362. [CrossRef]
24. De, P.; Hu, Y.; Rahman, M.S. Product-oriented web technologies and product re-turns: An exploratory study. *Inf. Syst. Res.* **2013**, *24*, 998–1010. [CrossRef]
25. Ofek, E.; Katona, Z.; Sarvary, M. "Bricks and clicks": The impact of product re-turns on the strategies of multichannel retailers. *Mark. Sci.* **2011**, *30*, 42–60. [CrossRef]
26. Xia, Y.; Xiao, T.; Zhang, G.P. The impact of product returns and retailer's service investment on manufacturer's channel strategies. *Decision Sci.* **2017**, *48*, 918–955. [CrossRef]
27. Bell, D.R.; Gallino, S.; Moreno, A. Offline showrooms in omnichannel retail: De-mand and operational benefits. *Manag. Sci.* **2018**, *64*, 1629–1651. [CrossRef]

28. Li, G.; Zhang, T.; Tayi, G.K. Inroad into omni-channel retailing: Physical showroom deployment of an online retailer. *Eur. J. Oper. Res.* **2020**, *283*, 676–691. [CrossRef]
29. Lee, D.H. An alternative explanation of consumer product returns from the post-purchase dissonance and ecological marketing perspectives. *Psychol. Mark.* **2015**, *32*, 49–64. [CrossRef]
30. Walsh, G.; Albrecht, A.K.; Kunz, W.; Hofacker, C.F. Relationship between online retailers' reputation and product returns. *Brit. J. Manage.* **2016**, *27*, 3–20. [CrossRef]
31. Cai, X.Y.; Li, J.B.; Dai, B.; Zhou, T. Pricing strategies in a supply chain with multi-manufacturer and a common retailer under online reviews. *J. Syst. Sci. Syst. Eng.* **2018**, *27*, 435–457. [CrossRef]
32. Batarfi, R.; Jaber, M.Y.; Aljazzar, S.M. A profit maximization for a reverse logistics dual-channel supply chain with a return policy. *Comput. Ind. Eng.* **2017**, *106*, 58–82. [CrossRef]
33. Guo, L.; Lin, L.; Sethi, S.P.; Guan, X. Return strategy and pricing in a dual-channel supply chain. *Int. J. Prod. Econ.* **2019**, *215*, 153–164. [CrossRef]
34. Letizia, P.; Pourakbar, M.; Harrison, T. The impact of consumer returns on the multichannel sales strategies of manufacturers. *Prod. Oper. Manag.* **2018**, *27*, 323–349. [CrossRef]
35. Zhang, Q.; Chen, J.; Chen, B.T. Information strategy in a supply chain under asymmetric customer returns information. *Transport. Res. Part E Logist. Transport. Rev.* **2021**, *155*, 102511. [CrossRef]
36. Ertekin, N.; Agrawal, A. How does a return period policy change affect multichannel retailer profitability? *Manuf. Serv. Oper. Manag.* **2021**, *23*, 210–229. [CrossRef]
37. Davis, S.; Gerstner, E.; Hagerty, M. Money back guarantees in retailing: Matching products to consumer tastes. *J. Retailing* **1995**, *71*, 7–22. [CrossRef]
38. Guo, L. Service cancellation and competitive refund policy. *Market. Sci.* **2009**, *28*, 901–917. [CrossRef]
39. Chen, J.; Grewal, R. Competing in a supply chain via full-refund and no-refund customer returns policies. *Int. J. Prod. Econ.* **2013**, *146*, 246–258. [CrossRef]
40. McWilliams, B. Money-back guarantees: Helping the low-quality retailer. *Manag. Sci.* **2012**, *58*, 1521–1524. [CrossRef]
41. Nasiry, J.; Popescu, I. Advance selling when consumers regret. *Manag. Sci.* **2012**, *58*, 1160–1177. [CrossRef]
42. Chen, B.T.; Chen, J. Compete in price or service?—A study of personalized pricing and money back guarantees. *J. Retailing* **2017**, *93*, 154–171. [CrossRef]
43. Li, M.; Liu, Y.C. Beneficial product returns in supply chains. *Prod. Oper. Manag.* **2021**, *30*, 3849–3855. [CrossRef]
44. Mondal, C.; Giri, B.C. Analyzing strategies in a green e-commerce supply chain with return policy and exchange offer. *Comput. Ind. Eng.* **2022**, *171*, 108492. [CrossRef]
45. Wang, J.; He, S.L. Optimal decisions of modularity, prices and return policy in a dual-channel supply chain under mass customization. *Transport. Res. Part E Logist. Transport. Rev.* **2022**, *160*, 102675. [CrossRef]
46. De Roover, F.E. Early examples of marine insurance. *J. Econ. Hist.* **1945**, *5*, 172–200. [CrossRef]
47. John, A.H. The London Assurance company and the marine insurance market of the eighteenth century. *Economica* **1958**, *25*, 126–141. [CrossRef]
48. Geng, S.D.; Li, W.L.; Qu, X.F.; Chen, L.R. Design for the pricing strategy of return-freight insurance based on online product reviews. *Electron. Commer. Res. Appl.* **2017**, *25*, 16–28. [CrossRef]
49. Marotta, A.; Martinelli, F.; Nanni, S.; Orlando, A.; Yautsiukhinb, A. Cyber-insurance survey. *Comput. Sci. Rev.* **2017**, *24*, 35–61. [CrossRef]
50. Lin, J.X.; Zhang, J.L.; Cheng, T.C.E. Optimal pricing and return policy and the value of freight insurance for a retailer facing heterogeneous consumers with uncertain product values. *Int. J. Prod. Econ.* **2020**, *229*, 107767. [CrossRef]
51. Chen, Z.W.; Fan, Z.P.; Zhu, S.X. Extracting values from consumer returns: The role of return-freight insurance for competing e-sellers. *Eur. J. Oper. Res.* **2022**. [CrossRef]
52. Lin, J.L.; Choi, T.M.; Kuo, Y.H. Will providing return-freight-insurances do more good than harm to dual-channel e-commerce retailers? *Eur. J. Oper. Res.* **2022**. [CrossRef]
53. Fan, Z.P.; Chen, Z.W. When should the e-tailer offer complimentary return-freight insurance? *Int. J. Prod. Econ.* **2020**, *230*, 107890. [CrossRef]
54. Ren, M.L.; Liu, J.Q.; Feng, S.; Yang, A.F. Pricing and return strategy of online retailers based on return insurance. *J. Retail. Consum. Serv.* **2021**, *59*, 102350. [CrossRef]
55. Chen, Z.W.; Fan, Z.P.; Zhao, X. Offering return-freight insurance or not: Strategic analysis of an e-seller's decisions. *Omega* **2021**, *103*, 102447. [CrossRef]
56. Yang, G.Y.; Ji, G.J. The impact of cross-selling on managing consumer returns in omnichannel operations. *Omega* **2022**, *111*, 102665. [CrossRef]
57. Tsay, A.A.; Agrawal, N. Channel conflict and coordination in the e-commerce age. *Prod. Oper. Manag.* **2004**, *13*, 93–110. [CrossRef]
58. Zhang, P.; He, Y.; Zhao, X. "Preorder-online, pickup-in-store" strategy for a dual-channel retailer. *Transport. Res. Part E Logist. Transport. Rev.* **2019**, *122*, 27–47. [CrossRef]
59. Zhang, Z.; Song, H.M.; Shi, V.; Yang, S.L. Quality differentiation in a dual-channel supply chain. *Eur. J. Oper. Res.* **2021**, *290*, 1000–1013. [CrossRef]

60. Shulman, J.D.; Coughlan, A.T.; Savaskan, R.C. Optimal reverse channel structure for consumer product returns. *Mark. Sci.* **2010**, *29*, 1071–1085. [CrossRef]
61. Mandal, P.; Basu, P.; Saha, K. Forays into omnichannel: An online retailer's strategies for managing product returns. *Eur. J. Oper. Res.* **2021**, *292*, 633–651. [CrossRef]

 *processes*

Article

Coupling Coordination Analysis of Regional IEE System: A Data-Driven Multimodel Decision Approach

Yaliu Yang *, Fagang Hu, Ling Ding and Xue Wu

Business School, Suzhou University, Suzhou 234000, China
* Correspondence: yangyaliu@ahszu.edu.cn; Tel.: +86-18949944995

Abstract: Coordinating regional innovation–economy–ecology (IEE) systems is an important prerequisite for overall continuous regional development. To fully understand the coordination relationship among the three, this study builds a data-driven multimodel decision approach to calculate, assess, diagnose, and improve the regional IEE system. First, the assessment indicator system of the regional IEE system is established. Secondly, the range method, entropy weight method, and weighted summation method are employed to calculate the synthetic developmental level. Thirdly, a multimodel decision approach including the coupling degree model, the coordination degree model, and the obstacle degree model is constructed to assess the spatiotemporal evolution characteristics of the regional IEE system coupling coordination and diagnose the main obstacles hindering its development. Finally, the approach is tested using Anhui Province as a case study. The results show that the coupling coordination degree of the Anhui IEE system presents a stable growth trend, but the coupling degree is always higher than the coordination degree. The main obstacle affecting its development has changed from the original innovation subsystem to the current ecology subsystem. Based on this, some countermeasures are put forward. This study, therefore, offers decision support methods to aid in evaluating and improving the regional IEE system.

Keywords: regional IEE system; multimodel decision; coupling coordination; decision support methods

1. Introduction

Along with the fast development of the worldwide economy, new requirements have been put forward for economic growth [1,2], scientific and technological innovation [3,4], and ecological environment construction [5,6]. In this new era, all countries face the task of integrating the construction of ecological civilization and scientific and technological innovation with economic construction [7] and must move toward green and coordinated development [8]. Ecology links production and consumption, and the construction of ecological civilization is determined to a great extent by the economic structure and economic development mode. With continuous changes in cloud computing, artificial intelligence, big data, and other modern information technologies, regional innovation has become the main force promoting regional economic [9] and ecological civilization construction [10]. Therefore, local governments and in-depth academic studies are very concerned about the important issues of how to coordinate relationships among the three systems of regional innovation, regional economy, and regional ecology, and exploring the high-quality development model promoted by regional coordinated development. In particular, scholars have conducted several studies examining the interaction between two of the three systems: between regional innovation and regional economy [11], between regional innovation and regional ecology [12,13], and between regional economy and regional ecology [14].

Numerous studies report on the connection between regional innovation and regional economy, however, most scholars begin with the angle of technological innovation [15] to analyze how regional innovation impacts and promotes regional economic growth [16]. Scholars generally agree that innovation is the basic motivity and source of economic

Citation: Yang, Y.; Hu, F.; Ding, L.; Wu, X. Coupling Coordination Analysis of Regional IEE System: A Data-Driven Multimodel Decision Approach. *Processes* **2022**, *10*, 2268. https://doi.org/10.3390/pr10112268

Academic Editor: Bhavik Bakshi

Received: 20 September 2022
Accepted: 31 October 2022
Published: 3 November 2022

growth and development activities [17,18]. Antonioli et al. [19] emphasized that innovation and institutional change could contribute to a rapidly developing economy. Adak [20] explored this linkage between scientific and technological innovation and economic growth, and analyzed the path from technology introduction, to the number of patents, to economic development. Liang et al. [21] discussed the co-evolution principle between technological innovation and regional economic development and studied the co-evolution law and crucial influencing properties between them through the use of the dynamical coupling model and geographical detector method. Cheng and Wang [22] explored the multiplier effect of science and technology innovation on community economic development through empirical data analysis from seaside cities. Chen and Zhang [23] applied this coupling coordination degree model and geographical detector for conducting the multi-scale evaluation of the coupling coordinative degree between innovation and economic development, taking resource-based cities as an example. This model was also used by Tian et al. [24] to reveal the synergistic condition between green innovation efficiency and economic development. Pradhan et al. [25] discussed the relationship between innovation and economic growth in Euro area countries, while Qamruzzaman and Jianguo [26] pointed out that there is a long-term cointegration relationship between financial innovation and economic growth in South Asia. Furthermore, Pala [27] selected 25 developing countries as research objects to explore the impact of innovation on economic growth.

Research has produced rich findings on the interaction between the regional economy and regional ecology, and several mature theories and models have been formed and widely used, such as the coordinated development theory [28], correlation analysis between economy and environment [29,30], and economic energy environment impact model [31]. Peng et al. [32] analyzed the external and inherent connections between the ecological environment and economic development by means of the coupled coordination theory. Fan et al. [33,34] investigated the coupling state between the urban social economy and the ecological environment. Liu et al. [35] explored the coupling mechanism and space-time coordination between the development of the economy and ecological environment along the Yellow River Basin. Taking the Beijing-Tianjin-Hebei region as the research target, Liao et al. [36] also examined the coupling and coordination level between them. Shi et al. [37] also analyzed the coupling coordination and spatial-temporal heterogeneity between them by applying a geographically and temporally weighted regression model. Finally, Adami and Schiavon [38] elaborated on the nature of the interrelation between economy and ecology.

The research stream on the interplay relationship between regional innovation and regional ecology is growing. Liu et al. [39] analyzed the mechanism of new knowledge growth and explored theoretically the effects of technological innovation on ecological environment quality improvement. After the environmental Kuznets curve hypothesis [40,41] was proposed, several topics emerged, such as the relationships between economic growth and suspended solids [42], between economic growth and sulfur dioxide emissions [43], between economic growth and carbon dioxide emissions, [44] and between economic growth and air pollution [45]. Xia and Bing [46] explored the impact mechanism of environmental optimization on the improvement of regional innovation performance and noted that the former can promote the latter. Ke et al. [47] took advantage of a spatial measurement technique to present the spatial effect of innovation efficiency on ecological footprints. Moreover, Meirun et al. [48] explored green technology innovation's dynamic impact on Singapore's economic growth and carbon dioxide emissions. Brancalion [49] emphasized the importance of innovation in regional ecological restoration. Furthermore, Murphy and Gouldson [50] pointed out that ecological and economic benefits can be achieved through innovation.

With the continuous expansion of research, the coordinated research among the three is expanding, such as innovation-economy-education [51,52], economy-ecology-health [53,54], and ecology-innovation-climate change [55]. For instance, Wang and Tan [56] explored the coupling conditions among finance development, scientific and technical innovation, and economic growth. Zhang [57] explored the coupled and coordinated state of the regional economy, tourism, and ecological environment. Liu et al. [58,59] explored the space-coupled and coordinated degree of the energy-economy-environment system. Similarly, Yan et al. [60] described the coupled and coordinated state among the three by using Australian data from 2007 to 2016. Taking the Huai-Hai economic region as an example, Zheng et al. [61] examined the coupling relationship and its heterogeneity among technological innovation, industrial transformation, and environmental efficiency based on models such as the super efficiency relaxation measurement model and grey correlation analysis. Furthermore, Xu et al. [62] worked to identify methods for achieving regional coordinated development using aspects of technological innovation, industrial upgrading, the ecological environment, and their interactions. Huang et al. [63] explored the coupling research on technological innovation, talent accumulation, and the ecological environment. Yin et al. [64] used the coupling coordination model to determine the coordination degree among environmental regulations, scientific and technical innovation, and green development. Wu [65] measured and evaluated the regional technology-economy-ecology coordination relationship. Essentially, it can be seen that the interrelationship and interaction among regional innovation, economy, and ecology are viewed with consensus. However, relevant studies on the analysis of the coupling and coordination of the three factors are limited.

The above research findings indicate that scholars have mostly focused on researching the pairwise relationships among regional economy, regional innovation, and regional ecology. This research stream is relatively mature, which provides a certain foundation for studying the relationships among the three; however, room still exists for further in-depth research. First, few studies place the regional innovation, regional economy, and regional ecology systems into the same framework to quantitatively assess coupling and coordination among them. Second, the existing literature provides little discussion on the spatial and temporal evolution trend for coupling coordination development among the three systems; thus, it is necessary to further analyze its spatial and temporal evolution characteristics and heterogeneity. Third, proposed policy recommendations have been relatively broad, making it necessary to further identify obstacles blocking the coordinated development of the three systems, and subsequently, formulate targeted countermeasures and suggestions. Challenged by the above, this study builds a data-driven multimodel decision approach for calculating, assessing, diagnosing, and improving the regional innovation–economy–ecology (IEE) system to provide a scientific foundation and reference for harmony and the integrated planning of sustainable regional development.

This study is of great theoretical and applied significance. On its theoretical significance, first, the assessment indicator system of the regional IEE system coupling coordination development is established, providing researchers with a new study vision of the good coupling and orderly coordination of three subsystems. Second, a data-driven multimodel decision approach for regional IEE coupling coordination development is established, and the spatial and temporal evolution laws and factors affecting its development are identified, thereby enriching the assessment method system. Third, this study reveals the trend of coupling and coordination and the function mechanism of the regional IEE system, expanding the framework of coupling coordination theory. However, the practical significance must also be considered. First, this study takes Anhui Province as a case study. By evaluating the development level and coupling coordination degree of Anhui's IEE system, it explores the coupling and coordination development state, identifies its key obstacle factors, and comprehensively reveals the internal mechanism of Anhui's IEE system development. Second, it analyzes the spatio-temporal evolution characteristics of the coupling and coordinated development of Anhui's IEE system, explores the internal mechanism of its coupling and coordinated development, and then proposes targeted strategies to provide decision-making guidance for relevant government departments and enterprises. Third, it actively implements the sustainable development goals of the

United Nations, which provides an important reference for the realization of regional innovation-economic-ecological sustainable development goals.

The basic structure of the thesis is as follows: Part 2 elaborates on the research method. Part 3 displays a case study, in which Anhui Province is selected as a case to verify the method's effectiveness and feasibility, and then provides effective countermeasures. Lastly, Part 4 presents the conclusion and sums up the essay.

2. Method

This part of the paper mainly discusses our research methods, involving the calculation, assessment, and diagnosis methods of the coupling coordination development of the regional IEE system. Its contents include method flow, data collection, data processing, and data modeling.

2.1. Method Flow

Promoting sustainable regional development necessitates placing the three systems—regional innovation, regional economy, and regional ecology—into the same framework, constructing the calculation framework and assessment system for the three subsystems' coupled coordinated development from the systems theory perspective, and conducting a systematic, integrated, and collaborative reform from the global optimization perspective [66,67]. This is urgently needed for high-quality coordinated regional development. However, given the complicacy and diversification of the three systems' data indicators, research has faced challenges regarding how to apply efficacious methods to calculate, assess, diagnose, and improve the three systems' coupling coordination development level.

To address this challenge and improve regional coordination development levels, this study constructs the data-driven multimodel decision approach [68,69] for the calculation, assessment, diagnosis, and promotion of regional IEE system coupling coordination development. In the data collection stage, we used surveys and research methods to define the data range. The collected data primarily includes statistical data on regional innovation, the regional economy, and the regional ecology. Empirical analysis is used in data processing and data modeling. Data processing mainly uses the range method-entropy weight method-weighted summation method to obtain the data standardization value, indicator weight, and synthetic developmental level score. Data modeling is used to build the coupling degree model and the coordination degree model for calculating the regional IEE system. Then, the coupling degree and coordination degree are identified as well as the key obstacles blocking the system's development using the obstacle degree model. The innovative practice applies the integrated method to the case of Anhui Province, verifies the integrated method's effectiveness, obtains the assessment results, and proposes targeted policy recommendations. The method flow chart of the study is illustrated in Figure 1.

2.2. Data Collection

Data collection is the basis of scientific data analysis. The accuracy of collected data is directly related to the value of the data analysis results. This study uses survey research methods to extract the desired data from the huge data. First of all, the source of the data should be clarified. The data of this study are from public publications and second-hand data collection. The main advantages of this data source are convenience, cost-saving, and accurate data. Therefore, this study's sample data are mainly observed from government statistical data resources, such as the "Statistical Yearbook", every year. Secondly, the scope of data collection must be defined. Based on the element connotation of the regional IEE system and the scientific and quantifiable principle, this study identifies the main indicators that can reflect the three subsystems, constructs an indicator system for the coordinated development of the regional IEE system, and determines the sample data capacity. Finally, according to the scope of the data collection, the data are collected through the data collection channels of the statistical yearbook, and the collected data are counted and summarized in an EXCEL table to lay a data foundation for the next stage of data processing.

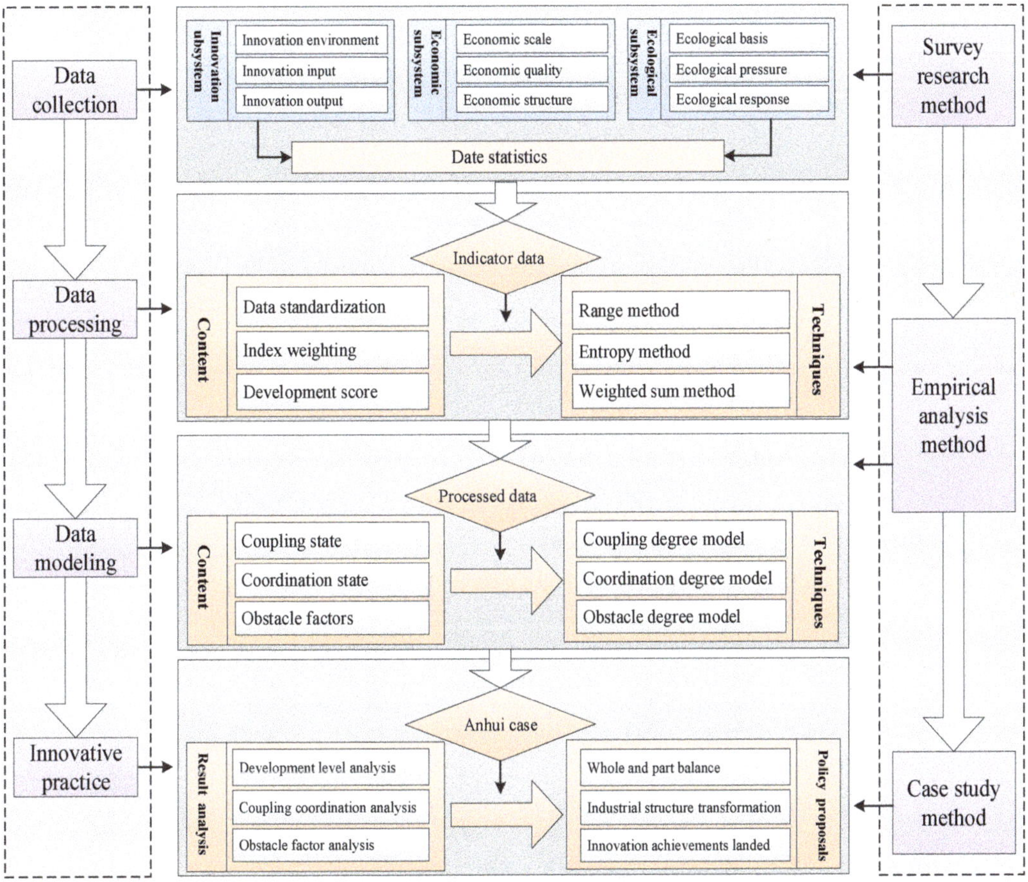

Figure 1. Method flow chart.

2.3. Data Processing

2.3.1. Range Method

This composite regional IEE indicator system includes three subsystems, with several criterion layers under the subsystem and several assessment indicators under the criterion layer because different indicators have different measurement units, which present non-comparability. Therefore, this study applies the range method for standardizing each indicator value to eliminate the measurement unit problem. The standardized calculation formula is:

$$X'_{ij} = \begin{cases} \frac{x_{ij} - \beta_j}{\alpha_j - \beta_j} \times 0.99 + 0.01 & x_j \text{ is a positive indicator} \\ \frac{\alpha_j - x_{ij}}{\alpha_j - \beta_j} \times 0.99 + 0.01 & x_j \text{ is a negative indicator} \end{cases} \tag{1}$$

where X_{ij} is the sample value, x'_{ij} is the standardized date value, and α_j and β_j are the highest and the lowest values of indicator j, respectively. The value range of x'_{ij} is between {0.01, 1}.

2.3.2. Entropy Weight Method

When calculating the comprehensive assessment score of each subsystem, it is necessary to weigh each indicator in the system. In view of the widespread use of the objective weight method-entropy weight method [70,71], we use this method to measure each indicator weight in each subsystem.

First, indicator weight R_{ij} is calculated using standardized data:

$$R_{ij} = \frac{X'_{ij}}{\sum_{i=1}^{t} X'_{ij}} \quad i = 1,2,3,\ldots t; j = 1,2,3,\ldots m \tag{2}$$

Let $k = \frac{1}{\ln(t)} > 0$, as the adjustment coefficient, and calculate indicator information entropy e_j:

$$e_j = -k \sum_{i=1}^{t} R_{ij} \ln R_{ij} \quad i = 1,2,3,\ldots t; j = 1,2,3,\ldots m \tag{3}$$

Determine indicator weight W_j:

$$W_j = \frac{1 - e_j}{m - \sum_{j=1}^{m} e_j} \quad \sum_{j=1}^{n} W_j = 1, j = 1,2,3,\ldots m \tag{4}$$

where W_j represents the weight coefficient of indicator j, and $0 < W_j < 1$, n represents the number of selected indicators. Using the weights of the indicators, we derive comprehensive assessment scores for the regional innovation (U_A), regional economy (U_B), and regional ecology (U_C) subsystems.

2.3.3. Weighted Summation Method

The weighted summation method is adopted in this study to determine the synthetic developmental level score of the regional innovation subsystem (U_A), the synthetic developmental level score of the regional economic subsystem (U_B), and the synthetic developmental level score of regional ecological subsystem (U_C). The specific formula is as follows:

$$U_A \text{ or } U_B \text{ or } U_C = \sum_{j=1}^{n} W_j \times X'_{ij} \tag{5}$$

where U_A, U_B, and U_C are the comprehensive assessment score of the regional innovation subsystem, regional economy subsystem, and regional ecology subsystem. W_j is the indicator weight of the response system and X'_{ij} is the standardized indicator value.

2.4. Data Modeling

The coupling coordination degree model [72] is a widespread application that incomprehensively calculates the degree of interaction, influence, and promotion between the interior factors of a complex system in a specific range. Coupling represents the interaction relationship [73], coordination is the constraint condition for the development process [74], and development is the ultimate goal. This model is often used to calculate and assess two systems [75,76], three systems [77,78], and multi-systems [71]. In this study, a multimodel decision approach of the regional IEE system is established, which includes the coupling degree model, coordination degree model, and obstacle degree model.

2.4.1. Coupling Degree Model

The coupling degree model can be applied to represent the influence degree of interaction of two or more systems. Referring to the research of relevant scholars [75,79], the three subsystems coupling degree model for the regional IEE system is derived as follows:

$$C = \frac{3 \cdot (U_A \cdot U_B \cdot U_C)^{\frac{1}{3}}}{U_A + U_B + U_C} \tag{6}$$

where C is the coupling degree score, with a value range from $\{0,1\}$. The higher the C value is, the stronger the interaction among the three subsystems and the better the coupling condition. Drawing on the views of relevant scholars [65,73], we divide the coupling degree into four stages. When $C \in [0, 0.3]$, it is in a stage of low-level coupling. When $C \in (0.3, 0.5]$, it is in the confrontation stage. When $C \in (0.5, 0.8]$, it is in the running stage. When $C \in (0.8, 0.1]$, it is in the high-level coupling stage.

2.4.2. Coordination Degree Model

Based on the result of the former model, we calculate the coordination degree of three systems and two systems. The coordination degree can clearly display the function degree size of interaction and influence among systems and can determine whether the systems develop harmoniously. This study represents the formula derivation of the coordination degree model.

$$D = \sqrt{C \times Y} \tag{7}$$

Among them, the coordination coefficients of two systems:

$$Y = \alpha U_A + \beta U_B \text{ or } \alpha U_A + \beta U_C \text{ or } \alpha U_B + \beta U_C \tag{8}$$

Three system coordination coefficients:

$$Y = a U_A + b U_B + c U_C \tag{9}$$

where Y is the synthetic developmental level score, a, b, c are the undetermined weights of the synthetic developmental level scores of the three systems, and α, β is the undetermined weight of the synthetic developmental level score of the two systems. Regional innovation, regional economy, and regional ecology are equally important. With reference to the practices of relevant scholars [32,80], the undetermined weight of the synthetic developmental level score of the three systems is $a = b = c = 1/3$, and the undetermined weight of the synthetic developmental level score of two systems is $\alpha = \beta = 1/2$. D is the coordination degree, which is in the range from 0 to 1. The classification of coordination degree D refers to the classification basis of relevant scholars [72,81], and the classification level is illustrated in Figure 2.

2.4.3. Obstacle Degree Model

This study uses the obstacle degree model [82,83] to diagnose and distinguish the major factors blocking the coupling coordinated development of the regional IEE system, which is helpful for the government when adopting targeted policies and measures to promote coordinated development. This is calculated as follows:

$$M_j = \left(1 - X'_{ij}\right) \times \left(P_i \times W_{ij}\right) \times 100 / \sum \left(1 - X'_{ij}\right) \times \left(P_i \times W_{ij}\right) \tag{10}$$

$$T_i = \sum M_{ij} \tag{11}$$

where X'_{ij} represents the standardized value of a single indicator x_{ij}, M_{ij} is the obstacle degree of the indicator j, and T_i represents the obstacle degree score of the subsystem i. P_i is the weight of subsystem i and W_{ij} is the weight of indicator j in system i.

To sum up, this study builds a coupling degree model, coordination degree model, obstacle degree model, and other technical means to measure the coupling, coordination, and obstacle degrees of the regional IEE system. This reveals the space-time evolution law of the coupling and coordinated development of the regional IEE system, clarifies the main factors affecting its development, and lays a solid foundation for promoting the sustainable development of the regional IEE system.

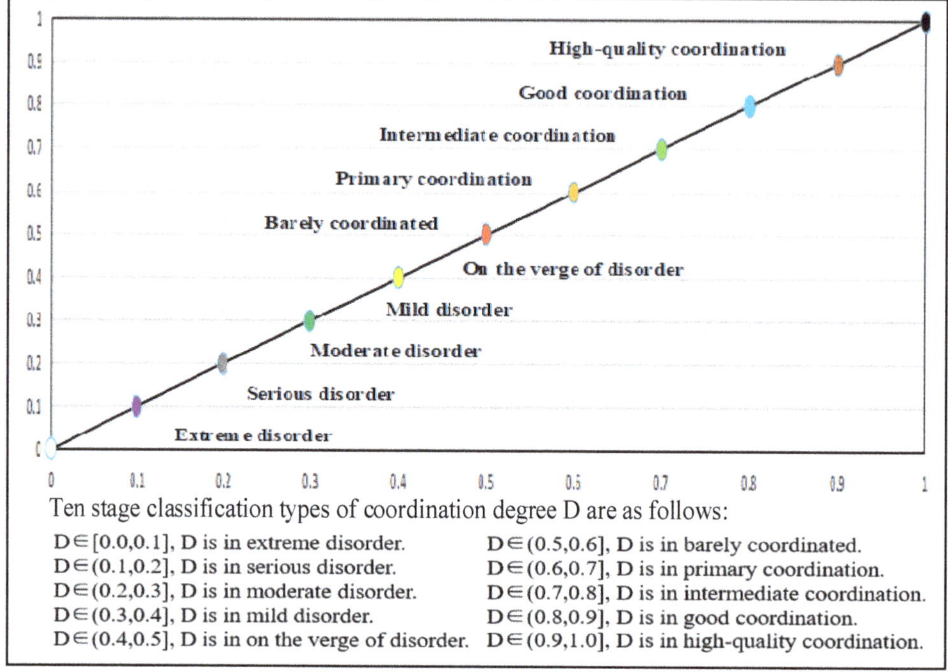

Ten stage classification types of coordination degree D are as follows:

$D \in [0.0, 0.1]$, D is in extreme disorder. $D \in (0.5, 0.6]$, D is in barely coordinated.

$D \in (0.1, 0.2]$, D is in serious disorder. $D \in (0.6, 0.7]$, D is in primary coordination.

$D \in (0.2, 0.3]$, D is in moderate disorder. $D \in (0.7, 0.8]$, D is in intermediate coordination.

$D \in (0.3, 0.4]$, D is in mild disorder. $D \in (0.8, 0.9]$, D is in good coordination.

$D \in (0.4, 0.5]$, D is in on the verge of disorder. $D \in (0.9, 1.0]$, D is in high-quality coordination.

Figure 2. Classification types of the coordination degree.

3. Case Study

We take Anhui Province as the object of study for validating the data-driven multi-model decision approach proposed and verifying the effectiveness of the approach. In 2014, Anhui's GDP broke through the 2 trillion mark for the first time, ushering in new breakthroughs in economic development. At the same time, 2014 was also the first year for Anhui to carry out the pilot work of building an innovative province. In addition, in 2014, Anhui Province carried out the construction of provincial ecological civilization pilot demonstration areas for the first time. Since then, Anhui's scientific and technological innovation, economic development, and ecological environment protection have ushered in a new stage, with accelerated progress and remarkable achievements. Therefore, 2014 is selected as the starting year of this study. At the same time, since the 2021 statistical yearbook has not been published yet, in view of the availability of data collection, this study chooses 2020 as the end year of the study. To sum up, the sample data used in this study are mainly from the statistical data resources of Anhui Province from 2014 to 2020, which is a good model and reference for regional research. In this section, we include the study area, results analysis, and discussion and managerial implications.

3.1. Study Area

Anhui Province is located in East China, between 114°54′–119°37′ E and 29°41′–34°38′ N, as shown in Figure 3. It is approximately 450 km wide from east to west and 570 km long from north to south. The land area is 139,400 km², which occupies 1.45% of the national territory. Anhui is an ecological resort with good mountains and water. It is the only province covered by the two national strategies of the Yangtze River Delta Yangtze Regional Integration and the Rise of the Central Region. Anhui is experiencing "three historical changes", from everything waiting to be done to a flourishing economy, a traditional agricultural province to a new industrial province, and innovation catching up to innovation leading. At present, how to deeply integrate regional innovation, economic development, and ecological construction, as well as actively explore innovation guidance, ecological priority, and high-quality economic development, have become important topics. Therefore, there is a need to calculate, assess, diagnose, and promote the coupling coordinated development of the Anhui regional IEE system.

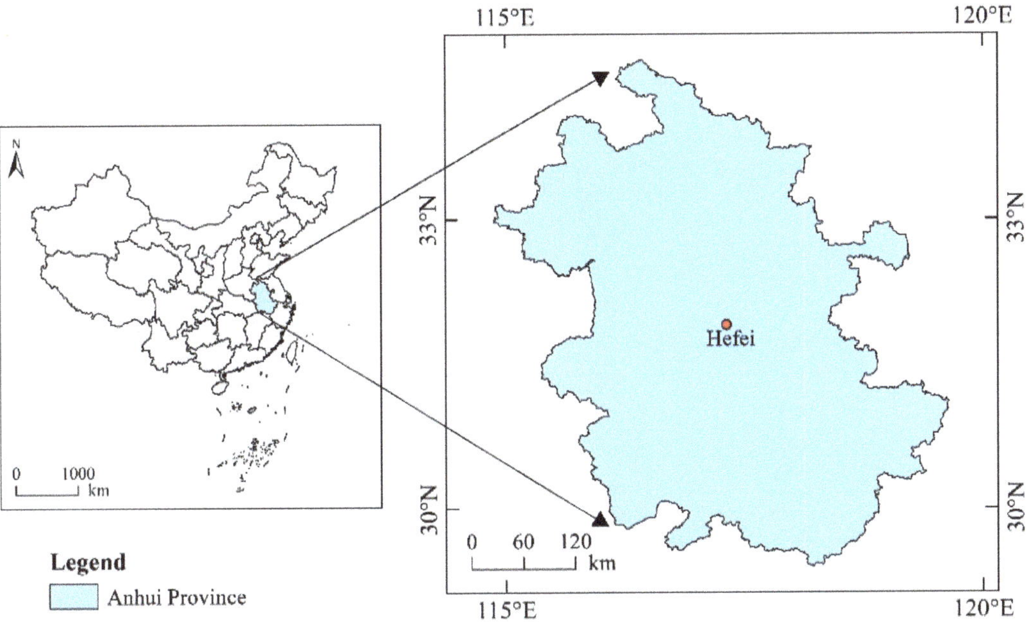

Figure 3. Administrative area map of the study area.

3.2. Results Analysis

3.2.1. Data Collection Results

Regional innovation, regional economy, and regional ecology are complex systems that affect and promote each other. The regional economy provides material demand guarantees for regional innovation and is restricted by the ecological environment and regional innovation level. Regional innovation is the first driving force leading development [84], as well as key to improving economic development quality [85] and promoting the transformation and upgrade of economic structures [86]. It also promotes ecological civilization construction by improving energy utilization efficiency and reducing energy consumption and three wastes discharge [87]. Further, regional eco-civilization construction is a strong base for increasing regional economic sustainability [6,10,88]. Therefore, the three subsystems—regional innovation, economy, and ecology—interact with, promote and support each other, and share a relationship of coupling coordination development.

Constructing the indicator system is a precondition and an important guarantee for calculation and assessment. This study follows the principles of scientific, hierarchical, and systematic indicator system construction and comprehensively considers the availability and persuasiveness of the indicator data. The indicator system constructed in this study includes 3 subsystems (system layer), 9 major criteria (criteria layer), and 35 basic indicators (indicator layer). The three subsystems include the innovation subsystem, regional economic subsystem, and ecological subsystem, and there is a coupling and coordination relationship among them [62,65].

The regional innovation subsystem refers to the regional innovation environment and innovation input and output [89,90]. The innovation environment refers to various software and hardware environments that affect the innovation of the innovation subject in the innovation process, such as a good policy environment, social environment, and cultural environment [91]. Compared with the existing research [92,93], this study increased the number of legal entities in cultural and related industries above the designated size (A11) [94] and the number of public library institutions (A12) [95] in the innovation environment criteria layer. Innovation input refers to the resources consumed for scientific research and technological innovation, mainly embodied in human and capital input [96]. Therefore, this paper selects the internal expenditure of R&D funds (A21), the number of college students per 10,000 population (A22), the full-time equivalent of R&D personnel (A23), and R&D personnel input/person (A24) to evaluate innovation input. Innovation output directly reflects the operation result of innovation, which is manifested as knowledge output and economic output [97]. Knowledge output is mainly manifested in scientific and technological achievements and patents. The economic output reflects the economic value created by innovation. This paper selects the technology market turnover (A31), number of registered scientific and technological achievements at or above the provincial or ministerial level (A32), invention patent authorization (A33), and patent application authorization (A34) to evaluate the innovation output.

For the regional economic subsystem, this study combines the current focus of regional economic development, referring to previous research [35], and expands the connotation of the regional economic subsystem from three aspects: economic scale, economic quality, and economic structure [98]. Economic scale is the material basis for economic development. This paper selects the GDP (B11), total retail sales of consumer goods (B12), total import and export trade (B13), and total investment in fixed assets (B14) to reflect the economic scale. Economic quality is reflected in economic benefits, social benefits, and ecological benefits. Compared with the existing research [99], per capita GDP (B21) and urban per capita disposable income (B23) have been added to relevant indicators of economic quality in this study. Economic structure refers to the composition and structure of the national economy, involving industrial structure, labor structure, and consumption structure [100]. This paper selects the proportion of employees in the tertiary industry (B31), the proportion of the added value of the tertiary industry in GDP (B32), and the urbanization rate (B33), to reflect the economic structure.

Drawing on the "China Sustainable Development Strategy Report" issued by the Chinese Academy of Sciences and relevant previous studies [59,101], the regional ecological subsystem is mainly assessed from three dimensions: the ecological basis, ecological pressure, and ecological response [102]. The ecological basis represents the ecological environment state and environmental change in a specific time period. This paper selects the percentage of forest cover (C11), urban per capita park green area (C12), area of nature reserve (C13), and per capita water resources (C14) to reflect the ecological foundation. Ecological pressure represents the load of human economic and social activities on the environment [103]. Compared with existing studies [104,105], this study adds industrial wastewater emissions (C21), industrial sulfur dioxide emissions (C22), and industrial solid waste output (C23) to the relevant indicators of ecological pressure. Ecological response refers to the actions and measures taken by human society to maintain the stability of the ecological environment. In this paper, the green coverage rate of urban built-up areas (C31), centralized sewage treatment rate (C32), the comprehensive utilization rate of general industrial solid waste (C33), and the harmless treatment rate of urban domestic garbage (C34) [106] are selected to evaluate the ecological response.

Finally, this study establishes a comprehensive assessment indicator system for the coupling coordination development of the regional IEE system, as described in Table 1. "Direction" refers to whether an indicator is a positive indicator or a negative indicator in Table 1. Positive indicators and negative indicators have different meanings. Positive indicators are represented by "+", and the larger the value, the better. Negative indicators are represented by "−", and the smaller the value, the better.

Table 1. Indicator system for the coordination development of the regional IEE system.

System Layer	Criteria Layer	Indicator Layer	Direction
A. Innovation subsystem	A1 Innovation environment	A11 Number of legal entities in cultural and related industries above designated size/unit	+
		A12 Number of public library institutions/unit	+
		A13 Public library collections per unit population/piece	+
		A14 Local financial expenditure on education/10^9 RMB	+
	A2 Innovation input	A21 Internal expenditure of R&D funds/million RMB	+
		A22 Number of college students per 10,000 population/person	+
		A23 Full-time equivalent of R&D personnel/person year	+
		A24 R&D personnel input/person	+
	A3 Innovation output	A31 Technology market turnover/million RMB	+
		A32 Number of registered scientific and technological achievements at or above the provincial or ministerial level/piece	+
		A33 Invention patent authorization/piece	+
		A34 Patent application authorization/piece	+
B. Economic subsystem	B1 Economic scale	B11 GDP/10^9 RMB	+
		B12 Total retail sales of consumer goods/10^9 RMB	+
		B13 Total import and export trade/million USD	+
		B14 Total investment in fixed assets/million RMB	+
	B2 Economic quality	B21 Per capita GDP/RMB	+
		B22 GDP energy intensity/tons of standard coal/million RMB	−
		B23 Urban per capita disposable income/RMB	+
		B24 Whole-society productivity/RMB/person	+
	B3 Economic structure	B31 Proportion of employees in the tertiary industry/%	+
		B32 Proportion of added value of the tertiary industry in GDP/%	+
		B33 Urbanization rate/%	+
C. Ecological subsystem	C1 Ecological basis	C11 Percentage of forest cover/%	+
		C12 Urban per capita park green area/m^2	+
		C13 Area of nature reserve/million hectares	+
		C14 Per capita water resources/m^3/person	+
	C2 Ecological pressure	C21 Industrial wastewater emissions/million tons	−
		C22 Industrial sulfur dioxide emissions/million tons	−
		C23 Industrial solid waste output/million tons	−
		C24 Application amount of agricultural chemical fertilizer (converted amount)/million tons	−
	C3 Ecological response	C31 Green coverage rate of urban built-up area/%	+
		C32 Centralized sewage treatment rate/%	+
		C33 Comprehensive utilization rate of general industrial solid waste/%	+
		C34 Harmless treatment rate of urban domestic garbage/%	+

The final assessment indicator system includes 3 subsystems (system layer), 9 major criteria (criteria layer), and 35 basic indicators (indicator layer). On this basis, we collect corresponding sample indicator data. All the indicator data used in this paper are from Anhui Statistical Yearbook (2014–2020) [107], China Statistical Yearbook (2014–2020), China Environmental Statistical Yearbook (2014–2020), and China Industrial Statistical Yearbook (2014–2020) [108].

3.2.2. Indicator Weight Analysis

In this study, the entropy weight method is used to determine the weight of each indicator. Before the entropy weight method [70] is adopted, we need to standardize the collected data to obtain the standardized value. The range method [71] is used in the standardization process, as shown in Formula (1). After using Formula (1) to obtain the standardized value, we use the entry weight method as shown in Formulas (2) and (3) to find the weight of each specific indicator in the subsystem. The weight of the criteria layer is obtained from the sum of its subordinate specific indicator weight. The weight determination process of each subsystem is the same. All calculation processes of obtained weight data resources were implemented in an EXCEL table. Finally, the weight of the Anhui IEE system coupling and coordinated development indicator system is shown in Table 2.

Table 2. The indicator system weight for the coordinated development of the regional IEE system.

System Layer	Criteria Layer	Weight	Indicator Layer	Weight
A. Innovation subsystem	A1 Innovation environment	0.194	A11 Number of legal entities in cultural and related industries above designated size/unit	0.037
			A12 Number of public library institutions/unit	0.035
			A13 Public library collections per unit population/piece	0.069
			A14 Local financial expenditure on education/10^9 RMB	0.053
	A2 Innovation input	0.444	A21 Internal expenditure of R&D funds/million RMB	0.076
			A22 Number of college students per 10,000 population/person	0.178
			A23 Full-time equivalent of R&D personnel/person year	0.103
			A24 R&D personnel input/person	0.088
	A3 Innovation output	0.362	A31 Technology market turnover/million RMB	0.105
			A32 Number of registered scientific and technological achievements at or above the provincial or ministerial level/piece	0.154
			A33 Invention patent authorization/piece	0.039
			A34 Patent application authorization/piece	0.064
B. Economic subsystem	B1 Economic scale	0.357	B11 GDP/10^9 RMB	0.091
			B12 Total retail sales of consumer goods/10^9 RMB	0.105
			B13 Total import and export trade/million USD	0.098
			B14 Total investment in fixed assets/million RMB	0.063
	B2 Economic quality	0.402	B21 Per capita GDP/RMB	0.091
			B22 GDP energy intensity/tons of standard coal/million RMB	0.111
			B23 Urban per capita disposable income/RMB	0.077
			B24 Whole-society productivity/RMB/person	0.123
	B3 Economic structure	0.242	B31 Proportion of employees in the tertiary industry/%	0.115
			B32 Proportion of added value of the tertiary industry in GDP/%	0.054
			B33 Urbanization rate/%	0.073

Table 2. *Cont.*

System Layer	Criteria Layer	Weight	Indicator Layer	Weight
C. Ecological subsystem	C1 Ecological basis	0.409	C11 Percentage of forest cover/%	0.000
			C12 Urban per capita park green area/m^2	0.084
			C13 Area of nature reserve/million hectares	0.250
			C14 Per capita water resources/m^3/person	0.075
	C2 Ecological pressure	0.345	C21 Industrial wastewater emissions/million tons	0.086
			C22 Industrial sulfur dioxide emissions/million tons	0.086
			C23 Industrial solid waste output/million tons	0.067
			C24 Application amount of agricultural chemical fertilizer (converted amount)/million tons	0.106
	C3 Ecological response	0.246	C31 Green coverage rate of urban built-up area/%	0.108
			C32 Centralized sewage treatment rate/%	0.070
			C33 Comprehensive utilization rate of general industrial solid waste/%	0.068
			C34 Harmless treatment rate of urban domestic garbage/%	0.000

As Table 2 illustrates, among the innovation subsystem's three criteria layers, the proportion of innovation input is the highest, reaching 0.444. The proportion of innovation environment is the lowest, with a weight value of 0.194. The proportion of innovation output is in the middle, with a weight value of 0.362. The number of college students per 10,000 population/person (A22) is the index with the largest weight in the innovation subsystem, and its weight value reaches 0.178. The scale of higher education is closely related to regional innovation. Then, in the economic subsystems' three criteria layers, the proportion of economic quality is the highest, reaching 0.402. The proportion of economic structure is the lowest, with a weight value of 0.242. The proportion of the economic scale is in the middle, with a weight value of 0.357. Whole-society productivity (B24) is the indicator with the largest weight in the economic subsystem, and its weight value reaches 0.123. It is evident that labor productivity is an important indicator to evaluate the level of regional economic development. Finally, in the three criteria layers of the ecological subsystem, the proportion of ecological foundation is the highest, reaching 0.409. The proportion of ecological response is the lowest, with a weight value of 0.246. The proportion of ecological pressure is in the middle, with a weight value of 0.345. The area of nature reserve/filling ecosystems (C13) has the largest weight in the ecosystem, and the area of nature reserve/filling ecosystems mainly reflects the stability and quality of regional ecosystems.

3.2.3. Synthetic Developmental Level Analysis

The development level and synthetic developmental level of each subsystem in Anhui Province from 2014 to 2020 are calculated using Equations (1)–(5), as illustrated in Figure 4.

As Figure 4 illustrates, the development level of the innovation subsystem follows a rapid upward trend, and the innovation environment is steadily optimized. Innovation input and output have been increasing since 2018 and 2017, respectively. This is highly consistent with the innovative development practices of Anhui Province. Anhui's regional innovation capacity has ranked first nationwide for ten consecutive years, and it is the province with the largest increase in the ranking of East China. The development level of the economic subsystem presents a strong uptrend, and the development level of the economic scale and economic quality criteria are the major drivers supporting the development of the economic subsystem. The development level of the ecological subsystem shows a rapid upward trend, with a good ecological foundation and strong ecological elasticity. This is consistent with Anhui's efforts to build an "ecological barrier" in the Yangtze River Delta and create a "green economy" growth point. On the whole, the synthetic developmental level of the innovation economy ecology shows a steady upward trend during the study period, resulting from the joint action of the three subsystems. However, the contribution degree of each subsystem is different. The contribution degree order is as follows: ecological subsystem > economic subsystem > innovation subsystem. It is thus clear that the main driving force of Anhui's synthetic developmental level has changed throughout the study period. Regarding the relationships among the three subsystems, the development level gradually trends toward coordination during the study period.

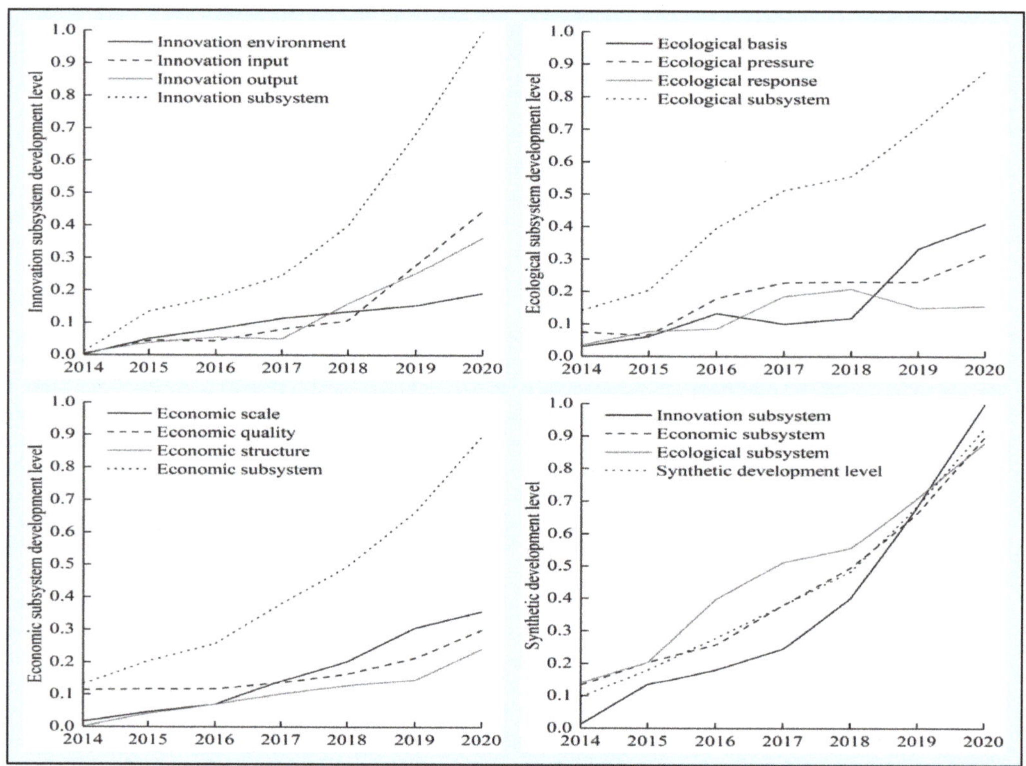

Figure 4. Development level and synthetic developmental level of each subsystem in Anhui.

3.2.4. Coupling Coordination Development Level Analysis

With regard to the development level data and the coupling coordination degree model, we use Excel 2018 software to calculate the coupling degree and coordination degree of the three systems and two systems of Anhui IEE. Table 3 illustrates the calculation results. On this basis, we describe the spatial and temporal evolution pattern of the coupling coordination of three systems and two systems from 2014 to 2020 (Figure 5) to analyze their spatial differences and evolution characteristics more intuitively.

As the data in Table 3 indicates, Y is the synthetic developmental level score; C is the coupling degree score; and D is the coordination degree. During the investigation period, the coupling degree among the three systems was at the high-level coupling stage, except for in 2014, when it was at the running-in stage. The coupling degrees of the innovation–economic and innovation–ecological subsystems were at a high-level coupling stage in all years except 2014, when they were at the running-in stage. The coupling degree of the economic ecological subsystem was at the high-level coupling stage in this investigation period. In addition, the coupling degree values of these high-level coupling stages are all above 0.927, which is a high level. However, a high coupling level does not mean a high coordination level. As Table 3 illustrates, in 2014–2019, the coordination degree of each system lagged behind the coupling degree. By 2020, the gap between the coordination degree and coupling degree was not apparent, and the gap value was below 0.06.

Table 3. Anhui Province IEE coupling and coordination degree calculation results.

Year	ABC			AB			AC			BC		
	Y	C	D	Y	C	D	Y	C	D	Y	C	D
2014	0.096	0.650	0.249	0.073	0.565	0.203	0.077	0.552	0.206	0.137	1.000	0.371
2015	0.181	0.982	0.421	0.169	0.979	0.407	0.169	0.979	0.407	0.204	1.000	0.451
2016	0.278	0.949	0.514	0.219	0.984	0.464	0.288	0.927	0.517	0.327	0.977	0.565
2017	0.379	0.956	0.602	0.312	0.976	0.552	0.378	0.935	0.595	0.446	0.989	0.664
2018	0.483	0.991	0.692	0.447	0.994	0.667	0.478	0.987	0.686	0.525	0.998	0.724
2019	0.686	1.000	0.828	0.674	1.000	0.821	0.697	1.000	0.835	0.687	0.999	0.828
2020	0.925	0.998	0.961	0.947	0.999	0.973	0.939	0.998	0.968	0.888	1.000	0.942

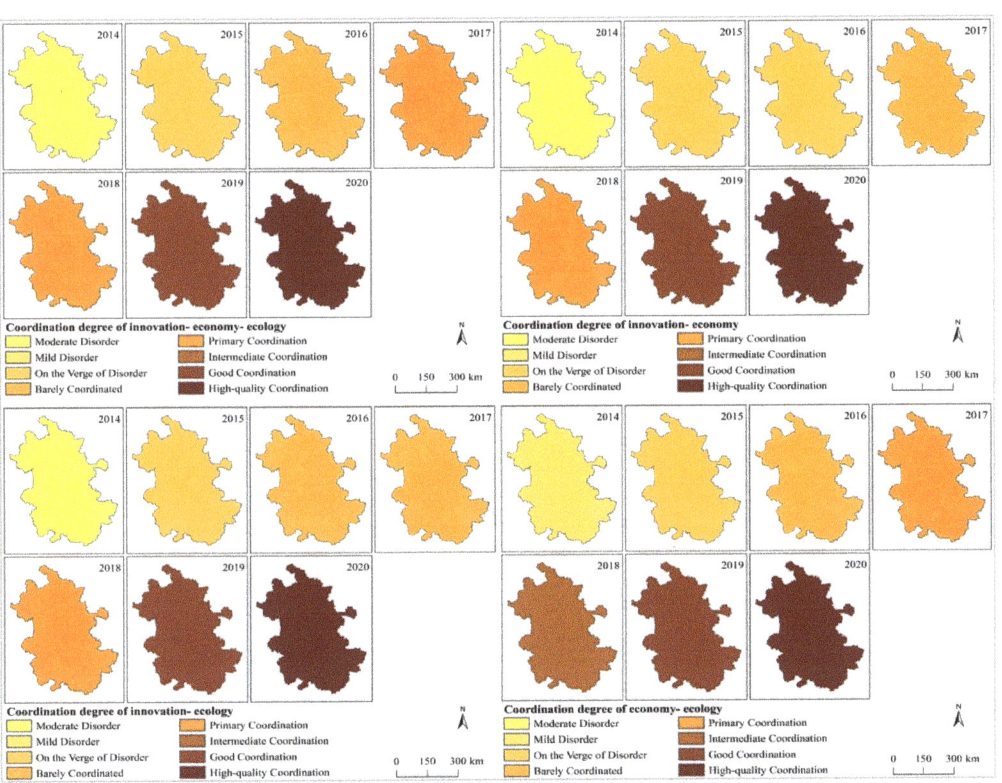

Figure 5. Spatial and temporal evolution of IEE coordination in Anhui Province.

Figure 5 illustrates the coordination development and evolution of three subsystems and two subsystems in Anhui Province. The IEE system indicates that the innovation–economy and innovation–ecology subsystems experienced the evolution track of "modern disorder-on the verge of disorder-barely coordinated-primary coordination-good coordination-high quality coordination". Over the course of the research, the level of coordination development improved consistently. The coordination degree of the economic ecological subsystem experienced the evolution track of "middle disorder-on the verge of disorder barely coordinated primary coordination-intermediate coordination-good coordination-high quality coordination". In addition, we find that, from 2014 to 2019, the coordination state of the two systems showed the following relationship: "economy-ecology > innovation–ecology > innovation–economy". In 2020, its coordination state showed the following relationship: "innovation–economy > innovation–ecology > economy–ecology". Anhui can be observed to shift from being driven by ecological factors to being driven by innovation, owing to Anhui's in-depth implementation of an innovation-driven development strategy, thereby promoting the transfer of science and technology into a real productive force and highlighting the regional innovation output benefits.

3.2.5. Obstacle Degree Analysis

By the use of the obstacle degree model, the obstacle degree score of the system, criterion, and indicator layers of the IEE system in Anhui Province during 2014–2019 is computed. The computed results are illustrated in Tables 4 and 5.

Table 4. Computed results of the obstacle degree of regional IEE in Anhui Province.

Year	2014	2015	2016	2017	2018	2019	2020
A1	7.07	5.81	5.21	4.30	3.84	4.25	1.08
A2	16.22	16.22	18.49	19.52	21.76	17.62	0.00
A3	13.10	13.16	14.15	16.71	13.08	11.65	0.00
B1	12.51	12.63	13.23	11.56	9.94	5.41	0.00
B2	10.61	11.63	13.14	14.21	15.32	20.05	45.15
B3	8.82	8.15	7.93	7.52	7.36	10.25	0.46
C1	13.96	14.15	12.78	16.57	18.78	8.28	0.00
C2	9.94	11.36	7.63	6.28	7.38	12.21	12.88
C3	7.77	6.90	7.43	3.33	2.55	10.30	40.42
A	36.39	35.19	37.85	40.53	38.68	33.51	1.08
B	31.94	32.40	34.30	33.29	32.61	35.70	45.62
C	31.66	32.41	27.85	26.17	28.71	30.79	53.31

Table 5. Main obstacles to coordinated IEE development in Anhui Province.

Year	Indicator Rankings									
	1		2		3		4		5	
	Factor	Obstacle Degree	Factor	Obstacle Degree	Factor	Obstacle Degree	Factor	Obstacle Degree	Factor	Obstacle Degree
2014	C13	9.13	A22	6.48	A32	5.50	B24	4.49	B31	4.19
2015	C13	9.70	A22	6.15	A32	6.08	B24	4.50	C31	4.34
2016	C13	10.71	A22	7.85	A32	6.95	B24	5.08	B31	4.52
2017	C13	12.45	A22	9.32	A32	8.16	B24	5.30	B31	4.68
2018	C13	15.21	A22	11.33	B22	6.34	A32	5.92	B24	5.63
2019	B22	11.64	A22	10.36	B31	8.13	C14	7.86	C33	7.19
2020	B22	45.15	C31	21.18	C33	19.25	C23	12.88	A11	1.08

As Table 4 and Figure 6 show, at the system level, from 2014 to 2019, the largest obstacle blocking the coordinated development of Anhui's IEE system was the innovation subsystem. In 2020, the obstacles of the innovation subsystem gradually disappeared, and the ecological subsystem became the largest development obstacle, followed by the economic subsystem. The economic rise of the innovation region in Anhui also placed certain pressures on the ecological environment. It is an important task for regional economic planning to seek the coupling coordination development of three subsystems: innovation, economy, and ecology. At the criterion level, innovation input (A2), innovation output (A3), economic scale (B1), economic quality (B2), and ecological basis (C1) were the top five projects in terms of barrier degree from 2014 to 2018. In 2019, the economic scale (B1) and ecological basis (C1) were replaced in the top five by ecological pressure (C2) and ecological response (C3), respectively. In 2020, the barrier degree of economic quality (B2) to the coupling coordination development of complex systems reached 45.15, and the barrier degree of ecological response (C3) to the coupling coordination development of complex systems reached 40.42, becoming the main obstacle of Anhui's IEE system. These results are consistent with those for the system layer.

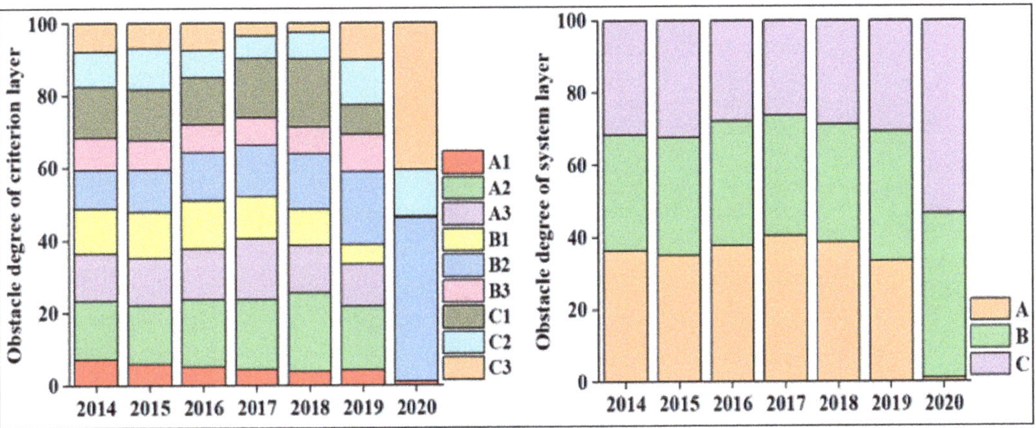

Figure 6. Change trends of the obstacle degrees of the system and criterion layers.

Finally, we sorted and ranked the obstacle degree results of 35 specific indicators in the indicator layer, and identified the top five major obstacle factors in Anhui Province from 2014 to 2020 (Table 5).

As Table 5 illustrates, from 2014 to 2017, the top four barrier factors remained unchanged: "C13 > A22 > A32 > B24." The fifth obstacle factor was C31 in 2015, and B31 in the other three years. The top five barrier factors in 2018, 2019, and 2020 were "C13 > A22 > B22 > A32 > B42", "B22 > A22 > B31 > C14 > C33", and "B22 > C31 > C33 > C23 > A11", respectively. We find that the changing trend of the barrier factors in the indicator layer is consistent with the change in the criteria layer and the system layer. During the study period, the obstacle degree of innovation indicators gradually decreased, while the obstacle degree of economic and ecological indicators gradually increased. It is thus clear that the key to realizing the coordinated development of Anhui's IEE system lies in giving consideration to innovation and focusing on enhancing the sustainable harmony development level of the economic and ecological subsystems.

4. Discussion and Managerial Implications

Apart from the existing literature [30,48,109], this paper gains significant advantages. First, an assessment indicator system of the coupling coordination development of regional IEE systems is established, which enriches the theoretical model of the coupled and coordinated relationship between regional systems. Second, a data-driven dynamic integration model method for the coupling coordination relationship of regional IEE systems is established to calculate and assess the coupling coordination degree of three systems and two systems, which enriches the empirical model of the coupling coordination relationship between regional systems. Third, constructing the obstacle factor diagnosis and identification mechanism that affects its development helps to optimize the path of the coupling coordination development of the three systems.

Promoting the coupling coordination development of regional innovation, regional economy, and regional ecology is the internal requirement for sustainable regional development [110] and has important practical significance for regional eco-civilization construction and innovation driven-development. Combined with the above, we identify three managerial implications. First, realizing the good coordination development of regional innovation, the regional economy, and the regional ecosystem is key to promoting high-quality regional development. Objective quantitative comprehensive assessment of the coupling coordination development of the three systems in the specific region will help reveal the region's spatial and temporal evolution law. Second, from a systems theory and synergy theory perspective, it becomes essential to correctly understand the coordination development relationship of regional innovation, regional economy, and regional ecology, identify the obstacles blocking coordinated development, and formulate a systematic and complete countermeasure system with a comprehensive top-level design and rigorous and detailed supporting measures. Third, the three subsystems of regional innovation, regional economy, and regional ecology are interdependent and interactive. To realize the sustainable and coordinated development of the economy and ecological environment, support and guidance should be provided for innovation; innovation will be the only way to maintain a green economic growth pattern.

5. Conclusions

Realizing high-level coupling coordination development of the regional IEE system is a very urgent request for achieving regional high-quality, sustainable healthy development. Therefore, calculating, assessing, diagnosing, and improving its development level are urgent tasks. Accordingly, this study proposes a data-driven comprehensive assessment mechanism for the coupling coordination development of the regional IEE system.

This study builds a data-driven multimodel decision approach to calculate, assess, diagnose, and improve the regional IEE system. Then, the data-driven multimodel decision approach is applied to the case of Anhui. The main conclusions are as follows: (1) The synthetic development level of the Anhui IEE system presents a stable growth trend. The contribution degree order of the three subsystems is as follows: ecological subsystem > economic subsystem > innovation subsystem. (2) The coupling coordination degree of the Anhui IEE system presents a stable growth trend, but the coordination degree has failed to achieve synchronous development with the coupling degree and still lags behind the coupling degree. (3) In the process of the coupling coordination development of the Anhui IEE system, the main obstacle at the system layer has changed from the original innovation subsystem to the current ecology subsystem. The main obstacles in the criteria layer have been the economic quality (B2) and ecological response (C3). The GDP energy intensity (B22), the green coverage rate of urban built-up areas (C31), the comprehensive utilization rate of general industrial solid waste (C33), and industrial solid waste output (C23) are the main obstacles affecting the coordinated development of the Anhui IEE system coupling. This shows that it is necessary to further break down the obstacle factors that hinder the orderly progress and coordinated development of the three subsystems in the future, especially to strengthen the ecological environment governance.

Regarding the innovations this study provides, first, from the systems theory perspective, the coupling coordination mechanism among the three systems is revealed, and the assessment indicator system on the coupling coordination development of regional innovation, regional economy, and regional ecology is established, covering 3 systems, 9 criteria, and 35 basic indicators. Second, based on the viewpoint of coupling coordination theory, a data-driven, three-system coupling coordination model is established to reveal the spatial and temporal evolution law of regional IEE system coupling coordination. Third, from a global perspective, a data-driven obstacle model is established to diagnose the factors impeding the development of the IEE system, and then formulate scientific, objective, and reasonable optimization countermeasures.

This article provides a new method and idea for the study of the coupling and coordinated development of regional IEE systems and provides a decision-making basis for the formulation of government coordination policies and the transformation and upgrading of enterprises. However, the regional IEE system is a complicated and poly-level operating system, involving many factors. In this study, only representative and available data indicators are included in the assessment system. Thus, indicator selection may not be comprehensive, and some deficiencies exist. Therefore, in future research, more data would be collected to calculate and assess the IEE system more objectively and provide additional support for regional synergy development.

Author Contributions: Conceptualization, Y.Y. and F.H.; methodology, F.H.; software, L.D.; validation, F.H.; formal analysis, X.W.; investigation, L.D.; resources, L.D.; data curation, L.D.; writing—original draft preparation, Y.Y.; writing—review and editing, Y.Y.; visualization, X.W.; supervision, X.W.; project administration, F.H.; funding acquisition, Y.Y. All authors have read and agreed to the published version of the manuscript.

Funding: This work was supported in part by a youth project of the Anhui Social Science Planning Project, No. AHSKQ2021D37.

Informed Consent Statement: Not applicable.

Data Availability Statement: Not applicable.

Conflicts of Interest: The authors declare no conflict of interest.

References

1. Acheampong, A.O.; Boateng, E.; Amponsah, M.; Dzator, J. Revisiting the economic growth–energy consumption nexus: Does globalization matter? *Energy Econ.* **2021**, *102*, 105472. [CrossRef]
2. Lin, B.; Zhou, Y. Measuring the green economic growth in China: Influencing factors and policy perspectives. *Energy* **2022**, *241*, 122518. [CrossRef]
3. Sovacool, B.K.; Newell, P.; Carley, S.; Franzo, J. Equity, technological innovation and sustainable behaviour in a low-carbon future. *Nat. Hum. Behav.* **2022**, *6*, 326–337. [CrossRef] [PubMed]
4. Holroyd, C. Technological innovation and building a 'super smart' society: Japan's vision of society 5.0. *J. Asian Public Policy* **2022**, *15*, 18–31. [CrossRef]
5. Yang, Y.; Lu, H.; Liang, D.; Tian, P.; Xia, J.; Wang, H.; Lei, X. Ecological sustainability and its driving factor of urban agglomerations in the Yangtze River Economic Belt based on three-dimensional ecological footprint analysis. *J. Clean. Prod.* **2022**, *330*, 129802. [CrossRef]
6. Amirgaliev, N.A.; Askarova, M.; Opp, C.; Medeu, A.; Kulbekova, R.; Medeu, A.R. Water quality problems analysis and assessment of the ecological security level of the Transboundary Ural-Caspian Basin of the Republic of Kazakhstan. *Appl. Sci* **2022**, *12*, 2059. [CrossRef]
7. Boons, F.; Wagner, M. Assessing the relationship between economic and ecological performance: Distinguishing system levels and the role of innovation. *Ecol. Econ.* **2009**, *68*, 1908–1914. [CrossRef]
8. Liao, B.; Li, L. Spatial division of labor, specialization of green technology innovation process and urban coordinated green development: Evidence from China. *Sustain. Cities Soc.* **2022**, *80*, 103778. [CrossRef]
9. Gertler, M.S.; Wolfe, D.A.; Garkut, D. No place like home? The embeddedness of innovation in a regional economy. *Rev. Int. Political Econ.* **2000**, *7*, 688–718. [CrossRef]
10. Gai, M.; Wang, X.; Qi, C. Spatiotemporal evolution and influencing factors of ecological civilization construction in China. *Complexity* **2020**, *2020*, 8829144. [CrossRef]
11. Potts, T. The natural advantage of regions: Linking sustainability, innovation, and regional development in Australia. *J. Clean. Prod.* **2010**, *18*, 713–725. [CrossRef]
12. Wang, S.; Wang, X.; Lu, F.; Fan, F. The impact of collaborative innovation on ecological efficiency–empirical research based on China's regions. *Technol. Anal. Strateg. Manag.* **2021**, *33*, 242–256. [CrossRef]

13. Yasmeen, H.; Tan, Q.; Zameer, H.; Tan, J.; Nawaz, K. Exploring the impact of technological innovation, environmental regulations and urbanization on ecological efficiency of China in the context of COP21. *J. Environ. Manag.* **2020**, *274*, 111210. [CrossRef] [PubMed]
14. Wang, J.; Wei, X.; Guo, Q. A three-dimensional evaluation model for regional carrying capacity of ecological environment to social economic development: Model development and a case study in China. *Ecol. Indic.* **2018**, *89*, 348–355. [CrossRef]
15. Perroux, F. Economic space: Theory and applications. *Q. J. Econ.* **1950**, *64*, 89–104. [CrossRef]
16. Solow, R.M. Technical change and the aggregate production function. *Rev. Econ. Stat.* **1957**, *39*, 554–562. [CrossRef]
17. Sener, S.; Saridogan, E. The effects of science-technology-innovation on competitiveness and economic growth. *Procedia Soc. Behav. Sci.* **2011**, *24*, 815–828. [CrossRef]
18. Droste, N.; Hansjürgens, B.; Kuikman, P.; Otter, N.; Antikainen, R.; Leskinen, P.; Pitkänen, K.; Saikku, L.; Loiseau, E.; Thomsen, M. Steering innovations towards a green economy: Understanding government intervention. *J. Clean. Prod.* **2016**, *135*, 426–434. [CrossRef]
19. Antonioli, D.; Mazzanti, M. Towards a green economy through innovations: The role of trade union involvement. *Ecol. Econ.* **2017**, *131*, 286–299. [CrossRef]
20. Mehmet, A. Technological progress, innovation and economic growth; the case of Turkey. *Procedia Soc. Behav. Sci.* **2015**, *195*, 776–782.
21. Liang, L.; Wang, Z.B.; Luo, D.; Wei, Y.; Sun, J. Synergy effects and it's influencing factors of China's high technological innovation and regional economy. *PLoS ONE* **2020**, *15*, e0231335. [CrossRef] [PubMed]
22. Cheng, H.; Wang, B. Multiplier effect of science and technology innovation in regional economic development: Based on panel data of coastal cities. *J. Coast. Res.* **2019**, *94*, 883. [CrossRef]
23. Chen, Y.; Zhang, D. Multiscale assessment of the coupling coordination between innovation and economic development in resource-based cities: A case study of Northeast China. *J. Clean. Prod.* **2021**, *318*, 128597. [CrossRef]
24. Tian, Y.; Huang, P.; Zhao, X. Spatial analysis, coupling coordination, and efficiency evaluation of green innovation: A case study of the Yangtze River Economic Belt. *PLoS ONE* **2020**, *15*, e0243459. [CrossRef] [PubMed]
25. Pradhan, R.P.; Arvin, M.B.; Nair, M.; Bennett, S.E. The dynamics among entrepreneurship, innovation, and economic growth in the Eurozone countries. *J. Policy Model.* **2020**, *42*, 1106–1122. [CrossRef]
26. Qamruzzaman, M.; Jianguo, W. Nexus between financial innovation and economic growth in South Asia: Evidence from ARDL and nonlinear ARDL approaches. *Financ. Innov.* **2018**, *4*, 20. [CrossRef]
27. Pala, A. Innovation and economic growth in developing countries: Empirical implication of swamy's random coefficient model (RCM). *Procedia Comput. Sci.* **2019**, *158*, 1122–1130. [CrossRef]
28. Norgaard, R.B. Economic indicators of resource scarcity: A critical essay. *J. Environ. Econ. Manag.* **1990**, *19*, 19–25. [CrossRef]
29. Boulding, K.E. *The Economics of the Coming Spaceship Earth*; Routledge: New York, NY, USA, 1966; pp. 1–17.
30. Liu, C.; Cai, W.; Zhai, M.; Zhu, G.; Zhang, C.; Jiang, Z. Decoupling of wastewater eco-environmental damage and China's economic development. *Sci. Total Environ.* **2021**, *789*, 147980. [CrossRef]
31. Oliveira, C.; Antunes, C.H. A multi-objective multi-sectoral economy–energy–environment model: Application to Portugal. *Energy* **2011**, *36*, 2856–2866. [CrossRef]
32. Peng, B.; Sheng, X.; Wei, G. Does environmental protection promote economic development? From the perspective of coupling coordination between environmental protection and economic development. *Environ. Sci. Pollut. Res.* **2020**, *27*, 39135–39148. [CrossRef] [PubMed]
33. Fan, Y.; Wu, S.Z.; Lu, Y.T.; Zhao, Y.H. Study on the effect of the environmental protection industry and investment for the national economy: An input-output perspective. *J. Clean. Prod.* **2019**, *227*, 1093–1106. [CrossRef]
34. Fan, Y.P.; Fang, C.L.; Zhang, Q. Coupling coordinated development between social economy and ecological environment in Chinese provincial capital cities-assessment and policy implications. *J. Clean. Prod.* **2019**, *229*, 289–298. [CrossRef]
35. Liu, K.; Qiao, Y.; Shi, T.; Zhou, Q. Study on coupling coordination and spatiotemporal heterogeneity between economic development and ecological environment of cities along the Yellow River Basin. *Environ. Sci. Pollut. Res.* **2021**, *28*, 6898–6912. [CrossRef] [PubMed]
36. Liao, M.L.; Chen, Y.; Wang, Y.J.; Lin, M.S. Study on the coupling and coordination degree of high-quality economic development and ecological environment in Beijing-Tianjin-Hebei region. *Appl. Ecol. Environ. Res.* **2019**, *17*, 11069–11083. [CrossRef]
37. Shi, T.; Yang, S.; Zhang, W.; Zhou, Q. Coupling coordination degree measurement and spatiotemporal heterogeneity between economic development and ecological environment—Empirical evidence from tropical and subtropical regions of China. *J. Clean. Prod.* **2020**, *244*, 118739. [CrossRef]
38. Adami, L.; Schiavon, M. From circular economy to circular ecology: A review on the solution of environmental problems through circular waste management approaches. *Sustainability* **2021**, *13*, 925. [CrossRef]
39. Liu, W.W.; Sun, R.; Li, Q. Measurement of coupling degree between regional knowledge innovation and technological innovation: An empirical analysis based on provincial panel data in China during 2010-2014. *J. Interdiscip. Math.* **2017**, *20*, 125–139. [CrossRef]
40. Stern, D.I. Progress on the environmental Kuznets curve? *Environ. Dev. Econ.* **1998**, *3*, 173–196. [CrossRef]
41. Cole, M.A.; Rayner, A.J.; Bates, J.M. The environmental Kuznets curve: An empirical analysis. *Environ. Dev. Econ.* **1997**, *2*, 401–416. [CrossRef]

42. Solakoglu, E.G. The effect of property rights on the relationship between economic growth and pollution for transition economies. *East. Europ. Econ.* **2007**, *45*, 77–94. [CrossRef]
43. Bakhsh, K.; Akmal, T.; Ahmad, T.; Abbas, Q. Investigating the nexus among sulfur dioxide emission, energy consumption, and economic growth: Empirical evidence from Pakistan. *Environ. Sci. Pollut. Res.* **2022**, *29*, 7214–7224. [CrossRef] [PubMed]
44. Shikwambana, L.; Mhangara, P.; Kganyago, M. Assessing the relationship between economic growth and emissions levels in South Africa between 1994 and 2019. *Sustainability* **2021**, *13*, 2645. [CrossRef]
45. Jiang, M.; Kim, E.; Woo, Y. The relationship between economic growth and air pollution—A regional comparison between China and South Korea. *Int. J. Environ. Res. Public Health* **2020**, *17*, 2761. [CrossRef] [PubMed]
46. Xia, C.; Bing, Y. Strategic leadership, environmental optimisation, and regional innovation performance with the regional innovation system coupling synergy degree: Evidence from China. *Technol. Anal. Strateg. Manag.* **2022**, 1–14. [CrossRef]
47. Ke, H.; Dai, S.; Yu, H. Spatial effect of innovation efficiency on ecological footprint: City-level empirical evidence from China. *Environ. Technol. Innov.* **2021**, *I*, 101536. [CrossRef]
48. Meirun, T.; Mihardjo, L.W.W.; Haseeb, M.; Khan, S.A.R.; Jermsittiparsert, K. The dynamics effect of green technology innovation on economic growth and CO2 emission in Singapore: New evidence from bootstrap ARDL approach. *Environ. Sci. Pollut. Res.* **2021**, *28*, 4184–4194. [CrossRef]
49. Brancalion, P.H.S.; van Melis, J. On the need for innovation in ecological restoration1. *Ann. Mo. Bot. Gard.* **2017**, *102*, 227–236. [CrossRef]
50. Murphy, J.; Gouldson, A. Environmental policy and industrial innovation: Integrating environment and economy through ecological modernisation. *Ecol. Mod. Read.* **2020**, *31*, 275–294.
51. Aribi, I.; Hassen, L.B. Phases of economic development in an endogenous growth model with innovation and education. *Oradea J. Bus. Econ.* **2021**, *6*, 78–87. [CrossRef]
52. Wu, N.; Liu, Z.K. Higher education development, technological innovation and industrial structure upgrade. *Technol. Forecast. Soc. Chang.* **2021**, *162*, 120400. [CrossRef]
53. Jogova, M.; Song, J.E.S.; Campbell, A.C.; Warbuton, D.; Warshawski, T.; Chanoine, J.-P. Process evaluation of the Living Green, Healthy and Thrifty (LiGHT) web-based child obesity management program: Combining health promotion with ecology and economy. *Can. J. Diabetes* **2013**, *37*, 72–81. [CrossRef] [PubMed]
54. Pontzer, H.; Wood, B.M. Effects of evolution, ecology, and economy on human diet: Insights from hunter-gatherers and other small-scale societies. *Annu. Rev. Nutr.* **2021**, *41*, 363–385. [CrossRef]
55. Sovacool, B.K. Who are the victims of low-carbon transitions? Towards a political ecology of climate change mitigation. *Energy Res. Soc. Sci.* **2021**, *73*, 101916. [CrossRef]
56. Wang, R.; Tan, J. Exploring the coupling and forecasting of financial development, technological innovation, and economic growth. *Technol. Forecast Soc. Chang.* **2021**, *163*, 120466.
57. Zhang, F.; Sarker, M.N.I.; Lv, Y. Coupling coordination of the regional economy, tourism industry, and the ecological environment: Evidence from western China. *Sustainability* **2022**, *14*, 1654. [CrossRef]
58. Liu, Y.; Liu, W.; Yan, Y.; Liu, C. A perspective of ecological civilization: Research on the spatial coupling and coordination of the energy-economy-environment system in the Yangtze River Economic Belt. *Environ. Monit. Assess.* **2022**, *194*, 403. [CrossRef]
59. Liu, J.; Tian, Y.; Huang, K.; Yi, T. Spatial-temporal differentiation of the coupling coordinated development of regional energy-economy-ecology system: A case study of the Yangtze River Economic Belt. *Ecol. Indic.* **2021**, *124*, 107394.
60. Yan, X.; Chen, M.; Chen, M.Y. Coupling and coordination development of Australian energy, economy, and ecological environment systems from 2007 to 2016. *Sustainability* **2019**, *11*, 6568. [CrossRef]
61. Zheng, Z.; Zhu, Y.; Qiu, F.; Wang, L. Coupling relationship among technological innovation, industrial transformation and environmental efficiency: A case study of the Huaihai Economic Zone, China. *Chin. Geogr. Sci.* **2022**, *32*, 686–706. [CrossRef]
62. Xu, X.; Zhou, Y. Spatio-Temporal differentiation characteristics and determinants of coupling coordination degree in technological innovation-industrial upgrading-ecological environment system: A case study of Jiangsu Province, China. *Pol. J. Environ. Stud.* **2021**, *30*, 3341–3355. [CrossRef]
63. Huang, J.; Wang, J.; Tian, G. Research on the coupling of technological innovation, talent accumulation and ecological environment. *Ekoloji* **2018**, *27*, 1735–1742.
64. Yin, K.; Zhang, R.; Jin, X.; Yu, I. Research and optimization of the coupling and coordination of environmental regulation, technological innovation, and green development. *Sustainability* **2022**, *14*, 501. [CrossRef]
65. Wu, D. Research on regional technology—Economy—Ecology coordination evaluation system in China. In *1st International Symposium on Innovative Management and Economics (ISIME 2021)*; Atlantis Press: Paris, France, 2021; pp. 108–120.
66. Liu, Z.; Li, K.W.; Tang, J.; Gong, B.; Huang, J. Optimal operations of a closed-loop supply chain under a dual regulation. *Int. J. Prod. Econ.* **2021**, *233*, 107991. [CrossRef]
67. Liu, Z.; Zheng, X.X.; Li, D.F.; Liao, C.N.; Sheu, J.B. A novel cooperative game-based method to coordinate a sustainable supply chain under psychological uncertainty in fairness concerns. *Transp. Res. Part E Logist. Transp. Rev.* **2021**, *147*, 102237. [CrossRef]
68. Liu, C.; Gao, M.; Zhu, G.; Zhang, C.; Zhang, P.; Chen, J.; Cai, W. Data driven eco-efficiency evaluation and optimization in industrial production. *Energy* **2021**, *224*, 120170. [CrossRef]
69. Yang, Y.; Wu, X.; Liu, F.; Zhang, Y.; Liu, C. Promoting the efficiency of scientific and technological innovation in regional industrial enterprises: Data-driven DEA-Malmquist evaluation model. *J. Intell. Fuzzy Syst.* **2022**, *43*, 4911–4928. [CrossRef]

70. Li, J.; Fang, H.; Fang, S.; Siddika, S.E. Investigation of the relationship among university–research institute–industry innovations using a coupling coordination degree model. *Sustainability* **2018**, *10*, 1954. [CrossRef]

71. Wang, M.; Zhao, X.; Gong, Q.; Ji, Z. Measurement of regional green economy sustainable development ability based on entropy weight-topsis-coupling coordination degree—A case study in Shandong Province, China. *Sustainability* **2019**, *11*, 280. [CrossRef]

72. Dong, L.; Longwu, L.; Zhenbo, W.; Liangkan, C.; Faming, Z. Exploration of coupling effects in the economy–society–environment system in urban areas: Case study of the Yangtze River Delta urban agglomeration. *Ecol. Indicat.* **2021**, *128*, 107858. [CrossRef]

73. Xing, L.; Xue, M.; Hu, M. Dynamic simulation and assessment of the coupling coordination degree of the economy–resource–environment system: Case of Wuhan City in China. *J. Environ. Manag.* **2019**, *230*, 474–487. [CrossRef] [PubMed]

74. Sun, Y.; Zhao, T.; Xia, L. Spatial-temporal differentiation of carbon efficiency and coupling coordination degree of Chinese county territory and obstacles analysis. *Sustain. Cities Soc.* **2022**, *76*, 103429. [CrossRef]

75. Song, Q.; Zhou, N.; Liu, T.; Siehr, S.A.; Qi, Y. Investigation of a "coupling model" of coordination between low-carbon development and urbanization in China. *Energy Policy* **2018**, *121*, 346–354. [CrossRef]

76. Xiao, Y.; Wang, R.; Wang, F.; Huang, H.; Wang, J. Investigation on spatial and temporal variation of coupling coordination between socioeconomic and ecological environment: A case study of the Loess Plateau, China. *Ecol. Indicat.* **2022**, *136*, 108667. [CrossRef]

77. Hou, C.; Chen, H.; Long, R. Coupling and coordination of China's economy, ecological environment and health from a green production perspective. *Int. J. Environ. Sci. Technol.* **2022**, *19*, 4087–4106. [CrossRef]

78. Pan, Y.; Weng, G.; Li, C.; Li, J. Coupling coordination and influencing factors among tourism carbon emission, tourism economic and tourism innovation. *Int. J. Environ. Res. Public Health* **2021**, *18*, 1601. [CrossRef]

79. Li, Y.; Li, Y.; Zhou, Y.; Shi, Y.; Zhu, X. Investigation of a coupling model of coordination between urbanization and the environment. *J. Environ. Manag.* **2012**, *98*, 127–133. [CrossRef]

80. Yang, Y.; Wang, Y.; Zhang, Y.; Liu, C. Data-driven coupling coordination development of regional innovation EROB composite system: An integrated model perspective. *Mathematics* **2022**, *10*, 2246. [CrossRef]

81. Liu, C.; Zhang, R.; Wang, M.; Xu, J. Measurement and prediction of regional tourism sustainability: An analysis of the Yangtze River economic zone, China. *Sustainability* **2018**, *10*, 1321. [CrossRef]

82. Lei, X.; Robin, Q.; Liu, Y. Evaluation of regional land use performance based on entropy TOPSIS model and diagnosis of its obstacle factors. *Trans. Chin. Soc. Agric. Eng.* **2016**, *32*, 243–253.

83. Zhang, F.; Sun, C.; An, Y.; Luo, Y.; Yang, Q.; Su, W.; Gao, L. Coupling coordination and obstacle factors between tourism and the ecological environment in Chongqing, China: A multi-model comparison. *Asia Pac. J. Tour. Res.* **2021**, *26*, 811–828. [CrossRef]

84. Çalışkan, H.K. Technological change and economic growth. *Procedia Soc. Behav. Sci* **2015**, *195*, 649–654. [CrossRef]

85. Fang, Y. The impact of innovation efficiency factors on the level of regional economic development. *Open J. Soc. Sci.* **2020**, *8*, 343. [CrossRef]

86. Leyden, D.P. Public-sector entrepreneurship and the creation of a sustainable innovative economy. *Small Bus. Econ.* **2016**, *46*, 553–564. [CrossRef]

87. Wang, J.; Wang, C.; Yu, S.; Li, M.; Cheng, Y. Coupling coordination and spatiotemporal evolution between carbon emissions, industrial structure, and regional innovation of counties in Shandong Province. *Sustainability* **2022**, *14*, 7484. [CrossRef]

88. Xu, L.; Zhao, X.; Chen, J. Exploring the governance dilemma of nuclear wastewater in Fukushima: A tripartite evolutionary game model. *Ocean. Coast. Manag.* **2022**, *225*, 106220. [CrossRef]

89. Buesa, M.; Heijs, J.; Pellitero, M.M.; Baumert, T. Regional systems of innovation and the knowledge production function: The Spanish case. *Technovation* **2004**, *26*, 463–472. [CrossRef]

90. Zhao, S.L.; Cacciolatti, L.; Lee, S.H.; Song, W. Regional collaborations and indigenous innovation capabilities in China: A multivariate method for the analysis of regional innovation systems. *Technol. Forecast. Soc. Chang.* **2015**, *94*, 202–220. [CrossRef]

91. Efrat, K. The direct and indirect impact of culture on innovation. *Technovation* **2014**, *34*, 12–20. [CrossRef]

92. Shan, D. Research of the construction of regional innovation capability evaluation system: Based on indicator analysis of Hangzhou and Ningbo. *Procedia Eng.* **2017**, *174*, 1244–1251. [CrossRef]

93. Li, X. Empirical analysis on the change of China's regional innovation capability: Based on the viewpoint of innovation system. *Manage World* **2007**, *12*, 18–22.

94. Tonghui, D.; Junfei, C. Evaluation and obstacle factors of coordination development of regional water-energy-food-ecology system under green development: A case study of Yangtze River Economic Belt, China. *Stoch. Environ. Res. Risk Assess.* **2021**, 1–17. [CrossRef]

95. Polina, E.A.; Solovyeva, I.A. Methodology for comprehensive assessment of regional innovative development. *R-Economy* **2019**, *5*, 79–91. [CrossRef]

96. Romer, P.M. Endogenous technological change. *J. Political Econ.* **1990**, *5*, S71–S102. [CrossRef]

97. Sleuwaegen, L.; Boiardi, P. Creativity and regional innovation: Evidence from EU regions. *Res. Policy* **2014**, *43*, 1508–1522. [CrossRef]

98. Xu, M.Y.; Chen, C.T.; Deng, X.Y. Systematic analysis of the coordination degree of China's economy-ecological environment system and its influencing factor. *Environ. Sci. Pollut. Res.* **2019**, *26*, 29722–29735. [CrossRef]

99. Lanchun, L.; Wei, C.; Kaimo, G.; Fang, Y.; Yun, T.; Ke, Z. Research on the evaluation system construction of county innovation driven development-based on evaluation and measurement model. In *Artificial Intelligence in China*; Liang, Q., Wang, W., Mu, J., Liu, X., Na, Z., Eds.; Springer: Singapore, 2022; pp. 218–225.
100. Deng, M.; Chen, J.; Tao, F.; Zhu, J.; Wang, M. On the coupling and coordination development between environment and economy: A case study in the Yangtze River Delta of China. *Int. J. Environ. Res. Public Health* **2022**, *19*, 586. [CrossRef]
101. Hu, X.; Xu, H. A new remote sensing indicator based on the pressure-state-response framework to assess regional ecological change. *Environ. Sci. Pollut. Res.* **2019**, *26*, 5381–5393. [CrossRef]
102. Lai, S.; Sha, J.; Eladawy, A.; Li, X.; Wang, J.; Kurbanov, E.; Lin, Z.; Wu, L.; Han, R.; Su, Y.C. Evaluation of ecological security and ecological maintenance based on pressure-state-response (PSR) model, case study: Fuzhou city, China. *Hum. Ecol. Risk Assess.* **2022**, *28*, 734–761. [CrossRef]
103. Shmelev, S.E.; Shmeleva, I.A. Global urban sustainability assessment: A multidimensional approach. *Sustain. Dev.* **2018**, *26*, 904–920. [CrossRef]
104. Su, Y.; Liang, D.; Guo, W. Application of multiattribute decision-making for evaluating regional innovation capacity. *Math. Probl. Eng.* **2020**, *2020*, 1–20. [CrossRef]
105. Xu, K.; Loh, L.; Chen, Q. Sustainable innovation governance: An analysis of regional innovation with a super efficiency slack-based measure model. *Sustainability* **2020**, *12*, 3008. [CrossRef]
106. Neri, A.C.; Dupin, P.; Sanchez, L.E. A Pressure-State-Response Approach to Cumulative Impact Assessment. *J. Clean. Prod.* **2016**, *126*, 288–298. [CrossRef]
107. Anhui Provincial Bureau of Statistics. Available online: http://tjj.ah.gov.cn/ssah/qwfbjd/tjnj/index.html (accessed on 10 December 2021).
108. National Bureau of Statistics. Available online: http://www.stats.gov.cn (accessed on 10 December 2021).
109. Guo, Y.; Ding, H. Coupled and coordinated development of the data-driven logistics industry and digital economy: A case study of Anhui Province. *Processes* **2022**, *10*, 2036. [CrossRef]
110. Glavič, P.; Pintarič, Z.N.; Bogataj, M. Process design and sustainable development—A European perspective. *Processes* **2021**, *9*, 148. [CrossRef]

Article

Coupled and Coordinated Development of the Data-Driven Logistics Industry and Digital Economy: A Case Study of Anhui Province

Yuxia Guo [1] and Heping Ding [1,2,*]

[1]   Business School, Suzhou University, Suzhou 234000, China
[2]   Department of Center for International Education, Philippine Christian University, Manila 1004, Philippines
*   Correspondence: dingheping@ahszu.edu.cn; Tel.: +86-133-5557-2363

Abstract: The digital transformation of the logistics industry is the current trend of development. In order to promote the integrated development of the logistics industry (LI) and the digital economy (DE), we propose a data-driven method which can be used to measure, evaluate, and identify the coupled and coordinated development (CCD) of the LI and DE. On the basis of data collection, we use the entropy weight method to measure the comprehensive development level of the LI and DE. A coordination model is then used to evaluate their CCD level. Finally, an obstacle degree model (ODM) is used to identify the key factors inhibiting the coordinated development (CD) of the two. This method is then applied to gauge the integration development of the LI and DE in Anhui Province. The results show that energy consumption and the lack of logistics employees are the main obstacles to the development of the LI in Anhui Province. The main obstacles to the development of the DE are the low development level of the electronic communications equipment manufacturing industry and the limited digitization of enterprises. Accordingly, this study puts forward corresponding countermeasures and suggestions to provide decision support for the CCD of the LI and DE.

Keywords: coordination degree model; logistics industry; digital economy; industrial integration

1. Introduction

Citation: Guo, Y.; Ding, H. Coupled and Coordinated Development of the Data-Driven Logistics Industry and Digital Economy: A Case Study of Anhui Province. *Processes* **2022**, *10*, 2036. https://doi.org/10.3390/pr10102036

Academic Editor: Jui-Yuan Lee

Received: 10 September 2022
Accepted: 5 October 2022
Published: 9 October 2022

With the advancement of economic globalization and the rise of the network economy, the logistics industry (LI) has achieved unprecedented development. This industry assumes a significant part of promoting the flow of industrial factors, optimizing regional resource allocation, and improving production and operational efficiency [1]. It is the "link" connecting production, circulation, distribution, and consumption in large-scale socialized production [2]. During the prevention and control period of the global COVID-19 pandemic, the logistics industry applied digital technologies such as big data, the Internet of Things, and block chain. These technologies have helped the industry to match supply and demand more quickly and accurately, which has helped to maintain normal economic and social development. In addition, the world has entered the era of the digital economy (DE), which has a profound impact on world economic development and human life. A white paper on the global digital economy published by the China Academy of Information Technology in 2022 points out that, in 2021, the added value of the DE in 47 major countries in the world reached USD 38.1 trillion, of which, China's digital economy reached USD 7.1 trillion [3]. The DE is a new engine for promoting the optimization of the whole industrial chain and a key force in changing the global competitive pattern. It integrates with traditional industries, promotes the upgrading and transformation of traditional industries, and provides a sustainable driving force for economic development [4]. With the improvement of the DE, the logistics industry has gradually become intelligent and networked, achieving digital transformation and upgrading, more accurately meeting supply and demand, reducing costs, and increasing efficiency [5,6]. The digital transformation of the LI is conducive to the development of industrial digitization in the DE and it provides production factors for the development of the digital economy industry. The interaction and coordinated development (CD) of the LI and DE are conducive to the integration, transformation, and upgrading of regional industries, helping to achieve efficient and sustainable industrial development [7]. This is also the key to promoting the high-quality development of regional economies [8]. Therefore, exploring the coupling and coordinated development (CCD) of the LI and DE, and identifying obstacle factors, has become the focus of academic circles and government departments.

First, the research on the improvement of the LI and DE is rich in content and diversity in perspective. Scholars have studied the ecological efficiency [9], carbon emission efficiency [10], green logistics [11,12], and sustainable development [13] of the logistics industry. Under this new situation, the logistics industry cannot achieve high-quality sustainable development without applying information technology [14]. The application of digital technology helps improve the efficiency of the LI and promotes the progress of green logistics [15]. As early as 1996, Tapscott [16] proposed the concept of the DE, and Carlsson [17] emphasized that the DE is a new business form based on the Internet that uses data and information technology to change the production, business, and social activities of various industries [18,19]. Intelligent logistics is the product of the deep integration of the LI and DE [20]. By integrating technologies such as the Internet of Things, the LI has achieved leapfrog development in terms of its service and management modes [21], laying a foundation for industrial transformation and upgrading and high-quality development.

Second, research has centered around the CD between the LI and DE as the LI and DE are two complex industrial systems [7,21]. The coupling coordination degree model (CCDM) can be used to measure the mutual promotion and interaction degree between two or more systems [22]. This method has been widely used in logistics, economics, agriculture, manufacturing, and other fields, achieving fruitful results [23]. For example, research has been conducted on the coupling and coordination of the LI and regional economy [24,25], urbanization and agricultural ecology [26], logistics industry and agriculture [27], innovative development and economy [28], and logistics industry and manufacturing industry [29]. There are also studies on the CD of the three systems, such as the logistics industry, economy, and ecological environment [30] and the economy, society, and environment [31,32]. However, research on the integration and CD of the LI using the coupling and coordination model rarely involves coupling and coordination with the DE.

Finally, another branch of the literature researches the logistics industry and the construction of the digital economy index system. There are numerous factors involved in the LI and DE, and the existing literature constructs an index system from different angles to measure these factors. From the perspective of sustainable development, these studies examine the economic basis, scale benefit, innovation capacity, environmental effect, and social responsibility of the LI [13]. Other studies construct indicator systems based on the sustainable development of industrial logistics [33]. For instance, Lan et al. constructed a metropolitan logistics index system from the total fixed asset investment in logistics, civilian truck holdings, number of logistics employees, total length of postal routes, and other factors [25]. Furthermore, other scholars measure it from the perspective of input-output [7]. At present, there is no unified indicator system for measuring the digital economy. Liu et al. measured the Internet penetration rate, number of employees, Internet-related output, and development of digital inclusive finance [34], while Liao et al. constructed indicators from three dimensions: the digital economic infrastructure index, digital economic development index, and digital economic innovation index [8]. In contrast, Wei et al. constructed indicators from four aspects: digital industrialization, industrial digitization, digital governance, and data value [35]. Some scholars have added GDP indicators, believing the DE has integrated into social life [36].

Research by international scholars on the logistics industry and digital economy gives a significant reference for exploring the coupling and coordination relationship between the regional logistics industry and the digital economy. However, there are still some deficiencies, i.e., scholars have studied the relationship between the logistics industry and the digital economy, but only considered the impact of digitalization on freight logistics [6] or the interaction between the two [7], and the impact of the application of digital technology on Imperial Retail Logistics Efficiency [14]. However, research on the coupling and coordination of the two systems needs to be deepened as several shortcomings remain:

- Scholars have researched the relationship between the LI and DE, but most of them only consider the influence of the application of digital technology on the industrial upgrading of the LI in the context of the DE. Therefore, the research on their coupling coordination needs to be deepened.
- The LI and DE are two complex industrial systems, involving multi-dimensional indicators, multi-source factors, and multiple dimensions. Therefore, there is an urgent need to identify how to accurately measure and evaluate the two industrial systems to promote their coordinated and sustainable development.
- There are many factors that affect the CD of the LI and DE. Therefore, it is necessary to scientifically distinguish the pivotal obstacle factors to give a quantitative basis for putting forward policy recommendations to support the development of the LI and DE.

In order to remedy these difficulties, this paper uses the research ideas of Lan [25,37–39] for reference. This research is more systematic and comprehensive than the existing research [7,8,33,34] in terms of index system construction. To begin, we construct a logistics industry system by comprehensively considering the facilities' input, energy consumption, business volume output, and carbon and exhaust emissions of the LI. In addition, an indicator system of the DE should be constructed by comprehensively considering the infrastructure construction of the DE, digital industry, and industrial digitalization, as well as innovative development. We propose a data-driven method which can be used to measure, evaluate and identify the CCD of the LI and DE. This method is then applied to Anhui Province to verify its effectiveness. Anhui Province has obvious regional advantages and is located in the middle of China. The digital development of logistics is directly related to the high-quality industrial development of adjacent regions. We then put forward targeted countermeasures and suggestions according to the evaluation and identification results. This method is conducive to providing a quantitative basis for policy recommendations and aims to provide decision support for relevant practitioners and decision makers in the logistics and digital economy industries.

The structure of this paper is as follows: Section 2 contains the materials and methods, which introduces the data collection and processing, data model, and application. Section 3 is a case study, taking Anhui Province of China as an example, to measure, evaluate, and identify the CCD of the LI and DE system. Section 4 is the discussion and, finally, Section 5 is the conclusion.

2. Materials and Methods

2.1. Study Area

We take the development of the LI and DE in Anhui Province in the Yangtze River Delta as an example. Anhui Province is located in the middle of China, in the hinterland of the Yangtze River Delta, near the sea and adjacent to rivers, with obvious geographical advantages. The strength of Anhui's LI is directly related to the industrial integration and development of adjacent regions. The application of digital technology is profitable to promote the transformation of the LI from an extensive traditional mode to an intelligent mode, and it is conducive to meeting the region's economic and social development needs more efficiently. The two bring about sustainable development in the process of continuous action and together promote the high-quality development of the regional economy. According to 2021 statistics released by the Anhui Provincial Bureau of Statistics in March 2022, the added value of transportation, storage, and postal services in 2021 was 205.69 billion yuan, an increase of 8.2%. The annual cargo transportation volume was 4.01 billion tons, an increase of 7.2% over the previous year. The turnover of cargo transportation was 1102.39 billion tons/kilometer, an increase of 8%. The added value of information transmission, software, and information technology services was 96.4 billion yuan, an increase of 10.6%. The operating income of other for-profit service industries, represented by emerging industries such as Internet information technology and business services, increased by 18.4% [40].

Although the two major industries exhibit stronger development over the previous year, they are also facing challenges such as industrial integration and uncoordinated development. Based on the research on the interaction and influence between the LI and DE in Anhui Province, this paper puts forward suggestions to promote the CCD of the LI and DE in Anhui Province, which is of great significance to promoting the high-quality development of industrial integration in Anhui Province.

2.2. Research Framework

To improve the sustainable development level of the LI and DE, the digital transformation of the LI must be realized to better promote industrial integration and accelerate high-quality economic development. The data-driven method is proposed which is mainly embodies putting forward policy recommendations through data collection, processing, modeling, application, and other links, centering on the data of regional LI and DE development. In the face of complex regional LI and DE systems and multiple data sources, this method should be used to overcome the challenges of comprehensive and scientific indicator systems, objective data model processing, and effective policy recommendations. This method involves data collection, processing, modeling, and application to establish a coupled co-scheduling and obstacle degree model. Through this model, the degree of CD is evaluated, and the factors inhibiting the CD are identified. Compared with existing research methods [24,25,29], this method can provide a quantitative basis for policy quality. The method flow is shown in Figure 1:

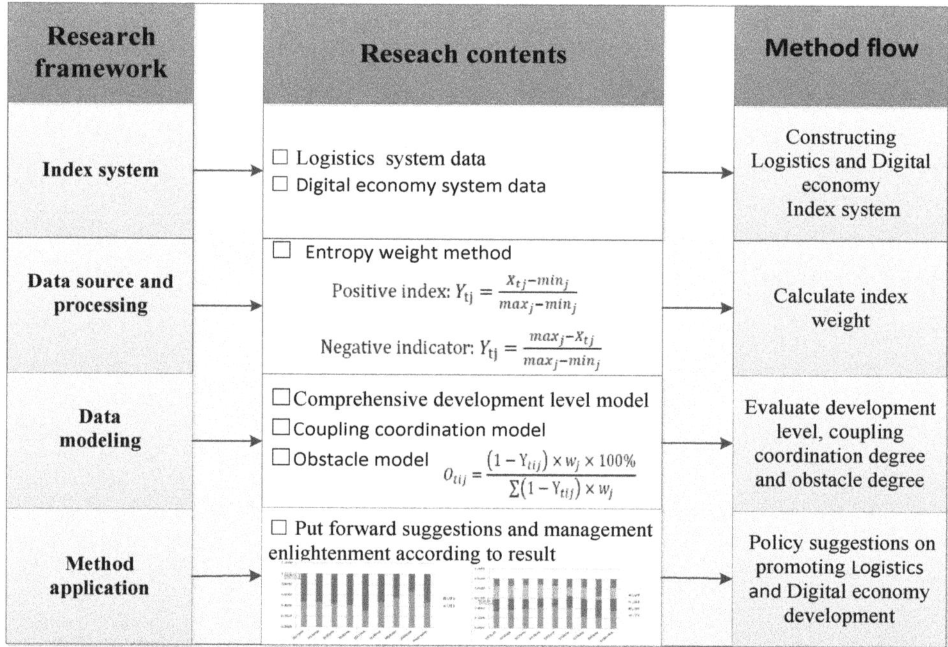

In the figure, the positive index is given by:

$$Y_{tj} = \frac{X_{tj} - min_j}{max_j - min_j}$$

Negative indicator:

$$Y_{tj} = \frac{max_j - X_{tj}}{max_j - min_j}$$

Obstacle model:

$$O_{tij} = \frac{(1 - Y_{tij}) \times w_j \times 100\%}{\sum(1 - Y_{tij}) \times w_j}$$

Figure 1. Method flow chart.

When the research object involves two or more systems that interact and influence each other, and there are many data sources, multiple dimensions, and more influencing factors, the method flow in Figure 2 can be used. However, the digital transformation of the LI is conducive to achieving cost reduction and increased efficiency, better serving related industries and accelerating the pace of industrial digital transformation. The development of the DE promotes the digitalization of the LI and requires the LI to optimize the allocation of resources for its development. The two complex systems interact and develop together. The specific data application process is shown in Figure 2:

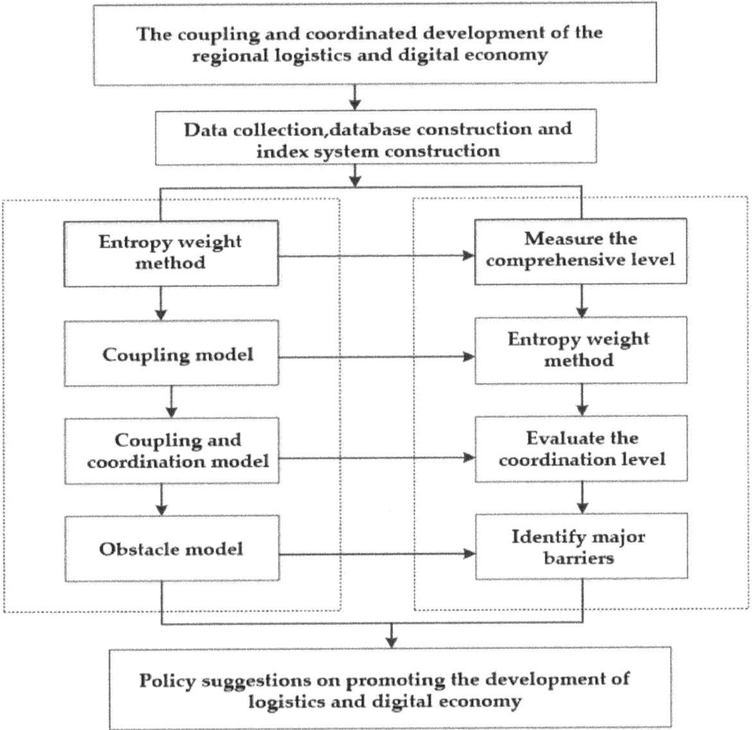

Figure 2. Data application diagram.

First, we collect data related to the development of the regional LI and DE to establish a database. Based on the principles of data systematization, objective rationality, scientificity, accessibility, and measurability, a development index system for the regional LI and DE is constructed. The index consists of six first-class indicators and 33 second-class indicators. Because the data are multi-source and include a wide range of information, the range method is used to make the data dimensionless.

Next, we use the entropy weight method to calculate the comprehensive development level of the LI and DE. The CCDM is then used to evaluate the CCD of the LI and DE system. The key obstacles to the CD of the regional LI and DE are identified through the obstacle degree model (ODM).

Finally, based on the quantitative research results, policy suggestions are put forward to provide a decision-making basis for relevant practitioners and managers to promote the coordinated and sustainable development of regional logistics industries and the digital economy.

2.3. Index System

Based on the research results regarding the development of the logistics industry [13,33] and digital economy [15–19,34–36], this study selects indicators that are scientific and quantifiable. From this, the index system, which includes six systems and 33 indicators, is constructed. This index is used to conduct comparative research on the CCD between systems.

The LI system indicators are constructed from the perspective of input-output. Compared with the indicators in previous studies [7,13,25], this study adds energy consumption and power consumption to the input indicators, reflecting the energy input in addition to infrastructure and talent. Correspondingly, carbon emissions and exhaust emissions are added to the output indicators. In addition to considering positive output, the negative output is also included, making the indicator system more comprehensive. Moreover, the index system for the digital economy comprehensively integrates the existing research [34–36]; that is, it comprehensively considers infrastructure, digital industry, digital innovation, and industrial digital level (see Table 1).

Table 1. Index system of the CCD of the LI and DE system.

System	Primary Index	Secondary Index	Symbol	Direction
Logistics system U1	Logistics industry input U11	Mileage of roads, railways, and waterways (km)	$x_{1\,1}$	+
		Fixed asset investment in the LI (RMB 100 mm)	$x_{1\,2}$	+
		Number of employees in the LI (10,000 people)	$x_{1\,3}$	+
		Postal and rural delivery routes (10,000 km)	$x_{1\,4}$	+
		Energy consumption of the LI (10,000 tons of standard coal)	$x_{1\,5}$	−
		Power consumption of the LI (100 million kWh)	$x_{1\,6}$	−
	Logistics industry output U12	Cargo turnover (100 million tons/km)	x_{17}	+
		Freight volume (10,000 tons)	$x_{1\,8}$	+
		Output value of the LI (RMB 100 mm)	$x_{1\,9}$	+
		Total amount of post and telecommunications business (RMB 10,000)	$x_{1\,10}$	+
		CO2 emission of the LI (10,000 tons)	$x_{1\,11}$	−
		Exhaust gas emissions of the LI (10,000 tons)	$x_{1\,12}$	−
Digital economy system U2	Digital infrastructure U21	Internet penetration rate (%)	x_{21}	+
		Total telecom business (RMB 100 mm)	x_{22}	+
		Mobile phone penetration rate (department/100 people)	x_{23}	+
		Number of domain names per capita (PCs)	x_{24}	+
		Number of pages per capita (PCs)	x_{25}	+
	Digital industry U22	Proportion of software business income to GDP (%)	x_{26}	+
		Proportion of income from software technology service industry in GDP (%)	x_{27}	+
		Proportion of operating income of computer electronic communication manufacturing industry in GDP (%)	x_{28}	+
		Fixed assets investment in the information service industry (RMB 100 mm)	x_{29}	+
		Number of employees in the software technology service industry (10,000 people)	$x_{2\,10}$	+
		Total profit of computer electronic communication manufacturing industry (RMB 100 mm)	$x_{2\,11}$	+
	Digital innovation U23	Employment in scientific research service industry (10,000 people)	$x_{2\,12}$	+
		Research expenditure (RMB 100 mm)	$x_{2\,13}$	+
		Total number of employees with bachelor's degree or above (people)	$x_{2\,14}$	+
		Number of patent applications per 10,000 people (PCS/10,000 people)	$x_{2\,15}$	+
		Proportion of output value of scientific research service industry in GDP (%)	$x_{2\,16}$	+
	Industrial digitization U24	Number of computers used by enterprises per 100 people (sets)	$x_{2\,17}$	+
		Number of websites per 100 enterprises (PCS)	$x_{2\,18}$	+
		E-commerce sales (RMB 100 mm)	$x_{2\,19}$	+
		Number of e-commerce enterprises	$x_{2\,20}$	+
		E-commerce purchase amount (RMB 100 mm)	$x_{2\,21}$	+

2.4. Data Source and Processing

Following Sun et al. [41], we combine the energy carbon emission coefficient and energy consumption given in "the 2006 IPCC guidelines for national greenhouse gas "inventories. The exhaust gas emission is the sum of the emissions of SO2, NOx, PM2.5, and PM10 by the LI in Anhui Province from 2013 to 2020. According to the primary energy consumption of the LI in the energy balance sheet of Anhui Province, the emission coefficient is calculated using the emission coefficient method. The emission coefficient is determined by referring to the EPA, AP-42, and Beijing emission coefficient [41,42] and combining the actual development of the LI in Anhui Province.

In this study, data on Anhui Province from 2013 to 2020 are taken as the statistical sample. The data on the regional logistics industry and the digital economy system are from the China Statistical Yearbook, China Internet development report, Anhui statistical yearbook, and China Energy Yearbook in the corresponding years. For some indicators,

2020 data are unavailable, therefore, we adopted the missing value processing method to complete the dataset [43].

We adopted the entropy weight method to weigh each index [44] in order to improve the accuracy of the results and avoid deviation of subjective weight. The specific calculation steps to prepare for measuring the comprehensive development level of the LI and DE are as follows:

First, the range standardization method is used to standardize the original logistics industry and digital economic data. Different algorithms are adopted for positive and negative indicators with different meanings. The specific calculation formulas are as follows:

$$\text{Positive index}: Y_{tj} = \frac{X_{tj} - min_j}{max_j - min_j} \tag{1}$$

$$\text{Negative indicator}: Y_{tj} = \frac{max_j - X_{tj}}{max_j - min_j}, \tag{2}$$

where X_{tj} and Y_{tj} represent the values before and after the dimensionless standardization of the j-th index data (t = 1,2,3, … , n; j = 1,2,3, … , m) of the logistics industry or the digital economy subsystem in the t year, respectively. To eliminate 0 and 1 after standardization, here, max_j represents 1.01 times the maximum value in the j-th index data of the logistics industry or the digital economic system during the study period. min_j represents 0.99 times the minimum value in the j-th index data of the logistics industry or the digital economic system during the study period. That is to say, $max_j = 1.01 * maxX_{tj}$; $min_j = 0.99 * minX_{tj}$.

The j-th index Y_{tj} of the specific gravity of P_{ij} is then calculated, as shown in Equation (3):

$$P_{tj} = \frac{Y_{tj}}{\sum Y_{tj}} \tag{3}$$

Next, the information entropy, e_j, of index j is calculated, as shown in Equation (4):

$$e_j = -k \sum P_{tj} \ln P_{tj}, \tag{4}$$

where, $k > 0$, $k = \left(\frac{1}{\ln n}\right)$, n refers to the year, $e_j \geq 0$.

Then, the information entropy redundancy g_j, of index j is calculated, as shown in Equation (5):

$$g_j = 1 - e_j \tag{5}$$

Finally, the weight, w_j, of indicator j is calculated, as shown in Equation (6):

$$w_j = g_j / \sum g_j \tag{6}$$

The entropy weight method is used to give weights to the indicators of each system objectively to prepare for the next calculation of the complete and thorough development level of the LI and DE system each year.

2.5. Data Model

2.5.1. Comprehensive Development Level Model

The complete and thorough development level of the regional LI and DE is calculated using the multi-objective linear weighting method. The specific calculation Equation (7) is as follows:

$$F_i = \sum_{j=1}^{m} w_j Y_{tj} \tag{7}$$

The higher the value of F_i, the higher the complete and thorough development level of the regional LI or DE.

2.5.2. Coupling Degree and CCDM

The concept of coupling originated from physics and refers to the phenomenon that two or more systems or motion forms affect each other through various interactions [37]. The coupling degree refers to the degree of mutual influence between systems or elements. The degree of interaction between the LI and DE, through their respective coupling elements, is defined as the degree of coupling between the LI and DE. Its magnitude indicates the degree of interaction between the two subsystems. Based on the above measurement of the complete and thorough development level of the regional LI and DE, we calculate the coupling degree of the regional LI and DE. The specific model is shown in Equation (8):

$$C = \frac{2\sqrt[2]{U_1 U_2}}{U_1 + U_2} \tag{8}$$

where, U_1 is the comprehensive level of the regional LI subsystem and U_2 is the comprehensive development level of the DE. C refers to the coupling degree between the regional LI and DE, and the value range is 0 to 1. The higher the value of C, the stronger the relationship between the regional LI and DE and the closer to orderly development. The smaller the value of C, the closer the two systems are to disordered development.

However, when the complete and thorough development degree of the regional LI and DE is relatively low and close, that is, when the development level of the two subsystems is not high, the coupling degree of the calculation is high. Therefore, to accurately reflect the interactive development level of the LI and DE, it is necessary to further build a CCMD. The CCDM is then used to evaluate the CCD of the regional LI and DE. For details, see Equations (9) and (10):

$$D = \sqrt{C \times T} \tag{9}$$

$$T = \alpha U_i + \beta U_i \tag{10}$$

In Equations (9) and (10), C is the coupling degree, and T is the comprehensive evaluation index used to reflect the overall level of the regional LI and DE. α, β is the specific weight of the two systems, which mainly measures the contribution of the LI and DE, of which, $\alpha + \beta = 1$. Based on the current situation of global economic development, the importance of the LI and DE to social and economic development is the same. As such, the weights are equal and the value is $\alpha = \beta = 1/2$. D is the coupled co-scheduling of the LI and DE, and the value is within {0,1}. The higher the value of D, the more coordinated the development levels of the LI and DE, and the closer the mutual promotion and coordination. The lower the value of D, the worse their coordination, and the weaker their mutual influence and coordination.

According to the calculated coupling degree and coupling co-scheduling and referring to existing relevant studies [26,37], the coordination status of the regional LI and DE is divided into the following levels (Table 2).

Table 2. The classification of CD and CCD.

C	Coupling Stage	Coupling Description	D	Coordination Level	U1 > U2	U1 < U2
(0.0, 0.3)	Low-level coupling	Gradually formed coupling	(0.0, 0.1)	Extreme disorder	Digital economy lag	Logistics lag
			(0.1, 0.2)	Severe disorder	Digital economy lag	Logistics lag
			(0.2, 0.3)	Moderate disorder	Digital economy lag	Logistics lag
(0.3, 0.5)	Antagonism	Development to a certain extent	(0.3, 0.4)	Mild disorder	Digital economy lag	Logistics lag
			(0.4, 0.5)	Near disorder	Digital economy lag	Logistics lag
			(0.5, 0.6)	Reluctantly coordinate	Digital economy lag	Logistics lag
(0.5, 0.8)	Run in	Good coupling development	(0.6, 0.7)	Primary coordination	Digital economy lag	Logistics lag
			(0.7, 0.8)	Intermediate coordination	Digital economy lag	Logistics lag
(0.8, 1.0)	High-level coupling	Promotes mutual development	(0.8, 0.9)	Good coordination	Digital economy lag	Logistics lag
			(0.9, 1.0)	Highly coordinated	Digital economy lag	Logistics lag

2.5.3. The ODM

To further identify the improvement direction of the CD of the regional LI and DE system, we need to identify specific influencing factors. The ODM is adopted to clarify the obstacle factors. The specific calculation steps are as follows:

$$O_{tij} = \frac{(1 - Y_{tij}) \times w_j \times 100\%}{\sum(1 - Y_{tij}) \times w_j} \tag{11}$$

$$O_{ti} = \sum O_{ij} \tag{12}$$

In Equation (12), O_{tij} is the obstacle degree of the j-th secondary index in the first index i of the LI or the DE system to the internal coupling and coordination of regional the LI or the DE. O_{ti} represents the obstacle degree of the i-th level I index in year t. Y_{tij} represents the standardized value of the j-th secondary index among the i-th primary index in the t-th year; $1 - Y_{tij}$ represents the deviation degree of the index; and w_j is the weight of the j-th index.

3. Results

3.1. Comprehensive Development Level

Based on the collected raw data for the LI and DE in Anhui Province, this study calculates the weight of each index of the logistics industry and digital economy system according to Equation (1) through (6), and calculates the comprehensive development level of the logistics industry and digital economy using Equation (7). The results are shown in Figure 3.

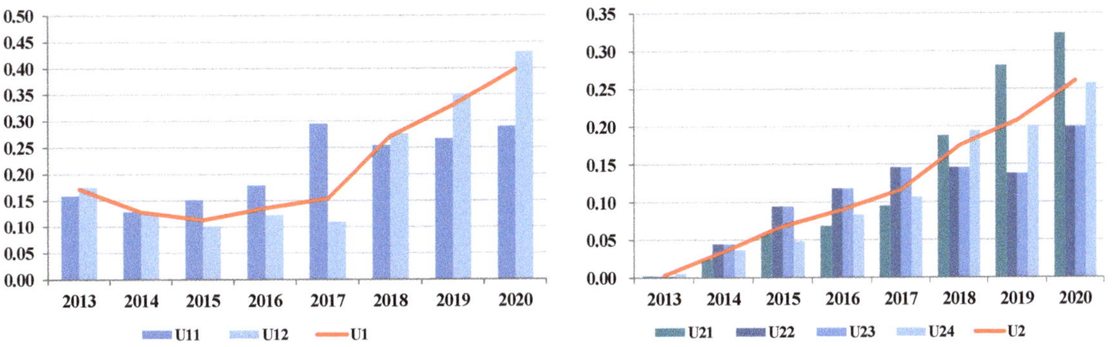

Figure 3. The comprehensive development level of Anhui Province's LI and DE.

Figure 3 shows that from 2013 to 2020, the development input (U11), output (U12), and comprehensive development level (U1) of the LI show a fluctuating growth trend; it also shows, from 2013 to 2020, the four first-level indicators of the DE system, i.e., digital infrastructure (U21), digital industry (U22), digital scientific and technological innovation (U23), industrial digitization (U24), and the comprehensive development level (U2), show a trend of continuous growth. From the perspective of the progress level of the LI subsystem, the comprehensive progress level declined in 2014 and 2015. At this stage, the LI was heavily taxed, which increased the cost of logistics, and the development of the LI was affected to a certain extent. From 2016 to 2020, there was an overall upward trend. The development of high-end technology in this stage opened up new space and provided new development opportunities for the LI, and the LI entered the stage of transformation and upgrading. In 2017, the input level of the LI was much higher than the output level, which is related to the 13th Five Year Plan of Anhui Province to increase the development of the LI. In 2018, the output level of the LI exceeded the input level, and by 2020, the output level reached 0.4313. This shows that the output efficiency of the logistics industry in Anhui Province has gradually improved. The application of new technologies such as big data and the Internet of Things in this period has made effective use of resource inputs. This improvement also reflects the increased demand for the LI in the global economy and network economy.

The overall comprehensive development level of the DE subsystem shows an upward trend. Although there are fluctuations in the level of the four first-level indicators: digital infrastructure (U21), digital industry (U22), digital scientific and technological innovation (U23), and industrial digitization (U24), the overall trend is still upward. Among these indicators, the digital infrastructure (U21) and the industrial digital level (U24) grew rapidly. Notably, the digital infrastructure construction level reached 0.3231 in 2020. This is inseparable from the implementation of national policies, increased digital infrastructure investment, and the strengthening of digital transformation and development of traditional industries in Anhui Province. However, the development level of the digital industry (U22) and digital technology innovation (U23) in Anhui Province was lower than that of digital infrastructure (U21) and industrial digitization (U24) in 2018 and beyond. This may have been affected by the global pandemic that occurred during this period, affecting the normal operation of innovative production inputs in relevant industries.

From the perspective of the development level of the LI system (U1) and the digital economy system (U2), the comprehensive development level of the logistics industry reached the highest level in 2020 (0.3981), while the development level of the DE reached a high of 0.2610. Therefore, the development level of the LI and DE in Anhui Province needs to be improved.

3.2. The Result of CCD

Data on the Anhui LI and DE are brought into the calculations of CD and CCD in Equation (8) through (10) to obtain the CD (C), comprehensive evaluation index (T), and coupling co-scheduling (D) of the Anhui logistics industry and digital economy system (Figure 4).

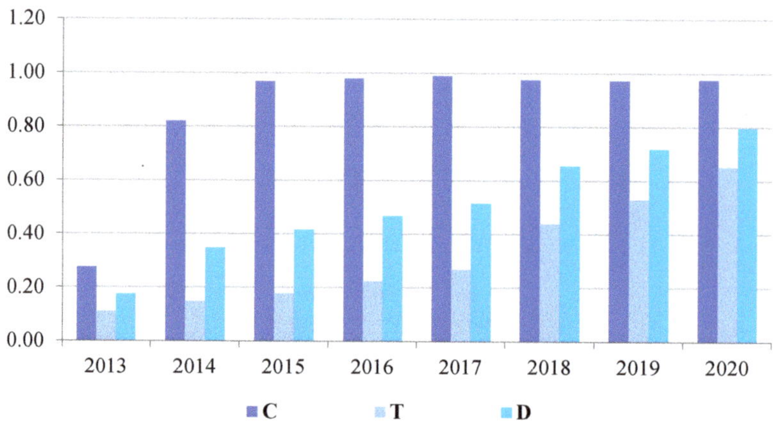

Figure 4. C, T, and D values of Anhui Province's LI and DE.

It can be observed in Figure 4 that the coupling degree (C) of LI in Anhui Province has been close to 1 since 2015, and reached a high of 0.9907 in 2017. However, the time-coupled co-scheduling (D) is 0.5163, not the highest. It should be noted that the development level of LI and DE in 2017 was low, and both need to be improved to increase their coordinated development level. The complete and thorough development index (T) and coupled co-scheduling (D) of the LI and DE in Anhui Province increased year by year from 2013 to

2020. This shows that the logistics industry and digital economy in Anhui Province are developing well, but the highest levels are no more than 0.7 and 0.8, respectively, so they still need to be strengthened.

According to the data in Figure 4, combined with the calculation results U1 and U2 of the complete and thorough development level, we can see the coupling and coordination type distribution between the logistics industry and the digital economic system in Anhui Province, as shown in Figure 5.

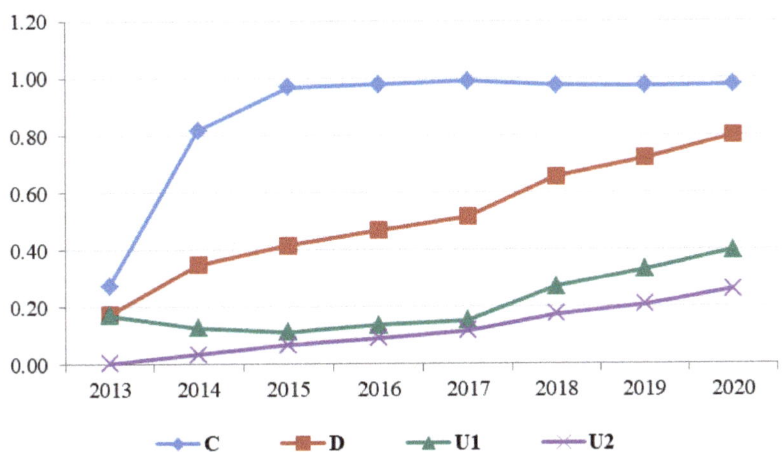

Figure 5. Coupling coordination type of the LI and DE in Anhui Province.

From Figure 5, combined with the level division in Table 2, we can note that the coupling degree between the LI and DE in Anhui Province was 0.2755 in 2013. This indicates a low-level coupling stage, and from this point, the LI and DE gradually formed a coupling. At this time, the logistics industry in Anhui Province had developed to a certain extent, while the digital economy had just started. From the above figure, it can be seen that U1 > U2 and the coupled co-scheduling (D) is 0.1753, which is a serious imbalance as the digital economy lags behind. After 2014, the CD between the LI and DE in Anhui Province is greater than 0.8, and even tends to 1 in the later stage, which indicates a highly coupled stage in which the two promote each other's orderly development. However, the level of coupled co-scheduling (D) at this stage is not high. In 2017, the coupling degree was as high as 0.9907, which is the highest in the study year. However, the coupled co-scheduling (D) was only 0.5163, which indicates a barely coordinated stage. Moreover, U1 > U2 showing that the digital economy lags slightly. Even though the D value reaches the highest level of 0.7985 in 2020, the coordination level is intermediate, U1 > U2 still exists, and the digital economy lags slightly.

To sum up, the CCD between the LI and DE systems in Anhui Province needs to be improved. The overall situation feature is "high coupling and low coordination". While promoting the respective development of the logistics industry and the digital economy, special attention should be paid to the CD of the LI and DE.

3.3. The Result of ODM

According to the ODM, Equations (11) and (12), the coupling and coordination barrier factors of the LI and DE in Anhui Province are calculated. As shown in Figure 6, the distribution of the barrier degree of the first-level indicators (U11-U24) of the LI and DE system in Anhui Province is as follows:

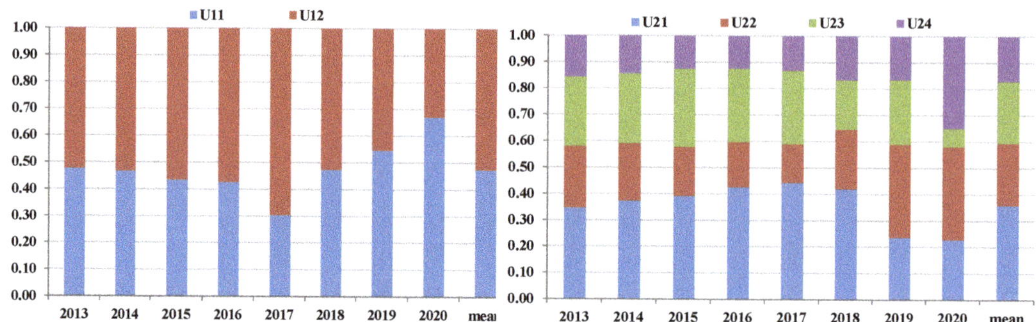

Figure 6. Results of obstacle degree of Anhui Province's LI and DE.

From Figure 6, we see the average U11 of the first level indicators of the LI in Anhui Province is slightly lower than U12. In 2017, the U12 of the LI in Anhui Province was higher. At that time, the social and economic development had higher requirements in terms of the service efficiency of the LI. By 2020, with the application of the Internet of Things, cloud computing, and other emerging technologies, the output efficiency of the LI was greatly improved. The barrier factor at this time is U11. From the perspective of the DE system of Anhui Province, the average ranking of the first-level indicators is U21 > U23 > U22 > U24. The main obstacle to the development of the DE is U21. However, increasing attention has been paid to the construction of a digital economic infrastructure after China proposed to increase the development of the DE. The DE foundation U21 < U22 in 2019 to 2020 is no longer the main obstacle. By 2020, the barrier degree of the first-level indicators of digital economic development is U22 > U24 > U21 > U23; therefore, U22 is the main obstacle factor.

In addition, to analyze the obstacle factors affecting the CCD of the LI and DE system in Anhui Province, we sort the obstacles of the third-level indicators according to the data in 2020, as shown in Figure 7:

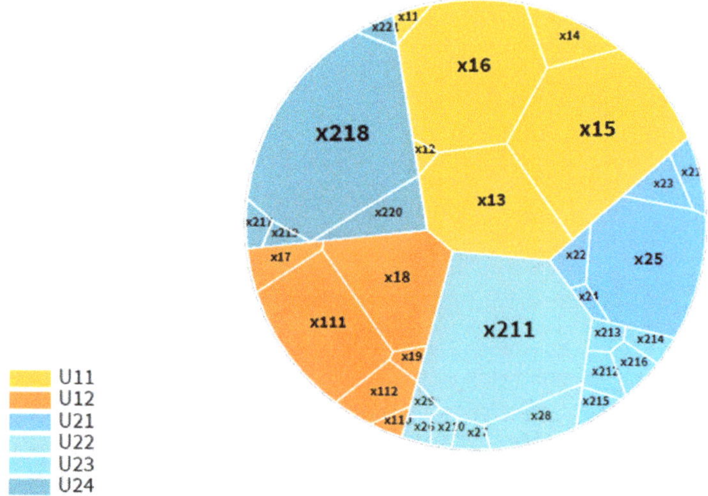

Figure 7. Ranked obstacle factors of Anhui Province's logistics industry and digital economy.

Figure 7 shows the distribution of the obstacle degree for each secondary index. First, in the LI subsystem, the barrier degree of the LI development input (U11) is greater than the barrier degree of the LI development output (U12). Among the secondary indicators of the logistics subsystem, x15 > x16 > x13 > x111 > x18 > x112 are the foremost obstacles. Among the indicators of the LI development investment (U11), energy consumption (x15), power consumption (x16), and the number of employees in the LI (x13) are the foremost development obstacles. This is mainly because the development of the LI in Anhui Province is facing problems such as high resource consumption and lack of logistics talent. For example, the total energy consumption of the LI in 2020 increased by 24.17% compared with that in 2013. The total power consumption of the LI in 2020 is about 2.41 times that of 2013. The number of employees in the LI decreased by 29,000 in 2020 compared with 2019. The key obstacles in the development output (U12) index of the LI are carbon dioxide emission (x111), freight volume (x18), and exhaust gas emission (x112). This shows that environmental pollution remains a major obstacle to the development of the LI and low-carbon emissions reduction is urgently needed. In addition, in 2020, the freight volume of Anhui Province was seriously affected by the COVID-19 pandemic.

Second, we can also see from Figure 7 that the barrier degree of the first-class indicators in the DE system of Anhui Province in 2020 is U22 > U24 > U21 > U23. The main obstacles for the secondary indicators in the DE system are: x218 > x211 > X25 > x28 > x220 > x23 > X212 > x216. In industrial digitization (U24), the number of websites owned by every 100 enterprises (x218) is the main obstacle factor, followed by the number of e-commerce enterprises (x220). This indicates the digitization level of enterprises in Anhui Province is still deficient. According to the statistical yearbook of Anhui Province in 2020, the number of websites owned by every 100 enterprises (x218) and the number of e-commerce enterprises (x220) decreased by 5% and 7.7%, respectively, compared with 2019. In the digital industry (U22), the total profit (x211) and total business volume (x28) of the computer, communication, and electronic equipment manufacturing industry are the main obstacles to development. In particular, the total profit of the computer, communication, and electronic equipment manufacturing industry (x211) is the main obstacle factor. According to the data in 2020, the index data decreased by 4.7% compared with that in 2019. On the one hand, the computer and electronic equipment manufacturing industry in Anhui Province needs further development; on the other hand, it has been substantially affected by the COVID-19 pandemic, which has caused great losses to the development of the manufacturing industry. In the digital infrastructure (U21), the number of web pages per capita (X25) is the most important obstacle factor, and it is also necessary to pay attention to the improvement of the mobile phone penetration rate (x23) per 100 people. Anhui is a populous province, and the number of web pages per capita needs to be further improved. In addition, although U23 ranks the lowest among the first-class indicators, attention should also be paid to the improvement of the number of employees in the scientific research and technical service industry (x212) and the proportion of the output value of the scientific research and technical service industry in GDP (x216).

4. Discussion

4.1. Policy Suggestions

The results of the CCD of the LI and DE in Anhui Province from 2013 to 2020 are this: the overall development level shows an upward trend. The development level of the LI system in Anhui Province is semi-U-shaped, and the development level of the DE is low, but it is growing rapidly. The LI and DE in Anhui Province show the feature of "high coupling and low coordination". The coordination degree needs to be further improved, and the development of the DE lags behind. The results of the obstacle degree calculation show that the main obstacles are the high energy consumption and power consumption of the LI as well as the urgent need for logistics talent. In addition, the development of the logistics industry also brings environmental pollution problems, such as carbon dioxide and exhaust gas emissions. Therefore, the following policy suggestions are put forward for the CD of the logistics industry and digital economy in Anhui Province:

1. Improve the comprehensive development level of the LI and DE in Anhui Province. CCD requires a high level of subsystem development to promote a high level of overall progress. In 2020, the overall development level of the LI in Anhui Province reached its highest level of 0.3981, and the overall development level of the DE reached its highest level of 0.2610. We can see that, although it reached the highest level in 2020, the comprehensive development level remains very low, and the development of the DE lags behind the development of the LI. Anhui Province should strengthen the development of its LI and DE, focus on the development of the DE because of its low starting point, and lay a foundation for the CD of the two at a higher level.

2. Promote the CD of the LI and DE in Anhui Province. From 2013 to 2020, the coupling degree between the LI and DE in Anhui Province was high, and from 2014 to 2020, it was almost 1. The LI and DE are developing in an orderly manner, but the CD of the two is low. In 2020, the highest level of coordinated dispatching was 0.7985, which is only at the intermediate coordination level. Moreover, the development of the digital economy lags behind, and there is still much room for progress. The world is seizing the opportunity of digital economic development to realize the digital transformation and development of traditional industries and win competitive advantages. Anhui Province should closely follow the pace of the times, strengthen the CD of the LI and DE, promote the digital transformation of the LI, and further realize the high-quality economic development of Anhui Province.

3. Overcome the obstacles to the CD of the LI and DE in Anhui Province. In 2020, the total energy consumption of the LI increased by 24.17% compared with 2013, and the number of logistics employees decreased by 29,000 compared with 2019. Efforts should be made to improve the output efficiency of the LI in Anhui Province, reduce energy consumption, reduce carbon dioxide and exhaust gas emissions, and realize the sustainable development of green logistics. Efforts should also be made to strengthen the education of high-level logistics practitioners to provide a talent guarantee for the development of the LI. In 2020, digital industrialization and industrial digitization were the main obstacles (U22 > U24 > U21 > U23). It is necessary to strengthen the application of new generation information technologies such as the Internet of Things, big data, and cloud computing. It is also necessary to promote the digital transformation of the logistics industry and help to increase the development level of industrial digitization in the digital economy. The development of the computer, communication, and electronic equipment manufacturing industry should be strengthened to win core competitiveness and promote the development of the digital industry. Finally, we should implement national policies and measures to promote the development of the logistics industry and digital economy, and we should increase investment in infrastructure construction.

4.2. Management Enlightenment

Compared with the existing literature [7,21,27–35], the following advantages and contributions are given: first, a more comprehensive evaluation index system for the CCD of regional LI and DE is established. The indicator system includes indicators reflecting the development input and output level of the LI, and it includes four subsystems reflecting the development level of the DE: digital infrastructure, digital industry, digital scientific and technological innovation, and industrial digitization, which are more systematic and scientific than indicators in previous studies. Second, using a data-driven method [39], we can more objectively and scientifically measure, evaluate, and identify the CCD level of the regional LI and DE so as to better serve relevant decision makers. Finally, the study puts forward some suggestions to promote the CD of the province's logistics industry and digital economy. Promoting the CCD of the LI and DE has important practical significance for sustainable and high-quality development. To sum up, some management implications are drawn:

(1) In order to achieve cost reduction, increase efficiency, and the high-quality development of LI, we should strengthen the digital transformation of the operation process of logistics and realize digital and intelligent development. The efficient development of the logistics industry has provided support for the development of the DE and is conducive to the development of digital industries such as information and communication equipment and infrastructure. Give full play to the interaction and interdependence of the two subsystems, which can effectively promote the sustainable development of industrial integration, and then promote the development of the regional economy.

(2) With the deep integration of the development of modern information technology and the LI, the mutual influence, interaction, and development of various factors between them can be clarified, which can more accurately formulate relevant countermeasures and better serve the decision-making of local governments and local related enterprises.

(3) The digital transformation of traditional industries is an essential a an adjustment facing the world. The LI lays a solid foundation for the integrated development of industry and the matching of resource supply and demand, ensuring sustained social and economic development. Realizing the digital transformation of the LI is conducive to accelerating the pace of industrial digitalization, which is crucial to promote the high-quality development of the regional economy. The CCD of the LI and DE should make a profound study.

5. Conclusions

Considering the acceleration of the global DE and the promotion of the digital transformation of traditional industries, we have entered a critical period for the LI. Logistics, as the leading and most basic component of economic development, is of great importance for realizing digital transformation. Promoting the CCD of the regional LI and DE is an inevitable requirement to promote the digital transformation of the regional logistics industry. This study is valuable in theory and practice. From the perspective of theoretical value, first, the evaluation index system of CCD for the LI and DE is constructed [45]. It takes into account the energy consumption, power consumption, and carbon dioxide and exhaust gas emissions of the logistics industry. In addition, it reflects the development level of the digital economy through four first-order dimension indicators. The implication of an integrated development index system for logistics and related industries is enriched. Second, a CCD model for the LI and DE is constructed. Because the model is based on development data for the logistics industry and digital economy, it provides a more scientific and objective way to quantitatively evaluate the CCD of the two. Third, through the correlation model, the factors that restrict the CCD of the logistics industry and the DE are identified. This provides a quantitative basis for the targeted design of policies for the development of the regional LI and DE. In terms of practical value, first, the research results can be used by managers of regional LI and DE-related enterprises to consider problems and make decisions from a quantitative perspective. Second, the findings provide direction for local governments to accurately form a planning strategy that promotes the development of the LI and DE. Third, new ideas for researchers focused on LI and DE are provided.

This paper evaluates and analyzes the CCD between the LI and DE and summarizes some rules and conclusions. However, due to the limitations of the current level of knowledge and objective conditions, there are still areas that need to be further improved in the research: first, in terms of the indicator system, the CCD of the LI and DE involves multiple indicator dimensions and multi-dimensional data, and the comprehensiveness of the indicator system needs to be further improved. Secondly, in terms of data collection and processing, due to the lack of data in 2021, future research should focus on timeliness. Finally, can the DE reduce the carbon emissions of the logistics industry? If so, how? These are the focus of future research.

Author Contributions: Conceptualization, Y.G. and H.D.; methodology, Y.G. and H.D. validation, H.D. and Y.G.; formal analysis, Y.G.; investigation, Y.G. and H.D.; resources, H.D.; data curation, Y.G.; writing—original draft preparation, Y.G.; writing—review and editing, Y.G. and H.D.; visualization, H.D. and Y.G.; supervision, Y.G. and H.D.; project administration, Y.G.; funding acquisition, Y.G. All authors have read and agreed to the published version of the manuscript.

Funding: This research is supported by the Major Project of Humanities and Social Sciences Research in Anhui Universities (NO.SK2021ZD0092), Anhui social science innovation and development research project (NO.2020CX093), the non-financial fund scientific research project of Suzhou University (NO.2022xhx149; NO.2022xhx150), and the Natural Science Foundation of Anhui Province (NO.2108085mg235).

Institutional Review Board Statement: Not applicable.

Informed Consent Statement: Not applicable.

Data Availability Statement: Not applicable.

Conflicts of Interest: The authors declare no conflict of interest.

References

1. Li, M.; Wang, J. The productivity effects of two-way FDI in China's logistics industry based on system GMM and GWR model. *J. Ambient Intell. Human. Comput.* **2021**, *5*, 1–5. [CrossRef]
2. Huang, Q.; Ling, J. Measuring embodied carbon dioxide of the logistics industry in China: Based on industry stripping method and input-output model. *Environ. Sci. Pollut. Res.* **2021**, *28*, 52780–52797. [CrossRef]
3. China Information and Communication Research Institute. *White Paper on Global Digital Economy*; China Academy of Communications: Beijing, China, 2022.
4. Xu, G.; Lu, T.; Liu, Y. Symmetric Reciprocal Symbiosis Mode of China's Digital Economy and Real Economy Based on the Logistic Model. *Symmetry* **2021**, *13*, 1136. [CrossRef]
5. Popkova, E.G.; Sergi, B.S. A Digital Economy to Develop Policy Related to Transport and Logistics. Predictive Lessons from Russia. *Land Use Policy* **2020**, *12*, 105083. [CrossRef]
6. Zhou, Z.; Wan, X. Does the Sharing Economy Technology Disrupt Incumbents? Exploring the Influences of Mobile Digital Freight Matching Platforms on Road Freight Logistics Firms. *Prod. Oper. Manag.* **2022**, *31*, 117–137. [CrossRef]
7. Guo, Y.; Mao, H.; Ding, H.; Wu, X.; Liu, Y.; Liu, H.; Zhou, S. Data-Driven Coordinated Development of the Digital Economy and Logistics Industry. *Sustainability* **2022**, *14*, 8963. [CrossRef]
8. Liao, X.; Yang, Z. Effect Measurement and Realization Path of Transformation and Upgrading of Digital Economy Enabling Manufacturing Industry in the Yangtze River Delta. *East China Econ. Manag.* **2021**, *35*, 22–30. [CrossRef]
9. Zheng, W.; Wang, J.; Mao, X.; Li, J. Ecological Efficiency Evaluation and Spatiotemporal Characteristics Analysis of the Linkage Development of the Logistics Industry and Manufacturing Industry. *Front. Energy Res.* **2022**, *9*, 709582. [CrossRef]
10. Yi, J.; Zhang, Y.; Liao, K. Regional Differential Decomposition and Formation Mechanism of Dynamic Carbon Emission Efficiency of China's Logistics Industry. *Int. J. Environ. Res. Public Health* **2021**, *18*, 13121. [CrossRef] [PubMed]
11. Yao, X.; Cheng, Y.; Zhou, L.; Song, M. Green efficiency performance analysis of the logistics industry in China: Based on a kind of machine learning methods. *Ann. Oper. Res.* **2022**, *1*, 727–752. [CrossRef]
12. Qin, W.; Qi, X. Evaluation of Green Logistics Efficiency in Northwest China. *Sustainability* **2022**, *14*, 6848. [CrossRef]
13. Ding, H.; Liu, Y.; Zhang, Y.; Wang, S.; Guo, Y.; Zhou, S.; Liu, C. Data-driven evaluation and optimization of the sustainable development of the logistics industry: Case study of the Yangtze River Delta in China. *Environ. Sci. Pollut. Res.* **2022**, *5*, 68815–68829. [CrossRef]
14. Klumpp, M.; Loske, D. Sustainability and Resilience Revisited: Impact of Information Technology Disruptions on Empirical Retail Logistics Efficiency. *Sustainability* **2021**, *13*, 5650. [CrossRef]
15. Li, J.; Wang, Q. Impact of the Digital Economy on the Carbon Emissions of China's Logistics Industry. *Sustainability* **2022**, *14*, 8641. [CrossRef]
16. Tapscott, D. *The Digital Economy: Promise and Peril in the Age of Networked Intelligence*; McGraw Hill: New York, NY, USA, 1996.
17. Carlsson, B. The Digital Economy: What is new and what is not? *Struct. Change Econ. D* **2004**, *15*, 245–264. [CrossRef]
18. Pan, W.R.; Xie, T.; Wang, Z.W.; Ma, L.S. Digital economy: An innovation driver for total factor productivity. *J. Bus. Res.* **2022**, *139*, 303–311. [CrossRef]
19. Sidorov, A.; Senchenko, P. Regional Digital Economy: Assessment of Development Levels. *Mathematics* **2020**, *8*, 2143. [CrossRef]
20. D'Amico, G.; Szopik-Depczyńska, K.; Dembińska, I.; Ioppolo, G. Smart and sustainable logistics of Port cities: A framework for comprehending enabling factors, domainsand goals. *Sustain. Cities Soc.* **2021**, *69*, 102801. [CrossRef]
21. Gan, W.; Yao, W.; Huang, S.; Liu, Y. A Study on the Coupled and Coordinated Development of the Logistics Industry, Digitalization, and Ecological Civilization in Chinese Regions. *Sustainability* **2022**, *14*, 6390. [CrossRef]

22. Bai, L.B.; Chen, H.L.; Gao, Q.; Luo, W. Project portfolio selection based on synergy degree of composite system. *Soft Comput.* **2018**, *22*, 5535–5545. [CrossRef]
23. Shi, T.; Yang, S.Y.; Zhang, W.; Zhou, Q. Coupling coordination degree measurement and spatiotemporal heterogeneity between economic development and ecological environment—Empirical evidence from tropical and subtropical regions of China. *J. Clean. Prod.* **2020**, *244*, 118739. [CrossRef]
24. Lan, S.L.; Zhong, R.Y. Coordinated development between metropolitan economy and logistics for sustainability. *Resour. Conserv. Recycl.* **2018**, *128*, 345–354. [CrossRef]
25. Lan, S.; Tseng, M.L. Coordinated Development of Metropolitan Logistics and Economy Toward Sustainability. *Comput. Econ.* **2018**, *52*, 1113–1138. [CrossRef]
26. Cai, J.; Li, X.P.; Liu, L.J.; Chen, Y.Z.; Wang, X.W.; Lu, S.H. CCD of new urbanization and agro-ecological environment in China. *Sci. Total Environ.* **2021**, *776*, 145837. [CrossRef]
27. Shu, H.; Hu, Y. Research on Cooperative Development of Agricultural Logistics Ecosphere Based on Composite System Coupling. *J. Stat. Inf.* **2020**, *35*, 62–71. Available online: https://kns.cnki.net/kcms/detail/detail.aspx?FileName=TJLT2 02012007&DbName=CJFQ2020 (accessed on 10 December 2020).
28. Chen, Y.; Zhang, D.N. Multiscale assessment of the coupling coordination between innovation and economic development in resource-based cities: A case study of Northeast China. *J. Clean. Prod.* **2021**, *318*, 128597. [CrossRef]
29. Su, J.; Shen, T.; Jin, S. Ecological efficiency evaluation and driving factor analysis of the coupling coordination of the logistics industry and manufacturing industry. *Environ. Sci. Pollut. Res.* **2022**, *9*, 62458–62474. [CrossRef] [PubMed]
30. Zhang, W.; Zhang, X.; Zhang, M.; Li, W. How to Coordinate Economic, Logistics and Ecological Environment? Evidences from 30 Provinces and Cities in China. *Sustainability* **2020**, *12*, 1058. [CrossRef]
31. Dong, L.; Longwu, L.; Zhenbo, W.; Liangkan, C.; Faming, Z. Exploration of coupling effects in the Economy-Society-Environment system in urban areas: Case study of the Yangtze River Delta Urban Agglomeration. *Ecol. Indic.* **2021**, *128*, 107858. [CrossRef]
32. Xu, M.X.; Hu, W.Q. A research on coordination between economy, society and environment in China: A case study of Jiangsu. *J. Cleaner Prod.* **2020**, *258*, 120641. [CrossRef]
33. Li, S. Study on the sustainable development evaluation of logistics enterprise for industry manufacturing based on data envelopment analysis. *Appl. Mech. Mater.* **2013**, *345*, 380–383. [CrossRef]
34. Liu, W.; Wang, Y. Effect and Mechanism of Digital Economy Empowering Urban Green and High-quality Development. *South China J. Econ.* **2022**, *8*, 73–91. [CrossRef]
35. Wei, L.; Hou, Y. Research on the Impact of China's Digital Economy on Urban Green Development. *J. Quant. Tech. Econ.* **2022**, *39*, 60–79. [CrossRef]
36. Brynjolfsson, E.; Collis, A. How Should We Measure the Digital Economy? *Harv. Bus. Rev.* **2019**, *6*, 140–150.
37. Ariken, M.; Zhang, F.; Liu, K.; Fang, C.L.; Kung, H.T. Coupling coordination analysis of urbanization and eco-environment in Yanqi Basin based on multi-source remote sensing data. *Ecol. Indic.* **2020**, *114*, 106331. [CrossRef]
38. Cheng, X.; Long, R.Y.; Chen, H.; Li, Q.W. Coupling coordination degree and spatial dynamic evolution of a regional green competitiveness system—A case study from China. *Ecol. Indic.* **2019**, *104*, 489–500. [CrossRef]
39. Liu, C.; Gao, M.; Zhu, G.; Zhang, C.; Zhang, P.; Chen, J.; Cai, W. Data driven eco-efficiency evaluation and optimization in industrial production. *Energy* **2021**, *224*, 120170. [CrossRef]
40. Anhui Provincial Bureau of Statistics. Anhui Statistical Yearbook. 2021. Available online: http://tjj.ah.gov.cn/ssah/qwfbjd/tjnj/index.html (accessed on 10 April 2022).
41. Sun, H.; Hu, X.; Nie, F. Spatio-temporal evolution and socio-economic drivers of primary air pollutants from energy consumption in the Yangtze River Delta. *China Environ. Manag.* **2019**, *11*, 71–78. [CrossRef]
42. Li, L.; Lei, Y.; Wu, S.; Huang, Z.; Luo, J.; Wang, Y.; Chen, J.; Yan, D. Evaluation of future energy consumption on PM2.5 emissions and public health economic loss in Beijing. *J. Clean. Prod.* **2018**, *187*, 1115–1128. [CrossRef]
43. Jiang, Y.; Lan, T.; Wu, L. A Comparison Study of Missing Value Processing Methods in Time Series Data Mining. In Proceedings of the IEEE International Conference on Computational Intelligence & Software Engineering, Wuhan, China, 11–13 December 2009; p. 12. [CrossRef]
44. Ren, F.; Yu, X. Coupling analysis of urbanization and ecological total factor energy efficiency—A case study from Hebei province in China. *Sustain. Cities Soc.* **2021**, *74*, 103183. [CrossRef]
45. Yang, Y.; Wang, Y.; Zhang, Y.; Liu, C. Data-Driven Coupling Coordination Development of Regional Innovation EROB Composite System: An Integrated Model Perspective. *Mathematics* **2022**, *10*, 2246. [CrossRef]

MDPI
St. Alban-Anlage 66
4052 Basel
Switzerland
www.mdpi.com

Processes Editorial Office
E-mail: processes@mdpi.com
www.mdpi.com/journal/processes